MXene-Filled Polymer Nanocomposites

MXenes are a new family of two-dimensional (2D) metal carbides, having properties such as metallic conductivity and hydrophilicity. Adding polymer binders/spacers between atomically thin MXene layers or reinforcing polymers with MXenes results in composite films that have excellent flexibility, good tensile and compressive strengths, and electrical conductivity. This book covers all advances in the field of MXene-filled polymer nanocomposites to date, illustrating fabrication and characterization, and specific properties like anti-healing, antifriction, and microwave absorption. It further covers potential applications like energy conversion, storage systems, antibacterial, and drug delivery.

The book features

- exclusive material on MXene-based polymer nanocomposites;
- properties and potential applications of polymers upon addition of MXenes;
- the effect of MXenes on various thermoplastic and elastomer polymers;
- a focus on the properties, fabrication methods, and applications of relevant polymer matrices; and
- extensive coverage of the role of MXenes in polymers.

This book is aimed at researchers, professionals, and graduate students in material science, polymer engineering, electronic materials, composites, chemical processing, chemical sciences, fire engineering, and biomedicine.

Emerging Materials and Technologies

Series Editor: Boris I. Kharissov

The *Emerging Materials and Technologies* series is devoted to highlighting publications centered on emerging advanced materials and novel technologies. Attention is paid to those newly discovered or applied materials with potential to solve pressing societal problems and improve quality of life, corresponding to environmental protection, medicine, communications, energy, transportation, advanced manufacturing, and related areas.

The series takes into account that, under present strong demands for energy, material, and cost savings, as well as heavy contamination problems and worldwide pandemic conditions, the area of emerging materials and related scalable technologies is a highly interdisciplinary field, with the need for researchers, professionals, and academics across the spectrum of engineering and technological disciplines. The main objective of this book series is to attract more attention to these materials and technologies and invite conversation among the international R&D community.

Gas Sensors
Manufacturing, Materials, and Technologies
Edited by Ankur Gupta, Mahesh Kumar, Rajeev Kumar Singh, and Shantanu Bhattacharya

Environmental Biotechnology
Fundamentals to Modern Techniques
Sibi G

Emerging Two Dimensional Materials and Applications
Edited by Arun Kumar Singh, Ram Sevak Singh, and Anar Singh

Advanced Porous Biomaterials for Drug Delivery Applications
Edited by Mahaveer Kurkuri, Dusan Losic, U.T. Uthappa, and Ho-Young Jung

Thermal Transport Characteristics of Phase Change Materials and Nanofluids
S. Harikrishnan and A.D. Dhass

Multidimensional Lithium-Ion Battery Status Monitoring
Shunli Wang, Kailong Liu, Yujie Wang, Daniel-Ioan Stroe, Carlos Fernandez, and Josep M. Guerrero

Scanning Probe Lithography
Fundamentals, Materials, and Applications
Yu Kyoung Ryu and Javier Martinez Rodrigo

Engineered Nanoparticles as Drug Delivery Systems
Nahid Rehman and Anjana Pandey

MXene-Filled Polymer Nanocomposites
Edited by Soney C. George, Sharika T. Nair, and Joice Sophia Ponraj

For more information about this series, please visit: www.routledge.com/Emerging-Materials-and-Technologies/book-series/CRCEMT

MXene-Filled Polymer Nanocomposites

Edited by Soney C. George, Sharika T. Nair,
and Joice Sophia Ponraj

CRC Press
Taylor & Francis Group
Boca Raton London New York

CRC Press is an imprint of the
Taylor & Francis Group, an **informa** business

Designed cover image: Shutterstock

First edition published 2023
by CRC Press
6000 Broken Sound Parkway NW, Suite 300, Boca Raton, FL 33487-2742

and by CRC Press
4 Park Square, Milton Park, Abingdon, Oxon, OX14 4RN

CRC Press is an imprint of Taylor & Francis Group, LLC

ISBN: 978-0-367-75979-7 (hbk)
ISBN: 978-0-367-75980-3 (pbk)
ISBN: 978-1-003-16497-5 (ebk)

DOI: 10.1201/9781003164975

Typeset in Times
by Apex CoVantage, LLC

Contents

Editor Biographies

Soney C. George is Dean of Research and Director, Centre for Nanoscience and Technology, Amal Jyothi College of Engineering, Kerala, India. He is a Fellow of the Royal Society of Chemistry and a recipient of KTU best researcher award. His paper in *International Hydrogen Energy* has been selected for the best citation award. He has published more than 200 papers in peer-reviewed national and international journals and conferences. He is also edited and contributed several books and chapters published by Elsevier, Wiley, Springer, and others. He has guided several PhD. and master's students. He is concentrating on structure—property relations in polymer-based nanomaterial. He has also successfully completed several central- and state-funded projects.

Sharika T. Nair works as Assistant Professor, St. Xavier's College Vaikom, Kerala, India. Her research areas include preparation of polypropylene/natural rubber-based blends and nanocomposites, their characterization, and their applications. She has published many research papers in international journals and presented papers in national and international Conferences. She has also published many book chapters. She has completed one UGC (New Delhi) funded project and guided several masters students.

Joice Sophia Ponraj is a researcher at Tyndall National Institute, Cork, Ireland. She obtained her Doctor of Philosophy from Anna University, Chennai, India. She was bestowed "Italian-Indian Bilateral Programme" hosted by UniFe, Ferrara, Italy and National Doctoral Fellowship by AICTE, India during her PhD. She was awarded TRIL fellowship three times by ICTP to work in IMEM-CNR, Parma, Italy during 2011 to 2014. She has worked as a postdoctoral fellow at Suzhou University, China, INL in Portugal, and as an assistant Professor in UIST, Ohrid, Republic of Macedonia. She bagged INSPIRE Faculty Award in India. Her research interests are two-dimensional materials, nanophotonics, and semiconducting device fabrication.

Contributors

Dana Susan Abraham
Department of Chemistry
Central University of Kerala
Periye, India

Shikha Agarwal
Synthetic Organic Chemistry Laboratory, Department of Chemistry
MLS University
Udaipur, India

Farhan J. Ahmad
Department of Pharmaceutics, School of Pharmaceutical Education & Research
New Delhi, India

Aruna Kumar Barick
Department of Chemistry
Veer SurendraSai University of Technology
Sambalpur, India

Bhavya Bhadran
PG Department of Chemistry
St. Thomas College
Pala, India

Margandan Bhagiyalakshmi
Department of Chemistry
Central University of Kerala
Periye, India

Vishal Chaudhary
Department of Physics, Bhagini Nivedita College
University of Delhi
New Delhi, India

Amal George Cheriyan
Mahatma Gandhi University
Kottayam, Kerala

Hugo Gajardoni de Lemos
Center of Engineering, Modeling and Applied Social Sciences
Federal University of ABC
Santo André (SP), Brazil

Sathish Chander Dhanabalan
Centre for Advanced Materials, Integrated-Inter-Department of LiWET Communications
Aaivalayam-Dynamic Integrated Research Academy and Corporations (A-DIRAC)
Coimbatore, India

Bibi Mary Francis
Centre for Advanced Materials, Integrated-Inter-Department of LiWET Communications
Aaivalayam-Dynamic Integrated Research Academy and Corporations (A-DIRAC)
Coimbatore, India

Divyani Gandhi
Synthetic Organic Chemistry Laboratory, Department of Chemistry
MLS University
Udaipur, India

Nupur Garg
Department of Pharmaceutics, School of Pharmaceutical Education & Research
New Delhi, India

Y. C. Goswami
Nano Research Group, School of Sciences
ITM University
Gwalior, India

S. Heera
University of Kerala
Palayam, Kerala

K. Jayanarayanan
Department of Chemical Engineering and Materials Science
Amrita Vishwa Vidyapeetham
Coimbatore, India

R. Jeeshma
Department of Chemistry
St. Joseph's College (Autonomous)
Calicut, India

Keloth Paduvilan Jibin
School of Chemical Sciences
Mahatma Gandhi University
Kottayam, India

Abitha Vayyaprontavida Kaliyathan
School of Chemical Sciences
Mahatma Gandhi University
Kottayam, India

Rajesh Kumar Manavalan
Institute of Natural Science & Mathematics
Ural Federal University
Ekaterinburg, Russia

Hanna Joseph Maria
School of Energy Materials
Mahatma Gandhi University
Kottayam, India

Patitapaban Mohanty
Department of Chemistry
Veer Surendra Sai University of Technology
Sambalpur, India

Pooja Mohapatra
Department of Chemistry
Veer Surendra Sai University of Technology
Sambalpur, Odisha, India

Priyaranjan Mohapatra
Department of Chemistry
Veer Surendra Sai University of Technology
Sambalpur, India,

Avinash R. Pai
International & Inter University Center for Nanoscience and Nanotechnology (IIUCNN)
Mahatma Gandhi University
Kottayam, India

Bhavya Pandey
Nano Research Group, School of Sciences
ITM University
Gwalior, India

Bigyan RanjanJali
Department of Chemistry
Veer Surendra Sai University of Technology
Sambalpur, India

N. Rasana
Department of Chemical Engineering and Material Sciences
Amrita School of Engineering
Coimbatore, India

Nivedita Raveendran
Centre for Advanced Materials, Integrated-Inter-Department of LiWET Communications
Aaivalayam-Dynamic Integrated Research Academy and Corporations (A-DIRAC)
Coimbatore, India

Rodrigo Mantovani Ronchi
Center of Engineering, Modeling and Applied Social Sciences
Federal University of ABC
Santo André (SP), Brazil

Pallishree Sahoo
Department of Chemistry
Veer Surendra Sai University of Technology
Sambalpur, India

Sydney Ferreira Santos
Center of Engineering, Modeling and Applied Social Sciences
Federal University of ABC
Santo André (SP), Brazil

P. Sarath Kumar
Amrita Vishwa Vidyapeetham
Coimbatore, India

Bismark Sarkodie
Key laboratory of ultrafine materials of the Ministry of Education
School of Materials Science and Engineering
Shanghai, China

Ayushi Sethiya
Synthetic Organic Chemistry Laboratory, Department of Chemistry
MLS University
Udaipur, India

Lipsa Shubhadarshinee
Department of Chemistry
Veer Surendra Sai University of Technology
Sambalpur, India

Sisanth Krishnan Sidharthan
Central Institute of Petrochemicals Engineering & Technology (CIPET)
Chennai, India

Jay Soni
Synthetic Organic Chemistry Laboratory, Department of Chemistry
MLS University
Udaipur, India

Joice Sophia
Centre for Advanced Materials, Integrated-Inter-Department of LiWET Communications
Aaivalayam-Dynamic Integrated Research Academy and Corporations (A-DIRAC)
Coimbatore, India

Rani mol Stephen
Department of Chemistry
St. Joseph's College (Autonomous)
Calicut, India

V. P. Swapna
Department of Chemistry
St. Joseph's College (Autonomous)
Calicut, India

Benjamin Tawiah
Kwame Nkrumah University of Science and Technology
Ghana, India

Sabu Thomas
Vice-Chancellor
Mahatma Gandhi University
Kottayam, India

Prajitha Velayudhan
School of Chemical Sciences
Mahatma Gandhi University
Kottayam, India

Mari Vinoba
Kuwait Institute for Scientific Research
Kuwait City, Kuwait

Preface

Over the years, polymer nanocomposites have been widely researched because filling nanoparticles into polymer matrix provides versatile routes for enhancing the processing and physical properties of polymers owing to the size effect of these tiny solid particles. The last twenty years have witnessed significant progress in the field of polymer nanocomposites, and they have shown fascinating properties compared to neat polymer. A variety of nanoparticles have been blended with polymers to form nanocomposites. Among these, MXenes have become a huge research interest in this decade.

MXenes are a relatively new and exciting class of two-dimensional (2D) materials, which offer a wide range of compositions and excellent properties. MXenes are known as 2D metal carbides or nitrides and have attracted great attention in recent years due to their high electrical conductivity, high hydrophilicity, chemical stability, and ultrathin 2D-sheet-like structure. MXene have drawn attention for many applications in the fields of sensors, supercapacitors, hydrogen storage, nanoelectronics, batteries, hybrid ion capacitors, and more.

Due to their special properties, MXenes have tremendous potential as advanced composite structures, especially those based on polymers due to a great affinity between macromolecules and the terminating groups of 2D MXenes. Adding polymer binders/spacers between atomically thin MXene layers or reinforcing polymers with MXenes results in composite films that have excellent flexibility, good tensile and compressive strengths, and electrical conductivity that can be adjusted over a wide range. Owing to their mechanical strength and impressive capacitive performance, these films have the potential to be used for structural energy storage devices, electrochemical actuators, and radiofrequency shielding, among other applications. To the best of the editor's knowledge, there is currently no single book that focuses exclusively on MXene-filled polymer nanocomposites.

The chapters of this book are organized as follows: Chapter 1 introduces MXene material by covering its preparation, properties, and applications. Chapter 2 provides an overview of MXene-filled polymer composites. This chapter examines emerging trends in MXene-filled polymer composites in research and development. Chapters 3 through 6 summarized the effect of MXene on the various properties of thermoplastic, elastomers, and thermoset polymers. All these chapters cover the preparation, properties, and applications of MXene-filled polymer composites at various structural levels. The thermal and crystallization behavior of MXene-filled polymer nanocomposites is discussed in Chapter 7. Chapters 8 through 14 examines the various applications of MXene-filled polymer nanocomposites, with these chapters covering antifriction performance, self-healing nature, electromagnetic interference shielding, sensing, energy storage, and biomedical applications of various MXene-filled polymer composites. Chapters 15 and 16 discuss the simulations and future perspective in MXene-filled polymer nanocomposites.

The book is the result of commitments by top researchers in the field of MXene with various backgrounds and expertise. This book introduces readers to the fundamental aspects, properties, fabrication methods, and the numerous applications of MXene and MXene-filled polymer nanocomposites with a strong focus on novel findings and technological challenges. This book is a one-stop reference for MXene materials and overviews up-to-date literature in the field of MXene-filled polymer nancomposites and applications. This book is unique in the sense that it covers all advances in the field of MXene-filled polymer nanocomposites to date. It serves as an invaluable guide to professionals, researchers, industrialists, graduate students, and senior undergraduates in the fields of polymer science and engineering, materials science, nano-devices, energy storage, flexible electronics, polymer chemistry, and nanoscience and who work in similar areas or are interested in the field of polymer blends and nanocomposites.

All chapters were contributed by renowned professionals from academic, industry, and government laboratories from various countries and were peer reviewed in accordance with the guidelines utilized elsewhere by top-rated polymer journals. The editors would like to thank all contributors for believing in this endeavor, sharing their views and precious time and obtaining documents. The editors would like to express their gratitude to the external reviewers whose contributions helped to improve the quality of this book. The editors are indebted to the organization for their selfless support in this venture. Finally, the editors also express their gratitude to CRC Press for its continuous encouragement.

1 Introduction

MXene—A Novel Two-Dimensional Material: Preparation, Properties, and Applications

Bibi Mary Francis, Joice Sophia Ponraj, Nivedita Lalitha Raveendran, Rajesh Kumar Manavalan, and Sathish Chander Dhanabalan

CONTENTS

1.1 INTRODUCTION

MXenes are two-dimensional (2D) layered materials that are obtained from their bulk parent material, MAX phases, by the selective etching of element 'A', where A stands for atoms such as aluminum or silicon that belong to group 13 to 16 in the periodic table (Anasori *et al.*, 2019). For example, the most common MXene is $Ti_3C_2T_x$, which is exfoliated from its 3D-MAX phase, Ti_3AlC_2, by selectively removing Al. The chemical formula for MXenes is $M_{n+1}X_nT_x$, in which M represents early transition metals; X represents carbon, nitride, or carbon nitride; T represents the functional groups –OH, –F and/or –O that attach to MXene during its etching process; and $n = 1, 2, 3$ (Naguib *et al.*, 2014; Abdelmalak, 2014). The in- and out-of-plane ordering of metal atoms as shown in Figure 1.1; different surface terminations and combinations of them and the possibility of forming solid solutions at the M and X sites lead to numerous 2D-MXenes with possible distinct properties (Jeon *et al.*, 2020; Anasori *et al.*, 2015; Yury Gogotsi and Qing Huang, 2021).

In other 2D materials, such as graphene, TMDs, and h-BN, it is the van der Waal's bond that holds different layers together, but in a MAX phase, the atom A binds with M and X via metallic bonding. Therefore, breaking the bond is a difficult task in a MAX phase via regular shear or other mechanical methods. However, chemical exfoliation is found to be an effective method to exfoliate

DOI: 10.1201/9781003164975-1

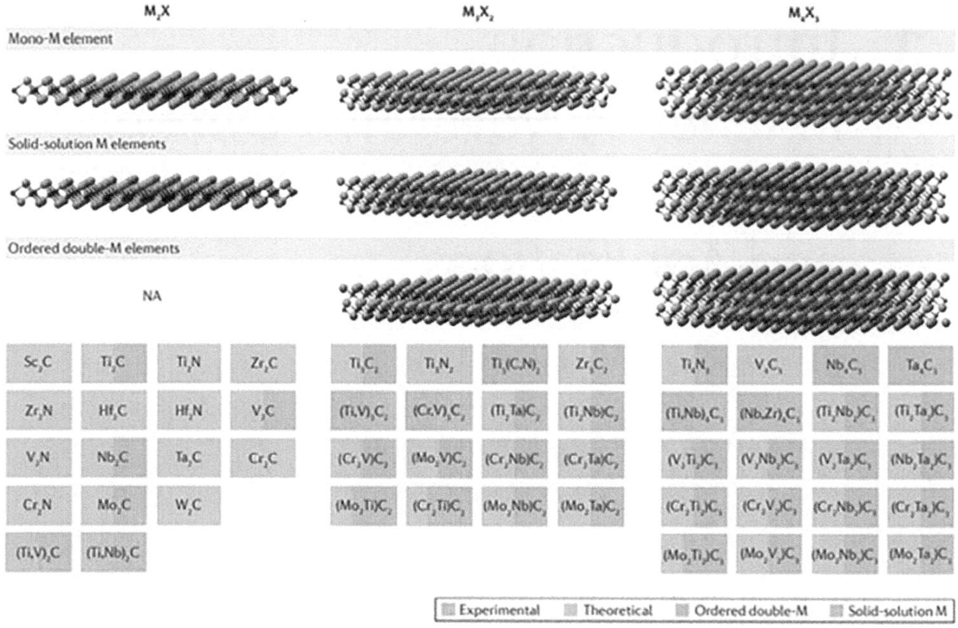

FIGURE 1.1 MXenes with three different formulae—M_2X, M_3X_2, and M_4X_3. The formula can contain single-M elements (e.g., Ti_2C or Ti_3C_2), a solid solution of minimum two different M elements (e.g., $(Ti,Nb)_4C_3$), and ordered two-M elements in which one type of transition metal element occupies the layers in the perimeter and the other one occupies M layers at the center (e.g., $(Mo_2V_2)C_3$).

MXene from its corresponding MAX phase without disrupting the MX bond, which is a mix of covalent, ionic, and metallic bonds (Naguib *et al.*, 2014). MXene synthesis is hence performed primarily using fluoride-based acid solutions such as hydrogen fluoride or a mix of hydrochloric acid and lithium fluoride (Anasori *et al.*, 2019). Chemical exfoliation allows selective etching and this gives MXenes an edge over the other two-dimensional materials considering the easiness for large-scale production (Anasori *et al.*, 2019; Jeon *et al.*, 2020). The surface termination attaches to the MXene during the exfoliation via a top-down method (Jeon *et al.*, 2020). It is possible to synthesize MXenes without surface termination using a bottom-up method. However, a bottom-up method for MXene synthesis needs to be explored better (Anasori *et al.*, 2019). The synthesis method has significant role in determining the lateral size, thickness, and defects in MXene (Jeon *et al.*, 2020).

MXenes have interesting electronic properties as they can be tuned by varying the elemental composition or the surface termination of MXene. Density functional theory (DFT) studies show that monolayers of bare MXene are metallic and have a high electron density in proximity of the Fermi level, in fact, higher than what is observed in its MAX phase (Naguib *et al.*, 2011, 2014; Shein and Ivanovskii, 2012; Khazaei *et al.*, 2013; Xie and Kent, 2013). However, like semiconducting materials, their Fermi levels are tunable with external stimuli (Gogotsi and Huang, 2021). Like other 2D materials, MXene also exhibits high electrical conductivity (~2.0×10^4 S cm^{-1} for $Ti_3C_2T_x$; Guo *et al.*, 2018), high surface area (external surface area of Ti_3C_2 MXene 2D flakes is found to be 61.56 m^2 g^{-1}; Rozmysłowska-Wojciechowska *et al.*, 2020), porosity (micropore surface area is found to be 31.52 m^2 g^{-1}; Rozmysłowska-Wojciechowska *et al.*, 2020), strength (for Ti_3C_2, in a perpendicular direction to the basal plane, Young's modulus is found to be about 80–100 GPa and tensile strength of a 40-nm thin nanoflake is found to be 670 MPa; Firestein *et al.*, 2020), hydrophilicity (~60-nm-thick membrane of MXene showed water flux in the range of 85.4 L m^{-2} h^{-1}; G. Liu *et al.*, 2018), high negative surface charge (ζ-potential for Ti_3C_2 is between −30 mV and −80 mV) and the ability for energy storage (Ma *et al.*, 2017; Akuzum *et al.*, 2018). The conductivity of MXene-Ti_3C_2

and its transition metal atom's redox activity has resulted in its wide application in batteries and supercapacitors (Gogotsi and Huang, 2021; VahidMohammadi *et al.*, 2021; Lukatskaya *et al.*, 2013; Pomerantseva and Gogotsi, 2017; Garg and Agarwal, 2020; C. J. Zhang *et al.*, 2018). Other important applications are electronic devices (e.g., field-effect transistors; Kim and Alshareef, 2020; Wang *et al.*, 2020), photovoltaics (Husam *et al.*, 2020; Yin *et al.*, 2021), polymer nanocomposite fillers (Zhang*et al.*, 2013; Naguib *et al.*, 2016a), photocatalysis (Ran *et al.*, 2017), and biomedical applications (George and Kandasubramanian, 2020; Sivasankarapillai *et al.*, 2020).

1.2 SYNTHESIS OF MXenes

Widely adopted synthesis method of MXenes are top-down methods in which a layer of A atoms is removed from the MAX phase, thus forming loosely packed 2D-natured-MXene layer. This is further reduced to a monolayer of MXene flakes (Naguib, Mashtalir, *et al.*, 2012). The metallic bond between M and A atoms is strong (compared to van der Waal's bond in other 2D materials) such that chemical or electrochemical selective etching that is aided by bonding energy difference and chemical reactivity of bonds in MAX precursors is needed (Naguib *et al.*, 2011, 2014; VahidMohammadi *et al.*, 2021). Different methods used for the synthesis of 2DMXenes are wet chemical etching using HF/fluoride salt and HCl/NH_4HF_2, non-HF etching using NaOH solutions at high temperatures, HCl via electrochemical methods, $ZnCl_2$ molten salt, iodine etching, and halogen etching. Finally, the scalable production of freestanding and stable $Ti_3C_2T_x$ Mxene is achieved via a blade-coating technique (VahidMohammadi *et al.*, 2021).

In the wet chemical etching method, fluoride ion containing strong acids, like hydrofluoric acid (HF) is employed to remove Al from its MAX phase, resulting in terminated, multilayer MXene powders (Gogotsi, 2013; Naguib *et al.*, 2011; Pomerantseva and Gogotsi, 2017). In the multilayered MXene, the 2D layers will be bonded by hydrogen bonds (Naguib *et al.*, 2011). Once the etching is over, next step is washing to remove the residual acid and attain a pH of around six. During washing, the etched solution is repeatedly centrifuged to segregate multilayer MXene from the acidic solution, after which decantation of acidic supernatant is done. Once the pH reaches the desired value, multilayered MXene flakes are collected using vacuum-assisted filtration and drying (Alhabeb *et al.*, 2017). Since the synthesis is carried out in an aqueous solution containing fluoride, MXenes are surface terminated by functional groups –O, –OH, and –F (Naguib *et al.*, 2011; Ahmed and Yi, 2017). The multilayered MXene is then delaminated into monolayer to a few-layered MXene using chemical intercalation using organic molecules like tetrabutylammonium hydroxide (TBAOH), dimethyl sulfoxide (DMSO), n-butylamine, or tetramethylammonium hydroxide (TMAOH) (Mashtalir *et al.*, 2013; Naguib *et al.*, 2015), and this is followed by sonication if required (Alhabeb *et al.*, 2017). The delaminated MXene flakes—$M_{n+1}X_nT_x$—will form a stable colloidal solution. During etching, the concentration of HF has an impact on the number of defects formed in MXene, and this affects the quality, as well as properties, of the synthesized Mxene (Sang *et al.*, 2016; Lipatov*et al.*, 2016a). Various MXenes were successfully etched from their MAX phases, Ti_3AlC_2, Ti_2AlC, V_2AlC, Nb_2AlC, and others, to name a few (Naguib *et al.*, 2021). The concentration of HF, temperature, and time of etching had to be varied for different MAX phases since the bond strength of transition metals with Al varied for each MAX phase (Alhabeb *et al.*, 2017; Naguib *et al.*, 2021). Additionally, tuning the etching conditions also help maintain a high yield and reduce the degradation of 2D MXene in acid (Naguib*et al.*, 2021). One of the major disadvantages of this method is that HF is a strong acid; hence, it is highly corrosive and poses risk to health. MILD, minimally intensive layer delamination, is a comparatively safe method. The MILD method uses a mixture of fluoride salts like LiF/NaF/KF and HCl acid (Ghidiu *et al.*, 2014). Another advantage of this method is that the fluoride salt–HCl mixture produces both HF that etches and an intercalant (Li ion if LiF is used) that delaminates side by side (Alhabeb *et al.*, 2017). The resulting MXene flakes exhibit claylike behavior due to the Li ions that are present between the flakes (VahidMohammadi *et al.*, 2021). It can be further delaminated into a single layer using sonication. MILD method offers better control over flake size and quality of Mxene (Lipatov *et al.*, 2016b; Choi *et al.*, 2018). Additionally, this method opened the research on electronic and optical research with its capacity for enhanced scalability and synthesis of MXene, especially $Ti_3C_2T_x$ (Shekhirev *et al.*, 2021). NH_4HF_2, ammonium bifluoride, have been used at room temperature for etching MXene from MAX phases (Halim *et al.*, 2014; Feng *et al.*, 2017). An Al atom from Ti3AlC2 is etched in both water and in polar organic solvent (N,N-dimethylformamide, N-methyl-2-pyrrolidone, dimethyl sulfoxide, and propylene carbonate) (Naguib *et al.*, 2021; Zhao*et al.*, 2020; Husmann *et al.*, 2020). During washing, an Al atom is extracted as AlF_3 and $(NH_4)_3AlF_6$. Due to the low solubilities of Al salts formed in organic solvents, a mix of HCl and propanol is used for washing. The clay for media of MXene thus delaminated into a mono- or a few layers using ultrasonication and stable colloids of 2DMXenes can be obtained in the same organic solvents (Zhao *et al.*, 2020). Two-dimensional $Ti_3C_2T_x$ has been produced from MAX phases using NH_4HF_2 (Zhao *et al.*, 2020; Husmann *et al.*, 2020; Halim *et al.*, 2014). A summary of wet chemical synthesis method is given in Figure 1.2.

Two-dimensional MXene–Ti_3C_2 has been synthesized from the MAX phase, Ti_3AlC_2 in choline chloride and deep eutectic solvents based on oxalic acid in the presence of NH_4F. The solid precursor and product at room temperature delivered a high purity of 98% in comparison with 95% seen in HF-etched Ti_3C_2 (J. Wu *et al.*, 2020).

Fluorine-free electrochemical etching using HCl has been demonstrated to produce Ti_2CT_x. In this method, porous Ti2AlC is used as electrodes in dilute HCl electrolyte. Electrochemical reaction results in the formation of layer of Ti2CTx on Ti2AlC electrode (Sun *et al.*, 2017). Ti2AlC will be etched into a three-layered structure consisting of CDC (carbide-derived carbon), MXene, and unetched MAX from outside to inside. The MXenes can be separated from this layer using bath sonication. These MXenes possess –Cl, –O, and –OH terminations (Sun *et al.*, 2017). Another fluoride-free etching has been done electrochemically to synthesize Ti_3C_2 using binary electrolyte (NH_4Cl/ $FeCl_3$ with NH_4OH). In this method, both cathode and anode used are Ti_3AlC_2 (Yang *et al.*, 2018). Chloride ions in the electrolyte enabled rapid etching by breaking Ti–Al bonds at the anode under ambient conditions (Yang *et al.*, 2018). The subsequent intercalation of ammonium hydroxide opened the edges of the anode that is etched, which led to the further etching of surfaces that were below.

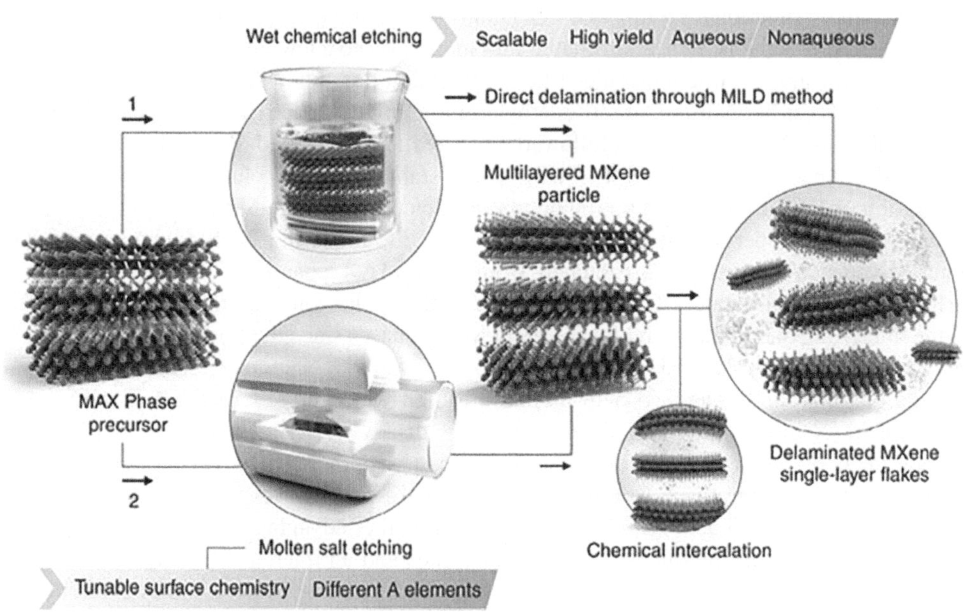

FIGURE 1.2 Schematic diagram of two methods of MXene synthesis—wet chemical and molten salt etching.

In this method, only –O and –OH terminations are found and fluorine termination is absent (Yang *et al.*, 2018). Multilayer $Ti_3C_2T_x$, which is free of fluorine termination, has been prepared using alkali (NaOH) treatment. The product was in powder form with 92 wt% purity (Li *et al.*, 2018). Molten salt ($ZnCl_2$, $CuCl_2$), and halogen (Br_2, I_2 and/or Cl_2 in cyclohexane) have also been used to etch MXenes (Mian Li *et al.*, 2019; Kamysbayev *et al.*, 2020; Shi *et al.*, 2021; Ali et al, 2021; Usman *et al.*, 2021).

Bottom-up synthesis method, chemical vapor deposition (CVD) has also been used to synthesize MXenes, such as α-Mo_2C, with lateral size of up to 100μm on a bilayer copper foil substrate above a foil of molybdenum (Xu *et al.*, 2015). A larger lateral size has been observed for MXene synthesized via this method, which was facilitated in analyzing their intrinsic properties (Gogotsi, 2015; Xu *et al.*, 2015). Other MXenes, such as tungsten carbide (WC) and tantalum carbide (TaC) have also been demonstrated using CVD (Xu *et al.*, 2015). Thus, a bottom-up method of synthesis is explored more for MXene monolayer synthesis.

1.3 PROPERTIES

MXenes are of exceptional interest in the world of 2D materials as they possess unique properties and characteristics that make them promising for a myriad of applications like energy storage, chemical sensing, electromagnetic interference (EMI) shielding, wireless communications, catalysis, and wearable electronics (Naguib *et al.*, 2012).

1.3.1 Physical, Chemical, and Mechanical Properties

One of the most notable features of MXenes that contributes to their unique properties is surface functionality. During the different processing steps, including intercalation and etching, there are these active functional groups that strongly couple to the surface of MXenes and have effects on their properties. Both the physical and chemical properties of an MXene are dependent on the host MAX phases, the processing, and postprocessing methods. Some of the properties, in lieu of the earlier argument, are discussed here. MXenes are environmentally sensitive materials, and the presence of water and oxygen facilitates their degradation, although vacuum filtration significantly counteracts the process (Double *et al.*, 2020). Due to the attached surface termination groups, MXenes are hydrophilic in nature and have water contact angle of about 21.5–35°(Ling *et al.*, 2014; Zhao *et al.*, 2017). It's been observed that F and O functionalization contributes to higher work functions while hydroxyl group tones it down (Khazaei *et al.*, 2015; Tahini and Smith, 2017). A work function ranging from 1.6 eV for $Sc_2C(OH)_2$ and 8.0 eV for Cr_2CO_2 has been reported (Khazaei *et al.*, 2015; Liu *et al.*, 2016). The use of Li^+ cation instead of HF during intercalation gives better electrical conductivity and modulates their structural response to relative humidity and hydration degree (Ghidiu *et al.*, 2014; Ghidiu *et al.*, 2016). Mechanical testing on MXenes have shown that they are stronger and stiffer than their MAX counterparts. The mechanical properties of MXenes depend on the surface terminations, and it is observed that MXenes with O-termination are stiffer in comparison with –F and –OH terminated ones (Zhan *et al.*, 2020). $Ti_3C_2T_x$ can support approximately 4000 times its own weight, and it can be raised by the formation of composites with polymer up to about 15,000 times its own weight (Ling *et al.*, 2014). $Ti_3C_2T_x$ has demonstrated a Young's modulus of about 330 GPa, which is higher than other solution-based 2D materials (Lipatov *et al.*, 2018). MXene composites with Polyvinyl alcohol (PVA), Polyacrylamide (PAM), and Ultra-high-molecular-weight-polyethylene (UHMWPE) have reported higher toughness and yield strength (Naguib *et al.*, 2016b; Zhang *et al.*, 2016; Mayerberger *et al.*, 2017). MXene composites with polymers have enhanced thermal, mechanical, and electrochemical properties. In Ti_3C_2Tx MXene/poly (vinylidene fluoride-trifluoro-ethylene-chlorofluoroethylene) (P[VDF-TrFE-CFE]) polymer composite (Tu *et al.*, 2018), a significantly high dielectric constant is observed, whereas $Ti_3C_2T_x$ MXene-based PVA hydrogel (Zhang *et al.*, 2018) exhibits good stretchability, asymmetrical strain sensitivity in addition to the ability for self-healing. Reinforcement of elastomers to improve their mechanical and functional properties is a frequent practice. Tensile tests done on $Ti_3C_2T_x$-reinforced nitrile butadiene rubber shows enhanced strength and increased electrical and thermal conductivity (Aakyiir *et al.*, 2020).

1.3.2 Electronic Properties

The majority of MXenes belong to the family of metals, semimetals, or semiconductors. Although almost all pristine MXenes are metallic, surface functionalization changes their nature to semiconducting mostly due to the shift in Fermi level or due to the modification of the crystal field around the transition metal in them (Khazaei *et al.*, 2017). The metallic conductivity originates from the highly aligned structure and high free-electron density in each MXene nanosheet paving the way for several electrical paths. The electrical properties of MXenes also depend on the lower concentrations of defects and larger size of flake, which promote higher conductivity (Come, *et al.*, 2012). This, in turn, is dependent on the synthesis, delamination, and sample preparation techniques. In the case of supported or freestanding films and compressed discs, both inter- and intra-flake resistance contribute to its conductivity, but for single-flake MXenes, electronic conduction is dependent on intra-flake and contact resistance. There is still room for in-depth investigation of MXene electronic conductivity that varies from less than 1 S cm^{-1} for pressed multilayered MXene disk to 1000s of S cm^{-1}, the highest being approximately 10,000 S cm^{-1} for $Ti_3C_2T_x$ thin films (Zhang *et al.*, 2017). The MXene conductivity is observed to be dependent on the etchants used during processing. HF–etchant results in smaller flakes with more defects, high –F terminations, and extremely low concentration of carriers at the Fermi level. However, significant improvement in conductivity is observed when a LiF–HCl mixture is used as the etchant (Ren *et al.*, 2016; Zhao *et al.*, 2016). Transparent thin MXene film samples tend to exhibit high conductivity, in which the possibility of finding well-aligned flakes with minimal inter-flake resistance occurs (Sang *et al.*, 2016). For a better understanding of the intrinsic transport properties of MXenes, it is imperative that studies must be conducted on singe flakes. However, the fabrication of single-flake MXenes of sufficient lateral size has been difficult. There are a few reports of successful studies on transport properties of mechanically exfoliated Ti_2CT_x (Lai *et al.*, 2015) and monolayer Field-effect transistors (FETs) (Sang *et al.*, 2016; Lipatov *et al.*, 2016b).

Another interesting electronic property of MXenes is its capability for EMI shielding effectiveness (SE). The good electrical conductivity, laminate structure, and ease of processibility with tunable surface chemistry attributes to excellent EMI SE. The films of Ti_3C_2 and its composites have shown EMI shielding capabilities comparable to metals and outperforming similar 2D materials. This outstanding performance of the MXene, $Ti_3C_2T_x$ is due to their high electrical conductivity (high electron density) and attenuation due to multiple reflections in the layered thin films (Iqbal *et al.*, 2020). Like other 2D materials, MXenes also exhibit anisotropy of electronic conductivity that is entirely dependent on the intrinsic van der Waals gap and stacked layers, and this makes it useful for multiple nanoelectronics devices.

1.3.3 Magnetic Properties

Although many MXenes have been theoretically predicted and experimentally synthesized, a majority of them are nonmagnetic, which limits their use in areas like spintronics. There are a few MXenes that are predicted to be intrinsically magnetic, like, Ti_2C and Cr_2C, of which a few are ferromagnetic and a few are antiferromagnetic (Gao *et al.*, 2016). In this quest to develop/discover magnetic MXenes, more attention is being given to half-metallic MXenes, since they have one spin channel as metallic and the other as semiconducting, resulting in a Fermi level with completely spin-polarized electrons. Although clear prediction of ferromagnetic and antiferromagnetic MXenes has been done, it is difficult to replicate the exact same result as the surface functionalization of these compounds during termination is still not completely controllable. A few examples of magnetic MXenes are Ti_2C, Ti_2N, Cr_2C, Cr_2N, and Mn_2C, among others (Ling *et al.*, 2014; He *et al.*, 2016; Wang, 2016; Xiong *et al.*, 2018). (See Figure 1.3.)

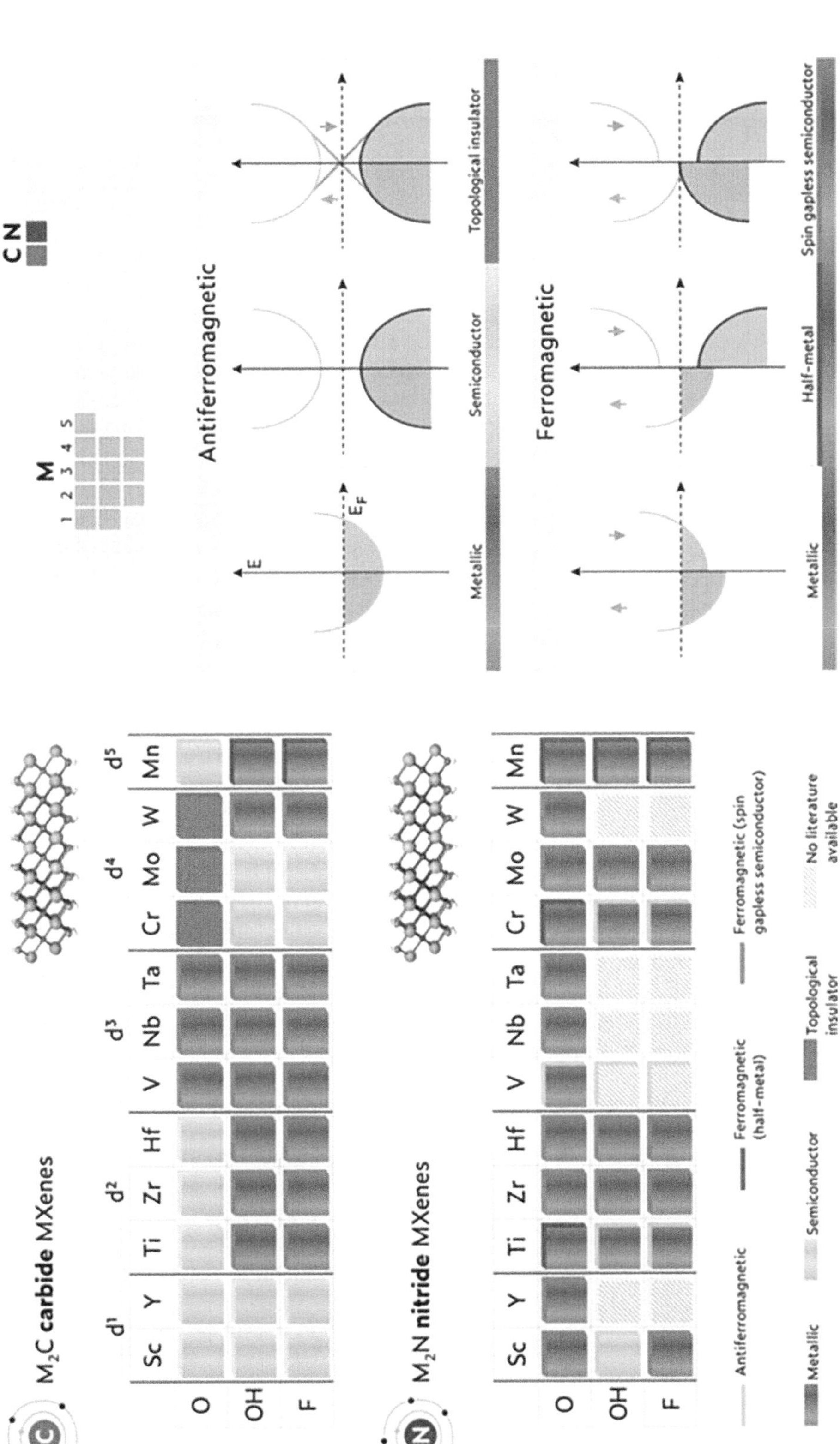

FIGURE 1.3 An outline of electronic and magnetic properties of M2CTx carbide MXenes and M2NTx nitride MXenes that are predicted theoretically. Here, M is an early-transition metal element and T_x is functional groups –O, –OH, or –F. The electronic band structure and ground-state magnetic configuration are shown.

1.3.4 Optical Properties

A wide range of optical properties, such as plasmonic properties and the nonlinear optical behavior of MXenes, have led the way to its application in saturable absorption for mode-locked lasing, plasmonic broadband absorber, surface-enhanced Raman scattering (SERS) substrate and photothermal therapy (Dong et al., 2018; Sarycheva *et al.*, 2017; Satheeshkumar *et al.*, 2016; Szuplewska *et al.*, 2019). Surface terminations are found to mostly affect the optical properties of MXenes (Wang *et al.*, 2017; Bai *et al.*, 2016).

The materials exhibit two types of plasmonic behavior—surface plasmon and bulk plasmon (Maier and Atwater, 2005). While bulk plasmons are oscillations of electrons seen deep inside the free carrier's body, surface plasmons (Figure 1.4) are electron oscillations at the surface at a metal–dielectric interface. Plasmonic excitation results in loss of energy due to the interactions of high-speed electrons within the solid (Maier and Atwater, 2005). The weak interlayer coupling of the MXene results in bulk plasmon energy that is not dependent on the thickness of the layers (El-Demellawi *et al.*, 2018). Additionally, $Ti_3C_2T_x$ MXene has been reported to exhibit high surface plasmon excitations (Mauchamp *et al.*, 2014). The surface plasmons can be observed at nanometer-scale and at mid-infrared spectral range, and this spectral range suitable for chemical-sensing applications (Ross Stanley, 2012). Depending on the functionalization, the surface plasmon wavelength of $Ti_3C_2T_x$ MXene can be varied or kept constant over a wide spectral range (6–2 μm for thicknesses with a spectral range between 5–45 nm). The variations can be made by controlling the thickness of MXene. Thus, the possibility to tune and enhance surface plasmons make MXenes promising materials for plasmonic applications (Mauchamp *et al.*, 2014). Additionally, plasmon energies of bare Ti_2C, Ti_2N, Ti_3C_2, and Ti_3N_2 are theoretically found to be 10.00, 11.62, 10.81, and 11.38 eV, respectively (Bai *et al.*, 2016; Khazaei *et al.*, 2017; Hantanasirisakul and Gogotsi, 2018).

MXenes exhibit photoluminescence in an aqueous solution, which is an important feature for biological applications (Wang *et al.*, 2020). Quantum dots (QDs) are developed using MXene as quantum confinement, and edge effects are enhanced with atomically thin quantum dots (Medintz *et al.*, 2005; Pan *et al.*, 2010). Mxene QDs with a photoluminescence property have been synthesized via hydrothermal method, which exhibited excitation-dependent photoluminescence behavior with a quantum yield of approximately 10% (Pan *et al.*, 2010; Xue *et al.*, 2017). Additionally, a $Ti_3C_2T_x$QD that exhibits white photoluminescence and fluorescence based on two photons has been reported to be synthesized using a solvothermal method, with a quantum yield of 9.36% (Lu *et al.*, 2019).

Nonlinear absorption of light by materials arises from nonlinear optical response of a material with the increase in intensity of illuminating light. This property has been applied in various fields like optical communications, sensing, lasers, and photonic devices (Wang *et al.*, 2020). Saturable absorbers (SAs) are important nonlinear optical devices used to convert continuous laser output to

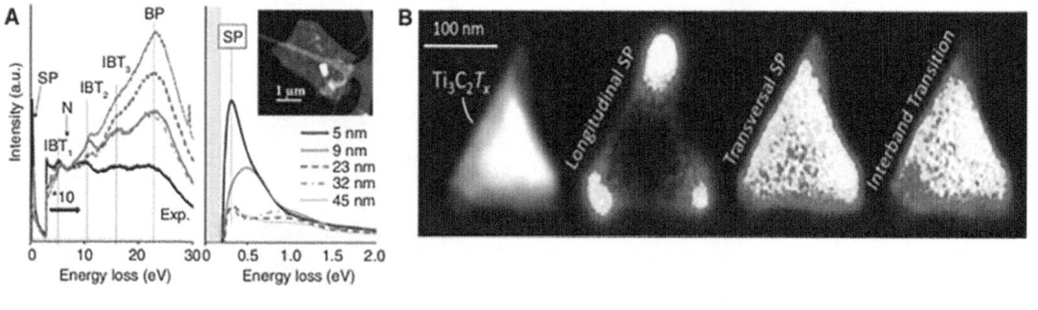

FIGURE 1.4 MXene surface plasmon. (A) Energy-loss spectra vs Intensity of $Ti_3C_2(OH)_2$ with various thicknesses (B) STEM-HAADF images of $Ti_3C_2T_x$ triangular flake displaying longitudinal SP, transversal SP, and interband transition SP.

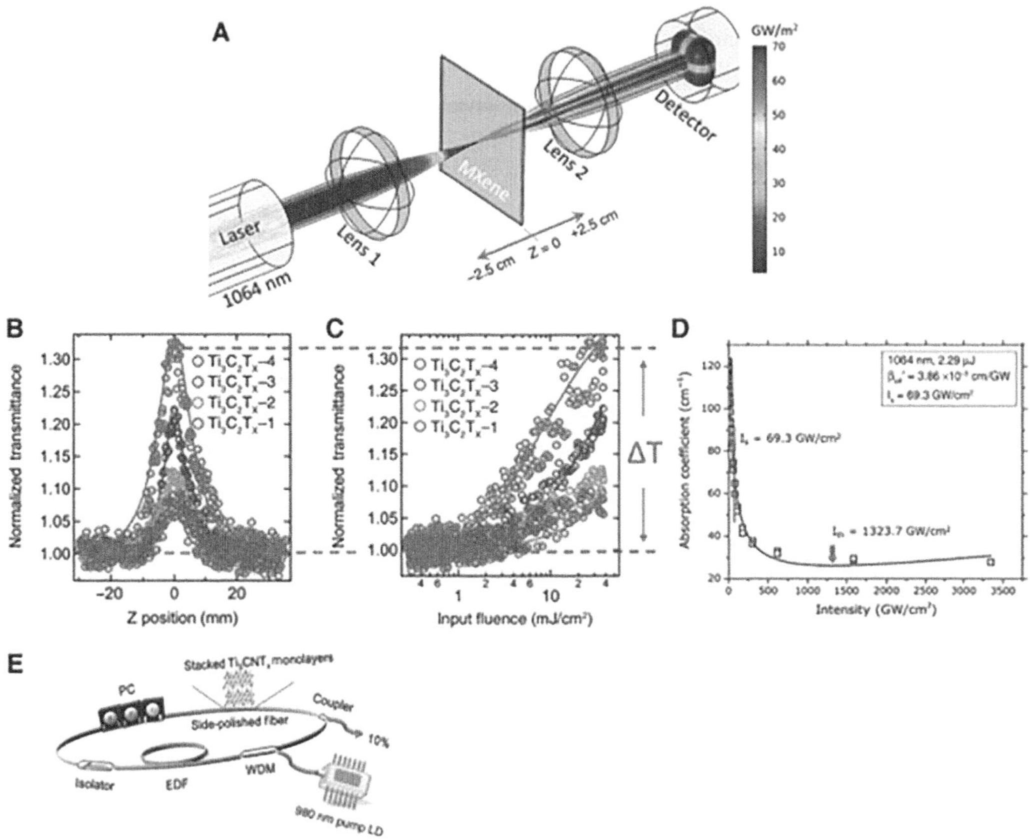

FIGURE 1.5 Nonlinear optical properties of MXene. (A) Z-scan measurement of an MXene thin film (B), (C) SA property of MXene, (D) Absorption coefficient vs laser intensity of MXene at 1064 nm (E) Ti_3CNT_x-based mode locker in ring-cavity erbium-doped fiber laser.

a train of laser pulses and are incorporated into fiber laser cavity (Ismail *et al.*, 2016). SAs have an absorption coefficient that decreases with increasing light intensity (Liu, 2009; Figure 1.5). The intensity-dependent optical transmission is a key factor in the SA (Yi *et al.*, 2019). The absorption of light by $Ti_3C_2T_x$ is reported as nonlinear, and $Ti_3C_2T_x$ exhibits saturable absorption for wavelengths from 800 to 1800 nm (Dong *et al.*, 2018). $Ti_3C_2T_x$ has been reported to exhibit broadband nonlinear optical response and has application in femtosecond laser generators (Jiang *et al.*, 2018). The optical properties of MXenes are a field that requires more extensive research for applications in photonic, optoelectronic, and plasmonic devices.

1.3.5 Rheological Properties

Rheological properties of materials in suspension help in understanding its application in solution processing, as well as manufacturing, methods. MXenes have been studied for viscous and visco-elastic properties using colloidal dispersions to high-loading slurries, which showed that a multilayer MXene suspension with about 70 wt.% are capable of fluidity (Akuzum *et al.*, 2018; Ghidiu *et al.*, 2014; Maleskiand Gogotsi, 2017). It is seen that MXene single-layer flakes possess high viscosity at extremely low concentrations and single-layer colloidal solutions exhibited partial elasticity even for the lowest concentration. Various applications have been demonstrated utilizing the rheological property of MXene (Akuzum *et al.*, 2018). A high aspect ratio and a negative surface charge result in complex hydrodynamics and electrostatic force between particles, and these aspects are found to have large impact on the rheology of material dispersions (Bergenholtz *et al.*, 2002; Russel, 1980). Single-layer and multilayer dispersions of MXene have been studied for various concentrations to understand rheological properties and demonstrated in various applications (C. [John] Zhang *et al.*, 2019; Yao *et al.*, 2020; Li *et al.*, 2020; Wang *et al.*, 2021).

1.4 APPLICATION

1.4.1 Electronic Applications

Metallic MXenes with high/low work functions (WFs) show enormous potential as Schottky barrier (SB)–free metal contacts to 2D semiconductors. It is predicted that –OH-terminated MXenes with ultralow WFs may form n-type barrier-free Schottky contacts with 2D semiconductors, like transition metal dichalcogenides and blue phosphorene (Zhan *et al.*, 2020). It was reported that there are greater chances for $Ti_3C_2T_x$ to form Ohmic contact, thereby improving carrier injection efficiency, due to its wide tunability of work function. Six MXenes—V_2CO_2, Mo_2CO_2, Cr_2CO_2, Cr_2NO_2, $V_4C_3O_2$, and V_2NO_2—are found to be useful as metal contacts to MoS_2 (You *et al.*, 2019). It paves the way for its application as electrode material in FETs (Lyu *et al.*, 2019).

An Ti_3C_2Tx MXene is used as contact electrodes in the fabrication of organic thin-film transistors (OTFTs) and complementary logic circuits (Lyu *et al.*, 2019). $Ti_3C_2T_x$ MXene-based electrodes in organic field-effect transistors (OFETs) demonstrated excellent carrier mobility of about 1 $cm^2 V^{-1}s^{-1}$ and on–off current ratio of about 10^7 for p-and n-type on flexible plastic substrates (Lyu *et al.*, 2019). The same MXene is also used as a contact for an Si–solar cell with better power conversion efficiency (PCE) and as Schottky junction heterostructures in self-driven photodetectors (Kang *et al.*, 2017; Fu *et al.*, 2019). There is also a recent report in which MXene is used in a perovskite solar cell to improve its work function and interface to enhance PCE (Agresti *et al.*, 2019).

A study of the mechanical properties of ultrathin MXene-aramid nanofiber (ANF) composite has proved to be efficient in EMI shielding. Even a lower MXene content of <10 wt% has been shown to improve the mechanical properties of pure ANF paper. A maximum of 197 MPa, 9.8% tensile strength and the fracture strain have been reported, respectively. With the addition of more MXene, it is seen that there is an increase in the electrical conductivity and EMI SE of a 17-μm-thick film to 170 S cm−1 and 33 dB, respectively (Xie *et al.*, 2019).

Five different MXenes, namely, Ti_2CT_x, $Ti_3C_2T_x$, Mo_2CT_x, Nb_2CT_x, and V_2CT_x are found to exhibit plasmonic features and are used in the fabrication of plasmonic photodetectors. Among the analyzed MXenes, Mo_2CT_x exhibited the best performance with excellent stability (Velusamy *et al.*, 2019).

MXene/polymer composites like hydroxyethyl cellulose (He *et al.*, 2020), PEDOT:PSS/PANI (R. Liu *et al.*, 2018; Y. Zhang *et al.*, 2019) with $Ti_3C_2T_x$MXene have been developed for EMI shielding applications. Other forms of MXene composites used for EMI applications are in the forms of hybrid structures (CNT/MXene/CNF (Cao *et al.*, 2019); PVDF/MXene/Ni (Wang *et al.*, 2019)), laminate structures (TiO_2-$Ti_3C_2T_x$/graphene (Xiang *et al.*, 2019), MXene–graphene–PVDF (Raagulan *et al.*, 2018), porous foams (Weng *et al.*, 2019; X. Wu *et al.*, 2020), and aerogels (Han *et al.*, 2019), all of which have found practical implementation in aerospace and military applications. Hydrogels based on $Ti_3C_2T_x$/PVA or CNT or PDAC (An *et al.*, 2018; Cai *et al.*, 2018; Y. Z. Zhang *et al.*, 2018) are developed and exhibits high sensitivity under strain, making it possible to sense vocal sounds of similar phonetics and finger motion sensing.

Several MXene-based sensors for sensing mechanical strain, biomolecules, and gases (ethanol, methanol, acetone, ammonia) have been developed. It is observed that MXene-based gas sensors show a high-selectivity response to hydrogen-bonding gases over acidic gases underlining the importance of the functional group (Lee *et al.*, 2017; Kim *et al.*, 2018). A piezoelectric sensor that is sensitive to external pressure and mechanical strain is developed, and it exhibits a gauge factor of 180.1–94.8 (0.19–0.82% strain) and 94.8–45.9 (0.82–2.13% strain) which is much higher when compared to the existing materials (Ma *et al.*, 2017).

This establishes the concept of using multiple MXenes for a range of 2D devices, like field-effect devices, Schottky junctions, ohmic contacts photodetectors, and memory devices, femtosecond pulsed laser, piezoresistive sensor, and contact electrodes for solar cells, among others.

1.4.2 Photonic Applications

Recent reports show that MXenes have been widely used for photonic-based applications. For example, V_2CT_x (T = F, OH, O) is used as conductive electrodes as it has absorption in the 500–2700-nm range with high conductivity (Ying *et al.*, 2018). MXenes like $Ti_3C_2T_x$ (Lin *et al.*, 2017), Nb_2CT_x (Lin *et al.*, 2017), $Ta_4C_3T_x$ (Dai *et al.*, 2017) modified by soybean phospholipid (SP) or polyvinylpyrrolidone (PVP) for improved colloidal stability, show high absorption coefficient in visible and near-infrared (NIR) regions, making them particularly suitable for photothermal therapy applications. A leaf–vein network-based MXene-coated flexible freestanding electrode is prepared, and its conductive and mechanical properties are found to be exemplary. The leaf-based electrode is used to fabricate a semi-transparent ultraviolet photodetector, which has a very clear photo response and outstanding flexibility, maintaining 90% of the photo response after being bent 1000 times (Chen *et al.*, 2020). A self-powered photodetector with an MXene–silicon van der Waals heterostructure is developed and tested (Zhang *et al.*, 2017). In another case, a $Ti_3C_2T_x$ transparent, flexible, conductive film with an optical figure of merit about 15, transmittance as high as 93%, and electrical conductivity of approximately 9880 S cm^{-1} has been developed for optoelectronic applications. MXene and noble metal (Ag, Au, Pd) nanoparticle hybrids have been fabricated for SERS application (Satheeshkumar *et al.*, 2016).

1.4.3 Biomedical Applications

The unique physicochemical properties of MXene make it convenient for its use in applications, such as photothermal therapy, anticancer treatment, and drug delivery. The main factors that must be considered when nanomaterials interact with living tissues are its biocompatibility and colloidal stability. It is imperative to gain a deeper understanding about neural dynamics, which would the pave way to treating brain diseases and injuries more effectively, and this is being made easier with the use of implantable microelectrodes. Simple and scalable flexible microelectrode arrays using MXene-based conductive inks have been developed for in vivo micro-electrocorticography recording and possessed remarkably high volumetric capacitance of 1500 F/cm^3 (Driscoll *et al.*, 2020). A multifunctional theranostic agent comprising of MnO_x/Ti_3C_2 composite MXene is developed and surface engineered with soybean phospholipid for efficient magnetic resonance (MR) and photoacoustic imaging-guided photothermal therapy (PTT) against cancer (Dai *et al.*, 2017). When Ti_3C_2 is loaded with doxorubicin (DOX), a fluorescent drug, and hyaluronic acid (HA), an effective tumor targeting nanomaterial is formed (Liu *et al.*, 2017). Ti_3C_2MXeneQDs are prepared, and it is seen that the photoluminescence emission maximum shifts from blue to red, and these could be used for in vitro bioimaging (Xue *et al.*, 2017). Another MXene that shows good promise in CT imaging is Ta_4C_3 due to their excellent biocompatibility and eco-friendly synthesis (Lin *et al.*, 2018; Z. Liu *et al.*, 2018). Similarly, MnO_x/Ti_3C_2 (Dai *et al.*, 2017) composites are also good candidates for MR imaging techniques.

Another important application is in the development of sensors to ensure environment and human safety. The major characteristics of an effective sensor are short response time, low limit of detection (LOD), high sensitivity, selectivity, low cost of production and wide linear range. A mediator-free nitrate biosensor with hemoglobin (Hb) immobilized on Ti_3C_2 MXenes as a protein-binding platform is reported (Liu *et al.*, 2015). They obtained direct energy transfer of Hb on the bare substrate and are used to detect the presence of nitrite in environmental water samples. There have been reports of a pH-sensitive Ti_3C_2QDs used to monitor intracellular pH and developed wearable fluorescent nano-sensors (Chen *et al.*, 2018). An Au/MXene composite biosensor is developed to detect enzymatic glucose (Rakhi *et al.*, 2016). A nonenzymatic glucose sensor using a porous nickel–cobalt-layered double hydroxide nanocomposite has been developed with a wide linearity range and low limit of detection (Menghui Li *et al.*, 2019). An MXene-based nanocomposite was used to detect dopamine from a conductive matrix of Pd/Pt nanoparticles with a DNA-adsorbed MXene surface exhibited excellent linearity and high selectivity against ascorbic acid, uric acid, and glucose (Zheng *et al.*, 2018). Hb immobilized on the MXene surface is also used to develop H_2O_2 mediator-free biosensor (Wang *et al.*, 2014). A Ti_3C_2-based FET biosensor for dopamine detection and neural activity monitoring is developed and had low LOD, and a wide linear range (Xu *et al.*, 2016). A wearable volatile organic compound sensor capable of detecting biomarkers (e.g., ammonia in lung diseases) are also developed by a group of researchers using MXene-based devices (Lee *et al.*, 2017). There are reports stating the use of MXene for removing from water bodies pollutants like heavy metal adsorption (Guo *et al.*, 2015, 2016; Ying *et al.*, 2015; Shahzad *et al.*, 2017), dye removal (Mashtalir *et al.*, 2014; Wong *et al.*, 2018), and biological sorption of bacteria (Xiao *et al.*, 2009; Jastrzebska *et al.*, 2017).

1.4.4 Electrochemical and Energy Storage Applications

Due to its role in developing sustainable energy, the engineering of efficient electrochemical energy storage is vital. To achieve this, the search for new materials and improvement of electrode architecture have been a research priority in recent years (Mathis *et al.*, 2019). However, conventionally seen charge storage mechanisms (ion intercalation and electric double-layer [EDL] capacitance) seen in electrochemical energy storage devices are not able to keep up with the demands of energy storage needs (Mathis *et al.*, 2019). Intercalation pseudo-capacitance that occurs via bulk redox reactions with ultrafast ion diffusion is emerging as an alternative mechanism in electrochemical energy (Mathis *et al.*, 2019; Okubo *et al.*, 2018). MXenes are promising intercalation pseudo-capacitor electrode materials due to their high electrical conductivity for a high current charge, a layered structure for ion intercalation, and complementary battery performance for large specific capacity (Okubo *et al.*, 2018). Other properties of MXenes like their ability for redox activity, high pseudo-capacitance rate, electrochemical stability in acidic and corrosive electrolytes, and good cycling stability make them preferable over other 2D materials (Okubo *et al.*, 2018). MXenes have already been demonstrated in various energy storage devices like batteries, fuel cells, and electrochemical capacitors (Lukatskaya *et al.*, 2013; Mashtalir *et al.*, 2013; Fei *et al.*, 2018; Huang *et al.*, 2019; Come *et al.*, 2012).

With the ever-increasing demand for energy and an urgency for implementing sustainable, green, and clean energy, renewable energy from sources like wind and sun are becoming vital. Since these energy sources are sporadic in nature, the role of energy storage devices is indispensable (Dusastre, 2010). High efficiency in energy conversion, comparatively long life-span, and low cost make electrochemical methods important technology for energy storage. The concept has been already demonstrated via lithium–ion batteries, lithium–sulfur batteries, sodium–ion batteries, potassium–ion batteries, and multivalent–ion batteries in many applications, such as portable devices and electric storage.

Two-dimensional layered materials (graphite, $LiCoO_2$, TiS_2), with their high anisotropic 2D structure for fast charge transfer and large interlayer space that allows fast ion intercalation/deintercalation, have proved to be good electrode materials in batteries (Aurbach *et al.*, 1999; Okumura *et al.*, 2020; Oh *et al.*, 2016). 2D MXenes possess tunability, high aspect ratio, high conductivity (9880 S/cm for $Ti_3C_2T_x$), and hydrophilicity (with contact angle in the range 21.5° to 35°), which makes them not only good electrodes but also a protective layer for metal anodes to prevent dendrite growth (Anasori*et al.*, 2019; Naguib *et al.*, 2014; Naguib *et al.*, 2021). MXene was initially used as a material for negative electrodes in Li–ion batteries (Naguib *et al.*, 2012). Nondelaminated Ti_2CT_x in 1M $LiPF_6$/ethylene carbonate (EC)–diethyl carbonate (DEC), delivered a specific capacity of 150–200 mAh/g at about 20 mA/g with good capacity retention (Naguib *et al.*, 2012). Similarly, nondelaminated Nb_2CT_x and V_2CT_xin a no-aqueous Li^+ electrolyte gave the specific capacities of 210–220 mAh/g (Naguib *et al.*, 2013).

Lithium-sulfur battery, with sulfur as active materials in cathode exhibits high specific capacity of around 1675 mAh/g (Ji *et al.*, 2009). But sulfide dissolution and deposition in electrolyte causes loss of sulfur and destroys specific density and cycle property of the battery (Bruce *et al.*, 2012; Liang *et al.*, 2015). In order to improve the performance of these batteries, new strategies are being developed, and one of them is finding the best material for building the anode of lithium–sulfur battery (Ji *et al.*, 2009; Liang *et al.*, 2015; D. W. Wang *et al.*, 2013). MXene sheets have been tested for the same, and sulfur atoms are seen to be bonding well with Mxenes (Lukatskaya *et al.*, 2013; Naguib *et al.*, 2012). Additionally, the high specific area of 2D MXenes provide sufficient area for electrochemical action and accommodation of sulfur (Liang *et al.*, 2015; Pang *et al.*, 2014; Tao *et al.*, 2014). It has been demonstrated that when a sulfur cathode is composited with Ti_2C or Ti_3C_2MXenes, it exhibits high performance (Zhao *et al.*, 2015). Additionally, considering the role of surface termination, it has been found that –OH-terminated MXene and its defective species are found to be efficient in trapping sulfur in lithium–sulfur batteries (Liang *et al.*, 2017).

High-power sodium–ion batteries have great potential in large-scale applications, and owing to the abundance and low cost of sodium, sodium–ion batteries are in focus (Wang *et al.*, 2015). The negative electrodes that are currently used in these batteries are made of materials such as hard carbon, TiO_2, phosphorous, and P_2-$Na_{0.66}[Li_{0.22}Ti_{0.78}]O_2$ suffer from trade-off between energy and power densities (Ponroucha and Goñiab, 2013; Oh *et al.*, 2014; Qian *et al.*, 2013; Y. Wang *et al.*, 2013). Additionally, a large volume change due to its reaction with large sodium ion leads to low cycle stability. New materials like MXenes with pseudocapacitive behavior as negative electrodes in sodium–ion batteries is found to resolve these problems (Wang *et al.*, 2015). $Ti_3C_2T_x$ MXene in non-aqueous Na^+electrolyte does not undergo any substantial structural change during the electrochemical reaction and has good capacity retention over 100 cycles (Kajiyama *et al.*, 2016). V_2CT_x has been demonstrated as positive electrode for sodium–ion capacitor. Electrochemical testing proved that both pseudocapacitive and diffusion-limited redox reactions occur in V_2CT_x. V_2CT_x/hard carbon sodium–ion capacitor is built, and it showed a maximum cell voltage of 3.5 V and a capacity of 50 mAh/g (Dall'Agnese *et al.*, 2015). In addition to Li–ion and sodium–ion batteries, MXenes are tested for potassium–ion batteries, multivalent–ion batteries, and others (Wu *et al.*, 2021; Ming *et al.*, 2021).

MXene-based electrodes have been tested for supercapacitors operating in aqueous electrolytes and nonaqueous electrolytes, as well as in hybrid, all-solid-state, and micro-supercapacitors (Anasori *et al.*, 2019). Other than pristine MXene electrodes, their composites have also been investigated to further enhance their electrochemical and mechanical performances (Dall'Agnese *et al.*, 2016; Ling *et al.*, 2014). $Ti_3C_2T_x$ electrodes are found to reversibly intercalate various cations like Na^+, K^+, $NH4^+$, Mg^{2+}, and Al^{3+} with a capacitance value of 440 Fcm^{-3}. Its performance over a wide range of charging rate in various alkaline and neutral electrolytes show capacitive-like electrochemical behavior (Lukatskaya *et al.*, 2013; Levi *et al.*, 2015). Meanwhile, in acidic electrolyte (1 M of H_2SO_4), an MXene clay electrode exhibited a higher capacitance of 900 Fcm^{-3} (Ghidiu *et al.*, 2014). Such a high value of capacitance is possible only with pseudocapacitive contributions from Ti atoms (Thomas *et al.*, 2021; Lukatskaya *et al.*, 2015). However, the limited potential range of MXene electrodes in aqueous electrolytes are a disadvantage (Anasori *et al.*, 2019). Using microporous electrode architecture, a capacitance of 210 Fg^{-1}at 10 $Vs^{-1}$and 100 Fg^{-1} at 40 Vs^{-1} rate and using MXene hydrogel by incorporating H_2SO_4 electrolyte in between MXene layers enabled capacitance of about 1500 Fcm^{-3} (Lukatskaya *et al.*, 2017). MXene electrodes have also been tested using carbon current collectors and potential window could be increased to 1V in 3M H_2SO_4 (Lukatskaya *et al.*, 2017).

Due to the absence of water electrolysis reaction, non-aqueous electrolytes offer higher potential window for MXene electrodes. A composite of MXene/carbon nanotube electrodes in 1 M of ethylmethylimidazole–triflouro-sulfonylimide gave a potential window of 1.8 V and a capacitance of 245 Fcm^{-3} (Dall'Agnese *et al.*, 2016; Figure 1.6). MXene electrodes, when used in neat ionic liquid electrolyte, showed a cell voltage of 3 V and capacitance of 75 $Fg^{-1}$with good power performance (Lin *et al.*, 2016). The key issue with MXene electrodes in a nonaqueous electrolyte is that it exhibits low capacitance, only in the range of 80 to 100 Fg^{-1}. These values, which are much less than what is seen in acidic electrolytes, is due to the absence of hydronium ions or protons of water molecules that facilitate redox pseudo-capacitance in transition metal (Conway *et al.*, 1997).

All-solid-state supercapacitors and micro-supercapacitors (MSCs), which are safer energy storage systems, are being developed for implementation in applications such as wearable devices and flexible displays (Lu *et al.*, 2014; Peng*et al.*, 2016; Hu *et al.*, 2017; Kurra *et al.*, 2016; Li *et al.*, 2016). MXene-based electrodes are used in all-solid-state MSCs for on-chip energy storage, which delivered an areal capacitance of 28 $mFcm^{-2}$ (Peng *et al.*, 2016). $Ti_3C_2T_x$ MXene ink is used for stamping coplanar MSCs on a flexible substrate, and this method of fabrication can result in low-cost, scalable, and fast manufacturing of MSCs (C. J. Zhang *et al.*, 2018; Figure 1.7).

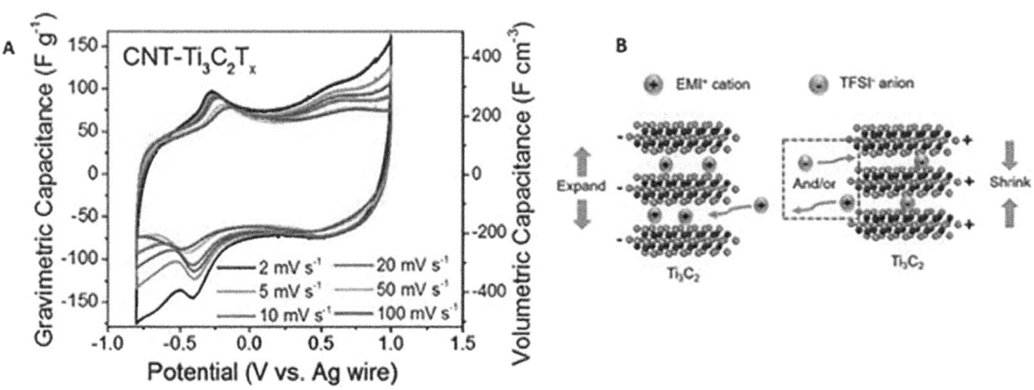

FIGURE 1.6 (a) Cyclic voltammetry plot and (b) charge storage mechanism of $Ti_3C_3T_x$ electrode in EMI–TFSI ionic liquid electrolyte.

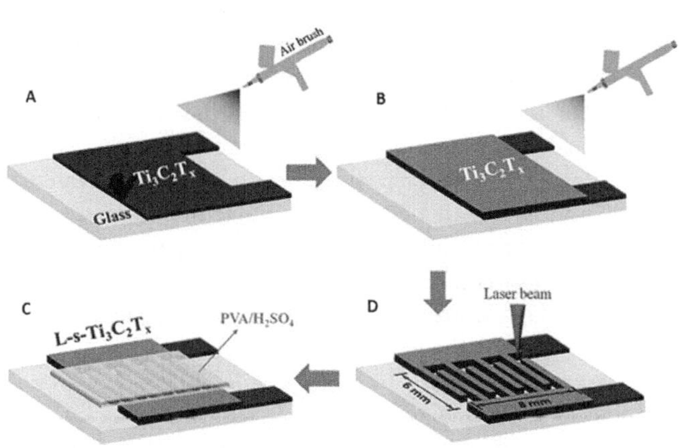

FIGURE 1.7 Illustration of $Ti_3C_2T_x$ micro supercapacitor fabrication (A) using spray coating: thin layer of $Ti_3C_2T_x$ is formed at the bottom as the current collector; (B) coating of an electroactive $Ti_3C_2T_x$ film; (C) laser cutting to obtain interdigital electrodes; and (D) painting of PVA/H_2SO_4 get electrolyte into interdigital $Ti_3C_2T_x$ to fabricate MSCs.

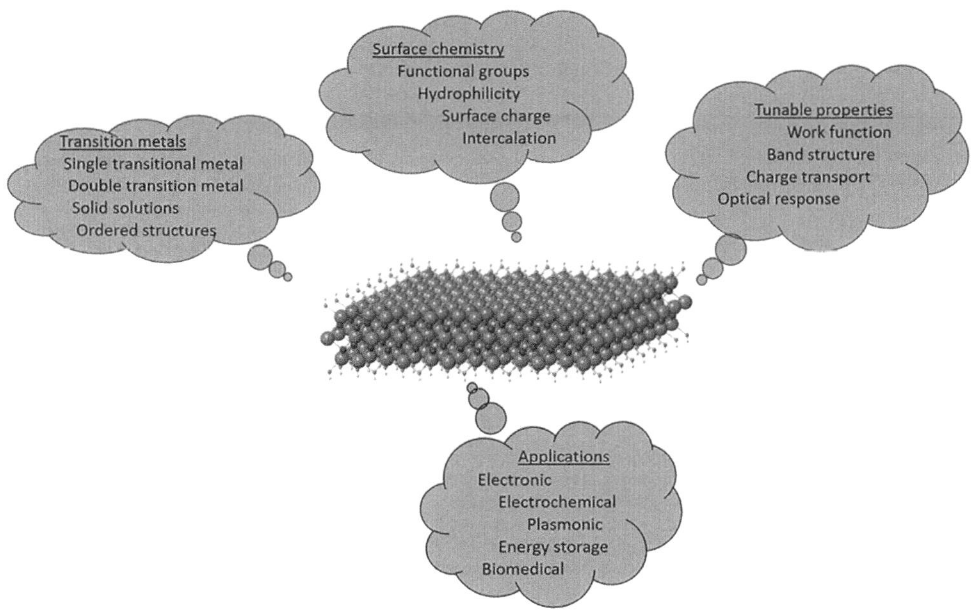

FIGURE 1.8 Schematic depiction of the most important properties and applications of MXenes.

These demonstrations show that MXene with its pseudocapacitive behavior can be used for energy harvesting and storage. Further study on optimization of MXenes using methods such as building nanocomposites or developing better electrode architecture can open exciting results in energy harvesting, conversion, and storage using electrochemical methods (Lukatskaya *et al.*, 2017; Figure 1.8).

1.5 CONCLUSION

The large family of MXenes has been found to have immense potential in various applications and is being extensively researched for the possibility of new properties, safer etching methods, and new functionalities. The outstanding electrochemical properties of MXenes like their ability for ion adsorption and diffusion have given these materials extra leverage in energy storage applications compared to other 2D materials. The biomedical applications of MXenes prove that they are biocompatible as well. MXetronics is an evolving field owing to the potential electronic and photonic applications of MXenes. While the research on its application in various applications is advancing, there are a few challenges that need to be overcome for its realization into scalable industrial applications. An environmentally friendly, safer, and scalable etching method is something that must be worked on since current methods involve concentrated acids and other chemicals. Understanding the surface chemistry, realizing new surface terminations, and the ability to control surface terminations can lead to various tunable properties and functionalities. Additionally, the possibility of MXene synthesis without surface termination is to be devised. Even though more than 60 MXenes have been predicted, Ti_3C_2 is the MXene that is widely studied for various applications. The need for extending studies to understand the stable etching, stability, properties, and functionalities of other MXenes are necessary for further advancement in MXenes. Other challenges like understanding magnetic properties of MXenes, defect engineering for better control of properties, improvement of chemical and thermal stability of MXenes, etching of MXenes from non-Al MAX phases, and learning the electromagnetic attenuation properties of MXenes are open to research for these materials, which have greatly progressed within a decade of its discovery.

REFERENCES

Aakyiir, M., Araby, S., Michelmore, A., Meng, Q., Amer, Y., Yao, Y., Li, M., Wu, X., Zhang, L. and Ma, J. (2020) 'Elastomer nanocomposites containing MXene for mechanical robustness and electrical and thermal conductivity', *Nanotechnology*, 31(31). doi: 10.1088/1361-6528/ab88eb.

Abdelmalak, M. N. (2014) *MXenes: A New Family of Two-Dimensional Materials and Its Application as Electrodes for Li-Ion Batteries*. Philadelphia, PA: Drexel University.

Agresti, A., Pazniak, A., Pescetelli, S., Di Vito, A., Rossi, D., Pecchia, A., Auf der Maur, M., Liedl, A., Larciprete, R., Kuznetsov, D. V. and Saranin, D. (2019) 'Titanium-carbide MXenes for work function and interface engineering in perovskite solar cells', *Nature Materials*, 18(11), pp. 1228–1234. doi: 10.1038/s41563-019-0478-1.

Ahmed, Sohail and Yi, Jiabao (2017) 'Two-dimensional transition metal dichalcogenides and their charge carrier mobilities in field-effect transistors', *Nano-Micro Letters*, 9(50). doi: 10.1007/s40820-017-0152-6.

Akuzum, B., Maleski, K., Anasori, B., Lelyukh, P., Alvarez, N. J., Kumbur, E. C. and Gogotsi, Y. (2018) 'Rheological characteristics of 2D titanium carbide (MXene) dispersions: A guide for processing MXenes', *ACS Nano*, 12(3), pp. 2685–2694. doi: 10.1021/acsnano.7b08889.

Alhabeb, M., Maleski, K., Anasori, B., Lelyukh, P., Clark, L., Sin, S. and Gogotsi, Y. (2017) 'Guidelines for synthesis and processing of two-dimensional titanium carbide (Ti3C2Tx MXene)', *Chemistry of Materials*, 29(18), pp. 7633–7644. doi: 10.1021/acs.chemmater.7b02847.

Ali, Jawaid, Hassan, Asra, Neher, Gregory, Nepal, Dhriti, Pachter, Ruth, Kennedy, W. Joshua, Ramakrishnan, Subramanian and Vaia, R. A. (2021) 'Halogen etch of Ti3AlC2 MAX phase for MXene fabrication', *ACS Nano*, 15(2), pp. 2771–2777. doi: 10.1021/acsnano.0c08630.

An, H., Habib, T., Shah, S., Gao, H., Radovic, M., Green, M. J. and Lutkenhaus, J. L. (2018) 'Surface-agnostic highly stretchable and bendable conductive MXene multilayers', *Science Advances*, 4(3), pp. 1–9. doi: 10.1126/sciadv.aaq0118.

Anasori, B., Xie, Y., Beidaghi, M., Lu, J., Hosler, B. C., Hultman, L., Kent, P. R., Gogotsi, Y. and Barsoum, M. W. (2015) 'Two-dimensional, ordered, double transition metals carbides (MXenes)', *ACS Nano*, 9(10), pp. 9507–9516. doi: 10.1021/acsnano.5b03591.

Anasori, B., Lukatskaya, M. R. and Gogotsi, Y. (2019) *2D Metal Carbides and Nitrides (MXenes)*. Switzerland: Springer Nature Switzerland AG. doi: 10.1007/978-3-030-19026-2.

Aurbach, D., Markovsky, B., Weissman, I., Levi, E. and Ein-Eli, Y. (1999) 'On the correlation between surface chemistry and performance of graphite negative electrodes for Li ion batteries', *Electrochimica Acta*, 45(1), pp. 67–86. doi: 10.1016/S0013-4686(99)00194-2.

Bai, Y., Zhou, K., Srikanth, N., Pang, J. H., He, X. and Wang, R. (2016) 'Dependence of elastic and optical properties on surface terminated groups in two-dimensional MXene monolayers: A first-principles study', *RSC Advances*, 6(42), pp. 35731–35739. doi: 10.1039/c6ra03090d.

Bergenholtz, J., Brady, J. F. and Vicic, M. (2002) 'The non-Newtonian rheology of dilute colloidal suspensions', *Journal of Fluid Mechanics*, 456, pp. 239–275. doi: 10.1017/S0022112001007583.

Bruce, P. G., Freunberger, S. A., Hardwick, L. J. and Tarascon, J. M. (2012) 'Li-O2 and Li-S batteries with high energy storage', *Nature Materials*, 11(1), pp. 19–29. doi: 10.1038/nmat3191.

Cai, Y., Shen, J., Ge, G., Zhang, Y., Jin, W., Huang, W., Shao, J., Yang, J. and Dong, X. (2018) 'Stretchable Ti3C2Tx MXene/carbon nanotube composite based strain sensor with ultrahigh sensitivity and tunable sensing range', *ACS Nano*, 12(1), pp. 56–62. doi: 10.1021/acsnano.7b06251.

Cao, W., Ma, C., Tan, S., Ma, M., Wan, P. and Chen, F. (2019) 'Ultrathin and flexible CNTs/MXene/cellulose nanofibrils composite paper for electromagnetic interference shielding', *Nano-Micro Letters*, 11(1), p. 72. doi: 10.1007/s40820-019-0304-y.

Chaudhuri, K., Alhabeb, M., Wang, Z., Shalaev, V. M., Gogotsi, Y. and Boltasseva, A. (2018) 'Highly broadband absorber using plasmonic titanium carbide (MXene)', *ACS Photonics*, 5(3), pp. 1115–1122. doi: 10.1016/j.pmatsci.2020.100757.

Chen, J., Li, Z., Ni, F., Ouyang, W. and Fang, X. (2020) 'Bio-inspired transparent MXene electrodes for flexible UV photodetectors', *Materials Horizons*, 7(7), pp. 1828–1833. doi: 10.1039/d0mh00394h.

Chen, X., Sun, X., Xu, W., Pan, G., Zhou, D., Zhu, J., Wang, H., Bai, X., Dong, B. and Song, H. (2018) 'Ratiometric photoluminescence sensing based on Ti3C2 MXene quantum dots as an intracellular pH sensor', *Nanoscale*, 10(3), pp. 1111–1118. doi: 10.1039/C7NR06958H.

Choi, G., Shahzad, F., Bahk, Y. M., Jhon, Y. M., Park, H., Alhabeb, M., Anasori, B., Kim, D. S., Koo, C. M., Gogotsi, Y. and Seo, M. (2018) 'Enhanced terahertz shielding of MXenes with nano-metamaterials', *Advanced Optical Materials*, 6(5), pp. 1–6. doi: 10.1002/adom.201701076.

Come, J., Naguib, M., Rozier, P., Barsoum, M. W., Gogotsi, Y., Taberna, P. L., Morcrette, M. and Simon, P. (2012) 'A non-aqueous asymmetric cell with a Ti2C-based two-dimensional negative electrode', *Journal of the Electrochemical Society*, 159(8), p. A1368. doi: 10.1149/2.003208jes.

Conway, B. E., Birss, V. and Wojtowicz, J. (1997) 'The role and utilization of pseudocapacitance for energy storage by supercapacitors', *Journal of Power Sources*, 66(1–2), pp. 1–14. doi: 10.1016/S0378-7753(96)02474-3.

Dai, C., Chen, Y., Jing, X., Xiang, L., Yang, D., Lin, H., Liu, Z., Han, X. and Wu, R. (2017) 'Two-dimensional tantalum carbide (MXenes) composite nanosheets for multiple imaging-guided photothermal tumor ablation', *ACS Nano*, 11(12), pp. 12696–12712. doi: 10.1021/acsnano.7b07241.

Dai, C., Lin, H., Xu, G., Liu, Z., Wu, R. and Chen, Y. (2017) 'Biocompatible 2D titanium carbide (MXenes) composite nanosheets for pH-responsive MRI-guided tumor hyperthermia', *Chemistry of Materials*, 29(20), pp. 8637–8652. doi: 10.1021/acs.chemmater.7b02441.

Dall'Agnese, Y., Rozier, P., Taberna, P. L., Gogotsi, Y. and Simon, P. (2016) 'Capacitance of two-dimensional titanium carbide (MXene) and MXene/carbon nanotube composites in organic electrolytes', *Journal of Power Sources*, 306, pp. 510–515. doi: 10.1016/j.jpowsour.2015.12.036.

Dall'Agnese, Y., Taberna, P. L., Gogotsi, Y. and Simon, P. (2015) 'Two-dimensional vanadium carbide (MXene) as positive electrode for sodium-ion capacitors', *Journal of Physical Chemistry Letters*, 6(12), pp. 2305–2309. doi: 10.1021/acs.jpclett.5b00868.

Dong, Y., Chertopalov, S., Maleski, K., Anasori, B., Hu, L., Bhattacharya, S., Rao, A. M., Gogotsi, Y., Mochalin, V. N. and Podila, R. (2018) 'Saturable absorption in 2D Ti3C2 MXene thin films for passive photonic diodes', *Advanced Materials*, 30(10), p. 1705714. doi: 10.1002/adma.201705714.

Driscoll, N., Maleski, K., Richardson, A. G., Murphy, B., Anasori, B., Lucas, T. H., Gogotsi, Y. and Vitale, F. (2020) 'Fabrication of Ti3c2 MXENE microelectrode arrays for in vivo neural recording', *Journal of Visualized Experiments*, 2020(156), pp. 1–9. doi: 10.3791/60741.

Dusastre, V. (2010) *Materials for Sustainable Energy*, Edited by Vincent Dusastre. Londan: Nature Publishing. doi: 10.1142/7848.

El-Demellawi, J. K., Lopatin, S., Yin, J., Mohammed, O. F. and Alshareef, H. N. (2018) 'Tunable multipolar surface plasmons in 2D Ti3C2 Tx MXene flakes', *ACS Nano*, 12(8), pp. 8485–8493. doi: 10.1021/acsnano.8b04029.

Fei, M., Lin, R., Deng, Y., Xian, H., Bian, R., Zhang, X., Cheng, J., Xu, C. and Cai, D. (2018) 'Polybenzimidazole/Mxene composite membranes for intermediate temperature polymer electrolyte membrane fuel cells', *Nanotechnology*, 29(3). doi: 10.1088/1361-6528/aa9ab0.

Feng, A., Yu, Y., Wang, Y., Jiang, F., Yu, Y., Mi, L. and Song, L. (2017) 'Two-dimensional MXene Ti3C2 produced by exfoliation of Ti3AlC2', *Materials and Design*, 114, pp. 161–166. doi: 10.1016/j.matdes.2016.10.053.

Firestein, K. L., von Treifeldt, J. E., Kvashnin, D. G., Fernando, J. F., Zhang, C., Kvashnin, A. G., Podryabinkin, E. V., Shapeev, A. V., Siriwardena, D. P., Sorokin, P. B. and Golberg, D. (2020) 'Young's modulus and tensile strength of Ti3C2 MXene nanosheets as revealed by *in situ* TEM probing, AFM nanomechanical mapping, and theoretical calculations', *Nano Lett.*, 20(8), pp. 5900–5908. doi: 10.1021/acs.nanolett.0c01861.

Fu, H. C., Ramalingam, V., Kim, H., Lin, C. H., Fang, X., Alshareef, H. N. and He, J. H. (2019) 'MXene-contacted silicon solar cells with 11.5% efficiency', *Advanced Energy Materials*, 9(22), p. 1900180. doi: 10.1002/aenm.201900180.

Gao, G., Ding, G., Li, J., Yao, K., Wu, M. and Qian, M. (2016) 'Monolayer MXenes: Promising half-metals and spin gapless semiconductors', *Nanoscale*, 8(16), pp. 8986–8994. doi: 10.1039/C6NR01333C.

Garg, R., Agarwal, A. and Agarwal, M. (2020) 'A review on MXene for energy storage application: Effect of interlayer distance', *Materials Research Express*, 7(2). doi: 10.1088/2053-1591/ab750d.

George, S. M. and Kandasubramanian, B. (2020) 'Advancements in MXene-polymer composites for various biomedical applications', *Ceramics International*, 46(7), pp. 8522–8535. doi: 10.1016/j.ceramint.2019.12.257.

Ghidiu, M., Halim, J., Kota, S., Bish, D., Gogotsi, Y. and Barsoum, M. W. (2016) 'Ion-exchange and cation solvation reactions in Ti3C2 MXene', *Chemistry of Materials*, 28(10), pp. 3507–3514. doi: 10.1021/acs.chemmater.6b01275.

Ghidiu, M., Lukatskaya, M. R., Zhao, M. Q., Gogotsi, Y. and Barsoum, M. W. (2014) 'Conductive two-dimensional titanium carbide "clay" with high volumetric capacitance', *Nature*, 516(7529), pp. 78–81. doi: 10.1038/nature13970.

Gogotsi, Y. (2015) 'Chemical vapour deposition: Transition metal carbides go 2D', *Nature Materials*, 14(11), pp. 1079–1080. doi: 10.1038/nmat4386.

Guo, J., Peng, Q., Fu, H., Zou, G. and Zhang, Q. (2015) 'Heavy-metal adsorption behavior of two-dimensional alkalization-intercalated MXene by first-principles calculations', *Journal of Physical Chemistry C*, 119(36), pp. 20923–20930. doi: 10.1021/acs.jpcc.5b05426.

Guo, X., Zhang, X., Zhao, S., Huang, Q. and Xue, J. (2016) 'High adsorption capacity of heavy metals on two-dimensional MXenes: An ab initio study with molecular dynamics simulation', *Physical Chemistry Chemical Physics*, 18(1), pp. 228–233. doi: 10.1039/c5cp06078h.

Guo, Z., Gao, L., Xu, Z., Teo, S., Zhang, C., Kamata, Y., Hayase, S. and Ma, T. (2018) 'High electrical conductivity 2D MXene serves as additive of perovskite for efficient solar cells', *Small*, 14(47), pp. 1–8. doi: 10.1002/smll.201802738.

Halim, J., Lukatskaya, M. R., Cook, K. M., Lu, J., Smith, C. R., Näslund, L. Å., May, S. J., Hultman, L., Gogotsi, Y., Eklund, P. and Barsoum, M. W. (2014) 'Transparent conductive two-dimensional titanium carbide epitaxial thin films', *Chemistry of Materials*, 26(7), pp. 2374–2381. doi: 10.1021/cm500641a.

Han, M., Yin, X., Hantanasirisakul, K., Li, X., Iqbal, A., Hatter, C. B., Anasori, B., Koo, C. M., Torita, T., Soda, Y. and Zhang, L. (2019) 'Anisotropic MXene aerogels with a mechanically tunable ratio of electromagnetic wave reflection to absorption', *Advanced Optical Materials*, 7(10), p. 1900267. doi: 10.1002/adom.201900267.

Hantanasirisakul, K. and Gogotsi, Y. (2018) 'Electronic and optical properties of 2D transition metal carbides and nitrides (MXenes)', *Advanced Materials*, 30(52). doi: 10.1002/adma.201804779.

He, J., Lyu, P. and Nachtigall, P. (2016) 'New two-dimensional Mn-based MXenes with room-temperature ferromagnetism and half-metallicity', *Journal of Materials Chemistry C*, 4(47), pp. 11143–11149. doi: 10.1039/C6TC03917K.

He, P., Cao, M. S., Cai, Y. Z., Shu, J. C., Cao, W. Q. and Yuan, J. (2020) 'Self-assembling flexible 2D carbide MXene film with tunable integrated electron migration and group relaxation toward energy storage and green EMI shielding', *Carbon*, 157, pp. 80–89. doi: 10.1016/j.carbon.2019.10.009.

Hu, M., Li, Z., Li, G., Hu, T., Zhang, C. and Wang, X. (2017) 'All-solid-state flexible fiber-based MXene supercapacitors', *Advanced Materials Technologies*, 2(10). doi: 10.1002/admt.201700143.

Hu, Y., Fan, X. L., Guo, W. J., An, Y. R., Luo, Z. F. and Kong, J. (2019) 'Ordered double-M elements MXenes TiMC: Large in-plane stiffness and ferromagnetism', *Journal of Magnetism and Magnetic Materials*, 486. doi: 10.1016/j.jmmm.2019.165280.

Huang, L., Li, T., Liu, Q. and Gu, J. (2019) 'Fluorine-free Ti3C2Tx as anode materials for Li-ion batteries', *Electrochemistry Communications*, 104(April), p. 106472. doi: 10.1016/j.elecom.2019.05.021.

Husmann, S., Budak, Ö., Shim, H., Liang, K., Aslan, M., Kruth, A., Quade, A., Naguib, M. and Presser, V. (2020) 'Ionic liquid-based synthesis of MXene', *Chemical Communications*, 56(75), pp. 11082–11085. doi: 10.1039/d0cc03189e.

Iqbal, A., Sambyal, P. and Koo, C. M. (2020) '2D MXenes for electromagnetic shielding: A review', *Advanced Functional Materials*, 30(47), pp. 1–25. doi: 10.1002/adfm.202000883.

Ismail, M. A., Harun, S. W., Ahmad, H. and Paul, M. C. (2016) *Passive Q-switched and Mode-Locked Fiber Lasers Using Carbon-Based Saturable Absorbers*. London, UK: IntechOpen. doi: 10.5772/60690.

Jastrzebska, A., Karwowska, E., Basiak, D., Zawada, A., Ziemkowska, W., Wojciechowski, T., Jakubowska, D. and Olszyna, A. (2017) 'Biological activity and bio-sorption properties of the Ti2C studied by means of zeta potential and SEM', *International Journal of Electrochemical Science*, 12(3), pp. 2159–2172. doi: 10.20964/2017.03.06.

Jeon, J., Yang, Y., Choi, H., Park, J. H., Lee, B. H. and Lee, S. (2020) 'MXenes for future nanophotonic device applications', *Nanophotonics*, 9(7), pp. 1831–1853. doi: 10.1515/nanoph-2020-0060.

Ji, X., Lee, K. T. and Nazar, L. F. (2009) 'A highly ordered nanostructured carbon-sulphur cathode for lithium-sulphur batteries', *Nature Materials*, 8(6), pp. 500–506. doi: 10.1038/nmat2460.

Jiang, Xiantao, Liu, Shunxiang, Liang, Weiyuan, Luo, Shaojuan, He, Zhiliang, Ge, Yanqi, Wang, Huide, Cao, Rui, Zhang, Feng, Wen, Qiao, Li, Jianqing, Bao, Qiaoliang, Fan, Dianyuan and Zhang, H. (2018) 'Broadband nonlinear photonics in few-layer MXene Ti3C2Tx (T = F, O, or OH)', *Laser and Photonics Review*, 12(2), p. 1700229. doi: 10.1002/lpor.201700229.

Kajiyama, S., Szabova, L., Sodeyama, K., Iinuma, H., Morita, R., Gotoh, K., Tateyama, Y., Okubo, M. and Yamada, A. (2016) 'Sodium-ion intercalation mechanism in MXene nanosheets', *ACS Nano*, 10(3), pp. 3334–3341. doi: 10.1021/acsnano.5b06958.

Kamysbayev, V., Filatov, A. S., Hu, H., Rui, X., Lagunas, F., Wang, D., Klie, R. F. and Talapin, D. V. (2020) 'Covalent surface modifications and superconductivity of two-dimensional metal carbide MXenes', *Science*, 369(6506), pp. 979–983. doi: 10.1126/science.aba8311.

Kang, Z., Ma, Y., Tan, X., Zhu, M., Zheng, Z., Liu, N., Li, L., Zou, Z., Jiang, X., Zhai, T. and Gao, Y. (2017) 'MXene–silicon van der waals heterostructures for high-speed self-driven photodetectors', *Advanced Electronic Materials*, 3(9), p. 1700165. doi: 10.1002/aelm.201700165.

Khazaei, M., Arai, M., Sasaki, T., Chung, C. Y., Venkataramanan, N. S., Estili, M., Sakka, Y. and Kawazoe, Y. (2013) 'Novel electronic and magnetic properties of two-dimensional transition metal carbides and nitrides', *Advanced Functional Materials*, 23(17), pp. 2185–2192. doi: 10.1002/adfm.201202502.

Khazaei, M., Arai, M., Sasaki, T., Ranjbar, A., Liang, Y. and Yunoki, S. (2015) 'OH-terminated two-dimensional transition metal carbides and nitrides as ultralow work function materials', *Physical Review B—Condensed Matter and Materials Physics*, 92(7). doi: 10.1103/PhysRevB.92.075411.

Khazaei, M., Ranjbar, A., Arai, M., Sasaki, T. and Yunoki, S. (2017) 'Electronic properties and applications of MXenes: A theoretical review', *Journal of Materials Chemistry C*, 5(10), pp. 2488–2503. doi: 10.1039/c7tc00140a.

Kim, H. and Alshareef, H. N. (2020) 'MXetronics: MXene-enabled electronic and photonic devices', *ACS Materials Letters*, 2(1), pp. 55–70. doi: 10.1021/acsmaterialslett.9b00419.

Kim, S. J., Koh, H. J., Ren, C. E., Kwon, O., Maleski, K., Cho, S. Y., Anasori, B., Kim, C. K., Choi, Y. K., Kim, J. and Gogotsi, Y. (2018) 'Metallic Ti3C2Tx MXene gas sensors with ultrahigh signal-to-noise ratio', *ACS Nano*, 12(2), pp. 986–993. doi: 10.1021/acsnano.7b07460.

Kurra, N., Ahmed, B., Gogotsi, Y. and Alshareef, H. N. (2016) 'MXene-on-paper coplanar microsupercapacitors', *Advanced Energy Materials*, 6(24), pp. 1–8. doi: 10.1002/aenm.201601372.

Lai, S., Jeon, J., Jang, S. K., Xu, J., Choi, Y. J., Park, J. H., Hwang, E. and Lee, S. (2015) 'Surface group modification and carrier transport properties of layered transition metal carbides (Ti2CTx, T: –OH, –F and –O)', *Nanoscale*, 7(46), pp. 19390–19396. doi: 10.1039/C5NR06513E.

Lee, E., VahidMohammadi, A., Prorok, B. C., Yoon, Y. S., Beidaghi, M. and Kim, D. J. (2017) 'Room temperature gas sensing of two-dimensional titanium carbide (MXene)', *ACS Applied Materials and Interfaces*, 9(42), pp. 37184–37190. doi: 10.1021/acsami.7b11055.

Levi, M. D., Lukatskaya, M. R., Sigalov, S., Beidaghi, M., Shpigel, N., Daikhin, L., Aurbach, D., Barsoum, M. W. and Gogotsi, Y. (2015) 'Solving the capacitive paradox of 2D MXene using electrochemical quartz-crystal admittance and *in situ* electronic conductance measurements', *Advanced Energy Materials*, 5(1), pp. 1–11. doi: 10.1002/aenm.201400815.

Li, B., Ma, K., Lu, S., Liu, X., Ma, Z., Zhang, L., Wang, X. and Wang, S. (2020) 'Structural health monitoring for polymer composites with surface printed MXene/ink sensitive sensors', *Applied Physics A: Materials Science and Processing*, 126(10), pp. 1–11. doi: 10.1007/s00339-020-03979-4.

Li, Hongyan, Hou, Yang, Wang, Faxing, Lohe, Martin R., Zhuang, Xiaodong, Niu L. and Feng, X. (2016) 'Flexible all-solid-state supercapacitors with high volumetric capacitances boosted by solution processable MXene and electrochemicallyexfoliated graphene', *Advanced Energy Materials*, 7(4), p. 1601847. doi: 10.1002/aenm.201601847.

Li, M., Fang, L., Zhou, H., Wu, F., Lu, Y., Luo, H., Zhang, Y. and Hu, B. (2019) 'Three-dimensional porous MXene/NiCo-LDH composite for high performance non-enzymatic glucose sensor', *Applied Surface Science*, 495, p. 143554. doi: 10.1016/j.apsusc.2019.143554.

Li, M., Lu, J., Luo, K., Li, Y., Chang, K., Chen, K., Zhou, J., Rosen, J., Hultman, L., Eklund, P. and Persson, P. O. (2019) 'Element replacement approach by reaction with lewis acidic molten salts to synthesize nanolaminated MAX phases and MXenes', *Journal of the American Chemical Society*, 141(11), pp. 4730–4737. doi: 10.1021/jacs.9b00574.

Li, T., Yao, L., Liu, Q., Gu, J., Luo, R., Li, J., Yan, X., Wang, W., Liu, P., Chen, B. and Zhang, W. (2018) 'Fluorine-free synthesis of high-purity Ti3C2Tx (T=OH, O) via alkali treatment', *Angewandte Chemie—International Edition*, 57(21), pp. 6115–6119. doi: 10.1002/anie.201800887.

Liang, X., Hart, C., Pang, Q., Garsuch, A., Weiss, T. and Nazar, L. F. (2015) 'A highly efficient polysulfide mediator for lithium-sulfur batteries', *Nature Communications*, 6, pp. 1–8. doi: 10.1038/ncomms6682.

Liang, X., Rangom, Y., Kwok, C. Y., Pang, Q. and Nazar, L. F. (2017) 'Interwoven MXene nanosheet/carbon-nanotube composites as Li–S cathode hosts', *Advanced Materials*, 29(3), pp. 1–7. doi: 10.1002/adma.201603040.

Lin, H., Gao, S., Dai, C., Chen, Y. and Shi, J. (2017) 'A two-dimensional biodegradable niobium carbide (MXene) for photothermal tumor eradication in NIR-I and NIR-II biowindows', *Journal of the American Chemical Society*, 139(45), pp. 16235–16247. doi: 10.1021/jacs.7b07818.

Lin, H., Wang, X., Yu, L., Chen, Y. and Shi, J. (2017) 'Two-dimensional ultrathin MXene ceramic nanosheets for photothermal conversion', *Nano Letters*, 17(1), pp. 384–391. doi: 10.1021/acs.nanolett.6b04339.

Lin, H., Wang, Y., Gao, S., Chen, Y. and Shi, J. (2018) 'Theranostic 2D tantalum carbide (MXene)', *Advanced Materials*, 30(4), p. 1703284. doi: 10.1002/adma.201703284.

Lin, Z., Barbara, D., Taberna, P. L., Van Aken, K. L., Anasori, B., Gogotsi, Y. and Simon, P. (2016) 'Capacitance of Ti3C2Tx MXene in ionic liquid electrolyte', *Journal of Power Sources*, 326, pp. 575–579. doi: 10.1016/j.jpowsour.2016.04.035.

Ling, Z., Ren, C. E., Zhao, M. Q., Yang, J., Giammarco, J. M., Qiu, J., Barsoum, M. W. and Gogotsi, Y. (2014) 'Flexible and conductive MXene films and nanocomposites with high capacitance', *Proceedings of the National Academy of Sciences of the United States of America*, 111(47), pp. 16676–16681. doi: 10.1073/pnas.1414215111.

Lipatov, A., Alhabeb, M., Lukatskaya, M. R., Boson, A., Gogotsi, Y. and Sinitskii, A. (2016a) 'Effect of synthesis on quality, electronic properties and environmental stability of individual monolayer Ti3C2 MXene flakes', *Advanced Electronic Materials*, 2(12), p. 1600255. doi: 10.1002/aelm.201600255.

Lipatov, A., Alhabeb, M., Lukatskaya, M. R., Boson, A., Gogotsi, Y. and Sinitskii, A. (2016b) 'MXene materials: Effect of synthesis on quality, electronic properties and environmental stability of individual monolayer Ti 3 C 2 MXene flakes', *Advanced Electronic Materials*, 2(12). doi: 10.1002/aelm.201670068.

Lipatov, A., Lu, H., Alhabeb, M., Anasori, B., Gruverman, A., Gogotsi, Y. and Sinitskii, A. (2018) 'Elastic properties of 2D Ti3C2Tx MXene monolayers and bilayers', *Science Advances*, 4(6), p. 491. doi: 10.1126/sciadv.aat0491.

Liu, G., Shen, J., Liu, Q., Liu, G., Xiong, J., Yang, J. and Jin, W. (2018) 'Ultrathin two-dimensional MXene membrane for pervaporation desalination', *Journal of Membrane Science*, 548, pp. 548–558. doi: 10.1016/j.memsci.2017.11.065.

Liu, G., Zou, J., Tang, Q., Yang, X., Zhang, Y., Zhang, Q., Huang, W., Chen, P., Shao, J. and Dong, X. (2017) 'Surface modified Ti3C2 MXene nanosheets for tumor targeting photothermal/photodynamic/chemo synergistic therapy', *ACS Applied Materials and Interfaces*, 9(46), pp. 40077–40086. doi: 10.1021/acsami.7b13421.

Liu, H., Duan, C., Yang, C., Shen, W., Wang, F. and Zhu, Z. (2015) 'A novel nitrite biosensor based on the direct electrochemistry of hemoglobin immobilized on MXene-Ti3C2', *Sensors and Actuators, B: Chemical*, 218, pp. 60–66. doi: 10.1016/j.snb.2015.04.090.

Liu, J.-M. (2009) *Photonic Devices*. Cambridge, UK: Cambridge University Press. doi: 10.1017/CBO9780511614255.

Liu, R., Miao, M., Li, Y., Zhang, J., Cao, S. and Feng, X. (2018) 'Ultrathin biomimetic polymeric Ti3C2Tx MXene composite films for electromagnetic interference shielding', *ACS Applied Materials & Interfaces*, 10(51), pp. 44787–44795. doi: 10.1021/acsami.8b18347.

Liu, Y., Xiao, H. and Goddard, W. A. (2016) 'Schottky-barrier-free contacts with two-dimensional semiconductors by surface-engineered MXenes', *Journal of the American Chemical Society*, 138(49), pp. 15853–15856. doi: 10.1021/jacs.6b10834.

Liu, Z., Lin, H., Zhao, M., Dai, C., Zhang, S., Peng, W. and Chen, Y. (2018) '2D superparamagnetic tantalum carbide composite MXenes for efficient breast-cancer theranostics', *Theranostics*, 8(6), pp. 1648–1664. doi: 10.7150/thno.23369.

Lu, S., Sui, L., Liu, Y., Yong, X., Xiao, G., Yuan, K., Liu, Z., Liu, B., Zou, B. and Yang, B. (2019) 'White photoluminescent Ti 3 C 2 MXene quantum dots with two-photon fluorescence', *Advanced Science*, 6(9). doi: 10.1002/advs.201801470.

Lu, X., Yu, M., Wang, G., Tong, Y. and Li, Y. (2014) 'Flexible solid-state supercapacitors: Design, fabrication and applications', *Energy and Environmental Science*, 7(7), pp. 2160–2181. doi: 10.1039/c4ee00960f.

Lukatskaya, M. R., Bak, S. M., Yu, X., Yang, X. Q., Barsoum, M. W. and Gogotsi, Y. (2015) 'Probing the mechanism of high capacitance in 2D titanium carbide using *in situ* X-Ray absorption spectroscopy', *Advanced Energy Materials*, 5(15), pp. 2–5. doi: 10.1002/aenm.201500589.

Lukatskaya, M. R., Kota, S., Lin, Z., Zhao, M. Q., Shpigel, N., Levi, M. D., Halim, J., Taberna, P. L., Barsoum, M. W., Simon, P. and Gogotsi, Y. (2017) 'Ultra-high-rate pseudocapacitive energy storage in two-dimensional transition metal carbides', *Nature Energy*, 6. doi: 10.1038/nenergy.2017.105.

Lukatskaya, M. R., Mashtalir, O., Ren, C. E., Dall'Agnese, Y., Rozier, P., Taberna, P. L., Naguib, M., Simon, P., Barsoum, M. W. and Gogotsi, Y. (2013) Cation intercalation and high volumetric capacitance of two-dimensional titanium carbide. *Science*, 341(6153), pp. 1502–1505. doi: 10.1126/science.1241488.

Lyu, B., Kim, M., Jing, H., Kang, J., Qian, C., Lee, S. and Cho, J. H. (2019) 'Large-Area MXene electrode array for flexible electronics', *ACS Nano*, 13(10), pp. 11392–11400. doi: 10.1021/acsnano.9b04731.

Ma, Y., Liu, N., Li, L., Hu, X., Zou, Z., Wang, J., Luo, S. and Gao, Y. (2017) 'A highly flexible and sensitive piezoresistive sensor based on MXene with greatly changed interlayer distances', *Nature Communications*, 8(1), pp. 1–7. doi: 10.1038/s41467-017-01136-9.

Maier, S. A. and Atwater, H. A. (2005) 'Plasmonics: Localization and guiding of electromagnetic energy in metal/dielectric structures', *Journal of Applied Physics*, 98(011101). doi: 10.1063/1.1951057.

Maleski, Kathleen, Mochalin, Vadym N. and Gogotsi, Y. (2017) 'Dispersions of two-dimensional titanium carbide MXene in organic solvents', *Chemistry of Materials*, 29(4), pp. 1632–1640. doi: 10.1021/acs.chemmater.6b04830.

Mashtalir, O., Cook, K. M., Mochalin, V. N., Crowe, M., Barsoum, M. W. and Gogotsi, Y. (2014) 'Dye adsorption and decomposition on two-dimensional titanium carbide in aqueous media', *Journal of Materials Chemistry A*, 2(35), pp. 14334–14338. doi: 10.1039/c4ta02638a.

Mashtalir, O., Naguib, M., Mochalin, V. N., Dall'Agnese, Y., Heon, M., Barsoum, M. W. and Gogotsi, Y. (2013) 'Intercalation and delamination of layered carbides and carbonitrides', *Nature Communications*, 4, pp. 1–7. doi: 10.1038/ncomms2664.

Mathis, T. S., Kurra, N., Wang, X., Pinto, D., Simon, P. and Gogotsi, Y. (2019) 'Energy storage data reporting in perspective—guidelines for interpreting the performance of electrochemical energy storage systems', *Advanced Energy Materials*, 9(39). doi: 10.1002/aenm.201902007.

Mauchamp, V., Bugnet, M., Bellido, E. P., Botton, G. A., Moreau, P., Magne, D., Naguib, M., Cabioc'h, T. and Barsoum, M. W. (2014) 'Enhanced and tunable surface plasmons in two-dimensional Ti3 C2 stacks: Electronic structure versus boundary effects', *Physical Review B—Condensed Matter and Materials Physics*, 89(23). doi: 10.1103/PhysRevB.89.235428.

Mayerberger, E. A., Urbanek, O., McDaniel, R. M., Street, R. M., Barsoum, M. W. and Schauer, C. L. (2017) 'Preparation and characterization of polymer-Ti3C2Tx (MXene) composite nanofibers produced via electrospinning', *Journal of Applied Polymer Science*, 134(37), p. 45295. doi: 10.1002/app.45295.

Medintz, I. L., Uyeda, H. T., Goldman, E. R. and Mattoussi, H. (2005) 'Quantum dot bioconjugates for imaging, labelling and sensing', *Nature Materials*, 4(6), pp. 435–446. doi: 10.1038/nmat1390.

Ming, Fangwang, Liang, Hanfeng, Huang, Gang, Bayhan, Zahra and Alshareef, H. N. (2021) 'MXenes for rechargeable batteries beyond the lithium-ion', *Advanced Materials*, 33, p. 202004039. doi: 10.1002/adma.202004039.

Naguib, M., Barsoum, M. W. and Gogotsi, Y. (2021) 'Ten years of progress in the synthesis and development of MXenes', *Advanced Materials*, 2103393, p. 2103393. doi: 10.1002/adma.202103393.

Naguib, M., Come, J., Dyatkin, B., Presser, V., Taberna, P. L., Simon, P., Barsoum, M. W. and Gogotsi, Y. (2012) 'MXene: A promising transition metal carbide anode for lithium-ion batteries', *Electrochemistry Communications*, 16(1), pp. 61–64. doi: 10.1016/j.elecom.2012.01.002.

Naguib, M., Halim, J., Lu, J., Cook, K. M., Hultman, L., Gogotsi, Y. and Barsoum, M. W. (2013) 'New two-dimensional niobium and vanadium carbides as promising materials for li-ion batteries', *Journal of the American Chemical Society*, 135(43), pp. 15966–15969. doi: 10.1021/ja405735d.

Naguib, M., Mashtalir, O., Carle, J., Presser, V., Lu, J., Hultman, L., Gogotsi, Y. and Barsoum, M. W. (2012) 'Two-dimensional transition metal carbides', *ACS Nano*, 6(2), pp. 1322–1331. doi: 10.1021/nn204153h.

Naguib, M., Mochalin, V. N., Barsoum, M. W. and Gogotsi, Y. (2014) '25th anniversary article: MXenes: A new family of two-dimensional materials', *Advanced Materials*, 26(7), pp. 992–1005. doi: 10.1002/adma.201304138.

Naguib, M., Saito, T., Lai, S., Rager, M. S., Aytug, T., Paranthaman, M. P., Zhao, M. Q. and Gogotsi, Y. (2016a) 'All-solid-state lithium-ion batteries with TiS2 nanosheets and sulphide solid electrolytes', *Journal of Materials Chemistry A*, 4(26), pp. 10329–10335. doi: 10.1039/c6ta01628f.

Naguib, M., Saito, T., Lai, S., Rager, M. S., Aytug, T., Paranthaman, M. P., Zhao, M. Q. and Gogotsi, Y. (2016b) 'Ti3C2Tx (MXene)-Polyacrylamide nanocomposite films', *RSC Advances*, 6, pp. 72069–72073. doi: 10.1039/C6RA10384G.

Naguib, M., Unocic, R. R., Armstrong, B. L. and Nanda, J. (2015) 'Large-scale delamination of multi-layers transition metal carbides and carbonitrides "mXenes"', *Dalton Transactions*, 44(20), pp. 9353–9358. doi: 10.1039/c5dt01247c.

Naguib, Michael, Kurtoglu, Murat, Presser, Volker, Lu, Jun, Niu, Junjie, Heon, Min, Hultman, Lars, Gogotsi, Yury and barsoum, M. W. (2011) 'Two-dimensional nanocrystals produced by exfoliation of Ti3AlC2', *Advanced Materials*, 23(37), pp. 4248–4253. doi: 10.1002/adma.201102306.

Oh, Seung-Min, Hwang, Jang-Yeon, Yoon, C. S., Lu, Jun, Amine, Khalil, Belharouak, Illias and Sun, Y.-K. (2014) 'High electrochemical performances of microsphere C-TiO2 anode for sodium-ion battery', *ACS Appl. Mater. Interfaces*, 6(14), pp. 11295–11301. doi: 10.1021/am501772a.

Okubo, M., Sugahara, A., Kajiyama, S. and Yamada, A. (2018) 'MXene as a charge storage host', *Accounts of Chemical Research*, 51(3), pp. 591–599. doi: 10.1021/acs.accounts.7b00481.

Okumura, Toyoki, Takeuchi, Tomonari and kobayashi, H. (2020) 'All-solid-state batteries with LiCoO2-type electrodes: realization of an impurity-free interface by utilizing a cosinterable Li3.5Ge0.5V0.5O4 electrolyte', *ACS Applied Energy Materials*, 4(1), pp. 30–34. doi: 10.1021/acsaem.0c02785.

Pan, Dengyu, Zhang, Jingchun, Li, Zhen and Wu, M. (2010) 'Hydrothermal route for cutting graphene sheets into blue-luminescent graphene quantum dots', *Advanced Materials*, 22(6), pp. 734–738. doi: 10.1002/adma.200902825.

Pang, Q., Kundu, D., Cuisinier, M. and Nazar, L. F. (2014) 'Surface-enhanced redox chemistry of polysulphides on a metallic and polar host for lithium-sulphur batteries', *Nature Communications*, 5(1), pp. 3–10. doi: 10.1038/ncomms5759.

Peng, Y. Y., Akuzum, B., Kurra, N., Zhao, M. Q., Alhabeb, M., Anasori, B., Kumbur, E. C., Alshareef, H. N., Ger, M. D. and Gogotsi, Y. (2016) 'All-MXene (2D titanium carbide) solid-state microsupercapacitors for on-chip energy storage', *Energy*, pp. 3–10. doi: 10.1039/C6EE01717G.View.

Pomerantseva, E. and Gogotsi, Y. (2017) 'Two-dimensional heterostructures for energy storage', *Nature Energy*, 2(7), pp. 1–6. doi: 10.1038/nenergy.2017.89.

Ponroucha, A. and Goñiab, M. R. (2013) 'High capacity hard carbon anodes for sodium ion batteries in additive free electrolyte', *Electrochemistry Communications*, 27, pp. 85–88. doi: 10.1016/j.elecom.2012.10.038.

Qian, J., Wu, X., Cao, Y., Ai, X. and Yang, H. (2013) 'High capacity and rate capability of amorphous phosphorus for sodium ion batteries', *Angewandte Chemie—International Edition*, 52(17), pp. 4633–4636. doi: 10.1002/anie.201209689.

Raagulan, K., Braveenth, R., Jang, H. J., Seon Lee, Y., Yang, C. M., Mi Kim, B., Moon, J. J. and Chai, K. Y. (2018) 'Electromagnetic shielding by MXene-Graphene-PVDF composite with hydrophobic, lightweight and flexible graphene coated fabric', *Materials*, 11(10), p. 1803. doi: 10.3390/ma11101803.

Rakhi, R. B., Nayak, P., Xia, C. and Alshareef, H. N. (2016) 'Novel amperometric glucose biosensor based on MXene nanocomposite', *Scientific Reports*, 6(October), pp. 1–10. doi: 10.1038/srep36422.

Ran, J., Gao, G., Li, F. T., Ma, T. Y., Du, A. and Qiao, S. Z. (2017) 'Ti3C2 MXene co-catalyst on metal sulfide photo-absorbers for enhanced visible-light photocatalytic hydrogen production', *Nature Communications*, 8. doi: 10.1038/ncomms13907.

Ren, C. E., Zhao, M. Q., Makaryan, T., Halim, J., Boota, M., Kota, S., Anasori, B., Barsoum, M. W. and Gogotsi, Y. (2016) 'Porous two-dimensional transition metal carbide (MXene) flakes for high-performance Li-Ion storage', *ChemElectroChem*, 3(5), pp. 689–693. doi: 10.1002/celc.201600059.

Rozmysłowska-Wojciechowska, A., Mitrzak, J., Szuplewska, A., Chudy, M., Woźniak, J., Petrus, M., Wojciechowski, T., Vasilchenko, A. S. and Jastrzębska, A. M. (2020) 'Engineering of 2D Ti3C2 MXene surface charge and its influence on biological properties', *Materials*, 13(10), pp. 1–18. doi: 10.3390/ma13102347.

Russel, W. B. (1980) 'Review of the role of colloidal forces in the rheology of suspensions', *Journal of Rheology*, 24(3), pp. 287–317. doi: 10.1122/1.549564.

Sang, X., Xie, Y., Lin, M. W., Alhabeb, M., Van Aken, K. L., Gogotsi, Y., Kent, P. R., Xiao, K. and Unocic, R. R. (2016) 'Atomic defects in monolayer titanium carbide (Ti3C2Tx) MXene', *ACS Nano*, 10(10), pp. 9193–9200. doi: 10.1021/acsnano.6b05240.

Sarycheva, A., Makaryan, T., Maleski, K., Satheeshkumar, E., Melikyan, A., Minassian, H., Yoshimura, M. and Gogotsi, Y. (2017) 'Two-dimensional titanium carbide (MXene) as surface-enhanced raman scattering substrate', *Journal of Physical Chemistry C*, 121(36), pp. 19983–19988. doi: 10.1021/acs.jpcc.7b08180.

Satheeshkumar, E., Makaryan, T., Melikyan, A., Minassian, H., Gogotsi, Y. and Yoshimura, M. (2016) 'One-step solution processing of Ag, Au and Pd@MXene hybrids for SERS', *Scientific Reports*, 6(1), pp. 1–9. doi: 10.1038/srep32049.

Shahzad, A., Rasool, K., Miran, W., Nawaz, M., Jang, J., Mahmoud, K. A. and Lee, D. S. (2017) 'Two-dimensional Ti3C2Tx MXene nanosheets for efficient copper removal from water', *ACS Sustainable Chemistry and Engineering*, 5(12), pp. 11481–11488. doi: 10.1021/acssuschemeng.7b02695.

Shein, I. R. and Ivanovskii, A. L. (2012) 'Graphene-like titanium carbides and nitrides Tin+1Cn, Tin+1Nn (n = 1, 2, and 3) from de-intercalated MAX phases: First-principles probing of their structural, electronic properties and relative stability', *Computational Materials Science*, 65, pp. 104–114. doi: 10.1016/j.commatsci.2012.07.011.

Shekhirev, M., Shuck, C. E., Sarycheva, A. and Gogotsi, Y. (2021) Characterization of MXenes at every step, from their precursors to single flakes and assembled films. *Progress in Materials Science*, 120, p. 100757. doi: 10.1016/j.pmatsci.2020.100757.

Shi, H., Zhang, P., Liu, Z., Park, S., Lohe, M. R., Wu, Y., Shaygan Nia, A., Yang, S. and Feng, X. (2021) 'Ambient-stable two-dimensional titanium carbide (MXene) enabled by iodine etching', *Angewandte Chemie—International Edition*, 60(16), pp. 8689–8693. doi: 10.1002/anie.202015627.

Sivasankarapillai, V. S., Somakumar, A. K., Joseph, J., Nikazar, S., Rahdar, A. and Kyzas, G. Z. (2020) 'Cancer theranostic applications of MXene nanomaterials: Recent updates', *Nano-Structures and Nano-Objects*, 22, p. 100457. doi: 10.1016/j.nanoso.2020.100457.

Stanley, Ross. (2012) 'Plasmonics in the mid-infrared', *Nature Photonics*, 6, pp. 409–411. doi: 10.1038/nphoton.2012.161.

Sun, W., Shah, S. A., Chen, Y., Tan, Z., Gao, H., Habib, T., Radovic, M. and Green, M. J. (2017) 'Electrochemical etching of Ti2AlC to Ti2CT:X (MXene) in low-concentration hydrochloric acid solution', *Journal of Materials Chemistry A*, 5(41), pp. 21663–21668. doi: 10.1039/c7ta05574a.

Szuplewska, A., Kulpińska, D., Dybko, A., Jastrzębska, A. M., Wojciechowski, T., Rozmysłowska, A., Chudy, M., Grabowska-Jadach, I., Ziemkowska, W., Brzózka, Z. and Olszyna, A. (2019) '2D Ti 2 C (MXene) as a novel highly efficient and selective agent for photothermal therapy', *Materials Science and Engineering C*, 98, pp. 874–886. doi: 10.1016/j.msec.2019.01.021.

Tahini, H. A., Tan, X. and Smith, S. C. (2017) 'The origin of low workfunctions in OH terminated MXenes', *Nanoscale*, 9(21), pp. 7016–7020. doi: 10.1039/c7nr01601h.

Tao, Xinyong, Wang, Jianguo, Ying, Zhuogao, Cai, Qiuxia, Zheng, Guangyuan, Gan, Yongping, Huang, Hui, Xia, Yang, Liang, Chu, Zhang, Wenkui and Cui, Y. (2014) 'Strong sulfur binding with conducting magnéli-phase TinO2n–1 nanomaterials for improving lithium–sulfur batteries', *Nano Lett. 2014, 14, 9, 5288–5294*, 14(9), pp. 5288–5294. doi: 10.1021/nl502331f.

Thomas, S., Sarathchandran, C., Ilangovan, S. A. and Moreno-Pirajan, J. C. eds., (2021) *Handbook of Carbon-Based Nanomaterials Chapter 16*. 1st edn. United States: Elsevier.

Tu, S., Jiang, Q., Zhang, X. and Alshareef, H. N. (2018) 'Large dielectric constant enhancement in MXene percolative polymer composites', *ACS Nano*, 12(4), pp. 3369–3377. doi: 10.1021/acsnano.7b08895.

Usman, K. A. A., Qin, S., Henderson, L. C., Zhang, J., Hegh, D. and Razal, J. M. (2021) 'Ti3C2Tx MXene: From dispersions to multifunctional architectures for diverse applications', *Materials Horizons*, (August). doi: 10.1039/d1mh00968k.

VahidMohammadi, A., Rosen, J. and Gogotsi, Y. (2021) 'The world of two-dimensional carbides and nitrides (MXenes)', *Science (New York, N.Y.)*, 372(6547). doi: 10.1126/science.abf1581.

Velusamy, D. B., El-Demellawi, J. K., El-Zohry, A. M., Giugni, A., Lopatin, S., Hedhili, M. N., Mansour, A. E., Fabrizio, E. D., Mohammed, O. F. and Alshareef, H. N. (2019) 'MXenes for plasmonic photodetection', *Advanced Materials*, 31(32), pp. 1–10. doi: 10.1002/adma.201807658.

Wang, D. W., Zeng, Q., Zhou, G., Yin, L., Li, F., Cheng, H. M., Gentle, I. R. and Lu, G. Q. M. (2013) 'Carbon-sulfur composites for Li-S batteries: Status and prospects', *Journal of Materials Chemistry A*, 1(33), pp. 9382–9394. doi: 10.1039/c3ta11045a.

Wang, F., Yang, C., Duan, C., Xiao, D., Tang, Y. and Zhu, J. (2014) 'An organ-like titanium carbide material (MXene) with multilayer structure encapsulating hemoglobin for a mediator-free biosensor', *Journal of The Electrochemical Society*, 162(1), pp. B16–B21. doi: 10.1149/2.0371501jes.

Wang, G. (2016) 'Theoretical prediction of the intrinsic half-metallicity in surface-oxygen-passivated Cr2N MXene', *The Journal of Physical Chemistry C*, 120(33), pp. 18850–18857. doi: 10.1021/acs.jpcc.6b05224.

Wang, S.-J., Li, D.-S. and Jiang, L. (2019) 'Synergistic effects between MXenes and Ni chains in flexible and ultrathin electromagnetic interference shielding films', *Advanced Materials Interfaces*, 6(19), p. 1900961. doi: 10.1002/admi.201900961.

Wang, Shun, Du, Y.-L. and Liao, W.-H. (2017) 'Tunable band gap and optical properties of surface functionalized Sc2C monolayer', *Chinese Physics B*, 26(1).

Wang, X., Kajiyama, S., Iinuma, H., Hosono, E., Oro, S., Moriguchi, I., Okubo, M. and Yamada, A. (2015) 'Pseudocapacitance of MXene nanosheets for high-power sodium-ion hybrid capacitors', *Nature Communications*, 6, pp. 1–6. doi: 10.1038/ncomms7544.

Wang, Y., Lubbers, T., Xia, R., Zhang, Y. Z., Mehrali, M., Huijben, M. and Johan, E. (2021) 'Printable two-dimensional V 2 O 5/MXene heterostructure cathode for lithium-ion battery', *Journal of the Electrochemical Society*, 168(2), p. 020507. doi: 10.1149/1945-7111/abdef2.

Wang, Y., Xu, Y., Hu, M., Ling, H. and Zhu, X. (2020) 'MXenes: Focus on optical and electronic properties and corresponding applications', *Nanophotonics*, 9(7), pp. 1601–1620. doi: 10.1515/nanoph-2019-0556.

Wang, Y., Yu, X., Xu, S., Bai, J., Xiao, R., Hu, Y. S., Li, H., Yang, X. Q., Chen, L. and Huang, X. (2013) 'A zero-strain layered metal oxide as the negative electrode for long-life sodium-ion batteries', *Nature Communications*, 4, pp. 1–8. doi: 10.1038/ncomms3365.

Weng, C., Wang, G., Dai, Z., Pei, Y., Liu, L. and Zhang, Z. (2019) 'Buckled AgNW/MXene hybrid hierarchical sponges for high-performance electromagnetic interference shielding', *Nanoscale*, 11(47), pp. 22804–22812. doi: 10.1039/C9NR07988B.

Wong, S., Ngadi, N., Inuwa, I. M. and Hassan, O. (2018) 'Recent advances in applications of activated carbon from biowaste for wastewater treatment: A short review', *Journal of Cleaner Production*, 175, pp. 361–375. doi: 10.1016/j.jclepro.2017.12.059.

Wu, J., Wang, Y., Zhang, Y., Meng, H., Xu, Y., Han, Y., Wang, Z., Dong, Y. and Zhang, X. (2020) 'Highly safe and ionothermal synthesis of Ti3C2 MXene with expanded interlayer spacing for enhanced lithium storage', *Journal of Energy Chemistry*, 47, pp. 203–209. doi: 10.1016/j.jechem.2019.11.029.

Wu, X., Han, B., Zhang, H. B., Xie, X., Tu, T., Zhang, Y., Dai, Y., Yang, R. and Yu, Z. Z. (2020) 'Compressible, durable and conductive polydimethylsiloxane-coated MXene foams for high-performance electromagnetic interference shielding', *Chemical Engineering Journal*, 381, p. 122622. doi: 10.1016/j.cej.2019.122622.

Wu, Y., Sun, Y., Zheng, J., Rong, J., Li, H. and Niu, L. (2021) 'MXenes: Advanced materials in potassium ion batteries', *Chemical Engineering Journal*, 404, p. 126565. doi: 10.1016/j.cej.2020.126565.

Xiang, C., Guo, R., Lin, S., Jiang, S., Lan, J., Wang, C., Cui, C., Xiao, H. and Zhang, Y. (2019) 'Lightweight and ultrathin TiO2-Ti3C2TX/graphene film with electromagnetic interference shielding', *Chemical Engineering Journal*, 360, pp. 1158–1166. doi: 10.1016/j.cej.2018.10.174.

Xiao, F., Zhao, F., Zhang, Y., Guo, G. and Zeng, B. (2009) 'Ultrasonic electrodeposition of gold—platinum alloy nanoparticles on ionic liquid—chitosan composite film and their application in fabricating nonenzyme hydrogen peroxide sensors', *Journal of Physical Chemistry C*, 113(3), pp. 849–855. doi: 10.1021/jp808162g.

Xie, Fan, Jia, Fengfeng, Zhuo, Longhai, Lu, Zhaoqing, Si, Lianmeng, Huang, Jizhen, Zhang, M. and Ma, Q. (2019) 'Ultrathin MXene/aramid nanofiber composite paper with excellent mechanical properties for efficient electromagnetic interference shielding', *Nanoscale*, 11, pp. 23382–23391. doi: 10.1039/C9NR07331K.

Xie, Y. and Kent, P. R. C. (2013) 'Hybrid density functional study of structural and electronic properties of functionalized Tin+1Xn (X=C, N) monolayers', *Physical Review B—Condensed Matter and Materials Physics*, 87(23). doi: 10.1103/PhysRevB.87.235441.

Xiong, D., Li, X., Bai, Z. and Lu, S. (2018) 'Recent advances in layered Ti3C2Tx MXene for electrochemical energy storage', *Small*, 14(17), p. 1703419. doi: 10.1002/smll.201703419.

Xu, B., Zhu, M., Zhang, W., Zhen, X., Pei, Z., Xue, Q., Zhi, C. and Shi, P. (2016) 'Ultrathin MXene-micropattern-based field-effect transistor for probing neural activity', *Advanced Materials*, 28(17), pp. 3333–3339. doi: 10.1002/adma.201504657.

Xu, C., Wang, L., Liu, Z., Chen, L., Guo, J., Kang, N., Ma, X. L., Cheng, H. M. and Ren, W. (2015) 'Large-area high-quality 2D ultrathin Mo2C superconducting crystals', *Nature Materials*, 14(11), pp. 1135–1141. doi: 10.1038/nmat4374.

Xue, Qi, Zhang, Huijie, Zhu, Minshen, Pei, Zengxia, Li, Hongfei, Wang, Zifeng, Huang, Yang, Huang, Yan, Deng, Qihuang, Zhou, Jie, Du, Shiyu, HuangQing and Zhi, C. (2017) 'Photoluminescent Ti3C2 MXene quantum dots for multicolor cellular imaging', *Advanced Materials*, 29(15). doi: 10.1002/adma.201604847.

Yang, S., Zhang, P., Wang, F., Ricciardulli, A.G., Lohe, M. R., Blom, P. W. M., Feng, X. (2018) 'Fluoride-free synthesis of two-dimensional titanium carbide (MXene) using a binary aqueous system', *Angewandte Chemie—International Edition*, 57(47), pp. 15491–15495. doi: 10.1002/anie.201809662.

Yang, S., Zhang, P., Wang, F., Ricciardulli, A. G., Lohe, M. R., Blom, P. W. and Feng, X. (2020) 'All-optical modulator using mxene inkjet-printed microring resonator', *IEEE Journal of Selected Topics in Quantum Electronics*, 26(5), pp. 1–6. doi: 10.1109/JSTQE.2020.2982985.

Yi, Y., Sun, Z., Li, J., Chu, P. K. and Yu, X. F. (2019) 'Optical and optoelectronic properties of black phosphorus and recent photonic and optoelectronic applications', *Small Methods*, 3(10), pp. 1–19. doi: 10.1002/smtd.201900165.

Yin, L., Li, Y., Yao, X., Wang, Y., Jia, L., Liu, Q., Li, J., Li, Y. and He, D. (2021) 'MXenes for solar cells', *Nano-Micro Letters*, 13(1), pp. 1–17. doi: 10.1007/s40820-021-00604-8.

Ying, G., Kota, S., Dillon, A. D., Fafarman, A. T. and Barsoum, M. W. (2018) 'Conductive transparent V2CTx (MXene) films', *FlatChem*, 8, pp. 25–30. doi: 10.1016/j.flatc.2018.03.001.

Ying, Y., Liu, Y., Wang, X., Mao, Y., Cao, W., Hu, P. and Peng, X. (2015) 'Two-dimensional titanium carbide for efficiently reductive removal of highly toxic chromium(VI) from water', *ACS Applied Materials and Interfaces*, 7(3), pp. 1795–1803. doi: 10.1021/am5074722.

You, J., Si, C., Zhou, J. and Sun, Z. (2019) 'Contacting MoS2 to MXene: Vanishing p-type schottky barrier and enhanced hydrogen evolution catalysis', *The Journal of Physical Chemistry C*, 123(6), pp. 3719–3726. doi: 10.1021/acs.jpcc.8b12469.

Yury Gogotsi and Qing Huang (2021) 'MXenes: Two-dimensional building blocks for future materials and devices', *ACS Nano*, 15(4), pp. 5775–5780. doi: 10.1021/acsnano.1c03161.

Zhan, X., Si, C., Zhou, J. and Sun, Z. (2020) 'MXene and MXene-based composites: Synthesis, properties and environment-related applications', *Nanoscale Horizons*, 5(2), pp. 235–258. doi: 10.1039/c9nh00571d.

Zhang, C., Anasori, B., Seral-Ascaso, A., Park, S. H., McEvoy, N., Shmeliov, A., Duesberg, G. S., Coleman, J. N., Gogotsi, Y. and Nicolosi, V. (2017) 'Transparent, flexible, and conductive 2D titanium carbide (MXene) films with high volumetric capacitance', *Advanced Materials*, 29(36), p. 1702678. doi: 10.1002/adma.201702678.

Zhang, C., Kremer, M. P., Seral-Ascaso, A., Park, S. H., McEvoy, N., Anasori, B., Gogotsi, Y. and Nicolosi, V. (2018) 'Stamping of flexible, coplanar micro-supercapacitors using MXene inks', *Advanced Functional Materials*. doi: 10.1002/adfm.201705506.

Zhang, C. J., McKeon, L., Kremer, M. P., Park, S. H., Ronan, O., Seral-Ascaso, A., Barwich, S., Coileáin, C. Ó., McEvoy, N., Nerl, H. C. and Anasori, B. (2019) 'Additive-free MXene inks and direct printing of micro-supercapacitors', *Nature Communications*, 10(1), pp. 1–10. doi: 10.1038/s41467-019-09398-1.

Zhang, H., Wang, L., Chen, Q., Li, P., Zhou, A., Cao, X. and Hu, Q. (2016) 'Preparation, mechanical and anti-friction performance of MXene/polymer composites', *Materials & Design*, 92, pp. 682–689. doi: 10.1016/j.matdes.2015.12.084.

Zhang, Xiaodong, Xu, Jianguang, Wang, Hui, Zhang, Jiajia, Yan, Hanbing, Pan, Bicai, Zhou, Jingfang and Xie, Y. (2013) 'Ultrathin nanosheets of MAX phases with enhanced thermal and mechanical properties in polymeric compositions: Ti3Si(0.75)Al(0.25)C2', *Angewandte Chemie—International Edition*, 52(16), pp. 4361–4365. doi: 10.1002/anie.201300285.

Zhang, Y., Wang, L., Zhang, J., Song, P., Xiao, Z., Liang, C., Qiu, H., Kong, J. and Gu, J. (2019) 'Fabrication and Investigation on the Ultra-thin and Flexible Ti3C2Tx/co-Doped Polyaniline Electromagnetic Interference Shielding Composite Films', *Composites Science and Technology*, 183(September), p. 107833. doi: 10.1016/j.compscitech.2019.107833.

Zhang, Y. Z., Lee, K. H., Anjum, D. H., Sougrat, R., Jiang, Q., Kim, H. and Alshareef, H. N. (2018) 'MXenes stretch hydrogel sensor performance to new limits', *Science Advances*, 4(6), pp. 1–8. doi: 10.1126/sciadv.aat0098.

Zhao, M. Q., Torelli, M., Ren, C. E., Ghidiu, M., Ling, Z., Anasori, B., Barsoum, M. W. and Gogotsi, Y. (2016) '2D titanium carbide and transition metal oxides hybrid electrodes for Li-ion storage', *Nano Energy*, 30, pp. 603–613. doi: 10.1016/j.nanoen.2016.10.062.

Zhao, M. Q., Xie, X., Ren, C. E., Makaryan, T., Anasori, B., Wang, G. and Gogotsi, Y. (2017) 'Hollow MXene spheres and 3D macroporous MXene frameworks for na-ion storage', *Advanced Materials*, 29(37), p. 1702410. doi: 10.1002/adma.201702410.

Zhao, X., Radovic, M. and Green, M. J. (2020) 'Synthesizing MXene nanosheets by water-free etching', *Chem*, 6(3), pp. 544–546. doi: 10.1016/j.chempr.2020.02.013.

Zhao, Xiaoqin, Liu, Min, Chen, Yong, Hou, Bo, Zhang, Na, Chen, Binbin, Yang, Ning, Chen, Ke, Li, J. and An, L. (2015) 'Fabrication of layered Ti3C2 with an accordion-like structure as a potential cathode material for high performance lithium–sulfur batteries', *Journal of Materials Chemistry A*, 3, pp. 7870–7876. doi: 10.1039/C4TA07101H.

Zheng, J., Wang, B., Ding, A., Weng, B. and Chen, J. (2018) 'Synthesis of MXene/DNA/Pd/Pt nanocomposite for sensitive detection of dopamine', *Journal of Electroanalytical Chemistry*, 816(March), pp. 189–194. doi: 10.1016/j.jelechem.2018.03.056.

2 Overview of MXene-Based Polymer Nanocomposites

State of the Art and New Challenges

Prajitha V, Jibin K P, Sisanth K S, Abitha V K, Hanna J Maria, and Sabu Thomas

CONTENTS

2.1 INTRODUCTION

Novel two-dimensional (2D) MXene nanomaterial discovered at Drexel University in 2011 has garnered significant interest among researchers due to its 2D morphology and high thermal, electrical, and mechanical properties. MXene is generated through a particular drawing of mass ternary carbide MAX-stage forerunners. The general representation of MXene is $M_{n+1}X_nT_x$, where M denotes transition metal, including Sc, Y, Ti, Zr, Hf, V, Nb, Ta, Cr, Mo, or W X, represents carbon or nitrogen, and n lies between 1–3 (Naguib *et al.*, 2012). Figure 2.1 shows the different elements that are used for the formation of the MAX phase. The MAX phase is exfoliated by the careful engraving of the A-phase using hydrofluoric (HF) acid.

Compared to the solid M–X bond, the bond between M–A b is weak and reactive. It is possible to remove the A layer with a suitable etching reagent and forms MXene denoted as $M_{n+1}X_nT_x$,T_x (oxygen [=O], hydroxyl [–OH] and fluorine [–F]) is the surface-terminating functional groups forms during the etching process, and the thickness of the MXene layers be governed by the n value (Lipton *et al.*, 2019).

The HF-etching route to disassociate the bond between M–A in the compound $M_{n+1}AX_n$ with Al can be explained with the following chemical changes.

Ti_3C_2Tz, the most frequently investigated MXene to date, was combined to polydiallyldimethylammonium chloride (PDDA) and polyvinyl alcohol (PVA), equally hydrophilic polymers, in the first-ever report of MXene polymer nanocomposites in 2014 (Ling *et al.*, 2014). Ti_3C_2Tz was made by etching Ti_3AlC_2 in 50% HF for 18 hours at room temperature (RT) with constant stirring. The MXene multiple layered sheets were rinsed using water in anticipation that a pH of 6 was measured after etching. These fine particles were then dried out at room temperature for 24 hours

DOI: 10.1201/9781003164975-2

in advance being immersed in dimethyl sulfoxide (DMSO) for 18 hours at RT and subsequently isolated by adding H_2O and centrifugation. To this set apart, DMSO interpolated ML MXene residue, 300:1 water was added, and the solution was sonicated in flowing argon (Ar) for about 5 hours. From 2014 to 2021, there were about 150 papers published in this area. Each year, the number of manuscripts in MXene polymer nanocomposites doubles compared to the previous year, as depicted in Figure 2.1. The increased number of published articles shows how important these materials are to the polymer field. The citation in Figure 2.2 reveals the interest among other researchers in this area

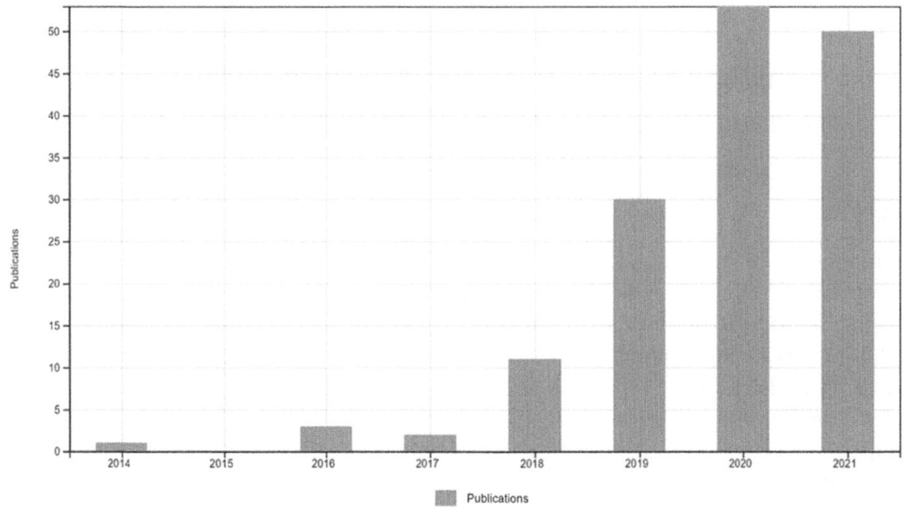

FIGURE 2.1 MXene nanocomposites publications and citations over the years starting from 2014.

Source: Web of Science.

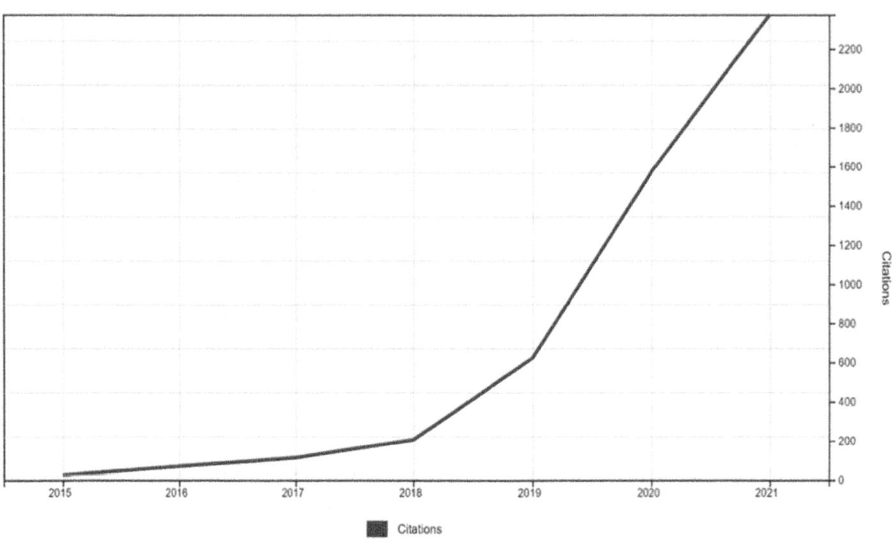

FIGURE 2.2 Citation report of MXene polymer nanocomposites articles.

Source: Web of Science.

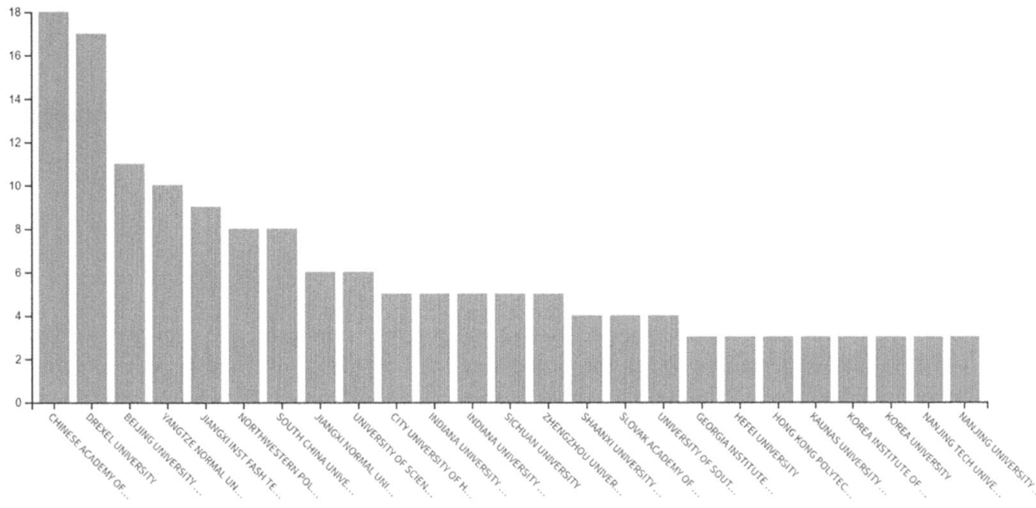

FIGURE 2.3 Top 25 institutions engaged in MXene polymer nanocomposites research.

and the potentiality of MAX phase materials in all fields of science, including electronic, environmental, energy storage, and water purification applications. We can see from Figure 2.2 that after 2019, a drastic change in citation and number of papers published because the difficulty in synthesis procedure of MAX phase materials is somewhat optimized in these years and more acceptable and reproducible synthetic production route is established.

MXene nanoparticles reinforced in the polymer matrix are hot areas where researchers are finding exciting results due to the extraordinary surface region, thermal, mechanical, and electrical applications of MXene (Chen *et al.*, 2021). MXene has unique properties due to its 2D structure, like one of them is high mechanical strength, and MXene has a 330 ± 30 GPa Young's modulus value. Because of the high electromagnetic shielding properties, the composites have been studied for electromagnetic interference (EMI) shielding applications. MXene/polymer composite has applications in innumerable arenas, such as energy expedients, membranes, radiation shielding, drug delivery, biosensors, and others (Zhang *et al.*, 2007; Carey *et al.*, 2019; Shi *et al.*, 2019; Guo *et al.*, 2020; Xiao *et al.*, 2020).

2.2 SYNTHESIS OF MXene/POLYMER NANOCOMPOSITES

The polymers are reinforced with a large variety of nanomaterials to improve mechanical strength, thermal, electrical, optical, barrier, and more. Layered MXene materials with high mechanical strength and conductivity, as a potential filler in polymers, allow the dispersion of nanomaterials, which has a significant role in improving the properties in polymer nanocomposites. There are several preparation techniques are used by scientists to improve distribution as well as properties. The interfacial thickness; secondary interactions, such as hydrogen bonding, electrostatic interaction, van der Waals forces; and others, even covalent bonding with polymer and MXene, are successfully synthesized by researchers. The synthesis techniques opted for are based on the type of polymers, water-soluble thermoplastic PVA, and MXene composites are synthesized by solution blending, solution casting, and other methods. The thermoplastic materials, like ultra-high molecular weight polyethene and polylactic acid polypropylene (PP), among others, are synthesized via melt blending (Ling *et al.*, 2014; Zhang *et al.*, 2016; Shi *et al.*, 2019; Huang *et al.*, 2020; Xue *et al.*, 2020; Tan *et al.*, 2021). Elastomeric/MXene nanocomposites are synthesized mainly via latex compounding and vacuum impregnation techniques used for the thermoset materials (Aakyiir*et al.*, 2021). Synthesis techniques also depend on the application of the prepared nanocomposites. Various methods are used for the synthesis of MXene/polymer nanocomposites:

- Solution blending
- Latex compounding
- Melt blending
- *In situ* polymerization

2.2.1 Solution Blending

Solution casting is a majorly used technique for MXene/polymer synthesis. The superior hydrophilicity of the MXene material simplifies the processing method for easily soluble polar polymers, such as polyvinylidene fluoride, polyurethane, and others, dissolved mainly in a dimethylformamide (DMF) or a PDDA water-soluble polymer like PVA. Easily dispersed MXene in such solvents mixed with the polymer solution and the composite prepared either via solution casting or vacuum filtration method. Various scientists also tried spin coating and dip coating of the dispersion. Among the solution blending process, PVA is more explored by scientists due to its solubility in water. Ling et al. fabricated PVA/MXene films via a vacuum-assisted filtration technique as the potential electronic application (Ling *et al.*, 2014), and Tan et al. explored the optical properties of drop cast PVA/MXene composites. They have theoretically calculated the optical absorption coefficient and extinction coefficient. They have a very high increase in conductivity, 30,000% higher than the neat PVA owing to hydrogen bond formation of MXene and PVA (Tan *et al.*, 2021). Wang *et al.* successfully cast Ti_3C_2Tx/epoxy compounds and revealed the EMI protective application for synthesized composites. The highest EMI shielding value is found in the 15 wt% compositions, with an ideal electrical conductivity of 105 S/m and an EMI shielding effectiveness of 41 dB (Wang *et al.*, 2019). Even though MXene is hydrophilic, getting good dispersion is a challenge, and to improve the distribution surface modification of MXene is opted by researchers using cetyltrimethylammonium bromide and an SiO_2 coating (Yu *et al.*, 2019; Wan *et al.*, 2020). Figure 2.4 shows the MXene/polymer

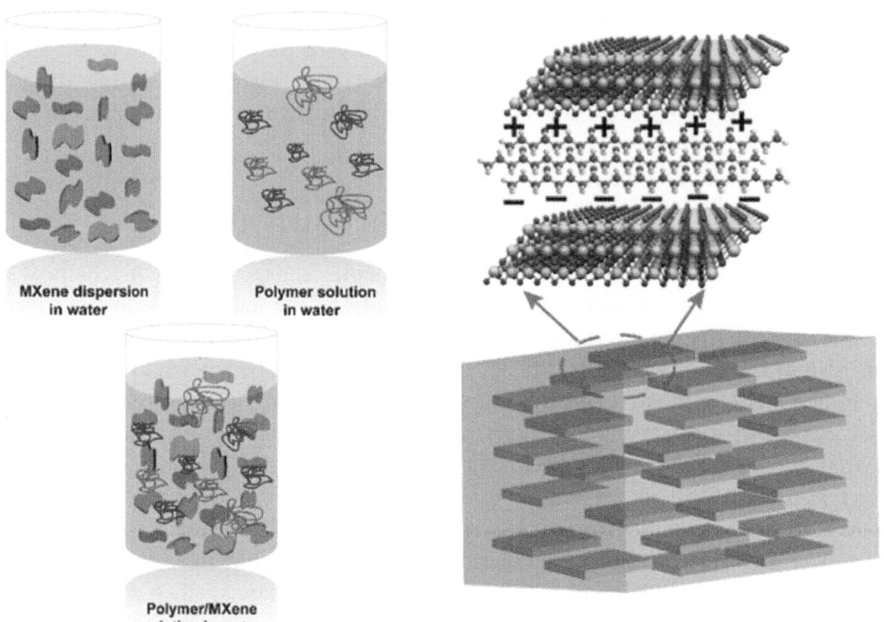

FIGURE 2.4 Schematic diagram of PVA/MXene solution before casting (Mirkhaniet *al.*, 2019).

TABLE 2.1
Solvent-Blended MXene/Polymer Composites

Sl. No	Mxene/Polymer Composites	Preparation Method/ Solvents	Applications/ Properties	References
1	Ti_3C_2Tx/PVA	Vacuum filtration Water	supercapacitor electrodes	(Ling *et al.*, 2014)
2	MXene/P(VDF-TrFE-CFE)	Solution casting DMF	Improved dielectric constant	(Tu *et al.*, 2018)
3	PVA/MXene nanocomposites	Vacuum-assisted filtration	Dielectric properties (ε' = 370.5, tan δ = 0.11) Electronic devices	(Mirkhani *et al.*, 2019)
4	MXene/Polypyrrole/ acrylic resin elastomer composite	Solvent casting Water and DMF	enhanced dielectric permittivity (493 at 100 Hz)	(Shao *et al.*, 2019)
5	MXene/PVA hydrogel	Gelatinization Water	Wearable Capacitive sensors	(Zhang *et al.*, 2019)
6	MXene-PDMS composites	Dip-coating	Wearable smart devices	(Song *et al.*, 2019)
7	PVA/MXene nanocomposites	Solution casting Water	Improved mechanical strength	(Pan *et al.*, 2020)
8	MXene/polyaniline fiber	Coating	Strain sensor	(Chao *et al.*, 2020)
9	MXene/P(VDF-TrFE)	Spin coating DMF	Pressure sensor	(Li *et al.*, 2020)
10	Ti_3C_2Tx MXene@SiO_2/ PVA composite	Solution coating Water	Dielectric constant of 27.2 Electronic application	(Wan *et al.*, 2020)
11	MXene/PVDF nanocomposites	Solvent-assisted mixing and compression molding Ethanol	48.47 ± 3.5 dB Loading concentration 22.55 vol% EMI shielding	(Rajavel *et al.*, 2020)
12	SiC-nanowire/MXene PVDF	Solution casting poly(diallyl dimethylammonium chloride, PDDA)	EMI shielding	(Ma *et al.*, 2021)
13	PVA/MXene films	Vacuum-assisted filtration	Pressure-sensing Artificial electronic skin	(Zhao *et al.*, 2021)

via solution blending, and Table 2.1 shows the various polymer/MXene composites synthesized via the solution blending process.

2.2.2 Latex Compounding

Aakyiir et al. synthesized MXene/nitrile butadiene rubber (NBR) composites via the emulsion-compounding technique and obtained good dispersion and an increase in mechanical strength. Using 14 vol% MXene, they got a 570% increase in Young's modulus (Aakyiir*et al.*, 2020). NBR/MXene nanocomposites were synthesized via the latex-stage-compounding method. The MXene suspension was mixed with NBR latex and coagulated using hydrochloric acid. The vulcanizing agents are combined with the rubber composite via two-roll mill mixing and compression molding with excellent electrical conductivity with a potential EMI shielding application (Aakyiir *et al.*, 2021).

Guo et al. used a saponification process between hydroxyl and carboxyl groups on serine to modify MXene nanoflakes and create nanostructured Ti_3C_2MXenes/rubber-based supramolecular elastomer (NMSE) with outstanding sensing capability (Guo *et al.*, 2020).

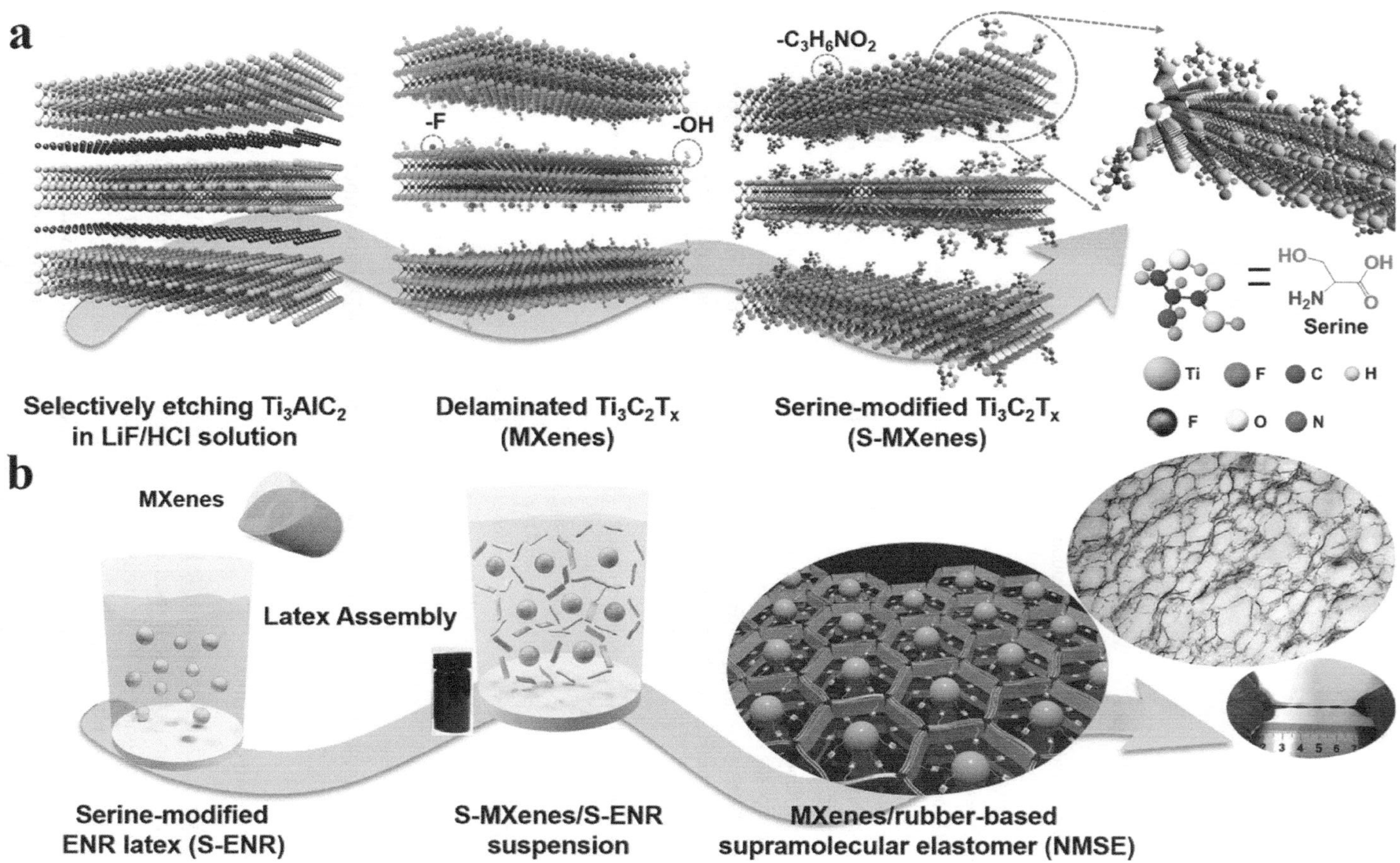

FIGURE 2.5 (a) The esterification reaction modifies the surface of MXene nanosheets by adding serine. (b) Using the latex assembly method, design a nanostructured MXene network in NMSE.(Guo *et al.*, 2020)

2.2.3 Melt Blending

In a twin-screw extruder, Gao et al. melt blended the polyethene glycol pretreated MXene/thermoplastic polyurethane (TPU) and extensively studied the filler dispersion, mechanical strength, and filler polymer interaction using rheological analysis. They have observed a strong hydrogen bonding interaction between PEG pretreated MXene and TPU chains, which results from good filler dispersion, and the morphological studies show no agglomeration at lower concentration of MXene, but at 1 wt%m small agglomerates are observed; at 0.5 wt% MXene, the tensile strength is increased upto 41.2% compared to neat TPU (Q. Gao *et al.*, 2020). Shi *et al.* (2019) successfully developedultra-thin 2D titanium carbide

(Ti_3C_2Tx)/PP nanocomposites with high mechanical strength via a melt blending technique. First, the MA-g-PP/Ti_3C_2Tx nanosheet powder is synthesized by using mechanical stirring of MXene powder and PP latex with the assistance of an oxygen-free fast aerationtactic. Then the MA-g-PP/Ti_3C_2Tx nanosheet powder melt blended with the PP granules to obtain better dispersion. The synthesis of the PP/MXene composites is shown in the Figure 2.6a, and the transmission electron microscopy (TEM) image of the composites shows the intercalated and exfoliated layers of MXene in the PP matrix and the scanning electron microscopy (SEM) images of surface fractured PP/MXene composite, which indicating interfacial solid adhesion with the composite shown in Figures 2.6band 2.6c, respectively. The excellent dispersion and the hydrogen bond interaction with the PP and theMA-g-PP/Ti3C2Tx nanosheet powder are responsible for the improved mechanical strength. The nanoconfinement by the hydrogen bond restricts the macromolecular chain movement and mechanical breakage and improves the mechanical strength, ductility, and modulus by 35.3%, 674.6%, and 102.2%, respectively; Figure 2.6d shows the model of mechanical failure process (Shi *et al.*, 2019).

2.2.4 *In Situ* Polymerization

In situ polymerization technique is a robust fabrication method in which the small monomers, and the MXene are dispersed together and allow the polymerization of the monomer with appropriate conditions. Polyaniline is a polymer that has gained significant interest *in situ* polymerization. The small molecule aniline and MXene dispersed in HCl are primarily mixed and polymerized with a suitable property for a potential gas sensor material (Zhao *et al.*, 2019). Ul Haq et al. synthesized PMMA/MXene/ZnO ternary compound via *in situ* bulk polymerization technique. Here, the MAX phase is Ti3C2Tx, which is added to the methyl methacrylate (MMA) monomer for the *in situ* bulk polymerization (Ul Haq *et al.*, 2020). The *in situ* click polymerization technique was explored by McDaniel et al. to synthesize uniformly dispersed in urethane/MXene composites. They created nanocomposite samples with a 5 wt% filler $Ti_3C_2T_z$concerning monomer and a polymer matrix with 1:1 composition of thiol and isocyanate groups ([SH]/[NCO] = 1). They also studied the reactions of synthesized MXene with urethane and prepared the composites. Figure 2.7 represents the complete synthetic route followed for the preparation of nanocomposite using urethane (McDaniel *et al.*, 2021)

2.3 PROPERTIES

Due to the 2D structure of the MXene material and its excellent surface properties with good conductivity and mechanical strength, incorporating this potential filler into the polymer matrix will improve its properties, which shows a prospective claim in energy storage, EMI shielding, membrane, sensors, and biomedical applications (Feng*et al.*, 2018; Cai *et al.*, 2020; Karahan *et al.*, 2020; Li *et al.*, 2020; Zhang *et al.*, 2020, 2021; Zheng *et al.*, 2021; Jaya Prakash and Kandasubramanian, 2021; Kim *et al.*, 2021; Kshetri *et al.*, 2021; Riazi, Taghizadeh and Soroush, 2021)

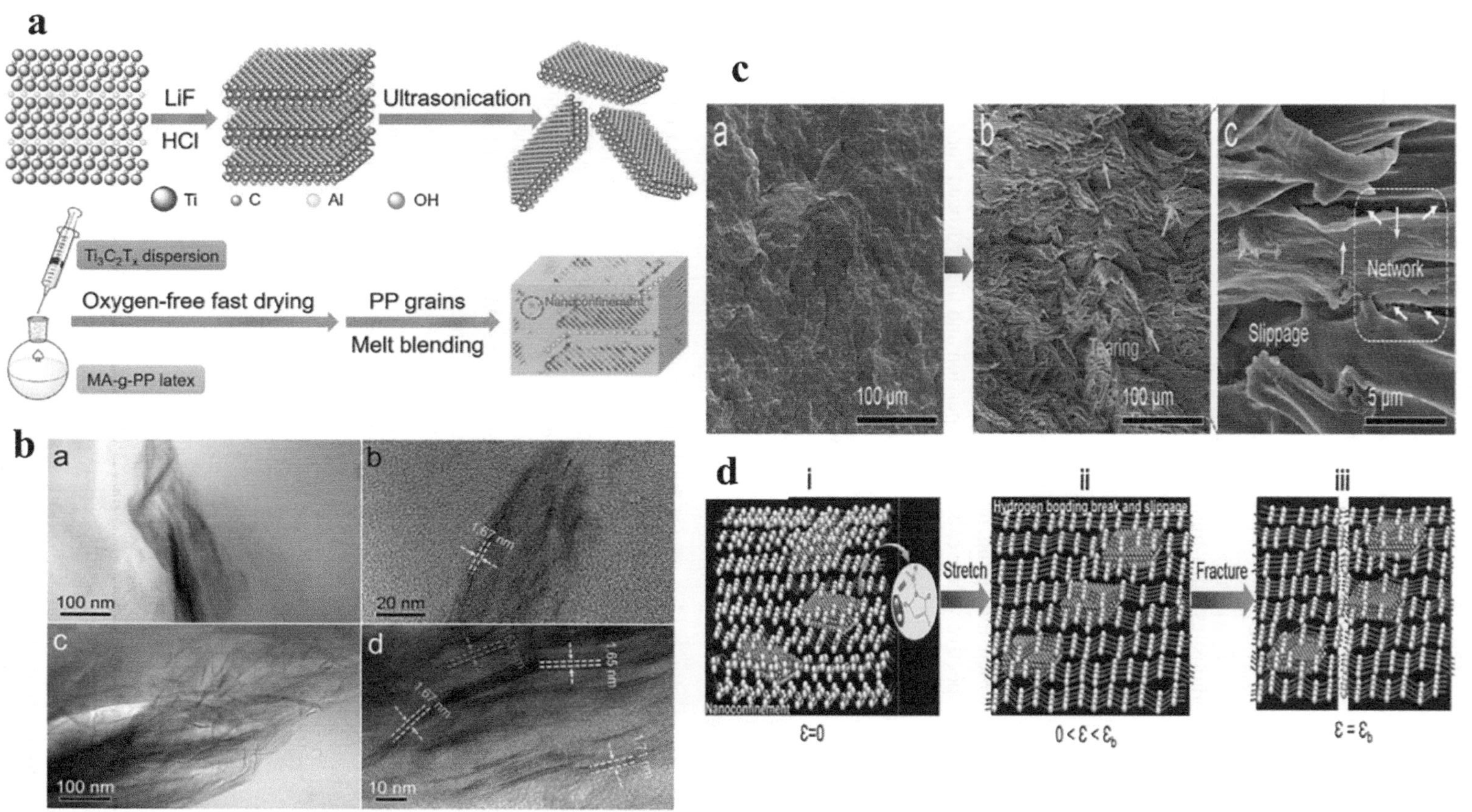

FIGURE 2.6 (a) Preparation of PP-g/MXene nanocomposites; (b) transmission electron microscopic images of PP-g/MXene nanocomposites 0.5 (a, b) and 2.0 (c, d); (c) SEM images of a fractured surface of PP-g/Ti3C2Tx NS-2.0, (d) diagrams of the mechanical failure process models of PP-g/Ti3C2Tx NS-2.0 (Shi *et al*., 2019).

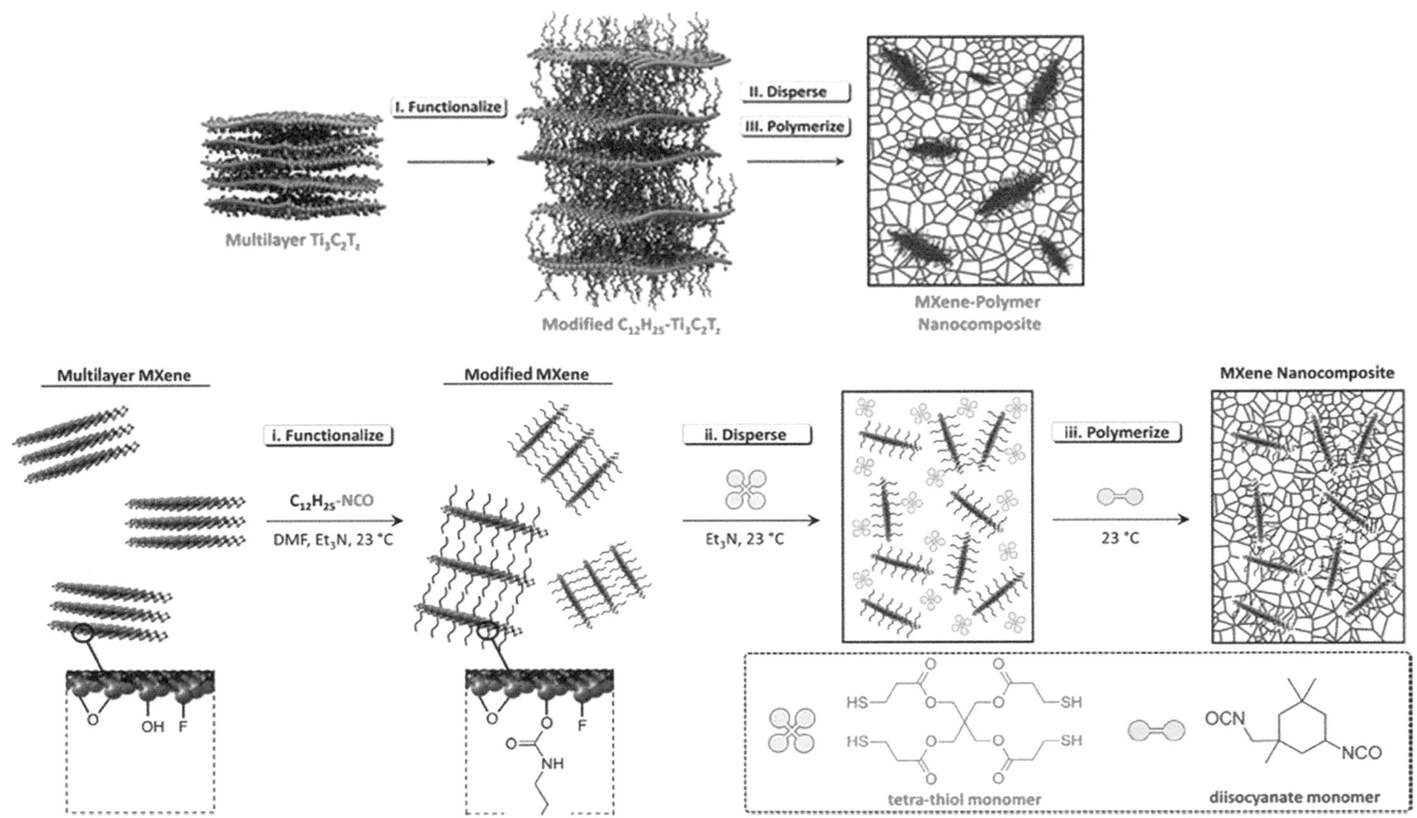

FIGURE 2.7 MXene thio urethane nanocomposites are made by (i) covalent functionalization and augmentation of multilayer Ti3C2Tz sheets via a reaction with dodecyl isocyanate, (ii) dispersion of customized Ti3C2Tz layers within a tetrafunctional thiol monomer, and (iii) *in situ* polymerization and nanocomposite establishment via the addition of a diisocyanate monomer (McDaniel *et al.*, 2021).

The increase in interfacial contact between the flakes and the surrounding polymer matrix is often attributed to the reinforcing effect on MXene nanocomposites. However, this interaction is rarely defined, and material qualities are often evidence of improved interfacial interactions/strength. The yield stress of nanocomposites rises when the filler loading of 2D materials, such as graphene, MXene, and layered silicates, among others, increases. This yield stress is a result of the filler's and hosts interfacial contact.

The mechanical strength of a polymer composite is very much crucial in every application. Ling *et al.* madeup the principal MXene/PVA nanocomposite, which improves mechanical properties and conductivity. The high elastic modulus (500 GPa) of MXene sheets can act as a sound reinforcement in the polymer matrix, and here the strength of the PVA/MXene composite improved to 91 ± 10 MPa. The enhanced mechanical strength indicates that effective stress transfer between the MXene layer, and the polymer matrix is taking action; also, the interfacial bonding between the matrix and MXene sheets are involved in the improvement in the mechanical strength of these nanocomposites (Ling *et al.*, 2014). Aakyiir *et al.* fabricated a stretchable material with a high electromagnetic shielding effectiveness of 49dB and elongation at a break of 170 ± 5.6%.

The MXene forms a good dispersion in the NBR matrix due to the surface modification of MXene with alkylamine and dispersed via latex compounding. The composite shows more ability to store electrical energy than the magnetic signals, so there is a rapid increase in permittivity. The electromagnetic shielding depends on the conductivity of the material here. The shielding mechanism depends on the absorption, and an increase in dielectric loss with the rise in MXene content is observed (Aakyiir *et al.*, 2021). The interfacial adhesion between the MXene and the polymer matrix ameliorates mechanical strength by effective load transfer to the MXene sheets. The surface energy and the hydrophilic functional groups present of the surface of MXene confirm good adhesion in the filler–matrix interface. TheMXene/epoxy coating showed high surface energy properties, and the interfacial adhesion is very high, which results in increased mechanical strength (Zukiene *et al.*, 2021). Biocompatibility is a significant concern in biomedical applications like artificial skin and pressure sensing applications. The hydrogen bond interaction between MXene and PVA provides a potentially conductive biomaterial for these applications. To monitor the signals, pressure sensors are placed near the tissue; for example, to observe the gastric peristalsis, the sensor is put to the stomach wall, so in vitro and in vivo studies are inevitable, and Zhao et al. fabricated the material can monitor cardiac motion also gastric peristalsis (Zhao *et al.*, 2021).Q. Song et al. constructed very flexibleFe_3O_4@Ti_3C_2Tx/DENR nanocomposites with excellent electric conductivity. Here, the conductive MXene layers display excellent interfacial adhesion with the natural rubber latex particles and form a segregated network structure that imparts high mechanical strength and conductivity and a 58-dB shielding efficiency (Song *et al.*, 2021). Table 2.2 shows some of the properties and applications of MXene/polymer composites.

MXenes' impermeable feature makes them ideal for increasing the barrier qualities of polymers, preventing gases and tiny molecules from passing through. When compared to their clean polymer equivalents, nanocomposites filled with layered silicates, and graphene have significantly enhanced barrier characteristics (Carey and Barsoum, 2021; Tan and Thomas, 2016; Wang *et al.*, 2018; Choudalakis and Gotsis, 2009). The improved barrier properties can be explained based on tortuous path creation. As a result of increases in volume fraction or aspect ratio of filler, the permeability will be decreased. Liu et al. studied the Ti_2CTz MXene membranes for isopropanol water mixture dehydration (Liu *et al.*, 2019), and Shamsabadi *et al.* studied the H_2, CO_2, CH_4 N_2 gas permeation usingTi_2CTz/polyurethane/Pebax(copolymer of polyamide and pole ether). They found good selectivity of CO_2/N_2 due to the solubility parameter values contributed by the functional group present in MXene sheets (Liu *et al.*, 2020; Yang *et al.*, 2020).

TABLE 2.2
MXene/Polymer Composites and Their Properties

Sl No	Polymer	Mxene	Properties/Applications	References
1	Thermoplastic polyurethane (TPU)	Ti_3C_2Tx	At an MXene loading value of 0.5 wt. %, tensile strength and elongation at break escalation by 41.2 wt%, and15.4 wt%, respectively.	(Gao *et al.*, 2020)
2	Polyvinylidene fluoride (PVDF)	Ti_3C_2Tx	Thermal conductivity is0.767 ± 0.034 W/mKand22.55 vol% filler concentration. 48.47 ± 3.5 dB EMI shielding effectiveness	(Rajavel *et al.*, 2020)
3	Epoxy	Ti_3C_2Tx	The electrical conductivity value of 105 S/m and EMI shielding effectiveness of 41 dB is the ideal values. 4.32 GPa Young's modulus and 0.29 GPa hardness	(Wang *et al.*, 2019)
4	Epoxy	Ti_3C_2Tx	Increase in young's modulus	(Sliozberg *et al.*, 2020)
5	Poly vinyl alcohol	Ti_3C_2Tx	Electrical conductivity of 1.4 x 106 S/m	(Tan *et al.*, 2021)
6	Nitrile butadiene rubber (NBR) latex	Ti_3C_2Tx	Electrical conductivity increased from 1.84×10^{-16} S.cm^{-1} to 9.0×10^{-05} S.cm$^{-1.}$	(Aakyiir *et al.*, 2020)
7	Natural rubber latex	Ti_3C_2Tx	Electrical conductivity value is6.18 × 10–4 S/m	(Yang *et al.*, 2021)
8	Polyurethane	Ti_3C_2Tx	PU/MX-0 has an electrical conductivity of 5983.5 203.6 S cm^{-1}, whereas PU/MX-20 has an SSEt of 33771.92 dB cm^2 g^{-1}.	(Liu *et al.*, 2020)
9	TPU	Ti_3C_2Tx-rGO hybrid	Excellent Flame retardant property	(Liu *et al.*, 2020)
10	P(VDF-TrFE-CFE)	Ti_3C_2Tx	Improved dielectric constant	(Tu *et al.*, 2018)

2.4 APPLICATIONS

The versatile surface chemistry of MXene and its exceptionally high electrical conductivity, mechanical strength, and thermal properties make MXene an incredibly suitable material that can be used in countless applications such as sensors, EMI shielding materials, and different applications in the biomedical field (Chen *et al.*, 2021). High electro-conductivity, extraordinary sensitivity, short detection limit, truncated hysteresis, and fast recovery after repeated usage are crucial properties for a potential sensor. The concentration of the surface group present in the MXene determines the performance of MXene/polymer composites (Riazi, Taghizadeh and Soroush, 2021).

MXene elastomer composites have potential use in self-healing flexible sensor applications. Guo *et al.* developed MXene/natural rubber nanocomposites with excellent self-healing properties. The serine-modified MXene developed an outstanding conductive network. The surface anchoring assemblies of MXene engage with the serine-grafted epoxidized natural rubber, forming interfacial supramolecular hydrogen bond interactions. They have developed a sensor that can sense minute human physiological motions and external moisture variations (Guo *et al.*, 2020). Luo *et al.* created an MXene/natural rubber composite that is highly conductive (1400 Sm^{-1}) and acceptable for EMI shielding (SE = 53.6 dB). Natural rubber's crosslinking improves reproducibility and stability, making it a viable candidate for use in foldable electronics (Luo *et al.*, 2019). The application of MXene/polymer composites, and the applications are tabulated in Table 2.3.

TABLE 2.3
Application of MXene/Polymer Nanocomposites

Sl. No	Mxene/Polymer Nanocomposite	Applications	References
1	PANI/Ti_3C_2Tx nanocomposites	Gas sensor	(Zhao *et al.*, 2019)
2	MXene/natural rubber nanocomposite	EMI shielding 53.6 dB 6.71 vol%	(Luo *et al.*, 2019)
3	Ti_3C_2Tx MXene/Epoxy nanocomposites	EMI shielding 41 dB 15 wt%	(Wang *et al.*, 2019)
4	Serine-modified MXene/Epoxidized natural rubber	Electronic sensor Speech, facial expression recognition, and pulse detection	(Guo *et al.*, 2020)
5	The hydrogel of modified polyacrylamide polymer and MXene-vinyl hybrid silica nanoparticle	Electronic skin sensors	(Cai *et al.*, 2020)
6	Composite of Ti_3C_2Tx MXene and PVDF polymer	EMI shielding At a thickness of 2 mm, 48.47 ± 3.5 dB corresponding to 22.55 vol % filler quantity	(Rajavel *et al.*, 2020)
7	MXene/Au/PET nanocomposite	Pressure sensor, wearable gadget For biological signal recognition Flexible piezoresistive sensor	(Gao *et al.*, 2020)
8	Cellulose Nanofibers/Ti_3C_2Tx MXene Aerogels/ Epoxy Nanocomposites	EMI shielding 74 dB, 1.38 vol%,	(Wang *et al.*, 2020)
9	Ti_3C_2Tx and P(VDF-TrFE) film	Piezoresistive Pressure sensors	(Li *et al.*, 2020)
10	MXene/polyanilinefiber nanocomposites	Ultrasensitive sensing	(Chao *et al.*, 2020)
11	MXene/EPDM rubber	EMI shielding 48 dB 2.7 wt%	(Lu *et al.*, 2020)
12	MXene/PU sensors	Wearable pressure sensor	(Zukiene *et al.*, 2021)
13	PDMS/MXene composite	E-skin triboelectric tactile sensor	(Cai *et al.*, 2021)

2.5 CONCLUSION

MXene is a versatile layered material that can be prepared from selective etching using strong acids, shows high mechanical strength and high electro conductive properties, and has potential applications in electromagnetic interference shielding, biomedical application, pressure sensing applications, and energy storage applications, among others. But the primary concern and difficulty in MXene research is its production on an industrial scale. If it fails in the industrial scale-up of synthesis, the applicability will also be a question mark, and its fate will be similar to graphene. Another foremost concern is its hindrance to oxidation; the ongoing research focuses on preventing the oxidation of MXene sheets. The cost and availability of MXene powders also limit its application potential to a greater extent. It is vital to find an alternative route for synthesizing the MAX phase materials without using hazardous HF acid. While comparing to graphene and carbon nanotube, the production cost of MAX phase materials is almost similar or slightly lower. One of the cheapest nanofillers available to make polymer composite is clay. MXene research in the future should focus on reducing the production cost of MXenes, and if the price of MXene is lower than that of clay, one can produce materials with high-end properties in a low-budget scheme.

We expect that as the playing arena of MXene polymer nanocomposites progresses and a countless emphasis is placed on scalable manufacturing approaches, the expansions will contribute the ways and means of making the MXene-based research and development more vibrant.

REFERENCES

Aakyiir, M. *et al.* (2020) 'Electrically and thermally conductive elastomer by using MXene nanosheets with interface modification', *Chemical Engineering Journal*. Elsevier, 397(February), p. 125439. doi: 10.1016/j.cej.2020.125439.

Aakyiir, M. *et al.* (2021) 'Stretchable, mechanically resilient, and high electromagnetic shielding polymer/MXene nanocomposites', *Journal of Applied Polymer Science*, 138(22). doi: 10.1002/app.50509.

Cai, Y. *et al.* (2020) 'Mixed-dimensional MXene-hydrogel heterostructures for electronic skin sensors with ultrabroad working range', *Science Advances*, 6(48). doi: 10.1126/sciadv.abb5367.

Cai, Y. W. *et al.* (2021) 'A flexible ultra-sensitive triboelectric tactile sensor of wrinkled PDMS/MXene composite films for E-skin', *Nano Energy*. Elsevier Ltd, 81(November 2020), p. 105663. doi: 10.1016/j.nanoen.2020.105663.

Carey, M. and Barsoum, M. W. (2021) 'MXene polymer nanocomposites: a review', *Materials Today Advances*. Elsevier Ltd, 9, p. 100120. doi: 10.1016/j.mtadv.2020.100120.

Carey, M. S. *et al.* (2019) 'Water transport and thermomechanical properties of Ti3C2T z MXene epoxy nanocomposites', *ACS Applied Materials and Interfaces*, 11(42), pp. 39143–39149. doi: 10.1021/acsami.9b11448.

Chao, M. *et al.* (2020) 'Wearable MXene nanocomposites-based strain sensor with tile-like stacked hierarchical microstructure for broad-range ultrasensitive sensing', *Nano Energy*, 78(July). doi: 10.1016/j.nanoen.2020.105187.

Chen, X. *et al.* (2021) 'MXene/Polymer nanocomposites: Preparation, properties, and applications', *Polymer Reviews*. Taylor & Francis, 61(1), pp. 80–115. doi: 10.1080/15583724.2020.1729179.

Choudalakis, G. and Gotsis, A. D. (2009) 'Permeability of polymer/clay nanocomposites: A review', *European Polymer Journal*. Elsevier Ltd, 45(4), pp. 967–984. doi: 10.1016/j.eurpolymj.2009.01.027.

Feng, Y. *et al.* (2018) 'An Ultrahigh Discharged Energy Density Achieved in an Inhomogeneous PVDF Dielectric Composite Filled with 2D MXene Nanosheets: Via Interface Engineering', *Journal of Materials Chemistry C*. Royal Society of Chemistry, 6(48), pp. 13283–13292. doi: 10.1039/c8tc05180a.

Gao, Q. *et al.* (2020) 'Mechanical, thermal, and rheological properties of Ti3C2Tx MXene/thermoplastic polyurethane nanocomposites', *Macromolecular Materials and Engineering*, 305(10), pp. 1–9. doi: 10.1002/mame.202000343.

Gao, Y. *et al.* (2020) 'Microchannel-confined MXene based flexible piezoresistive multifunctional micro-force sensor', *Advanced Functional Materials*, 30(11), pp. 1–8. doi: 10.1002/adfm.201909603.

Guo, Q. *et al.* (2020) 'Protein-inspired self-healable Ti3C2 MXenes/rubber-based supramolecular elastomer for intelligent sensing', *ACS Nano*, 14(3), pp. 2788–2797. doi: 10.1021/acsnano.9b09802.

Huang, H. *et al.* (2020) 'Synergistic effect of MXene on the flame retardancy and thermal degradation of intumescent flame retardant biodegradable poly (lactic acid) composites', *Chinese Journal of Chemical Engineering*. Elsevier BV, 28(7), pp. 1981–1993. doi: 10.1016/j.cjche.2020.04.014.

Jaya Prakash, N. and Kandasubramanian, B. (2021) 'Nanocomposites of MXene for industrial applications', *Journal of Alloys and Compounds*. Elsevier, 862, p. 158547. doi: 10.1016/j.jallcom.2020.158547.

Karahan, H. E. *et al.*(2020) 'MXene materials for designing advanced separation membranes', *Advanced Materials*, 32(29), pp. 1–23. doi: 10.1002/adma.201906697.

Kim, E. *et al.* (2021) 'MXene/polyurethane auxetic composite foam for electromagnetic interference shielding and impact attenuation', *Composites Part A: Applied Science and Manufacturing*. Elsevier Ltd, 147(April), p. 106430. doi: 10.1016/j.compositesa.2021.106430.

Kshetri, T. *et al.* (2021) 'Recent advances in MXene-based nanocomposites for electrochemical energy storage applications', *Progress in Materials Science*. Elsevier, 117(August), p. 100733. doi: 10.1016/j.pmatsci.2020.100733.

Li, L. *et al.* (2020) 'Hydrophobic and stable MXene-polymer pressure sensors for wearable electronics', *ACS Applied Materials and Interfaces*, 12(13), pp. 15362–15369. doi: 10.1021/acsami.0c00255.

Ling, Z. *et al.* (2014) 'Flexible and conductive MXene films and nanocomposites with high capacitance', *Proceedings of the National Academy of Sciences of the United States of America*, 111(47), pp. 16676–16681. doi: 10.1073/pnas.1414215111.

Lipton, J. *et al.* (2019) 'Mechanically strong and electrically conductive multilayer MXene nanocomposites', *Nanoscale*. Royal Society of Chemistry, 11(42), pp. 20295–20300. doi: 10.1039/c9nr06015d.

Liu, C. *et al.* (2020) 'Creating MXene/reduced graphene oxide hybrid towards highly fire safe thermoplastic polyurethane nanocomposites', *Composites Part B: Engineering*. Elsevier Ltd, 203(October), p. 108486. doi: 10.1016/j.compositesb.2020.108486.

Liu, G. *et al.* (2020) 'Polyelectrolyte functionalised ti2ct x mxene membranes for pervaporation dehydration of isopropanol/water mixtures', *Industrial and Engineering Chemistry Research*, 59(10), pp. 4732–4741. doi: 10.1021/acs.iecr.9b06881.

Liu, Guozhen *et al.* (2019) 'Two-dimensional Ti2CT: X MXene membranes with integrated and ordered nano-channels for efficient solvent dehydration', *Journal of Materials Chemistry A*, 7(19), pp. 12095–12104. doi: 10.1039/c9ta01507h.

Liu, Z. *et al.* (2020) 'Bioinspired ultra-thin polyurethane/MXene nacre-like nanocomposite films with synergistic mechanical properties for electromagnetic interference shielding', *Journal of Materials Chemistry C*, 8(21), pp. 7170–7180. doi: 10.1039/d0tc01249a.

Lu, S. *et al.* (2020) 'Flexible MXene/EPDM rubber with excellent thermal conductivity and electromagnetic interference performance', *Applied Physics A: Materials Science and Processing*. Springer Berlin Heidelberg, 126(7), pp. 1–12. doi: 10.1007/s00339-020-03675-3.

Luo, J. Q. *et al.*(2019) 'Flexible, stretchable and electrically conductive MXene/natural rubber nanocomposite films for efficient electromagnetic interference shielding', *Composites Science and Technology*. Elsevier, 182(July), p. 107754. doi: 10.1016/j.compscitech.2019.107754.

Ma, L. *et al.* (2021) 'Enhanced electromagnetic wave absorption performance of polymer/SiC-nanowire/MXene (Ti3C2Tx) composites', *Carbon*, 179, pp. 408–416. doi: 10.1016/j.carbon.2021.04.063.

McDaniel, R. M. *et al.*(2021) 'Well-dispersed nanocomposites using covalently modified, multilayer, 2D titanium carbide (MXene) and *In-Situ* "click" polymerisation', *Chemistry of Materials*, 33(5), pp. 1648–1656. doi: 10.1021/acs.chemmater.0c03972.

Mirkhani, S. A. *et al.*(2019) 'High dielectric constant and low dielectric loss via poly(vinyl alcohol)/Ti3C2Tx MXene nanocomposites', *ACS Applied Materials and Interfaces*, 11(20), pp. 18599–18608. doi: 10.1021/acsami.9b00393.

Naguib, M. *et al.* (2012) 'Two-dimensional transition metal carbides', *ACS Nano*, 6(2), pp. 1322–1331. doi: 10.1021/nn204153h.

Pan, Y. *et al.* (2020) 'Flammability, thermal stability and mechanical properties of polyvinyl alcohol nanocomposites reinforced with delaminated Ti3C2Tx (MXene)', *Polymer Composites*, 41(1), pp. 210–218. doi: 10.1002/pc.25361.

Rajavel, K. *et al.* (2020) '2D Ti3C2Tx MXene/polyvinylidene fluoride (PVDF) nanocomposites for attenuation of electromagnetic radiation with excellent heat dissipation', *Composites Part A: Applied Science and Manufacturing*. Elsevier Ltd, 129, p. 105693. doi: 10.1016/j.compositesa.2019.105693.

Riazi, H., Taghizadeh, G. and Soroush, M. (2021) 'MXene-based nanocomposite sensors', *ACS Omega*, 6(17), pp. 11103–11112. doi: 10.1021/acsomega.0c05828.

Shao, J. *et al.* (2019) 'A novel high permittivity percolative composite with modified MXene', *polymer*. Elsevier, 174(March), pp. 86–95. doi: 10.1016/j.polymer.2019.04.057.

Shi, Y. *et al.* (2019) 'Strengthening, toughing and thermally stable ultra-thin MXene nanosheets/polypropylene nanocomposites via nanoconfinement', *Chemical Engineering Journal*. Elsevier, 378(April), p. 122267. doi: 10.1016/j.cej.2019.122267.

Sliozberg, Y. *et al.* (2020) 'Interface binding and mechanical properties of MXene-epoxy nanocomposites', *Composites Science and Technology*. Elsevier Ltd, 192(January), p. 108124. doi: 10.1016/j.compscitech.2020.108124.

Song, D. *et al.* (2019) 'Hollow-structured MXene-PDMS composites as flexible, wearable and highly bendable sensors with wide working range', *Journal of Colloid and Interface Science*. Elsevier Inc., 555, pp. 751–758. doi: 10.1016/j.jcis.2019.08.020.

Song, Q. *et al.* (2021) 'Flexible, stretchable and magnetic Fe3O4@Ti3C2Tx/elastomer with supramolecular interfacial crosslinking for enhancing mechanical and electromagnetic interference shielding performance', *Science China Materials*, 64(6), pp. 1437–1448. doi: 10.1007/s40843-020-1539-2.

Tan, B. and Thomas, N. L. (2016) 'A review of the water barrier properties of polymer/clay and polymer/graphene nanocomposites', *Journal of Membrane Science*. doi: 10.1016/j.memsci.2016.05.026.

Tan, K. H. *et al.* (2021) 'Optical and conductivity studies of polyvinyl alcohol-MXene (PVA-MXene) nanocomposite thin films for electronic applications', *Optics and Laser Technology*. Elsevier Ltd, 136(October), p. 106772. doi: 10.1016/j.optlastec.2020.106772.

Tu, S. *et al.* (2018) 'Large dielectric constant enhancement in MXene percolative polymer composites', *ACS Nano*, 12(4), pp. 3369–3377. doi: 10.1021/acsnano.7b08895.

Ul Haq, Y. *et al.* (2020) 'Investigation of improved dielectric and thermal properties of ternary nanocomposite PMMA/MXene/ZnO fabricated by *in-situ* bulk polymerisation', *Journal of Applied Polymer Science*, 137(40), pp. 1–16. doi: 10.1002/app.49197.

Wan, W. *et al.* (2020) 'Enhanced dielectric properties of homogeneous Ti3C2Tx MXene@SiO2/polyvinyl alcohol composite films', *Ceramics International*, 46(9), pp. 13862–13868. doi: 10.1016/j.ceramint.2020.02.179.

Wang, K. *et al.* (2018) '3D graphene foams/epoxy composites with double-sided binder polyaniline interlayers for maintaining excellent electrical conductivities and mechanical properties', *Composites Part A: Applied Science and Manufacturing*, 110, pp. 246–257. doi: 10.1016/j.compositesa.2018.05.001.

Wang, L. *et al.* (2019) 'Fabrication on the annealed Ti3C2Tx MXene/Epoxy nanocomposites for electromagnetic interference shielding application', *Composites Part B: Engineering*. Elsevier Ltd, 171(April), pp. 111–118. doi: 10.1016/j.compositesb.2019.04.050.

Wang, L. *et al.* (2020) '3D shapeable, superior electrically conductive cellulose nanofibers/Ti 3 C 2 T x MXene Aerogels/Epoxy nanocomposites for promising EMI shielding', *Research*, 2020, pp. 1–12. doi: 10.34133/2020/4093732.

Xiao, Z. *et al.* (2020) *MXenes and MXenes-based composites*. Available at: http://link.springer.com/10.1007/978-3-030-59373-5.

Xue, Y. *et al.* (2020) 'Polyphosphoramide-intercalated MXene for simultaneously enhancing thermal stability, flame retardancy and mechanical properties of polylactide', *Chemical Engineering Journal*. Elsevier, 397(April), p. 125336. doi: 10.1016/j.cej.2020.125336.

Yang, K. *et al.* (2020) 'Triazine-based two-dimensional organic polymer for selective NO2 sensing with excellent performance', *ACS Applied Materials and Interfaces*, 12(3), pp. 3919–3927. doi: 10.1021/acsami.9b17450.

Yang, Y. D. *et al.* (2021) 'Natural rubber latex/MXene foam with robust and multifunctional properties', *E-Polymers*, 21(1), pp. 179–185. doi: 10.1515/epoly-2021-0017.

Yu, B. *et al.* (2019) 'Interface decoration of exfoliated MXene ultra-thin nanosheets for fire and smoke suppressions of thermoplastic polyurethane elastomer', *Journal of Hazardous Materials*, 374, pp. 110–119. doi: 10.1016/j.jhazmat.2019.04.026.

Zhang, A. *et al.* (2020) 'MXene-based nanocomposites for energy conversion and storage applications', *Chemistry—A European Journal*, 26(29), pp. 6342–6359. doi: 10.1002/chem.202000191.

Zhang, C. S. *et al.*(2007) 'Electromagnetic interference shielding effect of nanocomposites with carbon nanotube and shape memory polymer', *Composites Science and Technology*, 67(14), pp. 2973–2980. doi: 10.1016/j.compscitech.2007.05.011.

Zhang, F. *et al.* (2021) 'Ke Pi: InvestigationSurface functionalisation of Ti3C2Tx and its application in aqueous polymer nanocomposites for reinforcing corrosion protection', *Composites Part B: Engineering*, 217(April). doi: 10.1016/j.compositesb.2021.108900.

Zhang, H. *et al.* (2016) 'Preparation, mechanical and anti-friction performance of MXene/polymer composites', *Materials and Design*. Elsevier Ltd, 92, pp. 682–689. doi: 10.1016/j.matdes.2015.12.084.

Zhang, J. *et al.*(2019) 'Highly stretchable and self-healable MXene/polyvinyl alcohol hydrogel electrode for wearable capacitive electronic skin', *Advanced Electronic Materials*, 5(7), pp. 1–10. doi: 10.1002/aelm.201900285.

Zhao, L. *et al.* (2019) 'High-performance flexible sensing devices based on polyaniline/MXene nanocomposites', *InfoMat*, 1(3), pp. 407–416. doi: 10.1002/inf2.12032.

Zhao, L. *et al.* (2021) 'Highly-stable polymer-crosslinked 2D MXene-based flexible biocompatible electronic skins for in vivo biomonitoring', *Nano Energy*. Elsevier Ltd, 84(February), p. 105921. doi: 10.1016/j.nanoen.2021.105921.

Zheng, S. *et al.* (2021) 'Multitasking MXene inks enable high-performance printable microelectrochemical energy storage devices for all-flexible self-powered integrated systems', *Advanced Materials*, 33(10), pp. 1–10. doi: 10.1002/adma.202005449.

Zukiene, K. *et al.* (2021) 'Wettability of MXene and its interfacial adhesion with epoxy resin', *Materials Chemistry and Physics*, 257(July 2020). doi: 10.1016/j.matchemphys.2020.123820.

3 Processing, Morphology, Mechanical, and Electrical Properties, and Applications of Thermoplastic Polymer/MXenes Nanocomposites

Pallishree Sahoo, Lipsa Shubhadarshinee, Pooja Mohapatra, Patitapaban Mohanty, Bigyan Ranjan Jali, Priyaranjan Mohapatra, and Aruna Kumar Barick

CONTENTS

3.1 INTRODUCTION

Owing to the excellent exfoliation of graphene and its superior properties, two-dimensional (2D) nanomaterials have attracted a wide attention for interdisciplinary research in science and engineering as compared to the 3D nanomaterials [1]. These materials are a big family, including elemental substances (e.g., graphene, silicene, borophene, and black phosphorus) and compounds (e.g., transition metal sulfides, transition metal oxides, layered double hydroxide [LDH], and MXene) [2,3].

DOI: 10.1201/9781003164975-3

Recently, novel types of 2D nanomaterials, namely, "MXenes," have been surfaced and produced immense interest in the research community because of their unique nanostructured morphology and electronic features that broadening their exercise in various promising applications. MXenes are first discovered by the Gogotsi and Barsoum groups in 2011 [4]. They are an exceptional class of hydrophilic, electrically conducting 2D nanomaterials, and architecturally, they are 2D nanosheets of transition metal (Ti, V, Cr, Nb, etc.) carbides, nitrides, or carbonitrides [5–8]. MXenes are obtained by selectively chemical etching of the layers of *s*- and *p*-block elements (Al, Si, Sn, In, etc.) from their equivalent 3D MAX phases consist of layered ternary metal carbides, nitrides, and carbonitrides with a typical chemical composition of $M_{n+1}AX_n$(n: 1, 2, 3, 4), where M, A, and X symbolize *d*-block transition metals (such as Mo, Ti, Zr, Cr, etc.), main-group *s*- and *p*-block elements (mainly IIIA/IVA), and either C/N or both C and N atoms, respectively. More than 70 kinds of MAX phases have been reported in literature [9,10]; however, the well-known MXene forms only involved Ti_3C_2, Ti_2C, $(Ti_{0.5}, Nb_{0.5})_2C$, $(V_{0.5}, Cr_{0.5})_3C_2$, Ti_3CN, Ta_4C_3 [11], Nb_2C, V_2C [12], and Nb_4C_3 stoichiometric compositions [13]. Particularly, the exterior surfaces of the dispersed single sheets are usually functionalized with fluoride (–F), hydroxyl (–OH), and –O– polar functional groups at the time of chemical etching treatments.

MXenes have attracted tremendous attention from researchers due to their both outstanding physical and chemical properties [7], electrical conductivity [14], hydrophilicity, and flexible surface chemistry [15,16]. These properties make MXene a potential material for utilized for electrodes in energy storage devices [17,18], electrochemical supercapacitors [19,20], catalytic promoters [21,22], absorbents for heavy metal ions, materials for the photodegradation of dyes [23], biosensors [24,25], and anti-biofouling membranes [26,27]. Among the synthetic materials, MXene has shown the highest level of electromagnetic shielding (92 dB of electromagnetic interference [EMI] shielding effectiveness [SE]) and has far surpassed that of silver and copper [28] and higher charge storage capability compared to graphene supercapacitors [29].

Various MXene-based polymer nanocomposites have been investigated because of their superior physical properties like mechanical strength, electrical conductivity, and thermal conductivity. Until now, a number of review articles have been published regarding recent advances of MXenes covering their synthesis and modifications together with preparation, properties and application of polymer/MXene nanocomposites [30–40]. Jimmy and Kandasubramanian presented several polymer nanocomposites of MXene with a focus on their fabrication methods and future applications [30]. Gao *et al.* discussed the synthesis and properties of the MXenes and polymer/MXene nanocomposite membranes together with promising applications in various areas like filtration, EMI shielding, energy storage and conversion devices, wearable devices, and so on. [31]. Chen *et al.* reviewed the classification, preparation methods, performance, and applications such as biomedical, sensing, energy storage, conversion, and supply, electromagnetic absorption and shielding, catalysis, chemisorption, sewage purification, flame retardance, separation membranes, and other areas of polymer/MXene nanocomposites [32]. George and Kandasubramanian deliberated the exciting potential biomedical applications specifically antibacterial activity, photothermal cancer therapy, drug delivery, bioimaging, biosensing, and bone regeneration of polymer/MXene nanocomposites [33]. Aghamohammadi *et al.* assessed the mechanical, tribological, thermal, corrosion, and electrical properties of both thermoset and thermoplastic polymer/MXene nanocomposites [34]. Recently, He *et al.*, Gong *et al.*, and Kausar reviewed the preparation, properties, and application of polymer/MXene nanocomposites [35–37]. Carey and Barsoum appraised the production methods and properties particularly conductivity, mechanical behavior, interfacial strength, gas permeability, and the dimensional and thermal stability of polymer/MXene nanocomposites [38]. Riazi *et al.* investigated the reported advanced applications of polymer/MXene nanocomposites [39]. Song *et al.* evaluated the fabrication and progress of polymer/MXene nanocomposites for EMI shielding applications [40]. Nevertheless, a large number of articles are published on the polymer/MXene nanocomposites, the research advances are even in the initial stages and emerging performances need to be elaborately investigated for ultimate applications.

Several authors have reported that the mechanical, thermal, and electrical properties of polymer/MXene nanocomposites noticeably increase with MXene loading. Feng *et al.* synthesized MXene-based polyvinylidene fluoride (PVDF) nanocomposites and demonstrated enhanced electrical properties, such as the permittivity, AC conductivity, and breakdown strength of about 26 at 100 Hz, 10^{-11} Scm^{-1}, and approximately 350 MVm^{-1} at 100 Hz, respectively [41]. PVDF/MXene nanocomposites showed an approximately one fold improvement in thermal conductivity in comparison with the pristine PVDF, attended by a prominent improvement in the dynamic mechanical properties of the PVDF, for example, increased storage modulus, and shifting of the glass transition temperature (T_g) toward higher temperature at higher MXene loading [42]. Ling *et al.* prepared MXene-based polyvinyl alcohol (PVA) nanocomposite that could support 15,000 times its own weight [43]. Boota *et al.* added 8 wt.% polypyrrole (PPy) to MXene and found that the volumetric capacitance density was of 1000 Fcm^{-3}, which is twice that of the only MXene electrode [44], and by adding 0.15 wt.% MXene, Xu *et al.* reached 28 dB of EMI SE in a wave transparent PVA matrix [45]. The mechanical properties of ultra-high molecular weight polyethylene (UHMWPE)/MXene nanocomposites exhibited better mechanical properties, especially the surface hardness and creep behavior [46]. PVA/MXene nanocomposite nanofibers with low $Ti_3C_2T_x$ (0.14 wt%) loading exhibited a DC conductivity of 0.8×10^{-3} Scm^{-1} [47].

The main objective of the recent book chapter deals with the processing, characterization, properties, and application of thermoplastic polymer/MXene nanocomposites. The processing of thermoplastic polymer/MXene nanocomposites using solution blending, *in situ* method, and melt mixing are discussed, along with a description of the synthesized materials by means of different characterization techniques. In addition, the mechanical and electrical properties of the fabricated materials are highlighted with a depiction of different applications such as sensing, biomedical, energy storage, EMI shielding, filtration, and other fields to review the status of the thermoplastic polymer/MXene nanocomposites. Finally, this book chapter illustrates the challenges confronted by thermoplastic polymer/MXene nanocomposites and their future aspects, which is important for further research and development of thermoplastic polymer/MXene nanocomposites with outstanding material properties.

3.2 THERMOPLASTIC POLYMER/MXene NANOCOMPOSITES

Nanomaterials, such as nanoclays and various 2D nanomaterials, display more benefits in comparison with their 3D counterparts owing to their exceptionally large specific surface area and aspect ratios. The material performances such as mechanical, electrical, thermal, rheological, and barrier properties can be substantially improved by incorporating small amount of nanofillers at a concentration less than 2 wt% owing to the homogeneous dispersion of MXene nanosheets within the polymer matrix [48]. Excellent nanomaterials can extend noticeable improvements in electrical conductivity, catalytic activity, and sensing behavior. Reported polymer/MXene nanocomposites are typically made up of filled composites and complexes. The single-layered MXenes exhibit more available hydrophilic surfaces and superior compatibility with various polymers as compared to the multilayered MXenes. Hence, MXenes are normally delaminated that can be easily dispersed within the polymer matrix. Particularly, polymer/MXene complex are classified into laminated composite and structure [49–51], MXene-coated polymeric fiber or fabric composite [52,53], polymer/MXene composite fiber or fabric [54–56], and polymer/MXene composite aerogel, foam, or sponge [45,57,58]. The most studied are the filled composites, that is, thermoplastic [43,46,59,60] and thermosetting polymer [61,62] matrix, compared to the complexes. For example, a spin-coated MXene dispersed onto polycaprolactone (PCL) electrospun fiber membranes, which is a layered composite structure was reported by Zhou *et al.* [63]. The first-ever MXene-filled polymer nanocomposites was published in 2014 in which the $Ti_3C_2T_x$ MXenes were incorporated into the hydrophilic polydiallyl dimethyl ammonium chloride (PDDA) and PVA [43]. Different polymer/MXene nanocomposites are studied, which consist of hydrophilic polymers like PVA [43, 64–66], polyacrylic acid

(PAA) [67,68], and biopolymers [69,70], along with hydrophobic polymers like polyethylene (PE) [71], polyethylene oxide (PEO) [72], polypropylene (PP) [73], polystyrene (PS) [74], polyamide (PA) [75], and epoxies [76], among others. However, current literature focuses on the conductive polymers such as polyaniline (PANI) [77,78], polyvinyl pyrrolidone (PVP) [79,80], polypyrrole (PPy) [81], poly3,4-ethylenedioxy thiophene (PEDOT) [82], polystyrene sulfonate (PSS) [83], polyvinylidene difluoride (PVDF) [84], and polyvinylidene fluoride–trifluoroethylene (PVDF-TrFE) [85], and others.

3.2.1 Processing of Thermoplastic Polymer/MXene Nanocomposites

Polymer/MXene nanocomposites or complexes can be synthesized by using various methods, such as solution blending, melt blending, dry mixing/thermal pressing, emulsion mixing, *in situ* polymerization, ultrasonic mixing, and lamination stacking. Mostly MXene is prepared by solutions consisting of fluoride; the resulting MXene possess –F, –OH, and –O-terminated polar functional groups on its surface that provide good hydrophilicity, but dried MXene has poor hydrophilicity, and redispersion is also difficult [86]. Solution blending is the most widely used technology for the preparation of polymer/MXene composite. Usually, water is the most frequently used, and in some cases, polar solvents could also be used.

3.2.1.1 Solution Blending

Solution blending takes advantage of the excellent hydrophilicity of MXene and is easy to operate. It flexibly modulates the ratio of the components in the composites. In this method, first the dispersion of MXene and hydrophilic polymer matrix are carried out in a suitable common solvent in separate containers [87,88]. Then, both the solutions are blended together followed by evaporation to remove the solvents. Generally, this method is suitable for polymer either soluble in water or organic solvents such as PVA [89], polyvinyl chloride (PVC) [90], polyurethane (PU) [91], PEO [92], polyfluorene (PFS) [93], polyacrylamide (PAM) [94], PVDF [42], polyacrylates [95], cellulose [96], chitosan (CS) [97], polylactic acid (PLA) [98], etc. To disperse MXene, the suitable polar dispersant used are N,N-dimethyl formamide (DMF) [42], water [45], dimethylsulfoxide (DMSO) [99] or dimethylacetamide (DMAc) [100]. The MXene and polymer can be mixed in the same dispersant because the dispersant can also dissolve the polymer as well. Solution blending could also be carried out in different dispersants, but these dispersants need to be mutually soluble. Ling *et al.* have explored the preparation of a single-layer MXene-based PVA nanocomposite by blending the aqueous solution of MXenes and PVA [43]. For instance, Naguib *et al.* dispersed MXene with DMSO and dissolved PAM with water and then mixed the two mixtures or solutions to obtain the corresponding PAM/MXene nanocomposites [94]. At the molecular level, solution blending cannot achieve a true dispersion of MXene and polymer matrix, which is mainly shown in the case of weakly polar or nonpolar polymers. Therefore, some surface modification was adopted to enhance the dispersibility of MXene in dispersants, further increasing its uniform distribution throughout the polymer matrix. Before the fabrication of $Ti_3C_2T_x$-based UHMWPE nanocomposites, first the surface modification of MXenes is carried out. The surface modification can enhance the compatibility and dispersibility of $Ti_3C_2T_x$ in a UHMWPE matrix. Si *et al.* modified the surface of Ti_3C_2using long-chain cationic bromide, which improved its dispersibility in DMF, thereby improving the dispersion of Ti_3C_2in PS [101]. However, Yu *et al.* used cetyl trimethyl ammonium bromide (CTAB) to modify the surface of Ti_3C_2 and, thus, improve the dispersion of Ti_3C_2in TPU matrix [91]. Figure 3.1 shows the procedure for the synthesis of the PVC/MXene nanocomposites using a solution blending technique [90]. Some disadvantages such as substantial waste, difficulty of evaporating the solvent, environmental pollution, easy wrinkling during film casting, poor flatness, and insufficient mechanical strength in films are produced by solution blending, which hinder the practical application of this method.

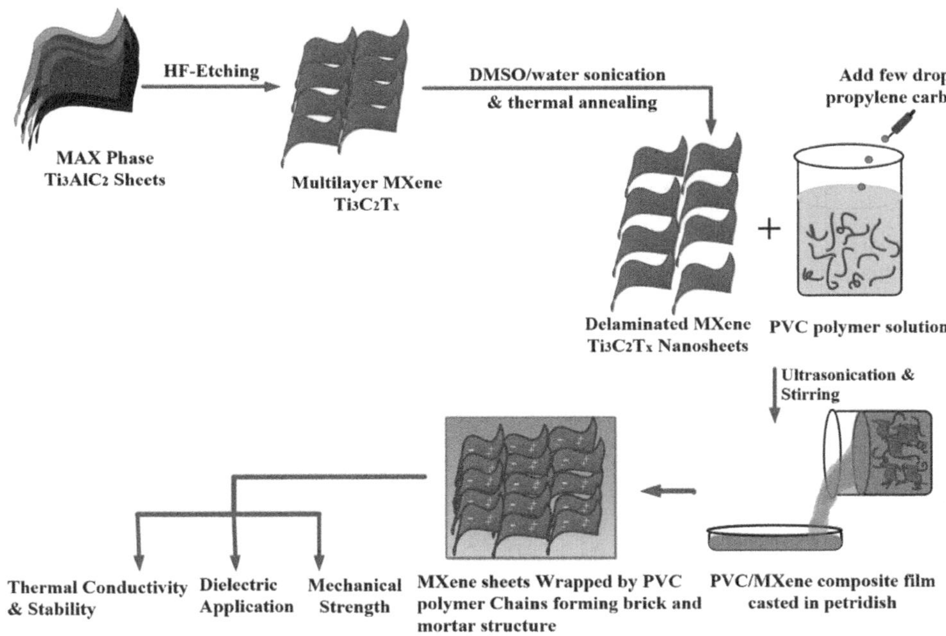

FIGURE 3.1 Schematic representation of solution blending technique for synthesis of PVC/MXene nanocomposites.

3.2.1.2 *In Situ* Method

The *in situ* method refers to the mixing of MXene nanosheets with small molecules, such as monomers, initiators, or curing agents, in presence of a filler, and then small molecules are *in situ* polymerized into macromolecules. Consequently, MXene is well distributed in the polymer hosts. The blending can significantly enhance the homogeneous distribution of MXene nanosheets within the polymer matrix. In this method, the polymer molecular chains are originated on the surfaces of the MXene nanosheets, which is an advantageous aspect for layered nanomaterials because the intercalation of monomers into nanolayers followed by a polymerization procedure to create enlargement and delamination into individual nanolayers. This type of blending is widely used to produce MXene-contained polymer nanocomposites, in the composites the polymers are thermosetting polymers containing cyclic or heterocyclic units, or linear macromolecules, which can be polymerized in mild conditions [75,102,103]. Epoxides, polydimethyl siloxane (PDMS), PPy, PANI, polydopamine (PDA), PAM, PAM/PVA, PEDOT, and further complex cyclopolymers can also be synthesized by *in situ* for preparing polymer/MXene nanocomposites to be used as electrodes, catalysis, shielding functional materials, and other purposes. Qin *et al.* successfully obtained PPy/MXene nanocomposites with 3D porous nanostructure via *in situ* electrodeposition method in pyrrole and MXene mixes [104]. Wang *et al.* synthesized PDA/$Ti_3C_2T_x$ composite film electrodes by single-step *in situ* technique [105]. Boota *et al.* and Tong *et al.* prepared PPy/$Ti_3C_2T_x$ composites by *in situ* polymerization as shown in Figure 3.2 [44,106]. All the previously mentioned approaches deal with the utilization of different solvents, while Carey *et al.* described the *in situ* bulk polymerization of polycaprolactam (ε-PCL) under inert conditions [75]. Here, 12-aminolauric acid and multilayer $Ti_3C_2T_x$ were intercalated with the ε-caprolactam monomer, which leads to the ring-opening polymerization of ε-caprolactam to prepare melt-blended nanocomposites, with 94% decrease in water vapor permeability. Linear polymers capable of being polymerized under mild conditions, such as acrylic monomers, olefinic monomers, polyesters, and polyamide precursor monomers, can also be polymerized *in situ* to prepare the corresponding nanocomposites.

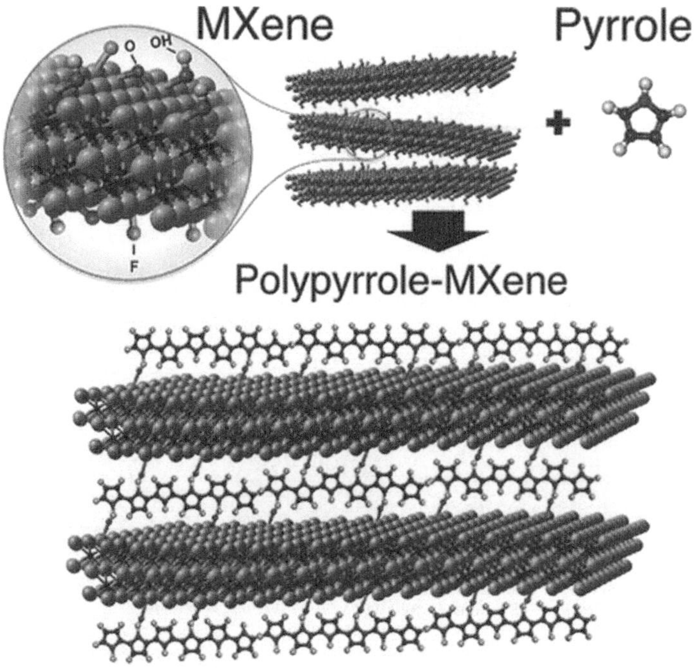

FIGURE 3.2 Schematic representation of *in situ* polymerization technique for synthesis of PPy/MXene nanocomposites.

3.2.1.3 Melt Blending

As compared to solution blending or *in situ* polymerization, which utilize a large amount of organic solvent, melt blending is considered more flexible for formulation, more economical, and more environmentally friendly. In this method, MXene added to the thermoplastic polymer matrix is usually done in the molten condition. This method includes extrusion, hot pressing, and injection molding of the polymer/MXene nanocomposites for uniform dispersion and directional orientation of the MXenes nanosheets within the polymer matrix. This method is especially appealing for industrial manufacturing of polymer/MXene nanocomposites for developing end-use products. MXenes can be easily supplied into the various methods of melt blending due to the satisfactorily high density of the MXenes of about 4.26 g/cm^3 [107] but only restricted to the thermoplastic polymers above the melting point. However, the melt processing needs to be carefully carried out at higher temperatures; otherwise, both polymers and MXenes undergo oxidative thermal degradation. Melt blending is discussed in the literature with regard to the fabrication of MXene-reinforced nonpolar thermoplastic polymers such as PP, UHMWPE [46], PS [73], PCL [67], linear low-density polyethylene (LLDPE) [71,108], and TPU [109,110]. The stepwise synthesis of melt blending of TPU/MXene composites is shown in Figure 3.3[110].

3.2.2 Characterization of Thermoplastic Polymer/MXene Nanocomposites

Tan *et al.* studied the Fourier transform infrared (FTIR) spectroscopy of neat PVA and PVA/MXene nanocomposites. The FTIR spectra of pure PVA and PVA/MXene nanocomposite with various concentration of MXene illustrates almost the same peaks, and no new peaks are observed as a result of the physical interactions existing between PVA and MXene nanosheets [111].

Figure 3.4 shows the scanning electron microscopy (SEM) morphology of the neat UHMWPE and UHMWPE/MXene nanocomposites, which displays several MXene nanosheets with lamellar

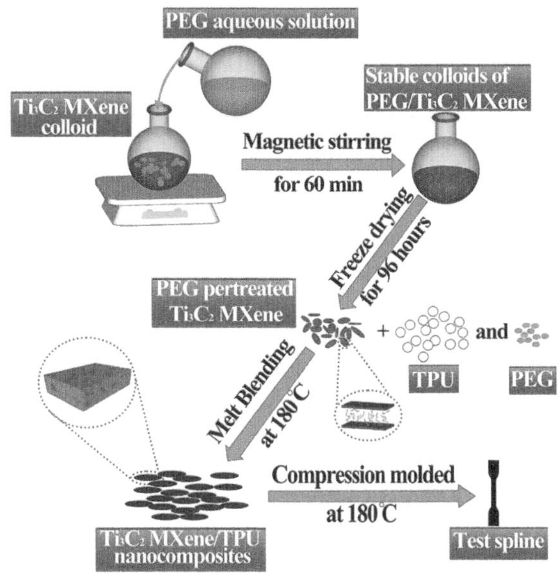

FIGURE 3.3 Schematic representation of melt blending for synthesis of TPU/MXene nanocomposites.

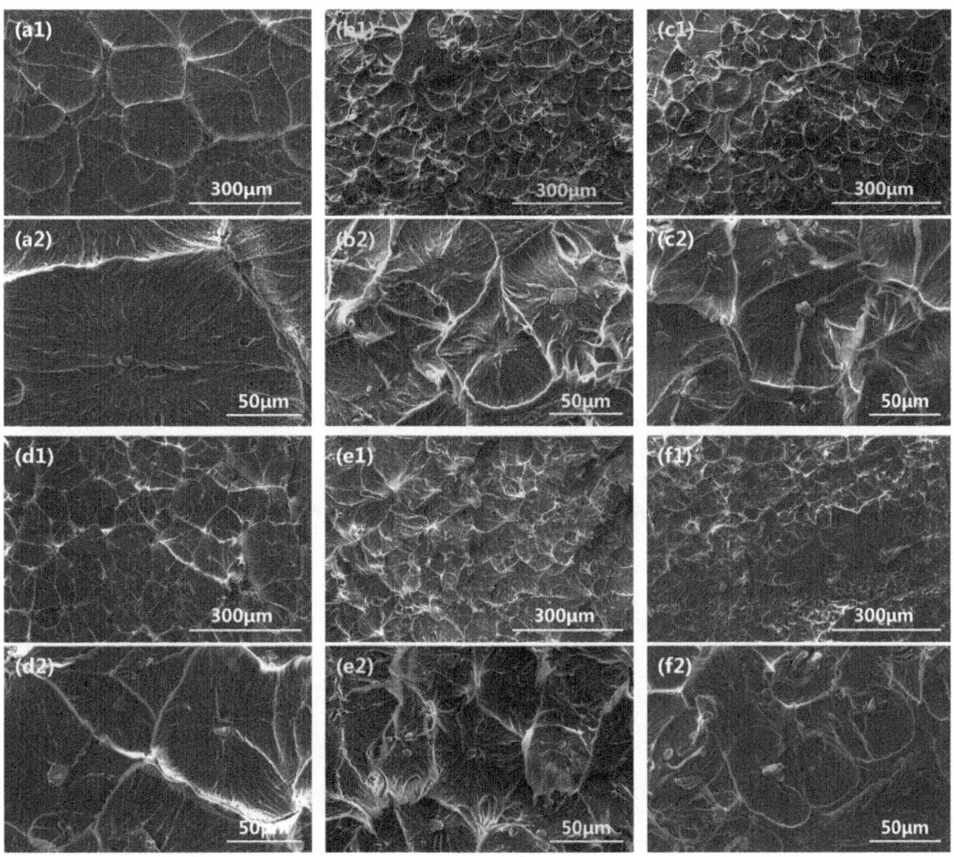

FIGURE 3.4 SEM microphotographs of the fractured surface of the UHMWPE/MXene nanocomposites: (**a**) 0, (**b**) 0.25, (**c**) 0.5, (**d**) 0.75, (**e**) 1, and (**f**) 2 wt.% of MXene loadings at 300 and 50 μm resolutions.

structures analogous to graphitic layers or layered silicate clays. The SEM surface shows the growth of spherulite structures of around 150–200 μm sizes with MXene at the nucleus. At higher loading of MXene, the creation of spherulites are more distinct because MXene acts as a nucleation center for UHMWPE to develop numerous amount of spherulites [30,46].

Lu *et al.* reported cross-sectional SEM micrographs of the pristine TPU and TPU/MXene nanocomposites. The pristine TPU displays a smooth and uniform surface morphology. The surface roughness of the TPU/MXene composites gradually improved with an increase in MXene concentration, which revealed an effective encompassing of the MXene nanosheets by the TPU matrix that developed analogic-segregated microstructures [112]. Moreover, the energy dispersive X-ray (EDX) spectroscopy for element mapping of O, Ti, N, and S distributions of MXene nanosheets within TPU phase in the designated area of the SEM image revealed that the MXene nanosheets are uniformly dispersed within the TPU matrix, confirming the commencement of the interconnected conductive network microstructures.

Figure 3.5 shows the SEM and transmission electron microscopy (TEM) morphology of the pristine PVDF and PVDF/MXene nanocomposites. The microstructure of the cryo-fractured surface of the neat PVDF exhibits a smooth surface (Figure 3.5a), whereas the surface roughness of the CTAB-modified MXene-based PVDF nanocomposites increases with increase in MXene loading as shown in Figure 3.5b–d, which informs the strong interfacial adhesion among MXene and PVDF matrix. The TEM images illustrate a little layer of the MXene nanosheets, which are homogeneously dispersed in the PVDF matrix as displayed in Figure 3.5e–h. At a higher MXene loading, few agglomerations are observed that implying the origination of the interconnected network structure [113].

Figure 3.6 shows the nanostructure morphology of pristine TPU and TPU/MXene nanocomposites using both SEM and TEM. Figures 3.6a and 3.6b show the SEM micrographs of cryo-fractured surface of neat TPU and 1.0 wt.% MXene-based TPU composites, respectively. Figure 3.6b show the distribution of the few layers of MXene nanosheets within the TPU matrix, which are well coated with TPU matrix, leading to the conclusion that the strong interphase interaction exists among the O and OH groups on the surface of the MXene, with a polar functional group present on the molecular backbone chain of the TPU matrix [110]. Figure 3.6c–f represent the low-and high-resolution TEM images of 1.0 wt.% MXene-loaded TPU composites. The low-magnification TEM image as shown in Figure 3.6c shows the black spots corresponds to the MXene nanolayers, which

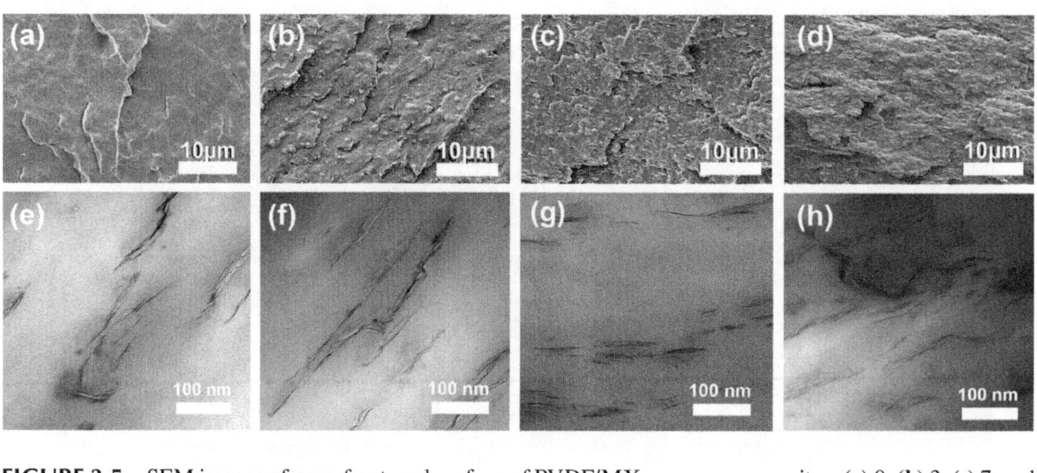

FIGURE 3.5 SEM images of cryo-fractured surface of PVDF/MXene nanocomposites: (**a**) 0, (**b**) 3, (**c**) 7, and (**d**) 10 wt.% of MXene loadings and TEM microphotographs of PVDF/MXene nanocomposites: (**e**) 3, (**f**) 5, (**g**) 7, and (**h**) 10 wt.% of MXene loadings.

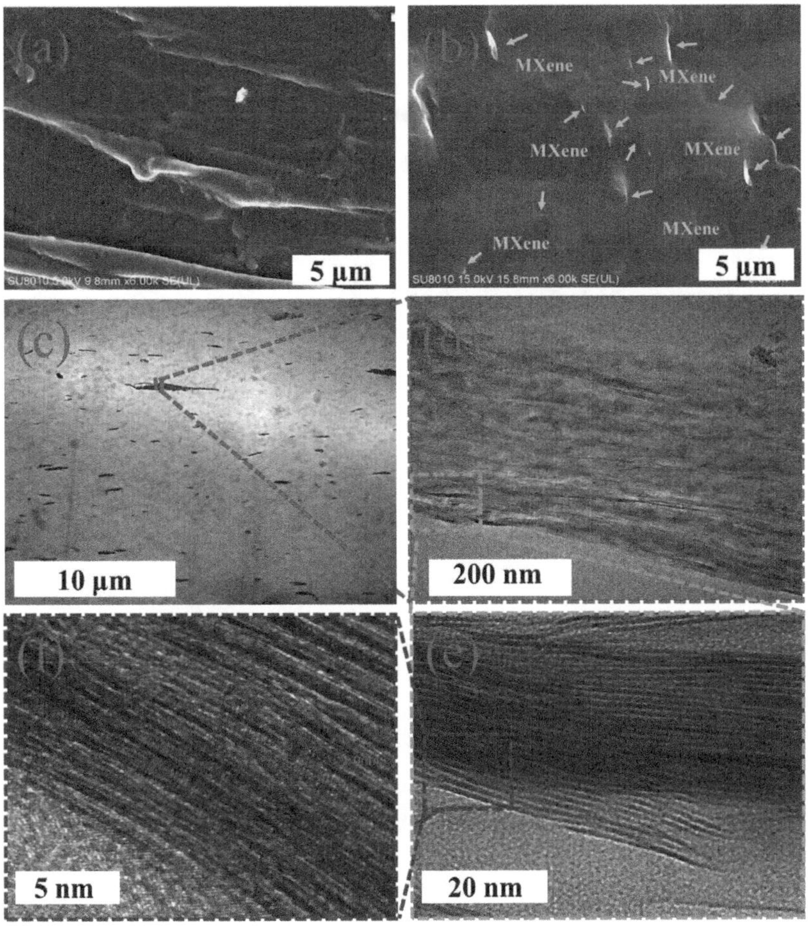

FIGURE 3.6 SEM micrographs of (**a**) neat TPU and (**b**) 1 wt.% MXene based TPU nanocomposites, and (**c–f**) TEM micrographs of 1 wt.% MXene based TPU nanocomposites.

are homogeneously dispersed throughout the TPU matrix without any large aggregation. The interlayer spacing of the MXene nanosheets is about 2.0 nm as observed from Figure 3.6f.

Figure 3.7 represents the polarized light microscope (PLM) photographs of neat PEO and PEO/MXene nanocomposites. The spherulites formed in PEO and 1 wt% MXene-based PEO nanocomposite are distinct and large in size whereas the size of the spherulites significantly decreases with an increase in MXene contents due to the improvement in the number of nucleation sites [92].

3.2.3 Properties of Thermoplastic Polymer/MXene Nanocomposites

3.2.3.1 Mechanical Properties of Thermoplastic Polymer/MXene Nanocomposites

The mechanical properties of the MXenes were theoretically evaluated by considering the thickness, surface termination, and composite effects through DFT [114,115] and MDs [116]. Figure 3.8a represents the molecular configurations of Ti_2C, Ti_3C_2, and Ti_4C_3 MXenes after equilibrium at 300 K. From Figure 3.8b, the in plane elastic modulus of Ti_2C, Ti_3C_2, and Ti_4C_3 MXenes are calculated as 597, 502, and 534 GPa, respectively, using molecular dynamics, which is noticeably higher than that of the graphene [116,31]. Moreover, the MXenes are inorganic fillers with good rigidity [114] and flexural properties [117].

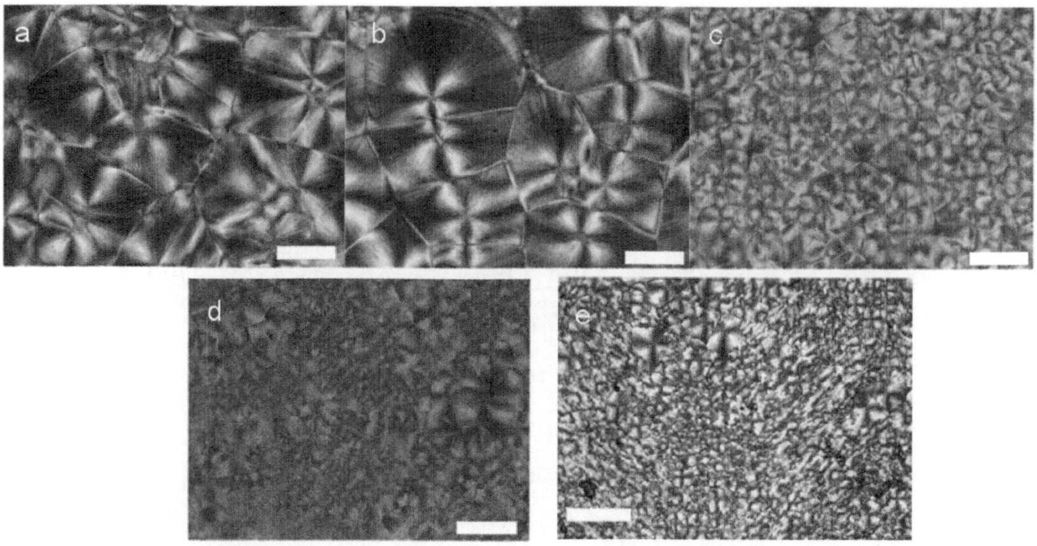

FIGURE 3.7 PLM images of PEO/MXene nanocomposites: (**a**) 0, (**b**) 1, (**c**) 2, (**d**) 3, and (**e**) 4 wt.% of MXene loadings.

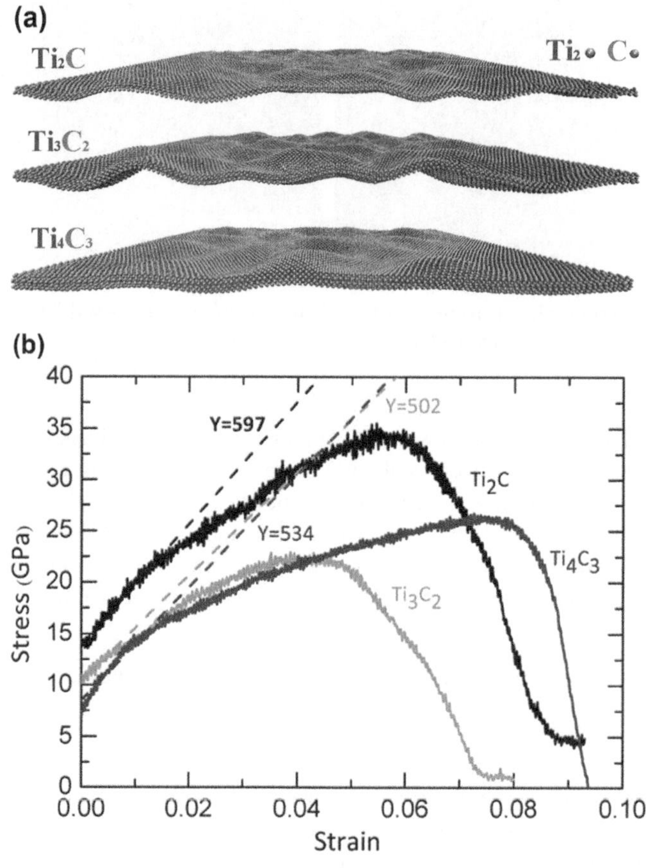

FIGURE 3.8 Modelling of mechanical behaviour of MXenes using DFT: (**a**) organizations of the different MXenes after equilibration at room temperature and (**b**) typical stress-strain curves of MXenes computed via MDs.

Some polymers having polar functional groups strongly interact with the MXenes because of the existence of terminal groups on the surface of MXenes. Consequently, the mechanical behaviors such as yield strength, elastic modulus, ultimate tensile strength, elongation at break, and hardness of the polymer matrix might be enhanced significantly. Ling *et al.* reported that the MXene nanosheets rolled with 10 wt% PVA can withstand about 15,000 times than the individual mass [43]. Hu *et al.* have shown that the mechanical properties of the MXene nanofilms substantially improved by the incorporation of 14 wt% chitosan [97]. Zhi *et al.* reported that the incorporation of MXene into the PU matrix improved the yield stress, hardness, and tensile strength by70%, 13%, and 20%, respectively, at 0.5 wt% MXene loading [118]. Sheng *et al.* reported that the mechanical strength and modulus of the 0.5 wt.% Ti_3C_2 MXene-based TPU were enhanced by 47.1% and 39.8%, respectively, in comparison with the neat TPU due to the homogeneous dispersion of the MXene in the TPU matrix along with strong chemical interactions among MXene and TPU molecular chains [110]. Zhang *et al.* reported that the tensile strength and modulus of UHMWPE were enhanced by the incorporation of Ti_3C_2 MXene. The ultimate tensile strength of the nanocomposite is increased by 8.6% with addition of 0.75 wt% Ti_3C_2 as shown in Figure 3.9a. The mechanism for the transfer of stress from UHMWPE to Ti_3C_2 MXene is displayed in Figure 3.9b, which acts as a vital factor for enhancement of the mechanical properties [46]. Liu *et al.* shown that the Young's modulus of Nafion film is improved about 23% by the addition of 2 wt.% of $Ti_3C_2T_x$ MXene [119]. Yi *et al.* developed 0.5 wt% stearic acid–modified $Ti_3C_2T_x$-based PLA nanocomposites with 131.6% improvement in

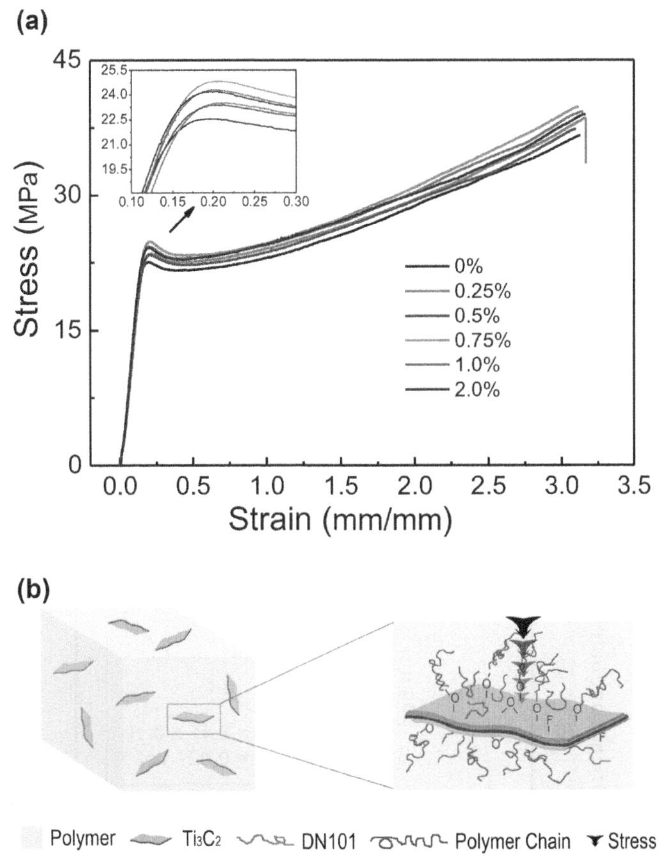

FIGURE 3.9 (**a**) Stress-strain curves for Ti_3C_2 MXene based UHMWPE nanocomposites with different MXene loading and (**b**) schematic representation of the mechanism for enhancement of the mechanical properties.

the elongation at break in comparison with the pristine PLA matrix [98]. Zhang *et al.* achieved that the incorporation of 0.3 wt% MXene into PMMA matrix improves the tensile strength and modulus of 89% and 397%, respectively [120]. Gao *et al.* have elaborated the various parameters of mechanical properties of different kinds of MXene-based polymer nanocomposites in tabular form [31].

3.2.3.2 Electrical Properties of Thermoplastic Polymer/MXene Nanocomposites

MXene shows excellent electrical conductivity (σ) of 9880 S/cm equal to that of graphene [121,122] and the conductivity were appreciably improved by the incorporation of MXene into the polymer matrices, which can be employed for electromagnetic interference shielding, stretchable strain sensors, humidity and volatile organic compounds (VOCs) sensors along with energy storage devices [38]. Mayerberger *et al.* obtained that the electrical conductivity of PEO/$Ti_3C_2T_x$ nanocomposites was enhanced by about 73.6% than that of pristine PEO [68]. Ling *et al.* observed that the conductivity of the PVA/MXene nanocomposites significantly increases with an increase in Ti_3C_2 loading [43]. Sobolčiak et al. studied the frequency dependence of the dc conductivity of MXene-based PVA nanofibers [47]. Naguib *et al.* synthesized $Ti_3C_2T_x$ MXene based PAM nanocomposites with significant improvement in electrical conductivity as shown in Figure 3.10a [94]. Sun *et al.* synthesized highly conducting PS/MXene nanocomposites with a percolation limit of 0.26 vol.% and conductivity of 1081 S/m [123]. Xu et al. prepared conductive PP/MXene nanocomposites with low percolation limit as represented in Figure 3.10b [124].

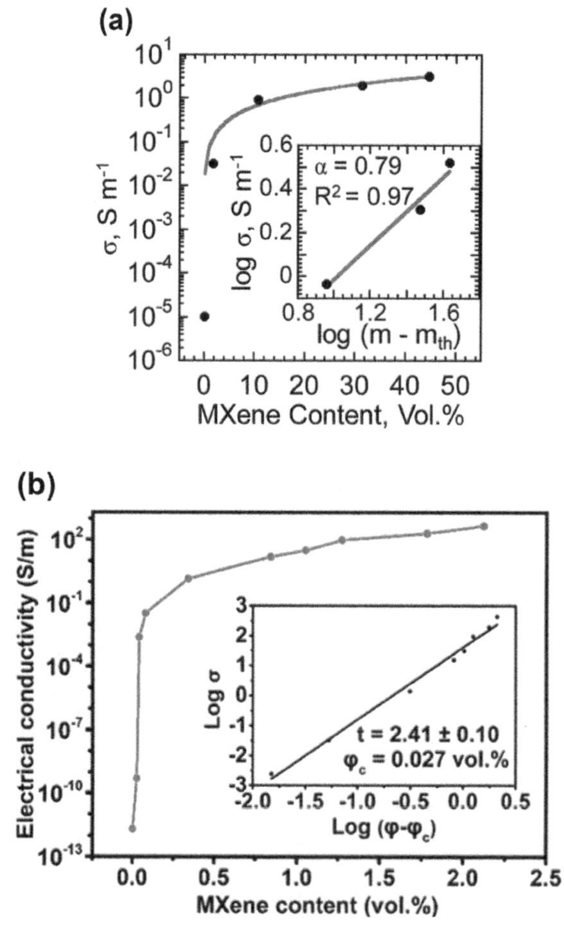

FIGURE 3.10 Variation of conductivity of (**a**) PAM and (**b**) PP nanocomposites with MXene loading.

Tu *et al.* perceived that the dielectric permittivity of the polyvinylidene fluoridetrifluoro-ethylene-chlorofluoroethylene (PVDF-TrFE-CFE) was significantly improved by the addition of MXene with a percolation limit of 15 wt% [125]. In addition, Mirkhani *et al.* examined that the dielectric constant of PVA/MXene nanocomposites prepared by vacuum-assisted filtration is noticeably higher in comparison with the solution casting methods [126].

Liu *et al.* reported that the proton conductivity of the Nafion/MXene nanocomposites is increased due to improvement in proton transportations by the incorporation of the MXene [119]. Fei *et al.* noted that the proton conductivity of the polybenzimidazole (PBI) is improved twofold in presence of the MXene [100]. The proton conductivity of the polymer/MXene nanocomposites is effectively improved due to the formation of long-range interconnected routes and enrichment of polar functional groups resulted from the interfacial interactions between the polymer matrix and MXene, which provides a large proton transfer rate and rapid mobility of protons with low resistance.

3.2.4 Application of Thermoplastic Polymer/MXene Nanocomposites

3.2.4.1 Sensing Applications

MXene is a suitable material for a broad range of highly sensitive and highly selective sensor applications [127–129] because of its superior conductivity [121,122,130], low bandgap [2], easy chemical modification [131], and rich active sites [132]. The polymer/MXene nanocomposites can be employed for both chemical (humidity [50,133,134] and VOCs [135]) and mechanical (strain [50,136] and pressure [58,129,137–140) sensors due to the effective synergistic effect imparted by high flexibility of the polymer and good sensing characteristics of the MXene. Kim *et al.* developed PAM/MXene sensor with improved linear and humidity sensing activities at both low and high humidity concentration [133]. Sajid *et al.* prepared PVA/MXene nanocomposites with humidity-sensing behavior [134]. An *et al.* described the ultra-fast humidity-sensing recovery and repeatability of MXene based polydiallyl dimethyl ammonium chloride (PDAC) nanocomposites [50]. Wang *et al.* fabricated a moisture sensor from PVA/MXene nanocomposites with response and recovery time of 0.9 and 6.3s, respectively [141].

Yuan *et al.* studied VOCs' sensing behavior of MXene-based PVA/polyethylene imine (PEI) nanocomposites. It exhibits low response and recovery times to detect different VOCs at concentrations ranging from parts per billion (ppb) to saturated levels because of the chemical adsorption of polar gases on the surface of the MXene through hydrogen bonding that is further supported by the large specific surface area of the highly porous PVA/PEI matrix [135]. Wang *et al.* developed a PANI/MXene nanocomposites based self-powered NH_3 gas sensor in real-time monitoring of the NH_3 concentration [142].

Shi *et al.* fabricated a bionic strain sensor with good sensitivity, high dynamic durability, and huge deformation from MXene-based PDA nanocomposites [136]. Yang *et al.* created an elastic strain sensor of MXene-based PU nanocomposite with excellent sensing range and ultra-sensitivity [139]. Zhang *et al.* reported a strain sensor of PVA/MXene nanocomposite hydrogel [60]. Chao *et al.* synthesized PANI/MXene nanocomposites based stretchable strain sensor with sensing range of 80% strain, sensitivity of 2369.1 for gauge factor, and detection limit of 0.1538% strain to monitor movements [143].

Guo *et al.* formed MXene based PLA nanocomposites with wearable pressure sensing properties of 11- and 25-ms response and recovery time, respectively. The PLA/MXene nanocomposites exhibit reliable response after 1000 compression cycles with a small detection limit, low power utilization, high sensitivity, and board detection range [129]. Hu *et al.* developed a wearable pressure sensor by using CS/MXene nanocomposite aerogels [137]. Li *et al.* constructed an excellent pressure sensor using PU/MXene nanocomposite foam [58]. Li *et al.* developed MXene-based PVDF-TrFE nanocomposites as a highly sensitive piezoresistive pressure sensor with response time of 16 ms [85]. Wang *et al.* also reported PVDF-TrFE/MXene nanocomposites with superior linear pressure sensors to detect the regular actions of a human [144]. Zhao *et al.* represented a very sensitive piezoelectricsensor based on PVDF/MXene nanocomposites with a high voltage sensitivity of 0.0480 V/N, a low recovery time of 3.1 ms, and a good stability of 10,000 s [145].

3.2.4.2 Biomedical Applications

Polymer/MXene nanocomposites can be suitable for various biomedical applications, such as drug delivery, antimicrobial activity, anticancer therapy, biosensing, bioimaging, and bone regeneration due to their good photothermal conversion efficiency (PCE), high selectivity, and fast stimuli sensitivity to tumor cells, electron sensitivity, higher antibacterial properties, unique surface reactivity, excellent hydrophilicity, good biocompatibility, and remarkable electrical and optical properties [33]. Rasool *et al.* illustrated that the PVDF/MXene nanocomposites membrane exhibit remarkable antibacterial action to both gram-positive *B. subtilis* and gram-negative *E. coli* bacterial strains and growth inhibition of 73% and 67%, respectively, along with a drastic fall in cell viability of the bacteria in comparison with the pristine PVDF [146]. Mayerberg *et al.* demonstrated the antibacterial behavior of the CS/MXene nanocomposites to gram-positive *S. aureus* and gram-negative *E. coli* bacteria having growth reduction of 62% and 95%, respectively [54].

Lin *et al.* reported photothermal therapy (PTT) of the polylactic-co-glycolic acid (PLGA)/MXene nanocomposites, with the destruction of tumorous cells happening at nearly 80°C in near-infrared (NIR) laser within 300 s [147]. Szuplewska *et al.* demonstrated the NIR laser-assisted annihilation of the cancerous cells using PEG-coated MXene nanoflakes with reasonable PTT efficiency and notable biocompatibility at broad span of measured concentrations of up to 37.5 μg/ml [148]. Feng *et al.* shown the PTT of the PVA/MXene nanoflakes at wide range of absorptions within NIR region having PCE of nearly 24.5% and 43.3% at 808 nm and 1064nm, respectively [149]. Lin *et al.* employed PVP/MXene nanosheets to study the *in vivo* PTT tumor treatment utilizing NIR laser sources, reporting an appreciable healing of the tumor cells within two days without reappearance and having a surviving period of 16 days [150].

Xing *et al.* evaluated the sustained and controlled release of the anticancer DOX drug by cellulose/MXene nanocomposite hydrogels, revealing a significant DOX releasing capability, which is further accelerated by irradiation of 808-nm light [151].

Rakhi *et al.* constructed a Nafion/MXene nanocomposites-based enzymatic glucose biosensor, which exhibited a linear amperometric response at concentrations of 0.1–18 mM, including sensitivity and detection limit of 4.2μAmM^{-1}cm^{-2}and 5.9μM (S/N = 3), respectively [152]. Chen *et al.* introduced biocompatible PLA/MXene nanocomposites for guided bone regeneration (GBR), which encouraged preosteoblast adhesion, proliferation, and osteogenic differentiation [153].

3.2.4.3 Energy Storage Applications

Polymer/MXene nanocomposites are extensively utilized for the fabrication of energy storage systems, for example, supercapacitors, rechargeable batteries, and energy conversions [30–32,35–38]. The conductive polymers, for example, polyfluorene (PF) derivatives [93], poly(3,4-ethylenedioxythiophene) (PEDOT) and its derivatives [102,154,155], PPy [44,104,156,157], PANI [78,158–162], PVA [43], PDAC [43], PDVF [41,163], and others, are generally used for energy storage devices. Boota *et al.* reported pseudocapacitance of the polar, nonpolar, and charged nitrogen functionalized PF derivatives, which displayed improvement in both volumetric and gravimetric capacitance with appreciable retention of capacitance both low and high scan rate along with good cycling stability up to 10^4 cycles that can be utilized for effective prolonged energy storage devices [93]. The capacitance of the PF derivatives is significantly high compared to both PVA/MXene [43] and PPy/MXene nanocomposites [44]. Chen *et al.* prepared PEDOT/MXene nanocomposites with good specific capacitance and broadened potential range of about1.8 V, along with promising supercapacitor properties [154]. Wu *et al.* prepared PPy/MXene nanocomposites that demonstrated an ultimate gravimetric capacitance of 184.36 Fg^{-1}at 2.0 mVs^{-1}and admirable cycling longevity to maintain the capacitance retention of 83.33% following 4 $\times 10^3$ charging/discharging cycles at a current density of 1.0 Ag^{-1} [156]. Li *et al.* demonstrated pseudocapacitor obtained from conducting PANI/MXene nanocomposites with a specific capacitance of 1632 $Fcm^{-3}$and a rate capacity of 827 Fcm^{-3}at $5\times10^3 mVs^{-1}$ [159]. Vahid Mohammadi *et al.* discussed the PANI/MXene nanocomposite electrodes with excellent specific and volumetric capacitance of 503 $Fg^{-1}$and 1682 Fcm^{-3}, respectively, as well as capacitance retention of 98.3% after 10^4 cycles [160]. Li *et al.* emphasized an electrochemical electrode based on PANI/MXene nanocomposites, showing volumetric capacitance of 563Fg^{-1}at 0.5$Ag^{-1}$and retention capacity of 84.72% and 95.15% by changing current density from 0.5–20Ag^{-1} and after 10^4 cycles, respectively. An asymmetric supercapacitor is developed by taking of PANI/MXene nanocomposites and activated carbon (AC) as positive and negative electrodes, respectively that achieved 22.67Wh/kg energy density at 0.217kW/kg power density [161].

3.2.4.4 Electromagnetic Interference Shielding and Absorption Applications

Polymer/MXene nanocomposites can be applied for EMI shielding due to a synergistic effect exhibited by highly conductive, hydrophilic, and chemical active MXene and good elastic, corrosion resistive, eco-friendly, and lightweight polymers [40,164]. Different polymer/MXene nanocomposites, such as PS [123], PP [124], elastomer [165], PEDOT:PSS [166–168], PVA [66,169], PAT–PANI–PpAP copolymer [170], PU [171–173], TPU [174], PVDF [175,178], and others, are taken to investigate the EMI shielding performance. Sun *et al.* developed PS/MXene nanocomposites achieving a superior EMI SE above 54 dB throughout the entire x-band frequency region, for example, the highest EMI SE of 62 dB at 1.9 vol.% MXene concentration, which is stated as the maximum reported value for conducting polymer nanocomposites [123]. Xu *et al.* explored the EMI shielding performance of the PP/MXene nanocomposites exhibiting EMI SE exceeding 60 dB throughout the complete X-band frequency range with the highest value of 66 dB at thickness of 1.93 mm for 2.12 vol.% MXene loading [124]. Aakyiir *et al.* reported highly flexible and mechanically tough elastomer/MXene nanocomposites with high EMI SE of 49 dB for a film 1 mm thick with a density of 1.25 gcm^{-3} at 19.6 vol.% of MXene loading. The high conductivity of MXene of $4350\pm125\ Scm^{-1}$ deliver spontaneous charge carriers in the polymer to take in EMI radiations [165]. Wan *et al.* and Liu *et al.* reported PEDOT:PSS/MXene nanocomposite films exhibiting EMI SE of 40.5 dB and 42.1 dB at about specimen thickness of 6.6 μm and 11.1 μm, respectively, and the former shows a specific EMI SE (SSE_t) of 19 497.8 $dBcm^2g^{-1}$. The total EMI SE (SE_T), along with both absorption (SE_A) and reflection (SE_R) components, is significantly increased with an increase in MXene loading [166,167]. Jin *et al.* studied the EMI shielding of 19.5 wt% MXene-filled PVA nanocomposite film, which showed 44.4 dB EMI SE at a 27.0-μm specimen thickness and 9343 $dBcm^2g^{-1}$specific EMI SE (SSE_t) [66]. Raagulan *et al.* synthesized MXene-loaded PAT–poly(*p*-aminophenol)–polyaniline (PAT–PANI–PpAP) copolymer nanocomposites and claimed 99.99% blocking of the incident radiations, with highest EMI SE range from 45.18 to 39.33 dB, but SSE_t, absolute effectiveness (SSE/*t*), and SE/*t*ranges are 27.99–40.49 $dBcm^3g^{-1}$, 236.45–721.25 $dBcm^2g^{-1}$, and 1.393–1.6 $dBmm^{-1}$, respectively [170]. Wang *et al.* investigated the EMI shielding properties of the PU/MXene nanocomposite aerogels and reported an EMI SE above 64.7 dB at a specimen thickness of 2 mm and a density of about 38.2 $mgcm^{-3}$. The typical SE_T and SE_A values enhanced from 32.5 dB to 64.7 dB and 23.6 dB to 55.6 dB with improvement in thickness from 0.5 to 2.0 mm, respectively [171]. Gao *et al.* conveyed remarkably high 50.7 dB EMI SE at a 52-μm specimen thickness and an SSE/t of 7276$dBcm^2g^{-1}$ [174]. Li *et al.* manufactured PVDF/MXene nanocomposite films with a maximum 19504.8$dBcm^2g^{-1}$ SSE/t at a 17-μm specimen thickness [175].

Wei *et al.* examined the microwave absorption behavior of the PANI/MXene nanocomposites, which showed the highest reflection loss of −56.30 dB at 13.80 GHz frequency and a 1.8-mm thickness as well as a relevant absorption bandwidth (>90%) extending from the X-band to the Ku-band GHz frequency region with the variation of thickness from 1.5 to 2.6 mm. The dependence of reflection loss on both frequency and thickness of the PANI/MXene nanocomposites and the suggested mechanism of microwave absorption [177].

3.2.4.5 Filtration Applications

Han *et al.* devised the filtration performance of the polyether sulfone (PES)/MXene nanocomposite membranes with superior flux value with a remarkable rejection of both organic dyes and inorganic salts at low applied pressure [99]. Ren *et al.* prepared PVDF/MXene nanocomposite membranes exhibited excellent water flux characteristic and considerable selective filtering of metal cations subjected to both dimension and charge. The water flux of the PVDF/MXene nanocomposites drastically decreased and filtration time rapidly increased with increase in thickness of the membrane. Moreover, the selectivity about the metal cation of sizes less than 4.5 Å and oxidation states between +2 to +4 [179].

Han *et al.* developed polyimide (PI)/MXene nanocomposite membranes with an outstanding rejection rate of the organic dyes at both ambient pressure and temperature [180]. Wu *et al.* reported the successful alcohol fluxes in both PEI/MXene and PDMS/MXene nanocomposites [181]. Moreover, Hao *et al.* demonstrated polar (amine ($-NH_2$) and carboxyl (–COOH) and nonpolar linear hydrocarbon chain ($-C_6H_6$ and $-C_{12}H_{26}$) functional groups in modified MXene-based polyacrylonitrile (PAN)/PEI and PAN/PDMS nanocomposite membranes with significant improvement in flux of organic solvents such as ethyl acetate, isopropanol, n-heptane, and toluene [182]. Xu *et al.* formulated chitosan/MXene nanocomposite membranes for pervaporation dehydration of organic solvents such as ethanol, ethyl acetate, and dimethyl carbonate [183]. Shen *et al.* investigated the CO_2 gas separation performance of the PEI/MXene nanocomposite membranes with permeation selectivity of 15.3 and 1.4 for CO_2/CH_4 and CO_2/H_2 gas mixtures, respectively [184].

3.2.4.6 Other Applications

In addition to the application of the polymer/MXene nanocomposites in the discussed fields, the novel nanostructures and excellent properties of the same can be suitable for various promising purposes, for example, rechargeable batteries [72,185], fuel cells [100,186,187], nanogenerators [52,188], flame retardancy [169, 189–193], and many more [34,194,195]. Luo *et al.* reported that the flame retardancy behavior of the TPU/PCS-MXene nanocomposites was appreciably improved, and the smoke emission capability was significantly reduced as the presence of the MXene endorses the labyrinth effect, which develops a compactness and concentration of residues through unevenly a nonaggregated distribution of MXenes within the TPU matrix. The time to ignition, peak heat release rate, total heat release, total smoke production, peak CO production rate, and peak CO_2 production rate of the 3wt% PCS-MXene-filled TPU nanocomposite was reduced by 35.13%, 66.66%, 21.01%, 27.72%, 52.24%, and 60.65%, respectively, in comparison with neat TPU [192].

3.3 FUTURE PERSPECTIVES AND CHALLENGES

This book chapter deals with the processing, morphology, mechanical and electrical properties, and application of polymer/MXene nanocomposites. This chapter discussed the important results of the current articles and proposed mechanisms for the influence of MXene on the properties of polymer nanocomposites. Reinforcement of MXene within the polymer substantially enhance the mechanical, electrical, and thermal properties due to the synergistic effect of the superior electrical conductivity, high specific surface area, exceptional nanostructured morphology, and distinct surface chemistry of the MXene nanosheets and superior flexibility, good mechanical strength, and easy processibility of the polymer matrices. Polymer/MXene nanocomposites can be successfully prepared through solution casting, melt blending, and *in situ* polymerization processes. The polymer/MXene nanocomposites has exhibited vast appealing projections that can be emerged as a potential substitute for sensor, biomedical, energy storage and conversion, EMI shielding and absorption, filtration, and many other applications.

Nonetheless, the research and development work on polymer/MXene nanocomposites are in their early stages, so industrial manufacturing and successive endue cannot be explored due to following two main challenges. Initially, exfoliation of the MXene nanolayers into single-layer nanosheets involves hazardous chemical methods, and achieving a homogeneous dispersion of the hydrophilic MXene nanosheets within the hydrophobic polymer matrix is a chaotic task. In spite of the tough challenges, the fabricated polymer/MXene nanocomposites can be suitable for various promising applications. Therefore, the research work on polymer/MXene nanocomposites can be further accelerated to accomplish with fruitful implementations in different application areas. A key drawback of MXene to be used for fabrication of polymer/MXene nanocomposites is its vulnerability to oxidation, which most affect the EMI shielding performance. Furthermore, immense attention can be given to the large-scale production of polymer/MXene nanocomposites using economically viable and environmentally friendly methodologies.

Advance research on polymer/MXene nanocomposites can be performed based on the following important approaches: (1) investigation of industrially feasible, cost effective, and eco-friendly techniques for fabrication of MXene nanosheets; (2) reducing the oxidation of MXene nanosheets for enhancing their chemical stability; (3) examining the origins for interphase adhesive interactions among polymer matrix and MXenes; and (4) advancement of multiperformance polymer/MXene nanocomposites.

REFERENCES

1. Lei, J.C., Zhang, X., Zhou, Z., Recent Advances in MXene: Preparation, Properties, and Applications, ***Frontiers of Physics***, 2015, **10**(3), *276–286.*
2. Zhu, J., Ha, E., Zhao, G., Zhou, Y., Huang, D., Yue, G., Hu, L., Sun, N., Wang, Y., Lee, L.Y.S. Xu, C., Wong, K.Y., Astruc, D., Zhao, P., Recent Advance in MXenes: A Promising 2D Material for Catalysis, Sensor and Chemical Adsorption, ***Coordination Chemistry Reviews***, 2017, **352**, *306–327.*
3. Salim, O., Mahmoud, K.A., Pant, K.K., Joshi, R.K., Introduction to MXenes: Synthesis and Characteristics, ***Materials Today Chemistry***, 2019, **14**, *100191.*
4. Naguib, M., Kurtoglu, M., Presser, V., Lu, J., Niu, J., Heon, M., Hultman, L., Gogotsi, Y., Barsoum, M.W., Two-Dimensional Nanocrystals Produced by Exfoliation of Ti_3AlC_2, ***Advanced Materials***, 2011, **23**(37), *4248–4253.*
5. Khazaei, M., Arai, M., Sasaki, T., Chung, C.Y., Venkataramanan, N.S., Estili, M., Sakka, Y., Kawazoe, Y., Novel Electronic and Magnetic Properties of Two-Dimensional Transition Metal Carbides and Nitrides, ***Advanced Functional Materials***, 2013, **23**(17), *2185–2192.*
6. Gao, Y., Wang, L., Zhou, A., Li, Z., Chen, J., Bala, H., Hu, Q., Cao, X., Hydrothermal Synthesis of TiO_2/Ti_3C_2 Nanocomposites with Enhanced Photocatalytic Activity, ***Materials Letters***, 2015, **150**, *62–64.*
7. Barsoum, M.W., The $M_{N+1}AX_N$ Phases: A New Class of Solids: Thermodynamically Stable Nanolaminates, ***Progress in Solid State Chemistry***, 2000, **28**(1–4), *201–281.*
8. Sun, Z., Music, D., Ahuja, R., Li, S., Schneider, J.M., Bonding and Classification of Nanolayered Ternary Carbides, ***PhysicalReviewB: Condensed Matter and Materials Physics***, 2004, **70**(9), *092102.*
9. Ronchi, R.M., Arantes, J.T., Santos, S.F., Synthesis, Structure, Properties and Applications of MXenes: Current Status and Perspectives, ***Ceramic International***, 2019, **45**(15), *18167–18188.*
10. Barsoum, M.W., MAX Phases: Properties of Machinable Ternary Carbides and Nitrides, Wiley-VCH, Weinheim, Germany, 2013.
11. Naguib, M., Mashtalir, O., Carle, J., Presser, V., Lu, J., Hultman, L., Gogotsi, Y., Barsoum, M.W., Two-Dimensional Transition Metal Carbides, ***ACS Nano***, 2012, **6**(2), *1322–1331.*
12. Naguib, M., Halim, J., Lu, J., Cook, K.M., Hultman, L., Gogotsi, Y., Barsoum, M.W., New Two-Dimensional Niobium and Vanadium Carbides as Promising Materials for Li-Ion Batteries, ***Journal of the American Chemical Society***, 2013, **135**(43), *15966–15969.*
13. Ghidiu, M., Naguib, M., Shi, C., Mashtalir, O., Pan, L.M., Zhang, B., Yang, J., Gogotsi, Y., Billinge, S.J., Barsoum, M.W., Synthesis and Characterization of Two-Dimensional Nb_4C_3 (MXene), ***Chemical Communications***, 2014, **50**(67), *9517–9520.*
14. Sarycheva, A., Polemi, A., Liu, Y., Dandekar, K., Anasori, B., Gogotsi, Y., 2D Titanium Carbide (MXene) for Wireless Communication, ***Science Advances***, 2018, **4**(9), *eaau0920.*
15. Al-Hamadani, Y.A., Jun, B.M., Yoon, M., Taheri-Qazvini, N., Snyder, S.A., Jang, M., Heo, J., Yoon, Y., Applications of MXene-based Membranes in Water Purification: A Review, ***Chemosphere***, 2020. **254**, *126821.*

16. Khazaei, M., Mishra, A., Venkataramanan, N.S., Singh, A.K., Yunoki, S., Recent Advances in MXenes: From Fundamentals to Applications, ***Current Opinion in Solid State and Materials Science***, 2019, **23**(3), *164–178.*
17. Come, J., Naguib, M., Rozier, P., Barsoum, M.W., Gogotsi, Y., Taberna, P.L., Morcrette, M., Simon, P., A Non-Aqueous Asymmetric Cell with a Ti_2C-Based Two-Dimensional Negative Electrode, ***Journal of the Electrochemical Society***, 2012, **159**(8), *A1368.*
18. Hu, J., Xu, B., Ouyang, C., Yang, S.A., Yao, Y., Investigations on V_2C and V_2CX_2 (X = F, OH) Monolayer as a Promising Anode Material for Li Ion Batteries from First-Principles Calculations, ***The Journal of Physical Chemistry C***, 2014, **118**(42), *24274–24281.*
19. Dall'Agnese, Y., Lukatskaya, M.R., Cook, K.M., Taberna, P.L., Gogotsi, Y., Simon, P., High Capacitance of Surface-modified 2D Titanium Carbide in Acidic Electrolyte, ***ElectrochemistryCommunications,*** 2014, **48**, *118–122.*
20. Xin, Y., Yu, Y.-X., Possibility of Bare and Functionalized Niobium Carbide MXenes for Electrode Materials of Supercapacitors and Field Emitters, ***Materials and Design***, 2017, **130**, *512–520.*
21. Xie, X., Chen, S., Ding, W., Nie, Y., Wei, Z., An Extraordinarily Stable Catalyst: Pt NPs Supported on Two-dimensional $Ti_3C_2X_2$ (X = OH, F) Nanosheets for Oxygen Reduction Reaction, ***Chemistry Communications***, 2013, **49**(86), *10112–10114.*
22. Gao, Y., Wang, L., Li, Z., Zhou, A., Hu, Q., Cao, X., Preparation of MXene-Cu_2O Nanocomposite and Effect on Thermal Decomposition of Ammonium Perchlorate, ***Solid State Science***, 2014, **35**, *62–65.*
23. Mashtalir, O., Cook, K.M., Mochalin, V.N., Crowe, M., Barsoum, M.W., Gogotsi, Y., Dye Adsorption and Decomposition on Two-Dimensional Titanium Carbide in Aqueous Media, ***Journal of Materials Chemistry A***, 2014, **2**(35), *14334–14338.*
24. Lorencova, L., Gajdosova, V., Hroncekova, S., Bertok, T., Blahutova, J., Vikartovska, A., Parrakova, L., Gemeiner, P., Kasak, P., Tkac, J., 2D MXenes as Perspective Immobilization Platforms for Design of Electrochemical Nanobiosensors, ***Electroanalysis***, 2019, **31**(10), *1833–1844.*
25. Lorencova, L., Bertok, T., Dosekova, E., Holazova, A., Paprckova, D., Vikartovska, A., Sasinkova, V., Filip, J., Kasak, P., Jerigova, M., Velic, D., Mahmoud, K.A., Tkac, J., Electrochemical Performance of Ti_3C_2Tx MXene in Aqueous Media: Towards Ultrasensitive H_2O_2 Sensing, ***Electrochimca Acta***, 2017, **235,** 471–479.
26. Zha, X.J., Zhao, X., Pu, J.H., Tang, L.S., Ke, K., Bao, R.Y., Bai, L., Liu, Z.Y., Yang, M.B., Yang, W., Flexible Anti-Biofouling MXene/Cellulose Fibrous Membrane for Sustainable Solar-Driven Water Purification, **ACS Applied Materials and Interfaces,** 2019, **11**(40), 36589–36597.
27. Rasool, K., Helal, M., Ali, A., Ren, C.E., Gogotsi, Y., Mahmoud, K.A., Antibacterial Activity of $Ti_3C_2T_x$ MXene, **ACS Nano**, 2016, **10**(3), *3674–3684.*
28. Shahzad, F., Alhabeb, M., Hatter, C.B., Anasori, B., Hong, S.M., Koo, C.M., Gogotsi, Y., Electromagnetic Interference Shielding with 2D Transition Metal Carbides (MXenes), ***Science***, 2016, **353**(6304), *1137–1140.*
29. Lukatskaya, M.R., Mashtalir, O., Ren, C.E., Dall'Agnese, Y., Rozier, P., Taberna, P.L., Naguib, M., Simon, P., Barsoum, M.W., Gogotsi, Y., Cation Intercalation and High Volumetric Capacitance of Two-Dimensional Titanium Carbide, ***Science***, 2013, **341**(6153), *1502–1505.*
30. Jimmy, J., Kandasubramanian, B., Mxene Functionalized Polymer Composites: Synthesis and Applications, ***European Polymer Journal***, 2020, **122**, *109367.*
31. Gao, L., Li, C., Huang, W., Mei, S., Lin, H., Ou, Q., Zhang, Y., Guo, J., Zhang, F., Xu, S., Zhang, H., MXene/Polymer Membranes: Synthesis, Properties, and Emerging Applications, ***Chemistry ofMaterials***, 2020, **32**(5), *1703–1747.*
32. Chen, X., Zhao, Y., Li, L., Wang, Y., Wang, J., Xiong, J., Du, S., Zhang, P., Shi, X., Yu, J., MXene/Polymer Nanocomposites: Preparation, Properties, and Applications, ***Polymer Review***, 2021, **61**(1), *80–115.*
33. George, S.M., Kandasubramanian, B., Advancements in MXene-Polymer Composites for Various Biomedical Applications, ***Ceramic International***, 2020, **46**(7), *8522–8535.*
34. Aghamohammadi, H., Amousa, N., Eslami-Farsani, R., Recent Advances in Developing the MXene/ Polymer Nanocomposites with Multiple Properties: A Review Study, ***Synthetic Metals***, 2021, **273**, *116695.*
35. He, S., Sun, X., Zhang, H., Yuan, C., Wei, Y., Li, J., Preparation Strategies and Applications of MXene-Polymer Composites: A Review, ***Macromol Rapid Commun***, 2021, **42**(19), *2100324.*
36. Gong, K., Zhou, K., Qian, X., Shi, C., Yu, B., MXene as Emerging Nanofillers for High-performance Polymer Composites: A Review, ***Composites Part B: Engineering***, 2021, **217**, *108867.*

37. Kausar, A., Polymer/MXene Nanocomposite – A New Age for Advanced Materials, ***Polymer-Plastics Technology and Materials***, 2021, **60**(13), *1377–1392*.
38. Carey, M., Barsoum, M.W., MXene Polymer Nanocomposites: A Review, ***Materials Today Advances***, 2021, **9**, *100120*.
39. Riazi, H., Nemani, S.K., Grady, M.C., Anasori, B., Soroush, M., Ti_3C_2 MXene–Polymer Nanocomposites and Their Applications, ***Journal of Materials Chemistry A***, 2021, **9**(13), *8051–8098*.
40. Song, P., Liu, B., Qiu, H., Shi, X., Cao, D., Gu, J., MXenes for Polymer Matrix Electromagnetic Interference Shielding Composites: A Review, ***CompositesCommunications***, 2021, **24**, *100653*.
41. Feng, Y., Deng, Q., Peng, C., Hu, J., Li, Y., Wu, Q., Xu, Z., An Ultrahigh Discharged Energy Density achieved in an Inhomogeneous PVDF Dielectric Composite Filled with 2D MXene Nanosheets *via* Interface Engineering, ***Journal of Materials Chemistry A***, 2018, **6**(48), *13283–13292*.
42. Cao, Y., Deng, Q., Liu, Z., Shen, D., Wang, T., Huang, Q., Du, S., Jiang, N., Lin, C.T., Yu, J., Enhanced Thermal Properties of Poly(Vinylidene Fluoride) Composites with Ultrathin Nanosheets of Mxene, ***RSC Advances***, 2017, **7**(33), *20494–20501*.
43. Ling, Z., Ren, C.E., Zhao, M.Q., Yang, J., Giammarco, J.M., Qiu, J., Barsoum, M.W., Gogotsi, Y., Flexible and Conductive MXene Films and Nanocomposites with High Capacitance, ***Proceedings of the National Academy of Sciences***, 2014, **111**(47), *16676–16681*.
44. Boota, M., Anasori, B., Voigt, C., Zhao, M.Q., Barsoum, M.W., Gogotsi, Y., Pseudocapacitive Electrodes Produced by Oxidant-Free Polymerization of Pyrrole between the Layers of 2D Titanium Carbide (MXene), ***Advanced Materials***, 2016, **28**(7), *1517–1522*.
45. Xu, H., Yin, X., Li, X., Li, M., Liang, S., Zhang, L., Cheng, L., Lightweight Ti_2CT_x MXene/Poly(vinyl alcohol) Composite Foams for Electromagnetic Wave Shielding with Absorption-Dominated Feature, ***ACS Applied Materials and Interfaces***, 2019, **11**(10), *10198–10207*.
46. Zhang, H., Wang, L., Chen, Q., Li, P., Zhou, A., Cao, X., Hu, Q., Preparation, Mechanical and Anti-friction Performance of MXene/Polymer Composites, ***Materials and Design***, 2016, **92**, *682–689*.
47. Sobolčiak, P., Ali, A., Hassan, M.K., Helal, M.I., Tanvir, A., Popelka, A., Al-Maadeed, M.A., Krupa, I., Mahmoud, K.A., 2D $Ti_3C_2T_x$ (MXene)-reinforced Polyvinyl Alcohol (PVA) Nanofibers with Enhanced Mechanical and Electrical Properties, ***Plos One***, 2017, **12**(8), *e0183705*.
48. Lipton, J., Weng, G.M., Alhabeb, M., Maleski, K., Antonio, F., Kong, J., Gogotsi, Y., Taylor, A.D., Mechanically Strong and Electrically Conductive Multilayer MXene Nanocomposites, ***Nanoscale***, 2019, **11**(42), *20295–20300*.
49. Yang, H., Dai, J., Liu, X., Lin, Y., Wang, J., Wang, L., Wang, F., Layered PVB/$Ba_3Co_2Fe_{24}O_{41}$/Ti_3C_2 MXene Composite: Enhanced Electromagnetic Wave Absorption Properties with High Impedance Match in a Wide Frequency Range, ***MaterialsChemistry and Physics***, 2017, **200**, *179–186*.
50. An, H., Habib, T., Shah, S., Gao, H., Patel, A., Echols, I., Zhao, X., Radovic, M., Green, M.J., Lutkenhaus, J.L., Water Sorption in MXene/Polyelectrolyte Multilayers for Ultrafast Humidity Sensing, ***ACS Applied Nano Materials***, 2019, **2**(2), *948–955*.
51. Yun, J., Echols, I., Flouda, P., Wang, S., Easley, A., Zhao, X., Tan, Z., Prehn, E., Zi, G., Radovic, M., Green, M.J., Lutkenhaus, J.L., Layer-by-Layer Assembly of Polyaniline Nanofibers and MXene Thin-Film Electrodes for Electrochemical Energy Storage, ***ACS Applied Materials and Interfaces***, 2019, **11**(51), *47929–47938*.
52. Jiang, C., Wu, C., Li, X., Yao, Y., Lan, L., Zhao, F., Ye, Z., Ying, Y., Ping, J., All-electrospun Flexible Triboelectric Nanogenerator Based on Metallic MXene Nanosheets, ***Nano Energy***, 2019, **59**, *268–276*.
53. Wang, Q.W., Zhang, H.B., Liu, J., Zhao, S., Xie, X., Liu, L., Yang, R., Koratkar, N., Yu, Z.Z., Multifunctional and Water-Resistant MXene-Decorated Polyester Textiles with Outstanding Electromagnetic Interference Shielding and Joule Heating Performances, ***Advanced Functional Materials***, 2019, **29**(7), *1806819*.
54. Mayerberger, E.A., Street, R.M., McDaniel, R.M., Barsoum, M.W., Schauer, C.L., Antibacterial Properties of Electrospun $Ti_3C_2T_z$ (MXene)/Chitosan Nanofibers, ***RSC Advances***, 2018, **8**, *35386–35394*.
55. Huang, X., Wang, R., Jiao, T., Zou, G., Zhan, F., Yin, J., Zhang, L., Zhou, J., Peng, Q., Facile Preparation of Hierarchical AgNP-Loaded MXene/Fe_3O_4/Polymer Nanocomposites by Electrospinning with Enhanced Catalytic Performance for Wastewater Treatment, ***ACS Omega***, 2019, **4**(1), *1897–1906*.
56. Gao, X., Li, Z.K., Xue, J., Qian, Y., Zhang, L.Z., Caro, J., Wang, H., Titanium Carbide $Ti_3C_2T_x$ (MXene) Enhanced PAN Nanofiber Membrane for Air Purification, ***Journal of Membrane Science***, 2019, **586**, *162–169*.
57. Liu, J., Zhang, H.B., Xie, X., Yang, R., Liu, Z., Liu, Y., Yu, Z.Z., Multifunctional, Superelastic, and Lightweight MXene/Polyimide Aerogels, ***Small***, 2018, **14**(45), *1802479*.

58. Li, X.P., Li, Y., Li, X., Song, D., Min, P., Hu, C., Zhang, H.B., Koratkar, N., Yu, Z.Z., Highly Sensitive, Reliable and Flexible Piezoresistive Pressure Sensors Featuring Polyurethane Sponge Coated with MXene Sheets, ***Journal of Colloid and Interface Science***, 2019, **542**, *54–62*.
59. Ronchi, R.M., Marchiori, C.F., Araujo, C.M., Arantes, J.T., Santos, S.F., Thermoplastic Polyurethane – $Ti_3C_2(T_x)$ MXene Nanocomposite: The Influence of Functional Groups upon the Matrix–Reinforcement Interaction, ***Applied Surface Science***, 2020, **528**, *146526*.
60. Zhang, Y.Z., Lee, K.H., Anjum, D.H., Sougrat, R., Jiang, Q., Kim, H., Alshareef, H.N., MXenes Stretch Hydrogel Sensor Performance to New Limits, ***Science Advances***, 2018, **4**(6), *eaat0098*.
61. Hatter, C.B., Shah, J., Anasori, B., Gogotsi, Y., Micromechanical Response of Two-Dimensional Transition Metal Carbonitride (MXene) Reinforced Epoxy Composites, ***Composites Part B: Engineering***, 2020, **182**, *107603*.
62. Sliozberg, Y., Andzelm, J., Hatter, C.B., Anasori, B., Gogotsi, Y., Hall, A., Interface Binding and Mechanical Properties of MXene-Epoxy Nanocomposites, ***Composites Science and Technology***, 2020, **192**, *108124*.
63. Zhou, Z., Panatdasirisuk, W., Mathis, T. S., Anasori, B., Lu, C., Zhang, X., Liao, Z., Gogotsi, Y., Yang, S., Layer-by-Layer Assembly of MXene and Carbon Nanotubes on Electrospun Polymer Films for Flexible Energy Storage, ***Nanoscale***, 2018, **10**(13), *6005–6013*.
64. Yang, Y., Liu, Y., Cai, X., Effects of Ultralow Concentration MXene (Nano-$Ti_3C_2T_x$) on the Electric and Physical Properties of Ternary Polyvinyl Alcohol Composites, ***Colloids Surfaces A: Physicochemical and Engineering Aspects***, 2021, **610**, *125929*.
65. Pan, Y., Fu, L., Zhou, Q., Wen, Z., Lin, C.T., Yu, J., Wang, W., Zhao, H., Flammability, Thermal Stability and Mechanical Properties of Polyvinyl Alcohol Nanocomposites Reinforced with Delaminated $Ti_3C_2T_x$ (MXene), ***Polymer Composites***, 2020, **41**(1), *210–218*.
66. Jin, X., Wang, J., Dai, L., Liu, X., Li, L., Yang, Y., Cao, Y., Wang, W., Wu, H., Guo, S., Flame-Retardant Poly(Vinyl Alcohol)/MXene Multilayered Films with Outstanding Electromagnetic Interference Shielding and Thermal Conductive Performances, ***Chemical Engineering Journal***, 2020, **380**, *122475*.
67. Li, K., Zou, G., Jiao, T., Xing, R., Zhang, L., Zhou, J., Zhang, Q., Peng, Q., Self-assembled MXene-based Nanocomposites via Layer-by-Layer Strategy for Elevated Adsorption Capacities, ***Colloids and Surfaces A: Physicochemical and Engineering Aspects***, 2018, **553**, *105–113*.
68. Mayerberger, E.A., Urbanek, O., McDaniel, R.M., Street, R.M., Barsoum, M.W., Schauer, C.L., Preparation and Characterization of Polymer-$Ti_3C_2T_x$ (MXene) Composite Nanofibers Produced via Electrospinning, ***Journal Applied Polymer Science***, 2017, **134**(37), *45295*.
69. Hu, D., Huang, X., Li, S., Jiang, P., Flexible and Durable Cellulose/MXene Nanocomposite Paper for Efficient Electromagnetic Interference Shielding, ***Composites Science and Technology***, 2020, **188**, *107995*.
70. Cao, W.T., Chen, F.F., Zhu, Y.J., Zhang, Y.G., Jiang, Y.Y., Ma, M.G., Chen, F., Binary Strengthening and Toughening of MXene/Cellulose Nanofiber Composite Paper with Nacre-Inspired Structure and Superior Electromagnetic Interference Shielding Properties, ***ACS Nano***, 2018, **12**(5), *4583–4593*.
71. Carey, M., Hinton, Z., Natu, V., Pai, R., Sokol, M., Alvarez, N.J., Kalra, V., Barsoum, M.W., Dispersion and Stabilization of Alkylated 2D MXene in Nonpolar Solvents and Their Pseudocapacitive Behavior, ***Cell Reports Physical Science***, 2020, **1**(4), *10042*.
72. Pan, Q., Zheng, Y., Kota, S., Huang, W., Wang, S., Qi, H., Kim, S., Tu, Y., Barsoum, M.W., Li, C.Y., 2D MXene-Containing Polymer Electrolytes for All-Solid-State Lithium Metal Batteries, ***Nanoscale Advances***, 2019, **1**, *395–402*.
73. Shi, Y., Liu, C., Liu, L., Fu, L., Yu, B., Lv, Y., Yang, F., Song, P., Strengthening, Toughing and Thermally Stable Ultra-Thin MXene Nanosheets/Polypropylene Nanocomposites via Nanoconfinement, ***Chemical Engineering Journal***, 2019, **378**, *122267*.
74. Zheng, Y., Chen, W., Sun, Y., Huang, C., Wang, Z., Zhou, D., High Conductivity and Stability of Polystyrene/MXene Composites with Orientation-3D Network Binary Structure, ***Journal of Colloid and Interface Science***, 2021, **595**, *151–158*.
75. Carey, M., Hinton, Z., Sokol, M., Alvarez, N.J., Barsoum, M.W., Nylon-6/$Ti_3C_2T_z$ MXene Nanocomposites Synthesized by *in Situ* Ring Opening Polymerization of ε-Caprolactam and Their Water Transport Properties, ***ACS Applied Materials and Interfaces***, 2019, **11**(22), *20425–20436*.
76. Monastyreckis, G., Mishnaevsky Jr, L., Hatter, C.B., Aniskevich, A., Gogotsi, Y., Zeleniakiene, D., Micromechanical Modeling of MXene-Polymer Composites, ***Carbon***, 2020, **162**, *402–409*.
77. Boota, M., Gogotsi, Y., MXene – Conducting Polymer Asymmetric Pseudocapacitors, ***Advanced Energy Materials***, 2019, **9**(7), *1802917*.

78. Xu, H., Zheng, D., Liu, F., Li, W., Lin, J., Synthesis of an MXene/Polyaniline Composite with Excellent Electrochemical Properties, ***Journal of Materials Chemistry A***, 2020, **8**(12), *5853–5858.*
79. Mao, H., Gu, C., Yan, S., Xin, Q., Cheng, S., Tan, P., Wang, X., Xiu, F., Liu, X., Liu, J., Huang, W., Sun, L., MXene Quantum Dot/Polymer Hybrid Structures with Tunable Electrical Conductance and Resistive Switching for Nonvolatile Memory Devices, ***Advanced Electronic Materials***, 2020, **6**(1), *1900493.*
80. Seroka, N.S., Mamo, M.A., Application of Functionalised MXene-Carbon Nanoparticle-Polymer Composites in Resistive Hydrostatic Pressure Sensors, ***SN Applied Sciences***, 2020, **2**, *413.*
81. Zhang, W., Ma, J., Zhang, W., Zhang, P., He, W., Chen, J., Sun, Z., A Multidimensional Nanostructural Design towards Electrochemically Stable and Mechanically Strong Hydrogel Electrodes, ***Nanoscale***, 2020, **12**(12), *6637–6643.*
82. Guan, X., Feng, W., Wang, X., Venkatesh, R., Ouyang, J., Significant Enhancement in the Seebeck Coefficient and Power Factor of p-Type Poly(3,4-ethylenedioxythiophene):Poly(styrenesulfonate) through the Incorporation of n-Type MXene, ***ACS Applied Materials and Interfaces***, 2020, **12**(11), *13013–13020.*
83. Bhatta, T., Maharjan, P., Cho, H., Park, C., Yoon, S.H., Sharma, S., Salauddin, M., Rahman, M.T., Rana, S.S., Park, J.Y., High-performance Triboelectric Nanogenerator based on MXene Functionalized Polyvinylidene Fluoride Composite Nanofibers, ***Nano Energy***, 2021, **81**, *105670.*
84. Seyedin, S., Zhang, J., Usman, K.A.S., Qin, S., Glushenkov, A.M., Yanza, E.R.S., Jones, R.T., Razal, J.M., Facile Solution Processing of Stable MXene Dispersions towards Conductive Composite Fibers, ***Global Challenges***, 2019, **3**(10), *1900037.*
85. Li, L., Fu, X., Chen, S., Uzun, S., Levitt, A.S., Shuck, C.E., Han, W., Gogotsi, Y., Hydrophobic and Stable MXene – Polymer Pressure Sensors for Wearable Electronics, ***ACS AppliedMaterials and Interfaces***, 2020, **12**(13), *15362–15369.*
86. Zhang, C., Ma, Y., Zhang, X., Abdolhosseinzadeh, S., Sheng, H., Lan, W., Pakdel, A., Heier, J., Nuesch, F., Two-Dimensional Transition Metal Carbides and Nitrides (MXenes): Synthesis, Properties, and Electrochemical Energy Storage Applications, ***Energy EnvironmentalMaterials***, 2020, **3**(1), *29–55.*
87. Ghaleb, Z.A.A., Jaafar, M., Rashid, A.A., Fabrication Methods of Carbon-based Rubber Nanocomposites and Their Applications, In Carbon-Based Nanofillers and Their Rubber Nanocomposites, Yaragalla, S., Mishra, R. K., Thomas, S., Kalarikkal, N., Maria, H.J., Eds., Elsevier, USA, 2019, *49–63.*
88. Woo, J.H., Kim, N.H., Kim, S.I., Park, O.K., Lee, J.H., Effects of the Addition of Boric Acid on the Physical Properties of MXene/Polyvinyl Alcohol (PVA) Nanocomposite, ***Composites Part B: Engineering***, 2020, **199**, *108205.*
89. Liu, R., Li, W., High-Thermal-Stability and High-Thermal-Conductivity $Ti_3C_2T_x$ MXene/Poly(vinyl alcohol) (PVA) Composites, ***ACS Omega***, 2018, **3**(3), *2609–2617.*
90. Mazhar, S., Qarni, A.A., Haq, Y.U., Haq, Z.U., Murtaza, I., Promising PVC/MXene Based Flexible Thin Film Nanocomposites with Excellent Dielectric, Thermal and Mechanical Properties, ***Ceramic International***, 2020, **46**(8), *12593–12605.*
91. Yu, B., Tawiah, B., Wang, L.Q., Yin Yuen, A. C., Zhang, Z.C., Shen, L.L., Lin, B., Fei, B., Yang, W., Li, A., Zhu, S.E., Hu, E.Z., Lu, H.D., Yeoh, G.H., Interface Decoration of Exfoliated MXene Ultra-Thin Nanosheets for Fire and Smoke Suppressions of Thermoplastic Polyurethane Elastomer, ***Journal of Hazardous Materials***, 2019, **374**, *110–119.*
92. Huang, Z., Wang, S., Kota, S., Pan, Q., Barsoum, M.W., Li, C.Y., Structure and Crystallization Behavior of Poly(Ethylene Oxide)/$Ti_3C_2T_x$ MXene Nanocomposites, ***Polymer***, 2016, **102**, *119–126.*
93. Boota, M., Pasini, M., Galeotti, F., Porzio, W., Zhao, M.Q., Halim, J., Gogotsi, Y., Interaction of Polar and Nonpolar Polyfluorenes with Layers of Two-Dimensional Titanium Carbide (MXene): Intercalation and Pseudocapacitance, ***Chemistry of Materials***, 2017, **29**(7), *2731–2738.*
94. Naguib, M., Saito, T., Lai, S., Rager, M. S., Aytug, T., Paranthaman, M. P., Zhao, M.Q., Gogotsi, Y., $Ti_3C_2T_x$ (MXene)–Polyacrylamide Nanocomposite Films, ***RSC Advances***, 2016, **6**(76), *72069–72073.*
95. Shao, J., Wang, J.W., Liu, D.N., Wei, L., Wu, S.Q., Ren, H., A Novel High Permittivity Percolative Composite with Modified Mxene, ***Polymer***, 2019, **174**, *86–95.*
96. Jiao, S., Zhou, A., Wu, M., Hu, H., Kirigami Patterning of MXene/Bacterial Cellulose Composite Paper for All-Solid-State Stretchable Micro-Supercapacitor Arrays, ***Advanced Science***, 2019, **6**(12), *1900529.*
97. Hu, C., Shen, F., Zhu, D., Zhang, H., Xue, J., Han, X., Characteristics of Ti_3C_2X–Chitosan Films with Enhanced Mechanical Properties, ***Frontiers inEnergy Research***, 2017, **4**, *41.*
98. Yi, Z., Yang, J., Liu, X., Mao, L., Cui, L., Liu, Y., Enhanced Mechanical Properties of Poly(Lactic Acid) Composites with Ultrathin Nanosheets of MXene Modified by Stearic Acid, ***Journal of AppliedPolymer Science***, 2020, **137**(17), *48621.*

99. Han, R., Ma, X., Xie, Y., Da, T., Zhang, S., Preparation of a New 2D MXene/PES Composite Membrane with Excellent Hydrophilicity and High Flux, ***RSC Advances***, 2017, **7**, *56204–56210.*
100. Fei, M., Lin, R., Deng, Y., Xian, H., Bian, R., Zhang, X., Cheng, J., Xu, C., Cai, D., Polybenzimidazole/Mxene Composite Membranes for Intermediate Temperature Polymer Electrolyte Membrane Fuel Cells, ***Nanotechnology***, 2018, **29**(3), *035403.*
101. Si, J.Y., Tawiah, B., Sun, W.L., Lin, B., Wang, C., Yuen, A.C.Y., Yu, B., Li, A., Yang, W., Lu, H.D., Chan, Q.N., Yeoh, G.H., Functionalization of MXene Nanosheets for Polystyrene towards High Thermal Stability and Flame Retardant Properties, ***Polymers***, 2019, **11**(6), *976.*
102. Qin, L., Tao, Q., El Ghazaly, A., Fernandez-Rodriguez, J., Persson, P.O., Rosen, J., Zhang, F., High-Performance Ultrathin Flexible Solid-State Supercapacitors Based on Solution Processable $Mo_{1.33}C$ MXene and PEDOT:PSS, ***Advanced Functional Materials***, 2018, **28**(2), *1703808.*
103. Kang, R., Zhang, Z., Guo, L., Cui, J., Chen, Y., Hou, X., Wang, B., Lin, C.T., Jiang, N., Yu, J., Enhanced Thermal Conductivity of Epoxy Composites Filled with 2D Transition Metal Carbides (MXenes) with Ultralow Loading, ***Scientific Reports***, 2019, **9**, *9135.*
104. Qin, L., Tao, Q., Liu, X., Fahlman, M., Halim, J., Persson, P.O., Rosen, J., Zhang, F., Polymer-MXene Composite Films Formed by MXene-facilitated Electrochemical Polymerization for Flexible Solid-state Microsupercapacitors, ***Nano Energy***, 2019, **60**, *734–742.*
105. Wang, H., Li, L., Zhu, C., Lin, S., Wen, J., Jin, Q., Zhang, X., *In Situ* Polymerized $Ti_3C_2T_x$/PDA Electrode with Superior Areal Capacitance for Supercapacitors, ***Journal of Alloys and Compounds***, 2019, **778**, *858–865.*
106. Tong, Y., He, M., Zhou, Y., Zhong, X., Fan, L., Huang, T., Liao, Q., Wang, Y., Hybridizing Polypyrrole Chains with Laminated and Two-Dimensional $Ti_3C_2T_x$ toward High-performance Electromagnetic Wave Absorption, ***Applied Surface Science***, 2018, **434**, *283–293.*
107. Gao, Q., Feng, M., Li, E., Liu, C., Shen, C., Liu, X., Mechanical, Thermal, and Rheological Properties of $Ti_3C_2T_x$ MXene/Thermoplastic Polyurethane Nanocomposites, ***Macromolecular Materials and Engineering***, 2020, **305**(10), *2000343.*
108. Cao X., Wu M., Zhou A., Wang Y., He X., Wang L., Non-isothermal Crystallization and Thermal Degradation Kinetics of MXene/Linear Low-Density Polyethylene Nanocomposites, ***e-Polymers***, 2017, **17**(5), *373–381.*
109. He L., Wang J., Wang B., Wang X., Zhou X., Cai W., Mu X., Hou Y., Hu Y., Song L., Large-scale Production of Simultaneously Exfoliated and Functionalized MXenes as Promising Flame Retardant for Polyurethane, ***Composites Part B: Engineering***, 2019, **179**, *107486.*
110. Sheng, X., Zhao, Y., Zhang, L., Lu, X., Properties of Two-Dimensional Ti_3C_2 MXene/Thermoplastic Polyurethane Nanocomposites with Effective Reinforcement via Melt Blending, ***Composites Science and Technology***, 2019, **181**, *107710.*
111. Tan, K.H., Samylingam, L., Aslfattahi, N., Saidur, R., Kadirgama, K., Optical and Conductivity Studies of Polyvinyl Alcohol-MXene (PVA-MXene) Nanocomposite Thin Films for Electronic Applications, ***Optics and Laser Technology***, 2021, **136**, *106772.*
112. Lu, J., Zhang, Y., Tao, Y., Wang, B., Cheng, W., Jie, G., Song, L., Hu, Y., Self-healable Castor Oil-based Waterborne Polyurethane/MXene Film with Outstanding Electromagnetic Interference Shielding Effectiveness and Excellent Shape Memory Performance, ***Journal of Colloid and Interface Science***, 2021, **588**, *164–174.*
113. Wu, W., Zhao, W., Sun, Q., Yu, B., Yin, X., Cao, X., Feng, Y., Li, R.K.Y., Qu, J., Surface Treatment of Two Dimensional MXene for Poly(Vinylidene Fluoride) Nanocomposites with Tunable Dielectric Permittivity, ***Composites Communications***, 2021, **23**, *100562.*
114. Kurtoglu, M., Naguib, M., Gogotsi, Y., Barsoum, M.W., First Principles Study of Two-Dimensional Early Transition Metal Carbides, ***MRS Communications***, 2012, **2**(4), *133–137.*
115. Guo, Z., Zhou, J., Si, C., Sun, Z., Flexible two-dimensional $Ti_{n+1}Cn$ (n = 1, 2 and 3) and Their Functionalized MXenes Predicted by Density Functional Theories, ***Physical Chemistry Chemical Physics***, 2015, **17**(23), *15348–15354.*
116. Borysiuk, V.N., Mochalin, V.N., Gogotsi, Y., Molecular Dynamic Study of the Mechanical Properties of Two-dimensional Titanium Carbides $Ti_{n+1}C_n$ (MXenes), ***Nanotechnology***, 2015, **26**(26), *265705.*
117. Borysiuk, V.N., Mochalin, V.N., Gogotsi, Y., Bending Rigidity of Two-Dimensional Titanium Carbide (MXene) Nanoribbons: A Molecular Dynamics Study, ***Computational Materials Science***, 2018, **143**, *418–424.*

118. Zhi, W., Xiang, S., Bian, R., Lin, R., Wu, K., Wang, T., Cai, D., Study of MXene-Filled Polyurethane Nanocomposites Prepared via an Emulsion Method, ***Composites Science and Technology***, 2018, **168**, *404–411*.
119. Liu, Y., Zhang, J., Zhang, X., Li, Y., Wang, J., $Ti_3C_2T_x$ Filler Effect on the Proton Conduction Property of Polymer Electrolyte Membrane, ***ACS AppliedMaterials and Interfaces***, 2016, **8**(31), *20352–20363*.
120. Zhang, X., Xu, J., Wang, H., Zhang, J., Yan, H., Pan, B., Zhou, J., Xie, Y., Ultrathin Nanosheets of MAX Phases with Enhanced Thermal and Mechanical Properties in Polymeric Compositions: $Ti_3Si_{0.75}Al_{0.25}C_2$, ***Angewandte Chemie International Edition***, 2013, **125**(16), *4457–4461*.
121. Zheng, S., Zhang, C., Zhou, F., Dong, Y., Shi, X., Nicolosi, V., Wu, Z., Bao, X., Ionic Liquid Pre-intercalated MXene Films for Ionogel-based Flexible Micro-supercapacitors with High Volumetric Energy Density, ***Journal of MaterialsChemistry A***, 2019, **7**(16), *9478–9485*.
122. Zhang, C., Anasori, B., Seral-Ascaso, A., Park, S., McEvoy, N., Shmeliov, A., Duesberg, G., Coleman, J., Gogotsi, Y., Nicolosi, V., Transparent, Flexible, and Conductive 2D Titanium Carbide (MXene) Films with High Volumetric Capacitance, ***Advanced Materials***, 2017, **29**(36), *1702678*.
123. Sun, R., Zhang, H.B., Liu, J., Xie, X., Yang, R., Li, Y., Hong, S., Yu, Z.Z., Highly Conductive Transition Metal Carbide/Carbonitride(MXene)@polystyrene Nanocomposites Fabricated by Electrostatic Assembly for Highly Efficient Electromagnetic Interference Shielding, ***Advanced FunctionalMaterials***, 2017, **27**(45), *1702807*.
124. Xu, M.K., Liu, J., Zhang, H.B., Zhang, Y., Wu, X., Deng, Z., Yu, Z.Z., Electrically Conductive $Ti_3C_2T_x$ MXene/Polypropylene Nanocomposites with an Ultralow Percolation Threshold for Efficient Electromagnetic Interference Shielding, ***Industrial & Engineering Chemistry Research***, 2021, **60**(11), *4342–4350*.
125. Tu, S., Jiang, Q., Zhang, X., Alshareef, H. N., Large Dielectric Constant Enhancement in MXene Percolative Polymer Composites, ***ACS Nano***, 2018, **12**(4), *3369–3377*.
126. Mirkhani, S.A., Shayesteh Zeraati, A., Aliabadian, E., Naguib, M., Sundararaj, U., High Dielectric Constant and Low Dielectric Loss via Poly(vinyl alcohol)/$Ti_3C_2T_x$ MXene Nanocomposites, ***ACS Applied Materials and Interfaces***, 2019, **11**(20), *18599–18608*.
127. Yang, Y., Shi, L., Cao, Z., Wang, R., Sun, J., Strain Sensors with a High Sensitivity and a Wide Sensing Range Based on a $Ti_3C_2T_x$ (MXene) Nanoparticle–Nanosheet Hybrid Network, ***Advanced Functional Materials***, 2019, **29**(14), *1807882*.
128. Kim, S.J., Koh, H.J., Ren, C.E., Kwon, O., Maleski, K., Cho, S.Y., Anasori, B., Kim, C.K., Choi, Y.K., Kim, J., Yury Gogotsi, Y., Jung, H.T., Metallic $Ti_3C_2T_x$ MXene Gas Sensors with Ultrahigh Signal-to-Noise Ratio, ***ACS Nano***, 2018, **12**(2), *986–993*.
129. Guo, Y., Zhong, M., Fang, Z., Wan, P., Yu, G., A Wearable Transient Pressure Sensor Made with MXene Nanosheets for Sensitive Broad-Range Human–Machine Interfacing, ***Nano Letters***, 2019, **19**(2), *1143–1150*.
130. Ciou, J., Li, S., Lee, P., Ti_3C_2 MXene Paper for the Effective Adsorption and Controllable Release of Aroma Molecules, ***Small***, 2019, **15**(38), *1903281*.
131. Shao, Y., Zhang, F., Shi, X., Pan, H., N-Functionalized MXenes: Ultrahigh Carrier Mobility and Multifunctional Properties, ***Physical Chemistry Chemical Physics***, 2017, **19**(42), *28710–28717*.
132. Zhan, X., Si, C., Zhou, J., Sun, Z., MXene and MXene-based Composites: Synthesis, Properties and Environment-related Applications, ***Nanoscale Horizons***, 2020, **5**(2), *235–258*.
133. Kim, H.B., Sajid, M., Kim, K.T., Na, K.H., Choi, K.H., Linear Humidity Sensor Fabrication Using Bi-layered Active Region of Transition Metal Carbide and Polymer Thin Films, ***Sensors and Actuators B: Chemical***, 2017, **252**, *725–734*.
134. Sajid, M., Kim, H.B., Siddiqui, G.U., Na, K.H., Choi, K.H., Linear Bi-layer Humidity Sensor with Tunable Response Using Combinations of Molybdenum Carbide with Polymers, ***Sensors and Actuators B: Physical***, 2017, **262**, *68–77*.
135. Yuan, W., Yang, K., Peng, H., Li, F., Yin, F., A Flexible VOCs Sensor Based on a 3D MXene Framework with a High Sensing Performance, ***Journal of MaterialsChemistry A,*** 2018, **6**(37), *18116–18124*.
136. Shi, X., Wang, H., Xie, X., Xue, Q., Zhang, J., Kang, S., Wang, C., Liang, J., Chen, Y., Bioinspired Ultrasensitive and Stretchable MXene-Based Strain Sensor via Nacre-Mimetic Microscale "Brick-and-Mortar" Architecture, ***ACS Nano***, 2019, **13**(1), *649–659*.
137. Hu, Y., Zhuo, H., Luo, Q., Wu, Y., Wen, R., Chen, Z., Liu, L., Zhong, L., Peng, X., Sun, R., Biomass Polymer-assisted Fabrication of Aerogels from MXenes with Ultrahigh Compression Elasticity and Pressure Sensitivity, ***Journal of MaterialsChemistry A***, 2019, **7**(17), *10273–10281*.

138. Ma, Y., Liu, N., Li, L., Hu, X., Zou, Z., Wang, J., Luo, S., Gao, Y., A Highly Flexible and Sensitive Piezoresistive Sensor Based on MXene with Greatly Changed Interlayer Distances, ***Nature Communications***, 2017, **8**, *1207.*
139. Yang, K., Yin, F., Xia, D., Peng, H., Yang, J., Yuan, W., A Highly Flexible and Multifunctional Strain Sensor Based on a Network-structured MXene/Polyurethane Mat with Ultra-high Sensitivity and a Broad Sensing Range, ***Nanoscale***, 2019, **11**(20), *9949–9957.*
140. Wang, K., Lou, Z., Wang, L., Zhao, L., Zhao, S., Wang, D., Han, W., Jiang, K., Shen, G., Bioinspired Interlocked Structure-induced High Deformability for Two-dimensional Titanium Carbide (MXene)/ Natural Microcapsule-Based Flexible Pressure Sensors, ***ACS Nano***, 2019, **13**(8), *9139–9147.*
141. Wang, D., Zhang, D., Li, P., Yang, Z., Mi, Q., Yu, L., Electrospinning of Flexible Poly(vinyl alcohol)/MXene Nanofiber-Based Humidity Sensor Self-powered by Monolayer Molybdenum Diselenide Piezoelectric Nanogenerator, ***Nano-Micro Letters***, 2021, **13**, *57.*
142. Wang, X., Zhang, D., Zhang, H., Gong, L., Yang, Y., Zhao, W., Yu, S., Yin, Y., Sun, D., *In Situ* Polymerized Polyaniline/MXene (V2C) as Building Blocks of Supercapacitor and Ammonia Sensor Self-powered by Electromagnetic-Triboelectric Hybrid Generator, ***Nano Energy***, 2021, **88**, *106242.*
143. Chao, M., Wang, Y., Ma, D., Wu, X., Zhang, W., Zhang, L., Wan, P., Wearable MXene Nanocomposites-based Strain Sensor with Tile-like Stacked Hierarchical Microstructure for Broad-range Ultrasensitive Sensing, ***Nano Energy***, 2020, **78**, *105187.*
144. Wang, S., Shao, H.Q., Liu, Y., Tang, C.Y., Zhao, X., Ke, K., Bao, R.Y., Yang, M.B., Yang, W., Boosting Piezoelectric Response of PVDF-TrFE via MXene for Self-powered Linear Pressure Sensor, ***Composites Science and Technology***, 2021, **202**, *108600.*
145. Zhao, Q., Yang, L., Ma, Y., Huang, H., He, H., Ji, H., Wang, Z., Qiu, J., Highly Sensitive, Reliable and Flexible Pressure Sensor Based on Piezoelectric PVDF Hybrid Film using MXene Nanosheet Reinforcement, ***Journal of Alloys and Compounds***, 2021, **886**, *161069.*
146. Rasool, K., Mahmoud, K.A., Johnson, D.J., Helal, M., Berdiyorov, G.R., Gogotsi, Y., Efficient Antibacterial Membrane Based on Two-dimensional $Ti_3C_2T_x$ (MXene) Nanosheets, ***Scientific Reports***, 2017, **7**, *1598.*
147. Lin, H., Wang, X., Yu, L., Chen, Y., Shi, J., Two-dimensional Ultrathin MXene Ceramic Nanosheets for Photothermal Conversion, **Nano Letters**, 2017, **17**(1), *384–391.*
148. Szuplewska, A., Kulpińska, D., Dybko, A., Jastrzębska, A.M., Wojciechowski, T., Rozmysłowska, A., Chudy, M., Grabowska-Jadach, I., Ziemkowska, W., Brzózka, Z., Olszyna, A., 2D Ti_2C (MXene) as a Novel Highly Efficient and Selective Agent for Photothermal Therapy, ***Materials Science and Engineering: C***, 2019, **98**, *874–886.*
149. Feng, W., Wang, R., Zhou, Y., Ding, L., Gao, X., Zhou, B., Hu, P., Chen, Y., Ultrathin Molybdenum Carbide MXene with Fast Biodegradability for Highly Efficient Theory-Oriented Photonic Tumor Hyperthermia, ***Advanced Functional Materials***, 2019, **29**(22), *1901942.*
150. Lin, H., Gao, S., Dai, C., Chen, Y., Shi, J., A Two-dimensional Biodegradable Niobium Carbide (MXene) for Photothermal Tumor Eradication in NIR-I and NIR-II Biowindows, ***Journal of the American Chemical Society***, 2017, **139**(45), *16235–16247.*
151. Xing, C., Chen, S., Liang, X., Liu, Q., Qu, M., Zou, Q., Li, J., Tan, H., Liu, L., Fan, D., Zhang, H., Two-dimensional MXene (Ti_3C_2)-integrated Cellulose Hydrogels: Toward Smart Three-Dimensional Network Nanoplatforms Exhibiting Light-induced Swelling and Bimodal Photothermal/Chemotherapy Anticancer Activity, ***ACS AppliedMaterials and Interfaces,*** 2018, **10**(33), *27631–27643.*
152. Rakhi, R.B., Nayak, P., Xia, C., Alshareef, H.N., Novel Amperometric Glucose Biosensor Based on MXene Nanocomposite, ***Scientific Reports***, 2016, **6**, *36422.*
153. Chen, K., Chen, Y., Deng, Q., Jeong, S.H., Jang, T.S., Du, S., Kim, H.E., Huang, Q., Han, C.M., Strong and Biocompatible Poly(Lactic Acid) Membrane Enhanced by $Ti_3C_2T_z$ (MXene) Nanosheets for Guided Bone Regeneration, ***Materials Letters***, 2018, **229**, *114–117.*
154. Chen, Z., Han, Y., Li, T., Zhang, X., Wang, T., Zhang, Z., Preparation and Electrochemical Performances of Doped MXene/Poly(3,4-ethylenedioxythiophene) Composites, ***Materials Letters***, 2018, **220**, *305–308.*
155. Gund, G.S., Park, J.H., Harpalsinh, R., Kota, M., Shin, J.H., Kim, T.I., Gogotsi, Y., Park, H.S., MXene/ Polymer Hybrid Materials for Flexible AC-Filtering Electrochemical Capacitors, ***Joule***, 2019, **3**(1), *164–176.*
156. Wu, W., Wei, D., Zhu, J., Niu, D., Wang, F., Wang, L., Yang, L., Yang, P., Wang, C., Enhanced Electrochemical Performances of Organ-like Ti_3C_2 MXenes/Polypyrrole Composites as Supercapacitors Electrode Materials, ***Ceramic International***, 2019, **45**(6), *7328–7337.*

157. Zhu, M., Huang, Y., Deng, Q., Zhou, J., Pei, Z., Xue, Q., Huang, Y., Wang, Z., Li, H., Huang, Q., Zhi, C., Highly Flexible, Freestanding Supercapacitor Electrode with Enhanced Performance Obtained by Hybridizing Polypyrrole Chains with Mxene, ***Advanced Energy Materials***, 2016, **6**(21), *1600969.*
158. Ren, Y., Zhu, J., Wang, L., Liu, H., Liu, Y., Wu, W., Wang, F., Synthesis of Polyaniline Nanoparticles Deposited on Two-dimensional Titanium Carbide for High-performance Supercapacitors, ***Materials Letters***, 2018, **214**, *84–87.*
159. Li, K., Wang, X., Li, S., Urbankowski, P., Li, J., Xu, Y., Gogotsi, Y., An Ultrafast Conducting Polymer@ MXene Positive Electrode with High Volumetric Capacitance for Advanced Asymmetric Supercapacitors, ***Small***, 2020, **16**(4), *1906851.*
160. VahidMohammadi, A., Moncada, J., Chen, H., Kayali, E., Orangi, J., Carrero, C.A., Beidaghi, M., Thick and Freestanding MXene/PANI Pseudocapacitive Electrodes with Ultrahigh Specific Capacitance, ***Journal of MaterialsChemistry A***, 2018, **6**(44), *22123–22133.*
161. Li, Y., Kamdem, P., Jin, X.J., Hierarchical Architecture of MXene/PANI Hybrid Electrode for Advanced Asymmetric Supercapacitors, ***Journal of Alloys and Compounds***, 2021, **850**, *156608.*
162. Zhou, J., Kang, Q., Xu, S., Li, X., Liu, C., Ni, L., Chen, N., Lu, C., Wang, X., Peng, L., Guo, X., Ding, W., Hou, W., Ultrahigh Rate Capability of 1D/2D Polyaniline/Titanium Carbide (MXene) Nanohybrid for Advanced Asymmetric Supercapacitors, ***Nano Research***, 2022, **15**, *285–295.*
163. Li, W., Song, Z., Zhong, J., Qian, J., Tan, Z., Wu, X., Chu, H., Nie, W., Ran, X., Multilayer-structured Transparent MXene/PVDF Film with Excellent Dielectric and Energy Storage Performance, ***Journal of MaterialsChemistry C***, 2019, **7**(33), *10371–10378.*
164. Iqbal, A., Sambyal, P., Koo, C.M., 2D MXenes for Electromagnetic Shielding: A Review, ***Advanced Functional Materials***, 2020, **30**(47), *2000883.*
165. Aakyiir, M., Kingu, M.A.S., Araby, S., Meng, Q., Shao, J., Amer, Y., Ma, J., Stretchable, Mechanically Resilient, and High Electromagnetic Shielding Polymer/MXene Nanocomposites, ***Journal of Applied Polymer Science***, 2021, **138**(22), *50509.*
166. Wan, Y.J., Li, X.M., Zhu, P.L., Sun, R., Wong, C.P., Liao, W.H., Lightweight, Flexible MXene/Polymer Film with Simultaneously Excellent Mechanical Property and High-performance Electromagnetic Interference Shielding, ***Composites Part A: Applied Science and Manufacturing***, 2020, **130**, *105764.*
167. Liu, R., Miao, M., Li, Y., Zhang, J., Cao, S., Feng, X., Ultrathin Biomimetic Polymeric $Ti_3C_2T_x$ MXene Composite Films for Electromagnetic Interference Shielding, ***ACS Applied Materials and Interfaces***, 2018, **10**(51), *44787–44795.*
168. Bora, P.J., Anil, A.G., Ramamurthy, P.C., Tan, D.Q., MXene Interlayered Crosslinked Conducting Polymer Film for Highly Specific Absorption and Electromagnetic Interference Shielding, ***Materials Advances***, 2020, **1**(2), *177–183.*
169. Wang, W., Yuen, A.C.Y., Long, H., Yang, W., Li, A., Song, L., Hu, Y., Yeoh, G.H., Random Nano-structuring of PVA/MXene Membranes for Outstanding Flammability Resistance and Electromagnetic Interference Shielding Performances, ***Composites Part B: Engineering***, 2021, **224**, *109174.*
170. Raagulan, K., Braveenth, R., Kim, B.M., Lim, K.J., Lee, S.B., Kim, M., Chai, K.Y., An Effective Utilization of MXene and Its Effect on Electromagnetic Interference Shielding: Flexible, Free-standing and Thermally Conductive Composite from MXene–PAT–Poly(p-aminophenol)–Polyaniline Co-polymer, ***RSC Advances***, 2020, **10**(3), *1613–1633.*
171. Wang, Y., Qi, Q., Yin, G., Wang, W., Yu, D., Flexible, Ultralight, and Mechanically Robust Waterborne Polyurethane/$Ti_3C_2T_x$ MXene/Nickel Ferrite Hybrid Aerogels for High-Performance Electromagnetic Interference Shielding, ***ACS Applied Materials and Interfaces***, 2021, **13**(18), *21831–21843.*
172. Liu, Z., Wang, W., Tan, J., Liu, J., Zhu, M., Zhu, B., Zhang, Q., Bioinspired Ultra-thin Polyurethane/ MXene Nacre-like Nanocomposite Films with Synergistic Mechanical Properties for Electromagnetic Interference Shielding, ***Journal of Materials Chemistry C***, 2020, **8**(21), *7170–7180.*
173. Yuan, W., Yang, J., Yin, F., Li, Y., Yuan, Y., Flexible and Stretchable MXene/Polyurethane Fabrics with Delicate Wrinkle Structure Design for Effective Electromagnetic Interference Shielding at a Dynamic Stretching Process, ***Composites Communications***, 2020, **19**, *90–98.*
174. Gao, Q., Pan, Y., Zheng, G., Liu, C., Shen, C., Liu, X., Flexible Multilayered MXene/Thermoplastic Polyurethane Films with Excellent Electromagnetic Interference Shielding, Thermal Conductivity, and Management Performances, ***Advanced Composites and Hybrid Materials***, 2021, **4**(2), *274–285.*
175. Li, Y., Zhou, B., Shen, Y., He, C., Wang, B., Liu, C., Feng, Y., Shen, C., Scalable Manufacturing of Flexible, Durable $Ti_3C_2T_x$ MXene/Polyvinylidene Fluoride Film for Multifunctional Electromagnetic

Interference Shielding and Electro/Photo-thermal Conversion Applications, ***Composites Part B: Engineering***, 2021, **217**, *108902*.

176. Wang, J., Yang, K., Wang, H., Li, H., A New Strategy for High-performance Electromagnetic Interference Shielding by Designing a Layered Double-percolated Structure in PS/PVDF/MXene Composites, ***Europrean Polymer Journal***, 2021, **151**, *110450*.
177. Wei, H., Dong, J., Fang, X., Zheng, W., Sun, Y., Qian, Y., Jiang, Z., Huang, Y., $Ti_3C_2T_x$ MXene/Polyaniline (PANI) Sandwich Intercalation Structure Composites Constructed for Microwave Absorption, ***Composites Science and Technology***, 2019, **169**, *52–59*.
178. Kumar, S., Kumar, P., Singh, N., Verma, V., Steady Microwave Absorption Behavior of Two-Dimensional Metal Carbide MXene and Polyaniline Composite in X-band, ***Journal of Magnetism and Magnetic Materials***, 2019, **488**, *165364*.
179. Ren, C. E., Hatzell, K. B., Alhabeb, M., Ling, Z., Mahmoud, K. A., Gogotsi, Y., Charge- and Size-Selective Ion Sieving Through $Ti_3C_2T_x$ MXene Membranes, ***The Journal of Physical Chemistry Letters***, 2015, **6**(20), *4026–4031*.
180. Han, R., Xie, Y., Ma, X., Crosslinked P84 Copolyimide/MXene Mixed Matrix Membrane with Excellent Solvent Resistance and Permselectivity, ***Chinese Journal of Chemical Engineering***, 2019, **27**(4), *877–883*.
181. Wu, X., Hao, L., Zhang, J., Zhang, X., Wang, J., Liu, J., Polymer-$Ti_3C_2T_x$ Composite Membranes to Overcome the Trade-off in Solvent Resistant Nanofiltration for Alcohol-based System, ***Journal of MembraneScience***, 2016, **515**, *175–188*.
182. Hao, L., Zhang, H., Wu, X., Zhang, J., Wang, J., Li, Y., Novel Thin-film Nanocomposite Membranes Filled with Multi-functional Ti3C2Tx Nanosheets for Task-specific Solvent Transport, ***Composites Part A: Applied Science and Manufacturing***, 2017, **100**, *139–149*.
183. Xu, Z., Liu, G., Ye, H., Jin, W., Cui, Z., Two-dimensional MXene Incorporated Chitosan Mixed-matrix Membranes for Efficient Solvent Dehydration, ***Journal of Membrane Science***, 2018, **563**, *625–632*.
184. Shen, J., Liu, G., Ji, Y., Liu, Q., Cheng, L., Guan, K., Zhang, M., Liu, G., Xiong, J., Yang, J., Jin, W., 2D MXene Nanofilms with Tunable Gas Transport Channels, ***Advanced Functional Materials***, 2018, **28**(31), *1801511*.
185. Chen, C., Boota, M., Xie, X., Zhao, M., Anasori, B., Ren, C. E., Miao, L., Jiang, J., Gogotsi, Y., Charge Transfer Induced Polymerization of EDOT Confined between 2D Titanium Carbide Layers, ***Journal of Materials Chemistry A***, 2017, **5**(11), *5260–5265*.
186. Zhang, X., Fan, C., Yao, N., Zhang, P., Hong, T., Xu, C., Cheng, J., Quaternary $Ti_3C_2T_x$ Enhanced Ionic Conduction in Quaternized Polysulfone Membrane for Alkaline Anion Exchange Membrane Fuel Cells, ***Journal of Membrane Science***, 2018, **563**, *882–887*.
187. Elancheziyan, M., Eswaran, M., Shuck, C.E., Senthilkumar, S., Elumalai, S., Dhanusuraman, R., Ponnusamy, V.K., Facile Synthesis of Polyaniline/Titanium Carbide (MXene) Nanosheets/Palladium Nanocomposite for Efficient Electrocatalytic Oxidation of Methanol for Fuel Cell Application, ***Fuel***, 2021, **303**, *121329*.
188. Zhang, Z., Yang, S., Zhang, P., Zhang, J., Chen, G., Feng, X., Mechanically Strong MXene/Kevlar Nanofiber Composite Membranes as High-performance Nanofluidic Osmotic Power Generators, ***NatureCommunications***, 2019, **10**, *2920*.
189. Li, L., Liu, X., Wang, J., Yang, Y., Cao, Y., Wang, W., New Application of MXene in Polymer Composites toward Remarkable Anti-dripping Performance for Flame Retardancy, ***Composites Part A: Applied Science and Manufacturing***, 2019, **127**, *105649*.
190. Yu, B., Yuen, A.C.Y., Xu, X., Zhang, Z.C., Yang, W., Lu, H., Fei, B., Yeoh, G.H., Song, P., Wang, H., Engineering MXene Surface with POSS for Reducing Fire Hazards of Polystyrene with Enhanced Thermal Stability, ***Journal of Hazardous Materials***, 2021, **401**, *123342*.
191. Zhang, L., Huang, Y., Dong, H., Xu, R., Jiang, S., Flame-Retardant Shape Memory Polyurethane/MXene Paper and the Application for Early Fire Alarm Sensor, ***Composites Part B: Engineering***, 2021, **223**, *109149*.
192. Luo, Y., Xie, Y., Geng, W., Dai, G., Sheng, X., Xie, D., Wu, H., Mei, Y., Fabrication of Thermoplastic Polyurethane with Functionalized MXene towards High Mechanical Strength, Flame-Retardant, and Smoke Suppression Properties, ***Journal of Colloid and Interface Science***, 2022, **606**, *223–235*.

193. Liu, L., Zhu, M., Shi, Y., Xu, X., Ma, Z., Yu, B., Fu, S., Huang, G., Wang, H., Song, P., Functionalizing MXene towards Highly Stretchable, Ultratough, Fatigue- and Fire-Resistant Polymer Nanocomposites, ***Chemical Engineering Journal***, 2021, **424**, *130338*.
194. Niranjana, J.P., Kandasubramanian, B., Nanocomposites of MXene for Industrial Applications, ***Journal of Alloys and Compounds***, 2021, **862**, *158547*.
195. Kshetri, T., Tran, D.T., Le, H.T., Nguyen, D.C., Hoa, H.V., Kim, N.H., Lee, J.H., Recent Advances in MXene-based Nanocomposites for Electrochemical Energy Storage Applications, ***Progress in Materials Science***, 2020, **117**, *100733*.

4 MXene–Elastomer Nanocomposites

Jeeshma R, Swapna V.P, and Ranimol Stephen

CONTENTS

4.1 INTRODUCTION

Two-dimensional (2D) transition metal carbides, nitrides and carbonitrides are generally called as MXenes. It is synthesized from a precursor MAX, where M is the early transition metal, A is group IIIA or IVA element and X is carbon and or nitrogen. Recently, material scientists have shown great interest in the study of different Mxenes and MXene-based nanocomposites because of their properties like electrical and thermal conductivity, high specific surface area and compatibility with polymer systems. The fine-tuning of size, surface chemistry and morphology of MXenes are possible, which makes it a promising nanomaterial for various applications. Moreover, the MXene surface is hydrophilic and can easily be dispersed in wide range of solvents. Introduced to the scientific community in 2011, MXene/polymer nanocomposite research is in an early stage [1]. Two-dimensional MXenes have good electrical conductivity and can be used in the electronic industry.

The importance of conductive elastomers arose from the demand for stretchable and flexible electronics. Elastomers can be made conductive by incorporating some conductive fillers such as

DOI: 10.1201/9781003164975-4

carbon nanotube, graphene, graphite, MXene and so on. Elastomer/MXene nanocomposites are promising materials for electromagnetic interference (EMI) shielding owing to the exceptional electrical conductivity of MXene [2]. Properties of nanocomposites greatly depend on the group present on the MXene surface, its dispersion in the polymer matrix and the processing conditions. MXene/polymer nanocomposite membranes exhibited high degree of water permeation and rejection of ions [3]. It can act as a flame retardant also, and its incorporation decreases the flammability of polymeric membranes [4]. The electrical conductivity of poly(vinyl alcohol) (PVA) is found to be 3000 times greater than pristine PVA in the presence of this nanomaterial [5]. MXene/poly(propylene) nanocomposites showed excellent EMI shielding [6]. From studies, it is revealed that MXene is an interesting nanomaterial for the fabrication of conductive flexible nanocomposites. This chapter focuses on the preparation, properties and applications of MXene/elastomer nanocomposites. It provides an insight into the different processing techniques, properties and applications of MXene/elastomer nanocomposites.

4.2 PREPARATION METHODS OF MXene/ELASTOMER NANOCOMPOSITES

4.2.1 Latex Compounding Method

Latex compounding is one of the effective and widely used methods to prepare MXene/elastomer nanocomposites. This method allows the preparation of nanocomposite at ambient conditions. In this method, the unmodified MXene (or modified MXene) solution is transferred to the neat elastomer latex of a specified amount, and it is mechanically stirred for to get uniform dispersion of MXene in the elastomer latex. Then the coagulation of MXene–elastomer latex is initiated by adding acid, and the resulting coagulant is washed thoroughly to remove the impurities and obtain pH 7. The wet product is allowed to dry in a ventilated oven followed by compounding using a two-roll mill. The composite is then vulcanized using a preheated-hydraulic press [7]. The preparation method of MXene/elastomer nanocomposite by latex compounding is summarized in Figure 4.1.

4.2.2 Freeze-Drying Method

MXene/elastomer nanocomposites can also be prepared by green and facile freeze-drying methods. The specified amount of MXene (Ti_3C_2) dispersion is gradually added to the elastomer latex containing additives and stirred mechanically at room temperature. The uniform MXene/elastomer suspension is directly dried via freeze-drying. It is then mixed in a two-roll mill, and the final rubber

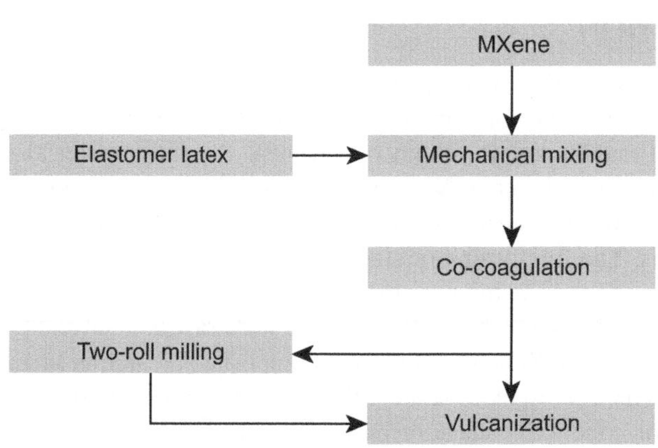

FIGURE 4.1 Schematics of MXene/elastomer nanocomposite preparation by latex compounding method.

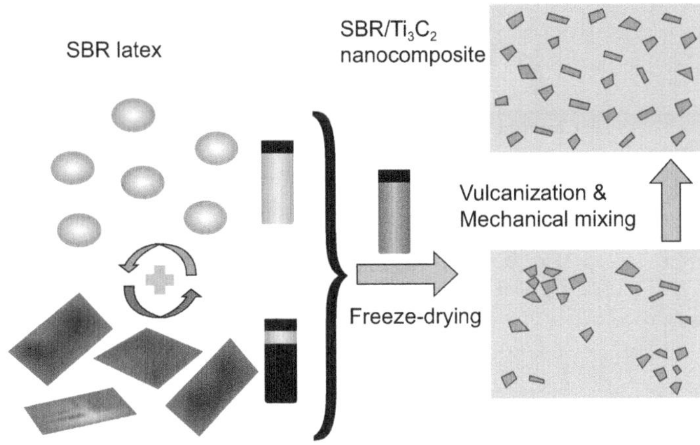

FIGURE 4.2 Schematics of styrene butadiene rubber (SBR)/MXene nanocomposite preparation by freeze-drying method [8].

Source: Reproduced with permission from RSC.

nanocomposite is obtained by compressing the compound at a specific temperature with a platen press for the optimum curing time. The preparation method of a styrene butadiene rubber (SBR)/MXene nanocomposite is shown in Figure 4.2 [8].

4.2.3 Solvent Mixing–Melt Blending Method

MXene/elastomer nanocomposite can be fabricated using the solvent mixing–melt blending method. One of the examples is the preparation of polyurethane/MXene-hybrid nanocomposite [9]. In this method, polymer is dissolved in a suitable solvent and ultrasonicated to get a homogeneous solution. MXene dispersed in ethyl alcohol is added to the polymer solution and sonicated. The prepared nanocomposite is precipitated in a nonsolvent and dried. It is then mixed using two-roll mill and hot-pressed to obtain elastomer/MXene nanocomposites.

4.2.4 Melt Blending Method

The melt blending method is also an effective method used to prepare MXene ($Ti_3C_2T_X$)/elastomer nanocomposite. The modified or unmodified MXene is blended with elastomer in a twin-screw extruder, where the extruder acts as a dispersion tool. The temperature and screw speed of the extruder have been optimized for the even dispersion of MXene in the elastomer. The extruded sample is then hot-pressed for further analysis [10,11].

4.2.5 Layer-by-Layer Deposition Method

MXene (Ti_3C_2)-coated elastomer can be synthesized by the layer-by-layer (LBL) deposition method, which involves the synthesis of Ti_3C_2 ultra-thin nanosheets and the coating solutions and the LBL deposition. The fundamental principle of the LBL deposition method includes the alignment of oppositely charged materials as multiple bilayers by utilizing the various interactive forces including van der Waals, hydrogen bonding, and electrostatic bonds. Lin and coworkers [12] successfully synthesized the Ti_3C_2/chitosan coated polyurethane foams (PUFs) by LBL deposition method for flame retardant application and the scheme is presented in Figure 4.3. They used Ti_3C_2 ultra-thin

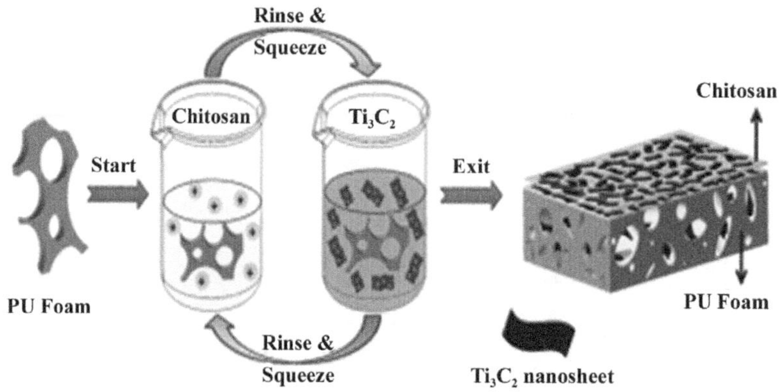

FIGURE 4.3 Schematic of the layer-by-layer assembled approach for construction of the Ti_3C_2/chitosan nanocoating on PUF [12].

Source: Reproduced with permission from Elsevier.

nanosheets, which are negatively charged, as the main component to prepare the fire retardant coating for PUF, and the chitosan is used as the positively charged partner combined with Ti_3C_2. The chitosan powder is dissolved in deionized water, and an aqueous suspension of MXene and sodium polyacrylate (PAANa) solution is adjusted by the diluted hydrochloric acid solution to pH 5 and pH 3, respectively. The LBL deposition is achieved by immersing the PUF into PAANa solution followed by the chitosan solution and MXene solution until the required numbers of bilayers are formed. This method is helpful to conserve the porous structure of elastomer foam, which is damaged during the loading of nanofillers.

4.2.6 Spin-Coating Method

The spin-coating method is effectively used for the synthesis of MXene/elastomer nanocomposites. First, the MXene dispersion in ethanol is prepared using an ultrasonication method. The MXene dispersion in ethanol is then dropped into the elastomer solution and mixed thoroughly by ultrasonication for 30 minutes to obtain uniform suspension. Next, the curing agents are added into the suspension, where the curing agent added should be one tenth of the elastomer mass, and the sample is stirred to produce a uniform dispersion. The produced mixture is spin-coated onto a nylon textile coated with Cu–Ni alloys substrate and cured in an oven at 80°C for almost 2 hours. This method was successively used for the fabrication of MXene/elastomer nanocomposites used for sensing applications [13].

4.2.7 Solution Casting Method

The solution casting is one of the simplest methods, which is used for the fabrication of MXene/elastomer nanocomposites. First, a homogeneous mixture of elastomer and MXene is prepared by taking the required mass fraction of elastomer and MXene and then adding a 10-wt% curing agent into the solution. The obtained mixture is poured onto the glass substrate with a required thickness and heated at 80°C for 2 hours. Finally, the MXene/elastomer composite film is detached from the glass surface and can be used for further applications. One of the examples is the formation of MXene/polydimethylsiloxane (PDMS) nanocomposite film by the casting method [14]. Figure 4.4 shows the fabrication procedures of the MXene/elastomer nanocomposite film by the casting method.

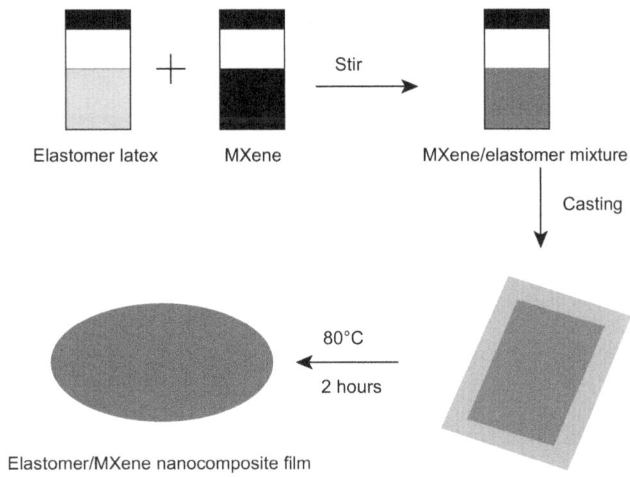

FIGURE 4.4 Schematic diagram for the fabrication of the elastomer/MXene nanocomposite film.

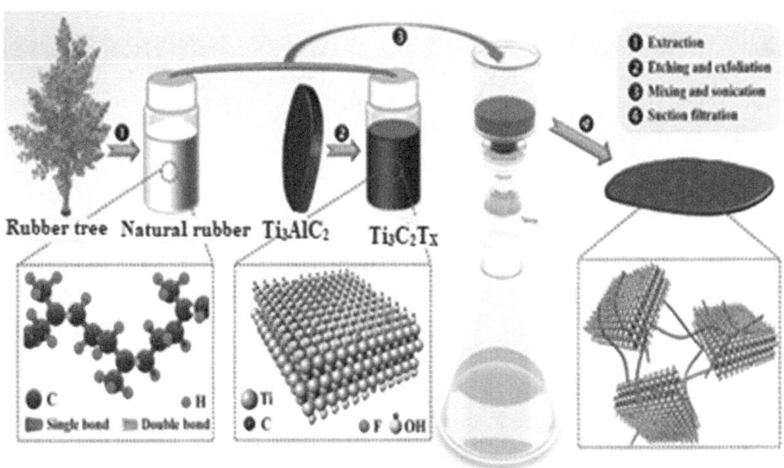

FIGURE 4.5 Schematic diagram of the preparation of MXene/NR nanocomposite [16]

Source: Reproduced with permission from Elsevier.

4.2.8 Vacuum-Assisted Filtration Method or Suction Filtration Method

MXene/elastomer nanocomposite films are also synthesized by a simple and cost-effective suction (vacuum-assisted) filtration method. In this method, the electrostatic repulsion forces between the similarly charged MXene and elastomer latex (both are negatively charged) ensure the distribution of the MXene sheets between the elastomer particles, which gives an interconnected 3D network [14,15]. First, a colloidal suspension of MXene of required concentration is prepared. Then, the MXene ($Ti_3C_2T_x$) suspension is mixed with elastomer latex under rapid stirring at room temperature, and then the mixture is sonicated for 10 minutes to obtain a uniform suspension. The prepared suspension is subjected to vacuum filtration to generate MXene/elastomer composite film. The films so obtained are then removed and allowed to air-dry at room temperature. The fabrication of the MXene/natural rubber nanocomposite [16] is illustrated in Figure 4.5.

4.2.9 Emulsion Method

The emulsion method is an important method used to prepare MXene/elastomer nanocomposite. For this, first, an MXene/ethanol dispersion is prepared by mixing the required quantity of MXene ($Ti_3C_2T_x$), ethanol and zirconia beads in a planetary milling container. The mixture is milled at 400r/min for 10 hours followed by sonicating at 25°C for 1 hour and centrifuged at 10,000r/min for 10 minutes. The obtained supernatant (MXene dispersion) is then added to the aqueous emulsion of elastomer (functionalized or nonfunctionalized), which can be prepared by stirring. The mixture is then sonicated to obtain uniform dispersion. Afterward, the MXene/elastomer emulsion is poured into a glass plate and dried for 2days at ambient temperature to obtain the composite film. One of the examples is the fabrication of MXene/polyurethane composite film using waterborne polyurethane and MXene dispersion in ethanol [17].

4.3 PROPERTIES OF MXene/ELASTOMER NANOCOMPOSITES

The crystals of MXenes possess overall a hexagonal closely packed structure, where M atoms are in closely packed arrays and X atoms reside in the octahedral interstitial sites [18–20]. To keep up structural stability the ordering of M atoms in MXenes could vary between M_2X, M_3X_2 and M_4X_3 [21]. The properties of MXene can be altered by varying its surface functionalities such as –O, –OH, –F as well as tuning the ratios of M or X elements. Due to the intriguing properties of MXene, it easily catches the attention of researchers for the improvement of various properties of polymers [22]. In this section, we discuss the properties of MXene/polymer composites, including electrical and thermal conductivity, dielectric properties, mechanical properties, thermal stability, flame retardancy and barrier properties

4.3.1 Conductivity

The electrical and thermal conductivity of pure elastomers are found to be lower. This drawback of the elastomers can be overwhelmed by fabricating their composite with conductive nanomaterials [23–25].

Conductive elastomer nanocomposites have attracted great attention in flexible sensors, electrostatic discharge environments and thermal interface management. Compared to other nanomaterials, MXenes have good electrical conductivity and have been increasingly used in flexible electronic devices (Table 4.1)[26]. Metallic conductivities of Ti_3C_2Tz MXenes are reported on the

TABLE 4.1
Electrical Conductivity of MXene and Other Materials

Nanomaterial	Electrical conductivity [S/m]
Carbon material	≈ 106 carbon nanotube (CNT)
	≈ 108 (graphene)
	1.6 ×105 (reduced graphene oxide [rGO] film)
Metal material	4.10 × 107 (Au bulk)
	6.3 × 107 (Ag bulk)
	5.96 × 107 (Cu bulk)
Conductive polymers	≈ 105
MXene family	2.4 × 106 (Ti3C2T*x* film)
	2.7 × 105 (Ti3CNT*x* film)
	1.6 × 105 (Ti2CT*x* film)
	9 × 104 (V2CT*x* film)

order of approximately 10,000 S/cm [27] and electrical conductivity in the range of approximately 2,000–6,500 S/cm [28]. The conductive MXene-based polymer nanocomposites are widely used as electromagnetic interference shielding materials [29], transparent conductors [28], energy storage and conversion applications [30–31], water purification [32] and sensors such as flexible strain sensors, pressure sensors, gas and selective molecule sensors [33–35].

Nitrile butadiene rubber (NBR) exhibits a thermal conductivity of 0.23 $W.m^{-1}K^{-1}$, while upon the dispersion of $Ti_3C_2T_x$MXene and allylamine-modified MXene (m-MXene) at 19.6 vol% in NBR, the thermal conductivity increases to 0.69 $W.m^{-1}K^{-1}$ and 1.01 $W.m^{-1}K^{-1}$, respectively, which is observed to be higher than the other reported elastomer nanocomposite system containing alumina, boron nitride, graphene and zinc oxide nanoparticle at the same or higher filler fraction. NBR/MXene and NBR/m-MXene nanocomposites, whose electrical conductivity increased upto 8.82×10^{-5} and 6.67×10^{-3} $S.cm^{-1}$, respectively, at 20.9 vol% filler loadings compared to the unfilled elastomer [7, 36]. SBR/Ti_3C_2Tx MXene nanocomposites exhibited a thermal conductivity of 0.477 $W.m^{-1}K^{-1}$, which is much higher than that of an SBR/reduced graphene oxide (3wt%) nanocomposite system (0. 265W. $m^{-1}K^{-1}$)[8]. Natural rubber latex (NRL)/MXene nanocomposite foam exhibits the electrical conductivity of 6.18 × 10−4 Sm^{-1}, which is two orders of magnitude higher than that of pure NRF [37].

4.3.2 Dielectric Properties

Elastomers with excellent dielectric properties are most suitable in the field of electromechanical transducers. Hyper-branched polysiloxane (HPSi)-modified $Ti_3C_2T_X$ sheets (HPSi-*d*-$Ti_3C_2T_X$) incorporated poly (dimethyl siloxane) (PDMS) composite system holds excellent dielectric properties. PDMS containing 1.40 vol.% HPSi-*d*-Ti_3C_2TX system exhibit a dielectric constant of 23.7 at 10^2 Hz, which is 8.5 times higher than that of neat PDMS [38]. Jena et al. fabricated $Ti_3C_2T_x$ MXene incorporated ethylene–vinyl acetate (EVA) nanocomposites and they found that an 8 wt% MXene-filler-incorporated EVA showed greater dielectric properties with low dielectric loss compared to all other weight percentages owing to their high electrical conductivity, good interaction between filler and matrix and well dispersion of $Ti_3C_2T_x$ in EVA matrix [39].

4.3.3 Mechanical Properties

Ti_3C_2 MXene possesses a higher Young's modulus among solution-processable 2D materials and has been observed to be 333 ± 30 GPa by the atomic force microscopy indentation technique [40], which is much higher than those of graphene with similar thickness [41]. These demonstrated that MXene materials are good reinforcements for polymeric composites. Theoretical molecular dynamic simulation study reports 502 GPa Young's modulus for MXene, but the experimental value is found to be lower owing to the surface functionalization and defect [42]. The good mechanical property of MXene facilitates their application in portable and wearable microelectronic systems like supercapacitors. The incorporation ofTi_3C_2Tx MXene (at 14.0 vol%) in elastomer, nitrile butadiene rubber, leads to 570% increase in Young's modulus. While both elongation at break and tensile strength record a 640% and 180% improvement at 3.9vol% and 7.5 vol% MXene, respectively, and then decreased [35]. NBR/MXene nanocomposites fabricated via latex compounding exhibited remarkable mechanical properties. At 14.0 vol% MXene-incorporated NBR exhibited a tensile strength of 25.94 ± 0.81 MPa, elongation at break of 170 ± 5.6% and Young's modulus of 15.85 ± 0.75 MPa [2]. MXene provides significant reinforcement to the natural rubber (NR) matrix. NR/Ti_3C_2 nanocomposite fabricated by vacuum-assisted filtration exhibited modulus and tensile strength 150 times and 7 times higher, respectively, than those of pure NR, at 6.71 vol%Ti_3C_2. Ti_3C_2-reinforcedNR by forming an interconnected network due to the electrostatic repulsion between NR and negatively charged Ti_3C_2 [15]. Compared with neat natural rubber latex foam (NRF), 2 and 3 phr of MXene containing NRF showed a 171% and 157% increase in tensile strength, respectively [37]. Polyurethane (PU)/MXene nanocomposites fabricated via an emulsion method showed a 70%,

20% and 13% improvement in yield stress, tensile strength and hardness, respectively, than pure PU at 0.5 wt% MXene loading [17]. It is interesting to observe that incorporation of MXene, increases rigidity and toughness of the polymer simultaneously. Sheng et al. prepared thermoplastic polyurethane (TPU)/MXene nanocomposites by a melt blending method and examined their properties. Their findings demonstrated that at 0.5 wt% MXene incorporation, the tensile strength and elongation at the break of TPU were increased by 47.1% (from 14 MPa of TPU to 20 MPa) and 17.5% (from 1580% of TPU to 1857%), respectively [11]. 3wt% MXene and melamine-cyanurate (Ti_3C_2Tx@ MCA) nanohybrid-integrated TPU nanocomposites exhibit 61.5 MPa of tensile strength, 175.4 ± 7.9 MJ m^{-3} of toughness and 588% of strain at failure [43]. Li et al. reported that in the presence of 2 phr Ti_3C_2, SBR nanocomposites showed higher tensile strength, which is higher than that observed in SBR nanocomposites with 2 phrgraphene [8]. Ti_3C_2MXenes/PDMS composite system modified by biomolecule exhibited a higher tensile strength at 10 wt% of filler loading, while elongation at the break decreased with the increase of filler loading from 2 wt% to 10 wt% [44].

4.3.4 Thermal Stability

Generally, the thermal stability and the dimensional stability of the nanocomposite are closely related to the thermal conductivity, thermal expansion, degradation and barrier properties of added nanofillers. MXenes possess good thermal conductivity compared to those of most semiconducting low-dimensional materials and metals. For example, 2.5 wt% of Ti_3C_2Tz MXene–incorporated PDMS nanocomposite showed a threefold improvement in thermal conductivity compared with pure PDMS [46]. SBR/Ti_3C_2nanocomposites fabricated by freeze-drying, and the mechanical mixing process showed higher thermal stability than neat SBR. In the presence of 4 phr Ti_3C_2, SBR

TABLE 4.2
Mechanical Properties of MXene/Elastomer Composites

MXene/Elastomer	Method	wt (%)	Thickness (μm)	Tensile Strength (MPa)	Strain (%)	Young's Modulus (GPa)	Ref
Ti_3C_2/NR	Vacuum-assisted filtration	0.58 ^{v}	–	3.11	935	0.4^e	15
		1.18^v	244	5.69	843	1.1^e	
		1.80^v	247	7.01	801	1.31^e	
		2.44^v	243	10.02	789	1.98^e	
		3.10^v	248	11.86	771	2.3^e	
		4.83^v	245	13.32	765	64.2^e	
		6.71^v	251	18.25	761	87.8^e	
Ti_3C_2/TPU	Hot press	0.125	250~300	17.3 ±1	1780 ±130		11
		0.25	250~300	18.0 ±1.2	1700 ±120		
		0.5	250~300	20.6 ±1.2	1850 ±100		
		1.0	250~300	17.3 ±0.8	1820 ±120		
A-MXenes/D-PDMS	Solution blending	2	1000	0.64 ±0.12	117 ±12	0.00130 ±0.0002	44
		6	1000	0.79 ±0.14	113 ±12	0.00151 ±0.0002	
		10	1000	1.81 ±0.22	81 ±6	0.004 ±0.0007	
CTAB-Ti_3C_2/TPU	Hot press	0.5	1000 ±300	50.8 ±1.1	599 ±45		45
		2.0	1000 ±300	50.1±0.8	614 ±18		
TBPC-Ti_3C_2/TPU	Hot press	0.5	1000 ±300	55.9 ±4.6	605 ±22		45
		2.0	1000 ±300	54.3 ±2.3	632 ±9		

Note: e and v, respectively, refer to elastic modulus (MPa) and vol% of MXene/elastomer nanocomposites.

nanocomposites showed initial decomposition at 150°C and maximum decomposition temperature of 160°C [8]. Thermoplastic polyurethane (TPU)/MXene nanocomposites fabricated via melt blending method exhibited an increased maximum degradation temperature of 20.3°C and the onset degradation temperature of 13.1°C at 0.5 wt% MXene loading [11].

4.3.5 Flame Retardancy

TPUs are extensively applied as lightweight materials in electronic device casings, footwear soles and shielding for wirings. Nevertheless, it is a highly flammable material and emits a lot of toxic gas (i.e., CO and HCN) and heavy smoke during combustion, which remains a major concern in industrial application. Thus, it is imperative to enhance the fire safety of TPU material. The introduction of flame retardants to TPU material is an effective way to improve its fire safety performance. MXene is a very promising flame retardant, and it reduces smoke and toxicity hazards in TPU significantly.

Liu et al. prepared a titanium carbide–reduced graphene oxide ($Ti_3C_2T_x$-rGO) hybrid and found that its addition in a TPU elastomer improve the thermal and fire-safe performances of TPU nanocomposite significantly. Incorporating 2.0wt% $Ti_3C_2T_x$-rGO led to a significant decrease in total smoke release and the smoke production rate of TPU nanocomposite by 54.0% and 81.2%, respectively. Moreover, this nanocomposite exhibited 46.2% and 54.1% distinct reductions in the total carbon monoxide yield and carbon monoxide production rate, respectively [9]. MXene and melamine cyanurate ($Ti_3C_2T_x$@MCA) nanohybrid integrated TPU nanocomposite exhibit 40% reduction in the peak-of-heat-release rate (PHRR) [43].

CTAB-modified Ti_3C_2 (MXene)–incorporated TPU exhibited reduction in PHRR by 51.2% and the production of smoke, CO_2 and CO by 57.1%, 49.7% and 39.4%, respectively [15]. TBPC-modified Ti_3C_2 (MXene)–incorporated TPU exhibited a reduction in PHRR by 52.2% and production of smoke, CO_2 and CO by 57.4%, 51.7% and 41.6%, respectively [45].

Ti_3C_2/chitosan-hybrid-coated PUF reduced the flammability and smoke release of PUF significantly. With respect to pure PUF, eight different number of bilayers (BL) coating decreased the PHRR by 57.2%, reduced the total heat release by 65.5%, reduced the smoke release by 71.1% and reduced the peak of CO_2 and CO production by 68.6% and 70.8%, respectively [12].

4.3.6 Barrier Properties

MXenes are impermeable particles, and their addition into polymers increases the gas barrier properties of polymers and resists the transport of gas molecules. Generally, impermeable nanoparticles in a polymer may limit the diffusion of larger gas molecules across the membrane due to the increase in the tortuous path of the gas molecule in the polymer matrix. A schematic representation of the tortuous path is presented in Figure 4.6. Shape, dispersion state, orientation and volume

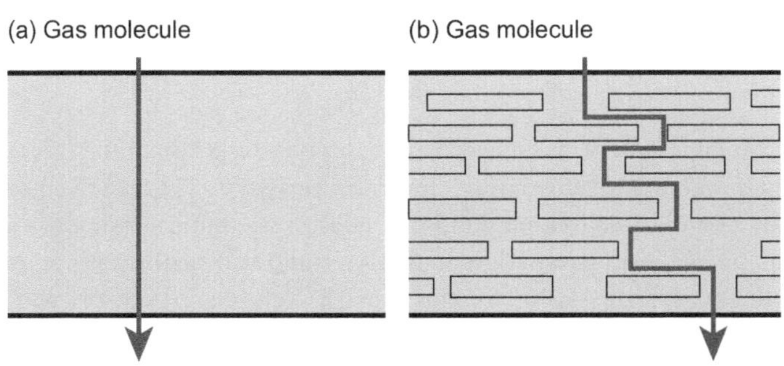

FIGURE 4.6 (a) Diffusion of gas molecule in pure polymeric film and (b) diffusion of gas molecule in polymer nanocomposites through the tortuous path.

fraction of nanofillers are the factors that determine the tortuosity of the path. Tortuosity ($\tau = L_1/L_2$) is the ratio of the diffusion path length of the permeating molecule in the nanocomposite membrane (L_1) to the diffusion path length of the permeating molecule in the pristine polymer (L_2) [47].

A nanoparticle-embedded polymer matrix either increases or decreases the gas barrier properties of a polymer, which depend on the nature of a nanofiller and the polymer–filler interaction. After introducing nanoparticles, polymers often exhibit high barrier properties by reducing free volume, constructing a tortuous path for permeating molecules and minimizing the permeation cross section in polymers. Here, nanofillers act as barrier, reducing the transport of gas molecules. In the reverse case, introducing particles such as silica and metal oxide decreases the gas barrier property of polymers by disrupting the polymer chain packing and creates voids at the polymer–particle interface or free volume pockets [48]. In 2020, Shamsabadi and coworkers carried out a gas permeability study onTi_3C_2Tz/Pebax/PU membrane using H_2, CO_2, N_2, and CH_4 gases, and they observed increased CO_2/N_2 selectivity due to the good adsorption of $CO_2$by the functional group (especially the hydroxyl group) on the $Ti_3C_2T_z$ surface [49].

4.4 APPLICATIONS

4.4.1 Sensors

Elastomer/MXene nanocomposites exhibit excellent stretchability and high conductivity due to synergistic effects. Hence, elastomer/MXene nanocomposites have attracted increasing interests in sensor fabrication. Because of the properties such as high conductivity, versatile surface chemistry and high aspect-ratio MXene nanomaterials have been used for sensor fabrication. These elastomer/MXene nanocomposite–based sensors have found application in electronic skin, electronic display equipment, human health and human body movements, wearable electronics and new generation robots [50–56]. High sensitivity, low fabrication prices, good stability, low detection limit, quick recovery for repetitive use, energy-efficient and a quick and preferably linear response are the important properties simultaneously required for a sensor for its practical application.

Flexible wearable sensors can detect an extensive range of human activities such as finger bending [57], arterial pulses [58], arm and leg movement [59] and vocal-muscle vibrations [60].

MXenes are highly promising candidates for the application of pressure and strain sensors. Pressure sensors are widely applied in the field of sports [61], electronic skin [62], healthcare monitoring [63] and speech recognition [64]. Most of the reported MXene/elastomer-based pressure sensors are observed to be piezoresistive type. Piezoresistive pressure sensors are always in high demand for wearable smart devices and human motion detection.

In a piezoresistive strain sensors, an applied pressure causes a change in resistivity. In the change in resistivity $\Delta R/R_0$, where ΔR indicates the difference between resistivity of the sensor's material before and after applying the strain and R is the resistivity of the sensor fabricated material at rest. The gauge factor (GF) is used to analyze sensitivity of strain sensors:

$$\frac{GF = \Delta R / R_0}{\Delta R / l_0},$$

Where L_0 is the original length of the sensor before applying a strain and ΔL is the absolute change in length. A high GF value indicates larger change in resistivity. Ti_3C_2 MXenes/PDMS composite system modified by biomolecule exhibit good potential in wearable sensing devices. It shows a GF of 3.6 and exhibits good tensile strength, conductivity and self-healing properties [44]. There are reports showing GF in the range 43–107 for elastomer/MXene-based sensors within the strain range of 0–10%. Guo and coworkers reported a good electronic sensor for intelligent sensing fabricated by assembling MXenenano-flakes modified with serine and a rubber-based supramolecular elastomer (NMSE). This NMSE-based sensor could precisely monitor minute human physiological motion and moisture variation [64], it is shown in Figure 4.7. Piezo resistive pressure sensors fabricated

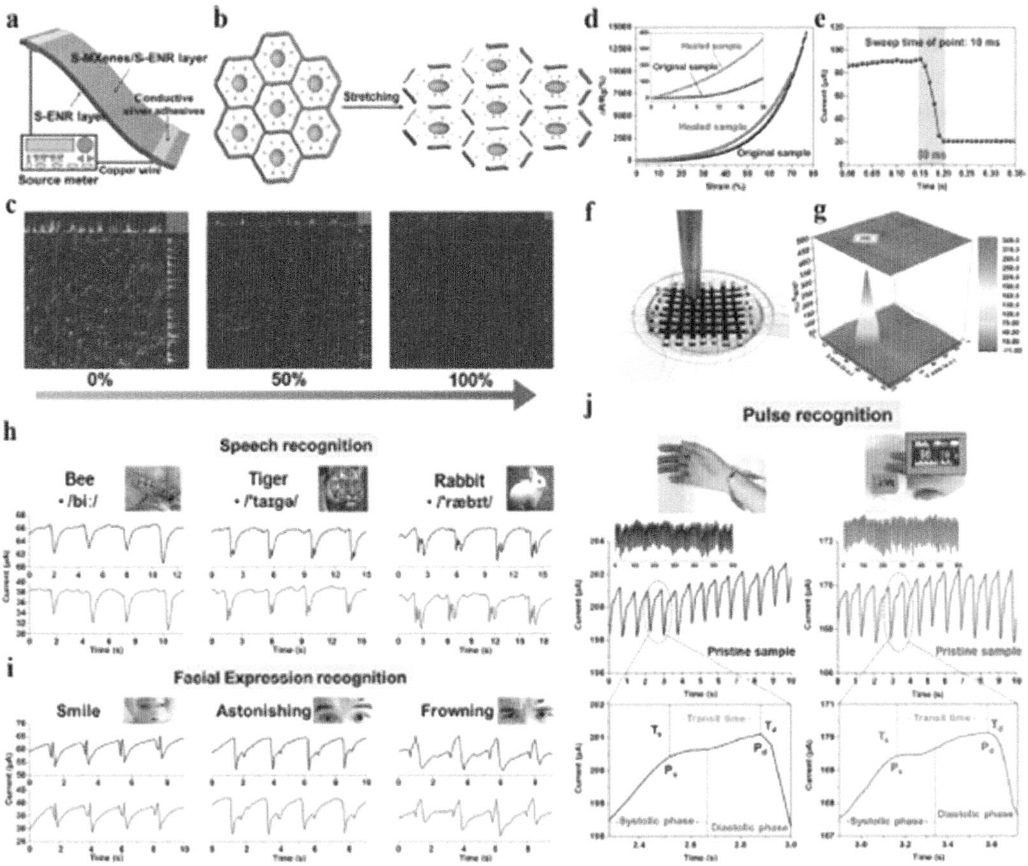

FIGURE 4.7 Sensing properties of modified MXene/epoxidized NMSE–based electronic sensor [64].

Source: Reproduced with permission from ACS.

from hollow-structured MXene–PDMS composite system exhibit superior properties, including a wide working range (0–180°), consistent performance with excellent stability (over 1000 cycles) and low detection limit (10 mg). Based on its excellent performance, hollow-structured MXene–PDMS composite sensors can detect facial muscle movement, swallowing, stereo sound and ultrasonic vibration monitoring, representing its potential application in wearable smart devices [65].

Li et al. prepared a flexible, sensitive, reliable and piezoresistive pressure sensors by coating PU/chitosan (CS) sponge with Ti_3C_2Tx (MXene) [66]. Recently, Zhang et al. developed s-MXene/elastomer (elastomer [Ecoflex series]) bilayer device having papillae-like arrays, here s-MXene is the 3D micro-structure composed of single-walled carbon nanotube (SWCNT) and Ti_3C_2Tx titanium carbide (MXene). The bilayer device was applied in flexible, high-precision and high-sensitivity (11.47 kPa–1) piezoresistive pressure sensors, and it was further used to detect microdroplet manipulation, as well as voice recognition [67].

It is challenging to fabricate a multifunctional, flexible, highly sensitive strain sensors with attractive properties, such as a wide sensing range and low detection limits for practical applications in physical care systems, smart robotics and wearable electronics. Generally, sensors would exhibit a trade-off effect between high sensitivity and wide sensing range during their application. Recently, MXene/polyurethane mat–based strain sensor were developed and exhibited superior performance for sensing application by achieving high sensitivity coupled with wide sensing range. MXene/polyurethane mats possess superior properties, including high sensitivity (GF upto 228), outstanding performance (more than 3200 cycles), sensing range upto 150%, low limit of detection (0.1%) and multifunctional detection. MXene/polyurethane-based sensors can have great potential to monitor subtle physiological signals (such as respiration and pulse waves), and hence, it could be applied for wearable health detectors and artificial electronic skin [68] (Figure 4.8).

TABLE 4.3
Elastomer/MXene-Based Sensors with Their Applications

Elastomer/MXenesystem	Application	Ref
Modified Ti_3C_2 MXenes/PDMS	Elastomeric wearable strain Sensors	44
Modified MXene/epoxidized natural rubber (NMSE)	Self-healable intelligent sensors and monitor minute human physiological motion and moisture variation	64
Hollow-structured MXene/PDMS	Pressure sensors and apply in wearable smart devices	65
Ti_3C_2TxMXene/polyurethane (PU)/chitosan (CS) sponge	Piezoresistive pressure sensors	66
s-MXene (SWCNT and Ti_3C_2Tx)/elastomer (Ecoflex series)	Pressure sensors and detect microdroplet manipulation as well as voice recognition	67
MXene/polyurethane mat	Strain sensors and apply in wearable health detectors and artificial electronic skin	68
Ti_3C_2 and polyurethane	Stretchable strain-sensing fabric	69
Ti_3C_2 and poly(dimethylsiloxane)	Skin conformal sensors for health Monitoring	70

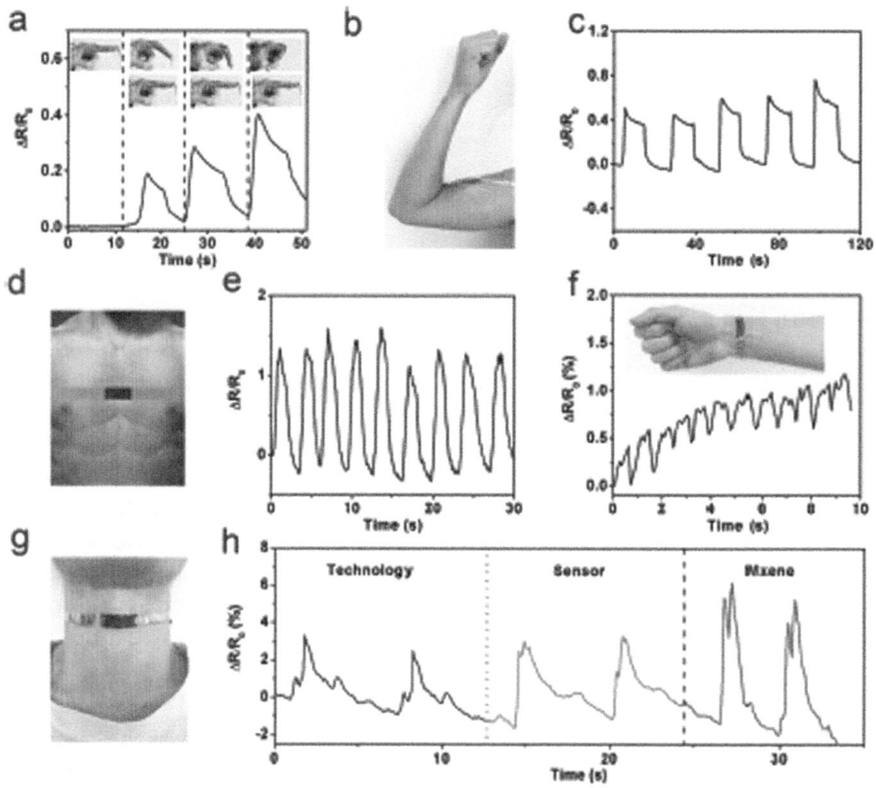

FIGURE 4.8 Physiological signal monitoring and physical motion detection of a network-M/P mat sensor [68].

Source: Reproduced with permission from RSC.

4.4.2 Electromagnetic Interference Shielding Materials

Globally, electromagnetic wave pollution has become one of the great challenges for the environment. Electromagnetic interference (EMI) shielding materials minimize or prevent the release of EMI signals considerably from electromagnetic devices into the environment. High-performance electromagnetic interference shielding materials are essential in flexible, foldable and wearable electronics. MXene sheets provide excellent EMI shielding performances to the polymers as compared with carbon nanomaterials incorporated into polymer nanocomposites. The challenging issue for the EMI shielding materials is to combine high EMI shielding performance together with excellent stretchability and mechanical robustness. Two-dimensional layered material have been proved to have exceptional EMI shielding performance along with excellent mechanical, electrical and thermal properties. The incorporation of MXene into elastomer enhances shielding performance as well as stretchability and mechanical robustness. NBR/MXene nanocomposites fabricated via latex compounding exhibited EMI shielding effectiveness of 49 dB at 19.6 vol%. Mechanical properties also found to increase remarkably by the addition of MXene to elastomers. In the presence of 14.0 vol% of MXene, NBR exhibited a tensile strength of 25.94 ± 0.81 MPa, elongation at break of 170 ± 5.6% and Young's modulus of 15.85 ± 0.75 MPa [2]. At 6.71 vol%, a $Ti_3C_2T_x$ MXene-doped NR nanocomposite system shows high flexibility, excellent EMI shielding performance of 53.6 dB and electrical conductivity of 1400 Sm^{-1}[15] as shown in Figure 4.9.

4.4.3 Other Applications

T.H. Chang and coworkers reported highly stretchable MXene/elastomer bilayer composite-based electrochemical electrode for bendable, stretchable supercapacitors. Recently, MXene-based composites have demonstrated a high potential for application in wearable and transparent devices. A 2D MXene nanosheet is found to be potential photothermal filler for various light-responsive applications. It possesses high thermal conductivity and 100% efficiency for light-to-heat conversion [72]. In 2019, Fan et al. developed a wearable, healable and transparent composite by integrating AgNP@MXene hybrid in PU. The AgNP@MXene–PU composite exhibits effective photothermal conversion and quick healing capabilities, and they applied this composite in a transparent, skin-mountable, and sun-powered warn coating to reveal its potential application for wearable and transparent devices [73].

4.5 CONCLUSION

Scientists are looking for promising materials to develop flexible electronic goods. Recently, Mxenes are found to be a good additive for polymers because of its distinctive properties. It is a 2D material with good electrical and thermal conductivity. Due to its large surface area, it is considered to be a good adsorbent and can be used for the removal of heavy metals. MXene/elastomer nanocomposites exhibited excellent electrical and thermal conductivity than carbon nanotube or graphene-filled systems. Therefore, these materials are considered to be favorable materials for EMI shielding and various sensor applications.

TABLE 4.4
EMI Shielding Effectiveness (SE) of Elastomer/MXene Nanocomposites

Elastomer/MXene System	Filler Fraction	Fabrication Method	EMI SE (dB)	Ref
1. NBR/MXene	19.6	Latex compounding	49	2
2. NR/MXene	6.71	Vacuum-assisted filtration	53.6	15
3. PDMS/MXene	1.1	Dip coating	43	71

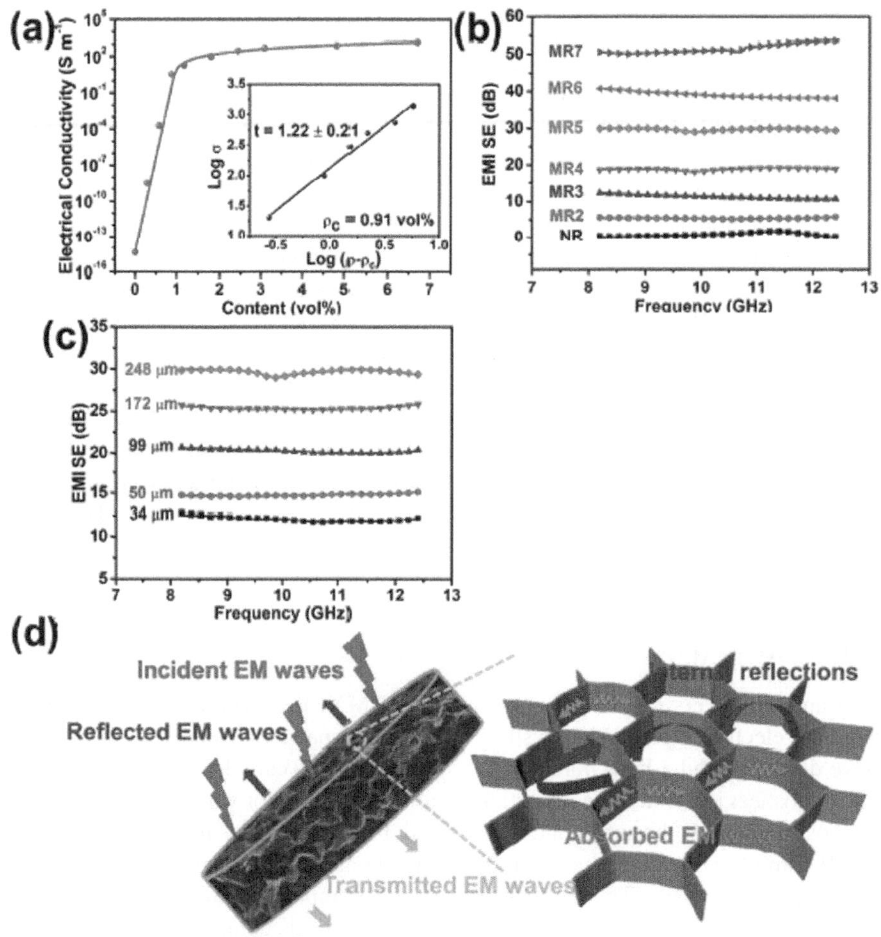

FIGURE 4.9 (a) Electrical conductivities and (b) Electromagnetic interference shielding effectiveness of different volume fraction of MXene loaded into MXene/rubber nanocomposites [15].

Source: Reproduced with permission from Elsevier.

REFERENCES

[1] Naguib M, Kurtoglu M, Presser V, Lu J, Niu J, Heon M, Hultman L, Gogotsi Y, Barsoum M W. Two-dimensional nanocrystals produced by exfoliation of Ti_3AlC_2. Adv Mater 2011:23;4248–4253.

[2] Aakyiir M, Kingu M A S, Araby S, Meng Q, Shao J, Amer Y, Ma J. Stretchable, mechanically resilient, and high electromagnetic shielding polymer/MXene nanocomposites. J Appl Polym Sci 2021:138.

[3] Mozafari M, Shamsabadi A A, Rahimpour A, Soroush M. Ion-selective MXene-based membranes: current status and prospects. Adv Mater Technol 2021:2001189.

[4] Ning H, Ma Z, Zhang Z, Zhang D, Wang Y. A novel multifunctional flame retardant MXene/nanosilica hybrid for poly(vinyl alcohol) with simultaneously improved mechanical properties. New J Chem 2021:45.

[5] Tan K H, Samylingam L, Aslfattahi N, Saidur R, Kadirgama K. Optical and conductivity studies of polyvinyl alcohol-MXene (PVA-MXene) nanocomposite thin films for electronic applications. Opt Laser Technol 2021:136;106772.

[6] Xu M, Liu, J, Zhang H, Zhang Y, Wu X, Deng Z, Yu Z. Electrically conductive $Ti_3C_2T_x$MXene/polypropylene nanocomposites with an ultralow percolation threshold for efficient electromagnetic interference shielding. Ind Eng Chem Res 2021:60;4342–4350.

[7] Aakyiir M, Yu H, Araby S, Ruoyu W, Michelmore A, Meng Q, Losic D, Choudhury N R, Ma J. Electrically and thermally conductive elastomer by using mxenenanosheets with interface modification. Chem Eng J 2020:397;125439.

[8] Li Q, Zhong B, Zhang W, Jia Z, Jia D, Qin S, Wang J, Razal J M, Wang X. Ti_3C_2MXene as a new nanofiller for robust and conductive elastomer composites. Nanoscale 2019:11;14712–14719.

[9] Liu C, Wu W, Shi Y, Yang F, Liu M, Chen Z, Yu B, Feng Y. Creating MXene/reduced graphene oxide hybrid towards highly fire safe thermoplastic polyurethane nanocomposites. Compos B Eng 2020:203;108486.

[10] Gao Q, Feng M, Li E, Liu C, Shen C, Liu X. Mechanical, thermal, and rheological properties of $Ti_3C_2T_x$MXene/thermoplastic polyurethane nanocomposites. Macromol Mater Eng 2020:305;2000343.

[11] Sheng X, Zhao Y, Zhang L, Lu X. Properties of two-dimensional Ti_3C_2MXene/thermoplastic polyurethane nanocomposites with effective reinforcement via melt blending. Compos Sci Technol 2019:181;107710.

[12] Lin B, Yuen A C Y, Li A, Zhang Y, Chen T B Y, Yu B, Lee E W M, Peng S, Yang W, Lu H, Chan Q N, Yeoh G H, Wang C H. MXene/chitosan nanocoating for flexible polyurethane foam towards remarkable fire hazards reductions. J Hazard Mater 2020:381;120952.

[13] He W, Sohn M, Ma R, Kang D J. Flexible single-electrode triboelectric nanogenerators with MXene/PDMS composite film for biomechanical motion sensors. Nano Energy 2020:78;105383.

[14] Cai Y, Zhang X, Wang G, Li G, Zhao D, Sun N, Li F, Zhang H, Han J, Yang Y. A flexible ultra-sensitive triboelectric tactile sensor of wrinkled PDMS/MXene composite films for E-skin. Nano Energy 2021:81;105663.

[15] Luo J, Zhao S, Zhang H B, Deng Z, Li L, Yu Z. Flexible, stretchable and electrically conductive MXene/natural rubber nanocomposite films for efficient electromagnetic interference shielding. Compos Sci Technol 2019:182;107754.

[16] Yang W, Liu J, Wang L, Wang W, Yuen A C Y, Peng S, Yu B, Lu H, Yeoh G H, Wang C. Multifunctional MXene/natural rubber composite films with exceptional flexibility and durability. Compos B Eng 2020:188;107875.

[17] Zhi W, Xiang S, Bian R, Lin R, Wu K, Wang T, Cai D. Study of MXene-filled polyurethane nanocomposites prepared via an emulsion method. Compos Sci Technol 2018:168;404–411.

[18] Wang X, Shen X, Gao Y, Wang Z, Yu R, Chen L. Atomic-scale recognition of surface structure and intercalation mechanism of Ti3C2X. J Am Chem Soc 2015:137;2715–2721.

[19] Karlsson L H, Birch J, Halim J, Barsoum M W, Persson P O Å. Atomically resolved structural and chemical investigation of single MXene sheets. Nano Lett 2015:15;4955–4960.

[20] Hong Ng VM, Huang H, Zhou K, et al. Recent progress in layered transition metal carbides and/or nitrides (MXenes) and their composites: synthesis and applications. J Mater Chem 2017:5(7);3039–3068.

[21] Halim J, Kota S, Lukatskaya M R, Naguib M, Zhao M-Q, Moon E J, Pitock J, Nanda J, May S J, Gogotsi Y, Barsoum M W. Synthesis and characterization of 2DMolybdenum Carbide (MXene). Adv Func Mater 2016:26;3118–3127.

[22] Gong K, Zhou K, Qian X, Shi C, Yu B. MXene as emerging nanofillers for high-performance polymer composites: A review. Composites Part B 2021:217;108867.

[23] Qiu A, Li P, Yang Z, Yao Y, Lee I, Ma J. A path beyond metal and silicon: polymer/nanomaterial composites for stretchable strain sensors. Adv Funct Mater 2019:29;1806306.

[24] Deng Z, Wang H, Ma P X, Guo B. Self-healing conductive hydrogels: Preparation, properties and applications. Nanoscale 2020:12;1224.

[25] Araby S, Meng Q, Zhang L, Kang H, Majewski P, Tang Y, Ma J. Electrically and thermally conductive elastomer/graphene nanocomposites by solution mixing. Polymer 2014:55;201–210.

[26] Ma C, Ma M G, Si C, Ji X-X, Wan P. Flexible MXene-based composites for wearable devices. Adv Funct Mater 2021:2009524.

[27] Shahzad F, Alhabeb M, Hatter C B, Anasori B, Hong S M, Koo C M, Gogotsi Y. Electromagnetic interference shielding with 2D transition metal carbides (MXenes). Science 2016:353;1137.

[28] Anasori B, Lukatskaya M R, Gogotsi Y. 2D metal carbides and nitrides (MXenes) for energy storage. Nat Rev Mater 2017:2;16098.

[29] Hantanasirisakul K, Zhao M, Urbankowski P, Halim J, Anasori B, Kota S, Ren C E, Barsoum M W, Gogotsi Y. Fabrication of $Ti_3C_2T_x$ mxene transparent thin films with tunable optoelectronic properties. Adv Electron Mater 2016:2;1600050.

[30] Mackanic D G, Chang T H, Huang Z, Cui Y, Bao Z. Stretchable electrochemical energy storage devices. Chem Soc Rev 2020:49;4466–4495.

[31] Pang J, Mendes R G, Bachmatiuk A, Zhao L, Ta H Q, Gemming T, Liu H, Liu Z, Rummeli M H. Applications of 2D MXenes in energy conversion and storage systems. Chem Soc Rev 2019:487;2.

[32] Ihsanullah I. MXenes (two-dimensional metal carbides) as emerging nanomaterials for water purification: Progress, challenges and prospects. Chem Engi J 2020:388:124340.

[33] Ma Y, Liu N, Li L, Hu X, Zou Z, Wang J, Luo S, Gao Y. A highly flexible and sensitive piezoresistive sensor based on MXene with greatly changed interlayer distances. Nat Commun 2017:8:1207.

[34] Cai Y, Shen J, Ge G, Y Zhang, Jin W, Huang W, Shao J, Yang J, Dong X. Stretchable $Ti_3C_2T_x$ MXene/carbon nanotube composite based strain sensor with ultrahigh sensitivity and tunable sensing range. ACS Nano 2018:12;56–62.

[35] Zhang Y, Lee K H, Anjum D H, Sougrat R, Jiang Q, Kim H, Alshareef H N. MXenes stretch hydrogel sensor performance to new limits. Sci Adv 2018:4;eaat009.

[36] Aakyiir M, Araby S, Michelmore A, Meng Q, Amer Y, Yao Y, Li M, Wu X, Zhang L, Ma J. Elastomer nanocomposites containing MXene for mechanical robustness and electrical and thermal conductivity. Nanotechnology 2020:31;315715.

[37] Yang Y D, Liu G X, Wei Y C, Liao S, Luo M C. Natural rubber latex/MXene foam with robust and multifunctional properties. Polymers 2021:21: 179–185.

[38] Wei L, Wang J W, Gao X H, Wang H Q, Wang X Z, Ren H. Enhanced dielectric properties of poly(dimethyl siloxane) bimodal network percolative composite with MXene. ACS Appl Mater Interfaces 2020:12:16805–16814.

[39] Jena D P, Anwar S, Parida R K, Parida B N, Nayak NC. Structural, thermal and dielectric behavior of two-dimensional layered $Ti_3C_2T_x$(MXene) filled ethylene–vinyl acetate (EVA) nanocomposites. J Mater Sci Materials in Electronics 2021:32;8081–8091.

[40] Lipatov A, Lu H, Alhabeb M, Anasori B, Gruverman A, Gogotsi Y, Sinitskii A. Elastic properties of 2D $Ti_3C_2T_x$ MXene monolayers and bilayers. Sci Adv 2018:4;eaat0491.

[41] Gao L, Li C, Huang W, Mei S, Lin H, Ou Q, Zhang Y, Guo J, Zhang F, Xu S, Zhang H. MXene/polymer membranes: Synthesis, properties and emerging applications. Chem Mater 2020:32;1703–1747.

[42] Carey M, Barsoum M W. MXene polymer nanocomposites: A review, Mater. Today Advan 2021:9;100120.

[43] Shia Y, Liu C, Duan Z, Yu B, Liu M, Song P. Interface engineering of MXene towards super-tough and strong polymer nanocomposites with high ductility and excellent fire safety. Chem Eng J 2020:399;125829.

[44] Zhang K, Sun J, Song J, Gao C, Wang Z, Song C, Wu Y, Liu Y. Self-healing Ti_3C_2MXenes/PDMS supramolecular elastomer based on small biomolecules modification for wearable sensor, ACS Appli Mater Interf 2020:12;45306–45314.

[45] Yu B, Tawiah B, Wang L, Yuen A C Y, Zhang Z, Shen L, Lin B, Fei B, Yang W, Li A, Zhu S, Hu E, Lu H, Yeoh G H, Interface decoration of exfoliated MXene ultra-thin nanosheets for fire and smoke suppressions of thermoplastic polyurethane elastomer. J Hazard Mater 2019:374;110–119.

[46] Wang D, Lin Y, Hu D, Jiang P, Huang X. Multifunctional 3D-MXene/PDMS nanocomposites for electrical, thermal and triboelectric applications. Compos Part A Appl Sci Manuf 2020:130;105754.

[47] Swapna V P, Abhisha V S, Stephen R. Polymer/POSS nanocomposite membranes for pervaporation. Elsevier 2020:201–229.eBookISBN: 9780128167854

[48] Abhisha V S, Swapna V P, Stephen R. Modern trends and applications of gas transport through various polymers. Elsevier 2017.eBook ISBN:9780128098851

[49] Shamsabadi A A, Isfahani A P, Salestan S K, Rahimpour A, Ghalei B, Sivaniah E, Soroush M. Pushing rubbery polymer membranes to be economic for CO_2 separation: embedment with $Ti_3C_2T_x$ MXene nanosheets. ACS Appl Mater Interfaces 2020:12;3984–3992.

[50] Yamamoto Y, Harada S, Yamamoto D, Honda W, Arie T, Akita S, Takei K. Printed multifunctional flexible device with an integrated motion sensor for health care monitoring. Sci Adv 2016:2;e1601473.

[51] Markvicka E J, Bartlett M D, Huang X, Majidi C, An autonomously electrically self-healing liquid metal-elastomer composite for robust soft-matter robotics and electronics. Nat Mater 2018:17;618–624.

[52] Lim H, Kim H S, Qazi R, Kwon Y, Jeong J, Yeo W. Advanced soft materials, sensor integrations, and applications of wearable flexible hybrid electronics in healthcare, energy, and environment. Adv Mater 2019:32;1901924.

[53] Kweon O Y, Lee S J, Oh J H. Wearable high-performance pressure sensors based on three-dimensional electrospun conductive nanofibers. NPG Asia Mater 2018:10;540–551.

[54] Wu X, Han Y, Zhang X, Zhou Z, Lu C. Large-area compliant, low-cost, and versatile pressure-sensing platform based on microcrack-designed carbon black@polyurethane sponge for human-machine interfacing. Adv Funct Mater 2016:26;6246–6256.

[55] He S, Sun X, Zhang H, Yuan C, Wei Y, Li J. Preparation strategies and applications of MXene-polymer composites: A review. Macromolecular Rapid Communications 2021:42;2100324.

[56] Ho D H, Choi Y Y, Jo S B, Myoung J M, Cho J H. Sensing with MXenes: Progress and prospects. Advan Mater 2021:33;2005846.

[57] Shi X, Wang H, Xie X, Xue Q, Zhang J, Kang S, Wang C, Liang J, Chen Y. Bioinspired ultrasensitive and stretchable MXene-based strain sensor via nacre-mimetic microscale "Brick-and-Mortar" architecture. ACS Nano 2019:13;649–659.

[58] Zhuo H, Hu Y, Chen Z, Peng X, Liu L, Luo Q, Yi J, Liu C, Zhong L.A carbon aerogel with super mechanical and sensing performances for wearable piezoresistive sensors. J Mater Chem A 2019:7;8092–8100.

[59] Yang Y, Shi L, Cao Z, Wang R, Sun J. Strain sensors with a high sensitivity and a wide sensing range based on a $Ti_3C_2T_x$ (MXene) nanoparticle–nanosheet hybrid network. Adv Funct Mater 2019:29;1807882.

[60] Li T, Chen L, Yang X, Chen X, Zhang Z, Zhao T, Li X, Zhang J. A flexible pressure sensor based on an MXene-textile network structure. J Mater Chem C 2019:7;1022–1027.

[61] Tao L, Zhang K, Tian H, Liu Y, Wang D, Chen Y, Yang Y, Ren T. Graphene-paper pressure sensor for detecting human motions. ACS Nano 2017:11;8790–8795.

[62] Bae G Y, Pak S W, Kim D, Lee G, Kim D H, Chung Y, Cho K. Linearly and highly pressure-sensitive electronic skin based on a bioinspired hierarchical structural array. Adv Mater 2016:28;5300–5306.

[63] Boutry C M, Nguyen A, Lawal Q O, Chortos A, Rondeau-Gagné S, Bao Z. A sensitive and biodegradable pressure sensor array for cardiovascular monitoring. Adv Mater 2015:27;6954–6961.

[64] Guo Q, Zhang X, Zhao F, Song Q, Su G, Tan Y, Tao Q, Zhou T, Yu Y, Zhou Z, Lu C. Protein-inspired self-healable Ti_3C_2MXenes/rubber-based supramolecular elastomer for intelligent sensing. ACS Nano 2020:14;2788–2797.

[65] Song D, Li X, Li X, Jia X, Min P, Yu Z. Hollow-structured MXene-PDMS composites as flexible, wearable and highly bendable sensors with wide working range. J Colloid Interf Sci 2019:555;751–758.

[66] Li X, Y. Li Y, Li X, Song D, Min P, Hu C, Zhang H, Koratkar N, Yu Z, Highly sensitive, reliable and flexible piezoresistive pressure sensors featuring polyurethane sponge coated with MXene sheets. J Colloid Interf Sci 2019:542;54–62.

[67] Zhang Y, Chang T, Jing L, Li K, Yang H, Chen P. Heterogeneous, 3D architecturing of 2D titanium carbide (MXene) for microdroplet manipulation and voice recognition. ACS Appl Mater Interfaces 2020:12;8392–8402.

[68] Yang K, Yin F, Xia D, Peng H, Yang J, Yuan W. A highly flexible and multifunctional strain sensor based on a network-structured MXene/polyurethane mat with ultra-high sensitivity and a broad sensing range. Nanoscale 2019:11;9949–9957.

[69] Seyedin S, Uzun S, Levitt A, Anasori B, Dion G, Gogotsi Y, Razal J M. MXene composite and coaxial fibers with high stretchability and conductivity for wearable strain sensing textiles. Adv Funct Mater 2020:30;1910504.

[70] Kedambaimoole V, Kumar N, Shirhatti V, Nuthalapati S, Sen P, Nayak M M, Rajanna K, Kumar S. Laser-induced direct patterning of free-standing Ti_3C_2-MXene films for skin conformal tattoo sensors. ACS sensors 2020:5;2086–2095.

[71] Hu D, Huang X, Li S, Jiang P. Flexible and durable cellulose/MXene nanocomposite paper for efficient electromagnetic interference shielding. Compos Sci Technol 2020:188;107995.

[72] Fan X, Ding Y, Liu Y, Liang J, Chen Y. Plasmonic $Ti_3C_2T_x$ MXene enables highly-efficient photothermal conversion for healable and transparent wearable device. ACS Nano 2019:13;8124–8134.

[73] Chang T, Zhang T, Yang H, Li K, Tian Y, Lee J Y, Chen P. Controlled crumpling of two-dimensional titanium carbide (MXene) for highly stretchable, bendable, efficient supercapacitors. ACS Nano 2018:12;8048–8059.

5 Green Polymer Nanocomposites with MXenes

Nanoth Rasana, Karingamanna Jayanarayanan, and P. Sarath Kumar

CONTENTS

5.1 INTRODUCTION

5.1.1 Green Biopolymer MXene Nanocomposites

MXenes are a novel two-dimensional (2D) nano-structured class of materials that offers various structural compositions, exciting metallic conductivity, electrical, mechanical, thermal properties and dispersibility [1–3]. These properties of MXenes could be exploited and used as nano-fillers in polymer nanocomposites [4]. Among these, MXene green polymer nanocomposites play a significant role owing to its biodegradability, nontoxicity, environmental-friendliness and low-cost synthesis techniques. Since the discovery of graphene, ultrathin 2D nanomaterials, such as phosphorene [5], antimonene [6–7] hexagonal boron nitride [8], layered metal oxides, transition metal disulfides [9] and layered double hydroxides [10], have gained great applicability owing to their compact, ordered structure and quantum confinement effects [11]. Very few studies have reported the incorporation of $Ti_3C_2T_x$ (titanium carbide) MXenes in various bio/green polymer matrices. The major preparation methodologies and the characterization of biodegradable polymers (poly(acrylic acid) (PAA) [12], poly (lactic acid) (PLA) [13], polyethylene oxide (PEO) [14], alginate/PEO nanofibers [12], poly(vinyl alcohol)(PVA) [12], cellulose [14], chitosan [15], polydopamine (PDA) [13], lignin biopolymer (L-DEA) [16] and reinforced MXenes were reported in literature. The major objective of this chapter is to investigate reported MXene-reinforced green nanocomposites synthesis, processing methodology, parameters, its sustainability, cost-effectiveness, tunable properties and applications.

5.2 SYNTHESIS METHODS OF GREEN MXene NANOCOMPOSITES

Different synthesis routes were discussed in the literature, and the method of exfoliation of MAX (M: early-transition metal, A: A group element, X is C or N) phase (polycrystalline nano-laminates of ternary carbides and nitrides) into MXene sheets, which maintain their properties and crystal

DOI: 10.1201/9781003164975-5

structure, is applied in most of the studies [17–22]. In addition to its adaptability, low-cost facile synthesis approaches using biodegradable sustainable polymers make it an obvious choice in various fields of materials research [23–29].

Elisa et al. [12] reported the synthesis route of MXene-reinforced biodegradable polymer nanocomposites. The authors have chosen sustainable biopolymers, such as PVA, PEO, PEO/alginate and PAA, for the reinforcement with MXenes. Initially, the etching of Ti_3AlC_2 (MAX phase) with lithium fluoride/hydrochloric acid (LiF/HCl) solution was carried out, and the colloidal form of MXene (Ti_3C_2) flakes was prepared by the sonication of Ti_3C_2 multilayers [30–31]. An amount of about 1 gm of Ti_3AlC_2 was gradually added into HCl solution (12M) along with 1 g of LiF and was stirred continuously for 1day at room temperature to obtain a colloidal solution. The obtained solution was then washed repeatedly with deionized water followed by continuous agitation, centrifugation and decanted till the black supernatant was visible (which shows an increase in the concentration of delaminated MXene in solution). In the last step of the first phase of preparation, the solution was sonicated in an ice bath under argon gas flow, followed by centrifuge for 1hour. In the second phase of preparation, fully dissolved polymeric solutions (PAA +DMF/H_2O, PEO +ethanol/H_2O, PVA + ethanol/H_2O) were added to 1 wt% of $Ti_3C_2T_x$ solutions and stirred for 12 hours at 23°C to achieve complete dissolution. To prepare PEO/alginate/$Ti_3C_2T_x$ solution, $Ti_3C_2T_x$/alginate was dissolved in glycerol/water and added to a PEO polymer diluted in water, and later, the $Ti_3C_2T_x$/PEO/alginate was further mixed for 24 hours. Then, $Ti_3C_2T_x$/polymer solution was electrospun to nanofiber mats by pumping at constant flow rates (250–750 μL/hr) under an applied voltage of (15–20kV) using a 21-gauge needle [12].

Huang et al. [32] explained the synthesis of MXene (Ti_3C_2)-reinforced biodegradable PLA composites. The authors also reported the production of crude Ti_3C_2 MXene using LiF/HCL solution. The homogeneous hydrofluoric acid solution was prepared by stirring LiF/HCl solution for 10 min and then added to powdered Ti_3AlC_2 MAX phase and was stirred at 38°C for 2days. The pH of supernatant was made neutral by washing the crude phase of Ti_3C_2 MXene multiple times with HCl, LiCl and deionized water. The precipitate was separated by centrifuging and sonicating in an ice bath and exfoliated, single layers of MXene under nitrogen flow, which was further centrifuged and freeze-dried. PLA biopolymer was kept in a vacuum oven at 80°C for 4 hours before melt processing owing to its hygroscopic nature. The dried intumescent flame retardant (IFR)/PLA biopolymer and Ti_3C_2 MXene were melt-blended in a Brabender Plasticorder maintaining the screw speed at 60 rpm for 6 min and chamber temperature at 180°C. The blended composite obtained after melt mixing is compression molded at 180°Cand 10MPa for 10 minutes [32]. Another work [33] reported the production of $Ti_3C_2T_x$/polydiallyl dimethyl ammonium chloride (PDDA) composites by a dropwise addition of an aqueous PDDA polymer solution into colloidal solution of MXene, and the mixture was magnetically stirred for 24 hours and then centrifuged at a high rpm of 3500 for 1 hour. A similar procedure was used to prepare PVA/MXene composites with PVA-MXene weight ratios as 60:40, 40:60, 20:80 and 10:90 [33]. Yi et al. [34] reported the modification of MXene sheets using stearic acid (SA) and proposed the synthesis route to prepare MXene/PLA composite films. The modification of MXene sheets using SA ($Ti_3C_2T_x$-SA) was prepared by mixing 0.5 g of ($Ti_3C_2T_x$-Na) with 2 g of SA in 80 ml of ethanol and allowed to react at 85°C for 24 hours under continuous magnetic agitation. After completion of the reaction, the mixture was ultrasonically treated under nitrogen atmosphere, the reaction product mixture was separated by centrifugation and the precipitation was washed by ethanol to remove unattached traces of SA from MXene–Na nanosheets. The product is then dried in a vacuum oven at 80°C for 12 hours to form $Ti_3C_2T_x$-grafted-SA nanosheets [34].

The SA-modified MXene (varied the content from 0.2–2 wt%) sheets are further embedded into PLA matrix by solution blending method (Figure 5.1). PLA granules were dissolved in chloroform to obtain PLA solution. $Ti_3C_2T_x$–SA suspension was slowly added to PLA solution, then magnetically stirred for 12 hours, and ultrasonically treated for 20 min. The mixture was allowed to evaporate slowly from Petri dishes at room temperature for 24 hours to form thin composite films [34].

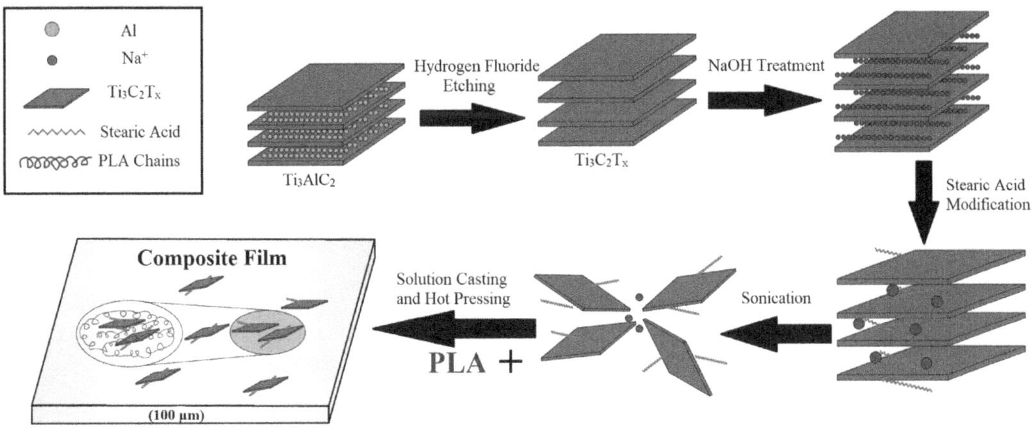

FIGURE 5.1 Synthesis method of stearic acid (SA) modified MXene with green biopolymer PLA nanocomposites [34].

In another work [35] on green/MXene composites, $Ti_3C_2T_x$ MXenes were synthesized from Ti_3AlC_2 MAX phase, and the cellulose-based filter paper was dipped in a colloidal solution of MXene. The M-filter composite papers with different dip-coating cycles (1,3,5,7 dip-coating cycle) were prepared and kept between PET plates for curing at 100°C for 2 hours [35]. In another study [15] on chitosan/MXene composites, MXene etched from MAX phase was mixed with ethanol, NH_3 (ammonia) and water, and the mixture was ultra-sonicated with modified polystyrene and washed and dried to MXene. Then the MXenes were treated with vinylimidazole to obtain MXene with imidazole brushes on the surface (QMXene-NH_2) and by solution casting dispersed in the chitosan matrix [15]. By altering the reaction time and acid content, diverse MXenes could be developed by HF (hydrofluoric acid) etching carried out from room temperature to various temperatures. The MXene nanosheets with terminal functional groups are then blended with green biopolymers using various processing techniques (listed in Table 5.1) like solution processing, melt blending, electrospinning, vacuum-assisted filtration, dip coating, polymerization and others [23,36–37]. Enhanced exfoliation, large-scale delamination of 2D MXenes with tailored thickness could be achieved from the previously mentioned synthesis routes.

5.3 CHARACTERIZATION OF GREEN MXene NANOCOMPOSITES

Several studies have reported the characterization techniques employed on MXene-reinforced biopolymer nanocomposites [38–41]. The reports revealed the relevance of MXene-based biopolymer nanocomposites and their broad application prospects in the field of material science [42–44].

The viscosity and conductivity of neat biopolymers and biopolymers loaded with 1 wt% MXene were investigated by Elisa et al. [12]. In their work, since the polymer/MXene solutions were electrospun viscosity was a prime factor in controlling the morphology of nanofibers. Low viscous polymer solutions could not produce uniform and continuous fibers and high viscous polymer solutions lead to difficulty in ejections while electrospinning [45–48]. With the addition of 1 wt% of $Ti_3C_2T_x$ MXene in PAA biopolymer, a slight variation in viscosity was observed. The viscosities of PEO and PVA incremented by 11.7% and 8.7%, respectively, with the incorporation of $Ti_3C_2T_x$. This enhancement in viscosity of PVA and PEO biopolymers is related to the presence of hydroxyl groups on $Ti_3C_2T_x$ and the hydrogen bonding effect between the biopolymers, which decides the rheological behavior of a polymer. On the other hand, in the case of $Ti_3C_2T_x$/alginate/PEO solutions, a decrement in viscosity by 10% was observed with regard to neat PEO, owing to the repulsive interactions

TABLE 5.1
Processing Techniques, Characterizations and Applications of MXene-Based Green Biopolymer Nanocomposites

Polymer/MXene	Synthesis/Processing	Characterization Methods	Applications	References
PVA, PEO, PAA/Ti_3C_2 MXene	Electrospun MXene/ polymer nanofibers	Morphology, Viscosity, Conductivity measurement, XRD, FTIR	Sensors, molecular electronics, tissue engineering	[12]
PLA/Ti_3C_2 MXene	Melt mixing in Brabender Plasticorder and then compression molded	XRD, AFM, XPS, CCTs LOI, SEM	Flame retardant applications	[32]
PDDA/PVA/MXene	Composite films via VAF (vacuum-assisted filtration)	Electrical conductivity, TEM, SEM	Energy storage devices	[33]
PLA/MXene-modified stearic acid (SA)	Solution casting	XRD, FTIR, TEM	Flame retardant and structural applications,	[34]
PDMS coated M-filter paper (cellulose)/$Ti_3C_2T_x$	Dip coating and curing	TEM, SEM, XRD, Electrical conductivity	Sensors, EMI (electro magnetic Interference)	[35]
Cellulose nanofiber (CNF)/MXene	Vacuum filtration approach	SEM, TEM, Mechanical properties measurement	EMI Shielding materials, wearable electronics and personal heating systems	[53]
Chitosan/MXene modified with Imidazole	Solution casting	TEM, SEM, XRD, TGA and Mechanical properties	Anion exchange membranes	[15]
PDA300 (Poly dopamine heat-treated at 300°C)/ MXene	Solution casting	XRD, SEM, TEM, lithium–ion diffusivity measurement	Lithium–ion batteries	[13]
$Ti_3C_2T_x$MXene/Alginate biopolymer	Solution casting	XRD, SEM, FTIR, XPS, Hg^{2+}adsorption measurements	Removal of heavy metal ions (mercuric ions)	[54]
$Ti_3C_2T_x$ MXene/Chitosan nanofibers	Electrospinning	XRD, SEM, TEM, antibacterial activity	Wound healing applications	[43]
Polydopamine/MXene	Polymerization	TEM, SEM, XRD, cycling stability	High-performance rechargeable batteries	[14]

among the polyanionic alginate and decreased interactions between PEO/alginate with the addition of MXenes [12]. Another major finding reported by this group of researchers is enhancement in the conductivity of polymers with the incorporation of MXenes. The enhancement in the conductivity of the polymeric solution could be achieved by adding ionic salt or by adjusting solvent ratios, which could produce low-diameter fibers due to the high number of charge carriers that increased the polymer jet stretching. The electrical conductivity of 1 wt% MXene in PEO solution enhanced by 74%, 34.6% in $Ti_3C_2T_x$/alginate/PEO solution and 6.2% increase in $Ti_3C_2T_x$/PVA solution with regard to neat polymeric solutions. With the incorporation of MXene in PAA biopolymer, the conductivity was reduced, which shows weak interactions between the MXene and PAA polymer. The morphological results from their study revealed the variation in the diameter of electrospun nanofibers with the incorporation of MXenes in polymers. The average diameter of PVA fiber increased by 32.9%, whereas there was a decrement in the average diameter of the PEO fiber by 20% and in the PAA nanofibers by 38.9% with the incorporation of MXenes. The reduction in fiber diameter suggests the applicability of such fibers in molecular electronics, sensors and tissue engineering due to the improvement in conductivity [49–52]. Fourier transform infrared (FTIR) analysis of electrospun fibers revealed the functional groups and vibration band corresponding to O–H, C–H and C–O stretch in MXene-reinforced polymer composites. A decrease in crystallinity was observed in PVA polymer owing to the hydrogen bonding between surface termination groups (hydroxyl) on MXene and biopolymer PVA, which limits the crystal growth of PVA. An enhanced nucleation rate and crystallization were reported for PEO, and the appearance of a new peak was observed in PAA spectra (confirmed by X-ray powder diffraction [XRD] analysis), which is indicative of the delamination of nanofiber mats [12].

Huang et al. [32] reported the atomic force microscopy and XRD analysis of Ti_3C_2 MXenes, which shows the exfoliated MXenes of a thickness of 1.55 nm and a diameter of 0.69 μm. The high aspect ratios and larger d spacing promoted a uniform distribution of MXene in the PLA matrix. The flame retardancy of PLA matrix could be enhanced by the incorporation of intumescent flame retardants (IFRs) are made up of piperazine pyrophosphate ($C_4H_{14}N_2O_7P_2$), phosphoric acid (H_3PO_4) and zinc oxide (ZnO). The uniform distribution of MXenes and flame retardants in PLA matrix was very well studied using SEM analysis and energy dispersive spectroscopy (EDS) mapping (elements Ti, P, N). The flammability of MXene/IFR/PLA composites was analyzed using UL-94 tests and LOI (limiting oxygen index) tests. The higher value of LOI indicates a strong flame retardant effect. PLA has low LOI value of 21.5% and is highly flammable with no rating (NR). Figure 5.2 demonstrates the LOI in percentage with respect to increase in MXene content in PLA matrix and its flame retardant effect [32].

The incorporation of 1 wt% of MXene in PLA/IFR (11 wt%) matrix enhanced the LOI to 160%, and the maximum heat release rate reduced by 64.6% (obtained by cone calorimeter tests [CCTs]), and in UL-94 tests, the sample achieved V-0 rating (Figure 5.3).

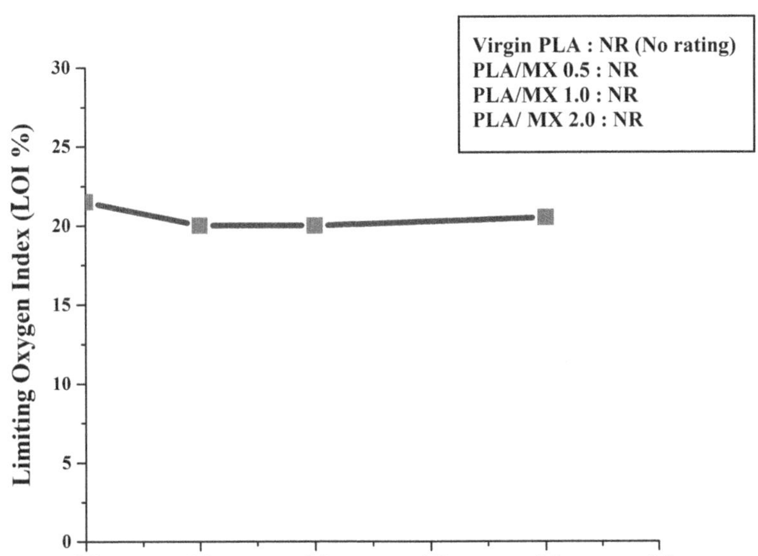

FIGURE 5.2 LOI values of PLA nanocomposite with increase in MXene loading [32].

Source: Reprinted with permission from Elsevier.

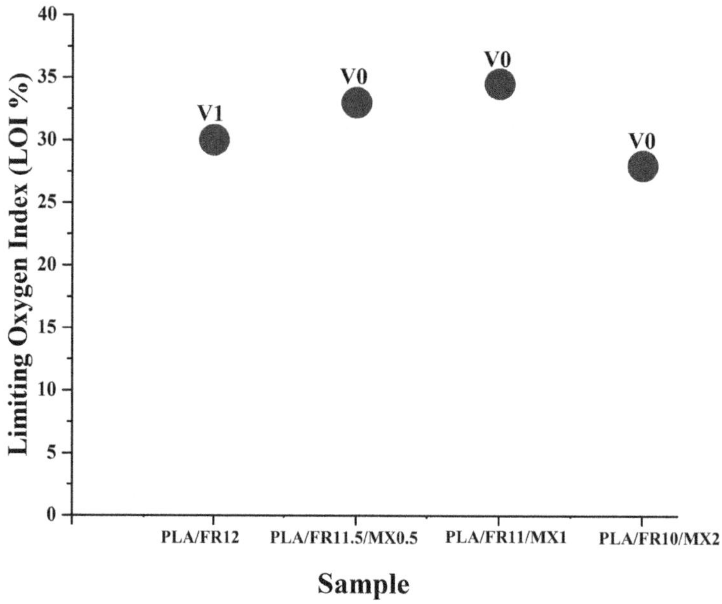

FIGURE 5.3 LOI of PLA nanocomposites incorporated with flame retardant and MXene [32].

Source: Reprinted with permission from Elsevier.

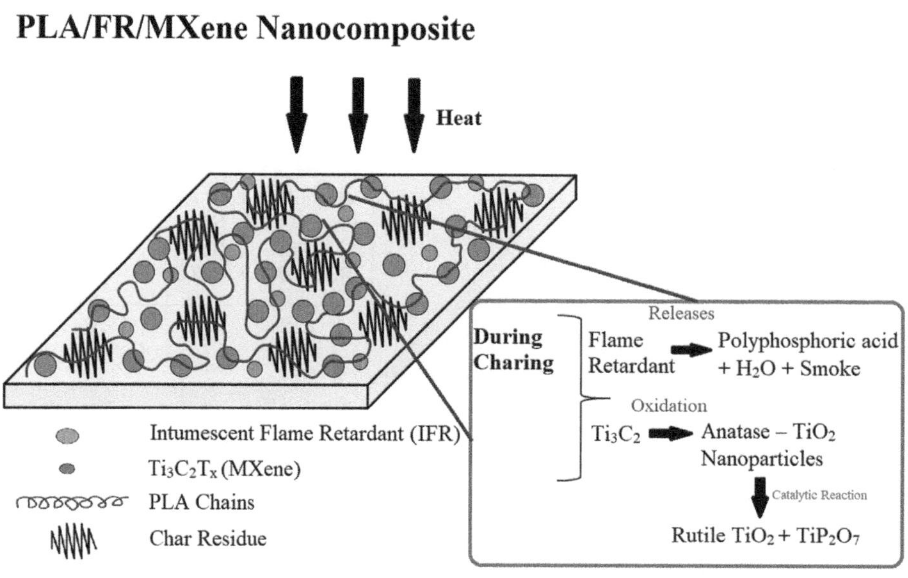

FIGURE 5.4 The mechanism of synergism of flame retardancy of biodegradable PLA composites incorporated with MXenes and IFR additive [32].

Source: Reprinted with permission from Elsevier.

Morphological results, CCT analysis, XPS and XRD analysis of MXene/PLA/IFR composites represented the nanosheet barrier effect of nano TiO_2 catalyst and 2D MXene sheets. Hence, the proposed PLA biopolymer composite system has the ability to enhance char formation and resists flame and fire hazard (Figure 5.4). It indicates that the addition of an IFR system can largely reduce the flame spread rate of composite system. This is largely due to the barrier effect of the char layer formed from the IFR (additive) system, which covers on MXene surface and decreases its wicking action. During the burning of the matrix and the generation of the char layer, the high-energy MXene surface is transformed into a carbonized rough surface, which leads to weakening of the wetting, spread and flow of interfacial PLA melt, thus reducing the actual flame spread rate. Furthermore, IFR retards the combustion of the composites through generating great amount of gases and developing an intact intumescent char shortly after the ignition. When part of IFR was substituted by MXene, there is a tremendous decline in the smoke production rate (SPR) and total smoke-release rate (TSR). The randomly distributed 2D MXene nanosheets provides a tortuous path effect that would effectively prolong the diffusion routes of degraded products during burning and hinder the release of gases.

Yu et al. [55] reported that nano-TiO_2 in MXene sheets promotes charring of polyurethane matrix during burning by blocking the release of volatiles. The compactness of char layer and degree of residue charring are improved owing to the crosslinking reaction promoted by MXene nanolayers [56]. The generation of pores and cracks during combustion could be obstructed, providing an improved barrier effect for heat and mass transfer [57].

Moreover, the inclusion of MXene nanosheets in PLA biopolymer with carbide content in Ti_3C_2 develops self-charring, promotes char formation and finally gets encapsulated in carbon residue. Owing to the high aspect ratio of MXene sheets like graphene [58] and other 2D materials, there is a delay in the release of flammable volatile products and offers a barrier to the transport of oxygen from the atmosphere. MXenes, along with IFR, exhibit excellent synergistic effects for enhancing the flame retardancy of PLA [32, 59–63]. In another study, MXene was incorporated into poly(diallyldimethylammonium) chloride (PDDA) and in a PVA biopolymer, which are neutral electrically [33, 64–65]. The authors reported excellent electrical conductivity, flexibility and controlled thicknesses [65]. XRD analysis revealed the ordered stacking of $Ti_3C_2T_x$ as evidenced by a sharp peak at 4.7°. With respect to pure $Ti_3C_2T_x$ films, there was a peak shift to lower angle side from 6.5°, which demonstrates the intercalation of PDDA molecules among MXene flakes. In the case of PDDA/MXene composites and PVA/MXene composites, there was a sharp decrease in electrical conductivity, the film thickness could be tailored with loading of the polymer and the films were flexible and freestanding. With the incorporation of polymer in MXenes, the distance between the flakes increases and they become less uniform. It is revealed that most of the $Ti_3C_2T_x$ films exist as single layers within the composite (verified by high-resolution transmission electron microscopy [HRTEM]) and the distance between the layers incremented with polymer loading. Furthermore, the conductivity (electrical) of the composites could be varied in a broad range from 22430 S/m to zero based on the composition of PVA [33]. Yi et al. [34] observed the characteristic X-ray diffraction peaks (9.5°, 19.2°, 34.0°, 36.8°, 39.0°, 41.8°, 48.5°, 52.4°, 56.6° and 60.2°) [66–68] of Ti_3AlC_2 crystalline MAX phase and confirmed its purity. However, after treatment with an HF solution, the diffraction peak corresponding to Al (39°) disappeared due to the removal of Al layer from MXene. Furthermore, lowered peak positions of $Ti_3C_2T_x$ indicated the larger interlayer spacing in $Ti_3C_2T_x$ than Ti_3AlC_2, and scanning electron microscopy (SEM) images revealed the lamellar structure of MXene, with each layer approximately 70nm thick [34]. FTIR analysis could confirm the presence of asymmetric and symmetric vibrations of C–H in $-CH_2$, indicating that the SA was attached to MXene, and the shift in diffraction peaks to new positions suggest that $Ti_3C_2T_x$ was successfully intercalated by SA and increased the spacing between the layers. Even after the strong washing in ethanol, it was observed that SA grafting remained on $Ti_3C_2T_x$, which due to the strong reaction between hydroxyl groups on MXene and carboxyl groups on SA.

The transmission electron microscopy (TEM) images of $Ti_3C_2T_x$ reported by Hu et al. [35] indicates a flat lamellar structure (average size of 0.58 μm) with wrinkles that could promote electrical conductivity. The special morphology [34] and ability to promote electrical conductivity conforms to another report by Ling et al. [33]. The SEM images of cellulose M-filter paper shows the presence of porous microstructure, which ranges from few to a dozen microns, that forms the basis for the electrically conductive pathways. With the incorporation of MXene in M-filter paper, cellulose fibers are coated with MXenes and interconnected MXenes are filled in the pores. As the number of dip-coating cycle increases, with rise in loading of $Ti_3C_2T_x$, MXene network becomes more and more compact and dense in comparison with M-filter/MXene composites with fewer dip cycles. This interconnected network of MXenes and cellulose fibers provides the flow path for charge carriers, causing an improvement in the electrical conductivity [69]. The morphology of PDMS-coated M-filter/MXene composites explains the tight bonding of MXenes with cellulose fibers [35]. This is a facile approach for the large-scale green production of M-filter/MXene green composites for the generation of flexible, multifunctional electromagnetic interference (EMI) materials. Zhou et al. [53] also reported that the incorporation of MXene along with cellulose nanofiber layers generates multilayered MXene/polymer films arranged alternatively that could be used as EMI films. A simple and efficient vacuum filtration approach was followed for the preparation of flexible, robust films of cellulose nanofiber and MXenes. SEM images revealed highly ordered lamella microstructure of carbon nanofiber and MXenes, and no gaps and exfoliations were observed at the interface of multilayered MXene and CNF (cellulose nanofiber) which indicates the flexibility and the reason for improved mechanical properties of the film. Cellulose nanofibers act as barriers that could prevent "zigzag" nano-crack growth [70] in MXene layers and show enhanced mechanical strength (112.5 MPa) and toughness of 2.7 MJ/m^3 compared to homogeneous CNF/MXene film and freestanding MXene film [53]. It was reported by Wang et al. [15] that imidazole functionalized MXene (QMXene-NH2) could be homogeneously distributed in the chitosan matrix. This constructs OH–conduction channels within the membrane and enhances mechanical and thermal stability of prepared composites [15].

Biodegradable modified PDA (polydopamine, PDA) biopolymers were prepared by Dong et al. [13] and then incorporated with highly conductive MXenes, imparting excellent rate capability, high capacity and good cycling stability. PDA materials were synthesized from polymerization reaction of dopamine monomers. Heat treatment of PDA was carried out at different temperatures (200°C, 300°C and 400°C). The authors proposed a simple heat treatment at 300°C, which promotes the superior electrochemical performance of PDA. SEM photographs of PDA reveals the spherical particles with diameters of 300–500 nm, and the polymer was treated at different temperatures: 200°C, 300°C and 400°C. In FTIR studies, compared with untreated PDA, heat treated PDA at 300°C shows the increase of carbonyl C atoms. The peaks that corresponds to catechol C atoms disappeared suggesting their oxidation. XRD analysis shows diffraction peaks from 9 to 6° for PDA/$Ti_3C_2T_x$ which indicates the increment in interlayer distance from 0.98 to 1.47 nm. The SEM images show that PDA/$Ti_3C_2T_x$ (heat-treated at 300°C) composite exhibited a layered morphology. The EDS elemental images reveal that PDA is uniformly distributed in $Ti_3C_2T_x$/PDA300 composite. The $Ti_3C_2T_x$/PDA (at 300°C) composite portrayed high specific capacities of 1190 $mAhg^{-1}$and 552 $mAhg^{-1}$ at 50 mA g^{-1} and 5 A g^{-1}, respectively, when used as an anode. The 2D layered lamella nanostructure and unsaturated carbon–carbon bonds in the PDA biopolymer, along with $Ti_3C_2T_x$ (conductive) MXenes, promote excellent lithium–ion storage capability [13]. A sandwich-structured composite was prepared by *in situ* polymerization of dopamine on the surface of $Ti_3C_2T_x$ MXenes to form ordered mesoporous polydopamine (MPDA)/$Ti_3C_2T_x$ by Li et al. [14]. The vertically oriented ordered MPDA layers with nanopores of approximately 20 nm provide a continuous diffusion channel, while $Ti_3C_2T_x$ layers guarantee a continuous electron flow path. Li et al. [14] also confirmed its high rate performance, cyclability and reversible capacity, which are the key characteristic requirements for high-performance rechargeable batteries as suggested by Dong et al. [13] in their work. Another study [71] revealed the lamellar layers and hydrophilic surface of MXenes functionalized with sulfonate polyelectrolyte brushes ($Ti_3C_2T_x$-SO_3H) via a facile precipitation-cum-polymerization approach. Sulphonated MXenes were used as nanofillers in chitosan biopolymer and sulfonated poly(ether ketone) (SPEEK). The report revealed the efficient proton transfer pathway, which connects with conduction channels in polymer phase. The incorporation of 10 wt% sulfonated MXene in polymer matrices enhanced the proton conductivity with respect to SPEEK and chitosan membrane by 144% and 66%, respectively [71].

In another study [54], $Ti_3C_2T_x$ MXene was prepared using a nontoxic facile approach by etching with NH_4F instead of HF acid. A novel green nanocomposite was synthesized by incorporating 2-D MXene and sodium alginate (SAG) biopolymer spheres. The MXene/Alginate biopolymers exhibit high adsorption behavior of Hg^{2+} (mercuric ions) owing to the porous nature of polymer spheres and functional moieties on biopolymer alginate, which was confirmed by FTIR, SEM, XRD and XPS analysis [54]. In the work reported by Elisa et al. [43] $Ti_3C_2T_x$ MXene flakes were incorporated within chitosan nano-biopolymer fibers via an electrospinning process. SEM micrographs of MXene/chitosan revealed the thick networks of uniform-diameter nonwoven fibers. The branching of fibers was not observed, which explains the proper balance of the electrical and surface tension forces of $Ti_3C_2T_x$/chitosan solution during electrospinning. The broadened and reduced peak intensity of XRD diffraction peaks suggests the interaction between the MXenes and chitosan nanofibers. TEM micrographs revealed the simultaneous visualization of chitosan and MXene flakes in the composite and suggest the two types of orientation as the embodiment of $Ti_3C_2T_x$ flakes within the fiber and the protrusion of MXene flakes [43].

5.4 MECHANICAL PROPERTIES OF GREEN POLYMER BASED MXene NANOCOMPOSITES

It was reported that $Ti_3C_2T_x$ films of approximately 3.3-μm thickness presented a Young's modulus of 3.5 ±0.01 GPa and tensile strength of 22 ± 2 MPa [33], which are comparable with graphene oxide (GO) [72] and carbon nanotube–based bucky paper [73]. At the same time, it was observed that MXene films possessed better conductivity. PVA film of 13 μm exhibits a tensile strength of 30 ±5 MPa. With the incorporation of 10 wt% of MXene in PVA, tensile strength enhanced by 34%. A further increase in MXene loading to 40 wt% in PVA/MXene films enhanced the Young's modulus to 3.7 ±0.02 GPa, the tensile strength to 91 ±10 MPa and a higher strain at break of 4 ±0.5% in comparison with other loadings of MXene. The general increase in strength and stiffness of these composite films indicates the efficient stress transfer from the matrix to MXene layers and suggestive of strong interfacial bonding between the MXenes and the PVA polymer [74]. The hydroxyl groups, which are termination groups on MXenes acts as a binder between the MXenes and PVA polymer. Reports suggested that MXene/PVA films rolled into cylinders by controlling the PVA–MXene ratio demonstrate enough tensile and compressive strength that they could be used for energy storage device applications [33,75]. The mechanical properties of pure-PLA and SA-modified PLA composites with different contents of MXenes 0.2, 0.5, 1 and 2 wt%) were reported by Yi et al.[34]. At low loading of 0.5 wt% of MXene it is observed that PLA composites exhibited excellent toughness, whereas it decreased when the filler content increased to 1 wt%. The elongation at break increased to 131.6%, approximately sixfold in comparison with pure PLA biopolymer. This increase in elongation at break is owing to two factors and could be due to the plasticizing effect of SA and the increased chain mobility of PLA or owing to the induced shear yielding caused by the incorporation of rigid fillers. Beyond 1 wt% of MXene toughness decreased due to the formation of aggregates. MXene flakes prevent the movement of polymer chains due to large specific surface area. The reported results indicate that even with a small wt% of MXenes ($Ti_3C_2T_x$), a significant increase in mechanical properties could be achieved. The percentage of crystallinity and the thermal stability of SA-modified MXene/PLA composites was also found to be very prominent even at low loadings of MXene, which shows the applicability of PLA biopolymer in novel applications [34]. Mechanical properties of chosen green polymer-based MXene nanocomposites are reported in Table 5.2.

It was reported [53] that freestanding MXene film shows a low tensile strength (3.5 MPa) and strain at the break of 0.6%, owing to the weak interaction between MXene sheets. For a uniform stress distribution, polymer binders like PVA and CNF were employed, which contribute to strong

TABLE 5.2
Tensile Properties of Green Biopolymer/MXene-Reinforced Nanocomposites

Composite Nomenclature	Mechanical Properties	References
PVA/$Ti_3C_2T_x$ $Ti_3C_2T_x$:40 wt% PVA film thickness: 12 μm	Tensile strength: 91 ± 10 MPa; Young's modulus: 3.7 ± 0.02MPa; Elongation at break: 4 ± 0.5MPa	[33]
PLA/$Ti_3C_2T_x$-g-SA (Stearic acid) $Ti_3C_2T_x$-g-SA: 0.5 wt%	Elongation at break: 131.6 ± 18%; Yield strength: 56.7 ±1.4M Pa; Young's modulus: 2.6 ± 0.1 GPa	[34]
Alternating Cellulose nanofiber/MXene CNF5@MXene4 film	Tensile strength: 112.5MPa, Toughness 2.7MJ/m^3, strain at break 4.3%	[53]
Chitosan/MXene-NH2 MXene: 7.5 wt%	Tensile strength: 41MPa; Strain at break: 0.045%	[15]
Chitosan/SPEEK (sulfonated (poly ether ketone))/$Ti_3C_2T_x$-SO_3H-5 wt%	Tensile strength: 39MPa 50% strain	[71]

interaction and facilitate stress transfer [76–78]. The homogeneous CNF-reinforced MXene film presented a tensile strength of 92.1 MPa and a strain at break of 2.2%, which is lower than free-standing pure CNF film. Hence, the authors proposed alternating a multilayered structure along with CNF layers. Multilayered CNF and MXene films with 50% CNF content undergo plastic yielding, and their strength and toughness are comparable with that of CNF film. With the incorporation of MXene layers varying from 1 to 5%, tensile strength, the toughness and strain at break were enhanced in the range of 86.6 to 112.5 MPa, 1.6 to. 2.7 MJ/m^3 and 3.1 to 4.3%, respectively. At the optimal content of CNF 5 layer and MXene 4 film, the achieved mechanical properties was equivalent to pure CNF film. It was also observed that at this optimal content of CNF5MXene4 film, it could withstand a weight of 500 g without failure representing the flexibility of alternate multilayered composite films [53]. It is found that QMXene-NH_2 (Imidazole-modified MXene) has a tensile strength of 27.5MPa, Young's modulus of 920.2MPa and strain at break of 5.5%. With the incorporation of QMXene-NH_2 in semi-crystalline chitosan biopolymer tensile strength raised to a maximum of 41 MPa, but elongation at break was slightly reduced [15]. The incorporation of sulfonated MXenes in SPEEK and chitosan–polymer matrices enhanced the mechanical strength significantly. It is reported that the tensile strengths of hybrid membranes obtained are greater than 10.4 MPa, which indicates the possibility to include sulfonated MXene in matrices, particularly in green biopolymers to enhance its structural stability [71].

5.5 APPLICATIONS OF GREEN POLYMER NANOCOMPOSITES WITH MXenes

A bioderived electrode material was made from PDA, which is very cost-effective and maintains sustainability by incorporating nanostructured $Ti_3C_2T_x$ MXene and could be used as an alternative in lithium–ion batteries. The electrochemical performance of biopolymer PDA could be improved in synergism with MXene and under a simple heat treatment at 300°C. The prepared composite exhibits enhanced rate capability (remaining 552 mAh g^{-1} at 5A g^{-1}) and retained 82% cycling stability even after 1000cycles and exhibited a high capacity of 1190 mAh g^{-1}, 50mA g^{-1}). The super-lithiation of the C–C (unsaturated) bonds in PDA in combination with the conductive substrate of MXene leads to the high performance of biopolymer MXene composite [13]. Li et al. [14] also proposed the applicability of green MXene composites for energy storage device applications owing to their cost-effectiveness, flexibility and environmental friendliness. For enhanced electrochemical performance the authors proposed a sandwich-layered composite structure of polystyrene-polyethylene oxide block copolymer, along with $Ti_3C_2T_x$ MXene, which promotes nanopore channels for

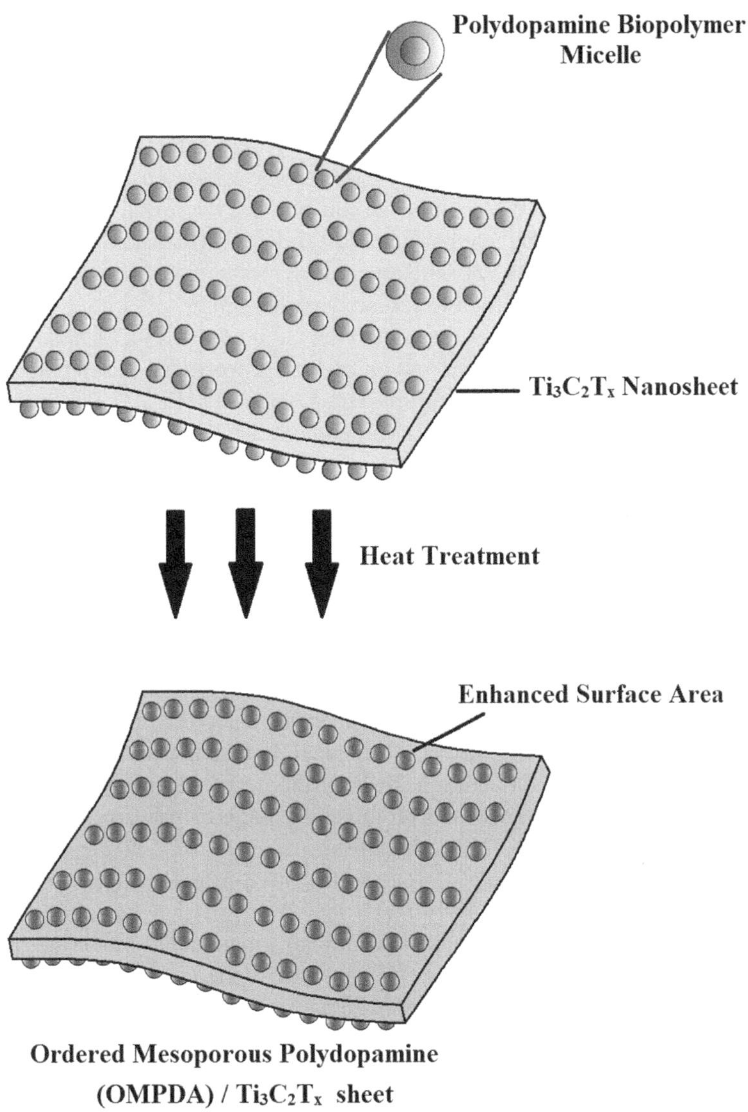

FIGURE 5.5 Ordered mesoporous structure of MXene nanosheets by direct polymerization and heat treatment of polydopamine biopolymer [14].

Source: Reprinted with permission from ACS publications.

diffusion of ions and continuous electron flow path. The ordered MPDA structure (Figure 5.5) with MXenes in the composite facilitates the ion transfer path, enhances the available surface area for lithium ions and demonstrates excellent rate performance, reversible capacity and cycling life. The ordered MPDA/$Ti_3C_2T_x$ electrode achieved a specific capacity of 1000 mA hg^{-1} at 50 mA g^{-1}. A high capacity of 430 mA h g^{-1} was also maintained after 600 cycles at 1A g^{-1}. The electrochemical performance of ordered MPDA/$Ti_3C_2T_x$ was evaluated using cyclic voltammetry (CV) via half-cell configuration.

From the cyclic voltametry analysis of ordered (MPDA)/MXene composite, the peak at 1.5 and 0.67 V is attributed to the addition of lithium to unsaturated nitrogen and enolization reaction of lithium with carbonyl oxygen [79–80]. The intercalation of Li+ ions into C_6aromatic rings corresponds to the generation of strong electrolyte interphase (SEI) film at the interface between the electrode and the electrolyte and was confirmed by voltametry analysis [81]. The redox properties of ordered MPDA/$Ti_3C_2T_x$ are highly reversible, which was indicated by overlapping of peaks in CV curves. The charge–discharge curves show the Li+ de intercalation and the low columbic efficiency at the initial cycles was increased to 95% from the third cycle. These observations suggest that generation of mesoporous structure in PDA composite can enhance the overall reversibility of reactions and capacity with lithium ions. The heterostructure of ordered MPDA and MXene corresponds to high rate capability and capacity [14,82]. The recent progress and advances in PDA/MXene composites was also discussed by other group of researchers [83–85].

In another study, Mayerberger et al. [12] reported the alignment of $Ti_3C_2T_x$ flakes within the fiber axis of biopolymer PVA and PAA polymer and observed the protruded flakes from the fibers. This is important for biomedical applications (antibacterial wound healing), where protruded MXene flakes could harm bacterial membranes [12]. The authors also reported the antibacterial properties of electrospun Ti_3C_2 MXene flakes with chitosan nanofibers for antibacterial wound dressing applications [43]. They demonstrated a reduction of 95% and 62%, respectively, of bacterial colony units against gram-negative *Escherichia coli* (*E.coli*) and gram-positive *Staphylococcus aureus* (*S.aureus*) with 0.75 wt% $Ti_3C_2T_x$- loaded nanofibers for 4-hour treatment. Cytotoxicity studies proves the nontoxicity of MXene/chitosan fibers. The average cell variability relative to the control specimen was about 85% at different test concentrations, which demonstrated the non-cytotoxicity of MXene/chitosan composite mats to bacterial cells over an exposure time of 72 hours [43].

The mechanism of the antibacterial nature of MXene flakes is related to the direct contact of bacterial cells with MXene flakes [43,86]. TEM analysis revealed the penetration of MXene flakes through bacterial membranes, causing direct mechanical destruction that also conforms with antibacterial activity of GO [87–89]. In addition, hydrophilic negatively charged surface termination functional groups (–F,–O, –OH) on MXenes promotes the agglomeration of bacteria, causing antibacterial activity. About 200 μg mL^{-1} of colloidal suspension of $Ti_3C_2T_x$ against *E. coli* indicates 99% bacterial inhibition zones [43,86,90–91]. Another study revealed the applicability of green MXene composites for the capture of heavy metal ions from waste water, owing to its large specific surface area, high porosities, oxygenated functional groups and unique internal structure of MXenes. The synthesized microspheres of MXene–SAG spheres (MX-SAG) composite under extreme pH conditions (0.5–1.0M HNO_3) demonstrated an adsorption capacity of 932.84 mg/g for the removal of mercuric ion (the highest among the reported adsorbents) and shows excellent reproducible properties [54]. The $Ti_3C_2T_x$ MXene is superior to GO, owing to the presence of termination surface groups, hydrophilic nature and chemical stability, where GO lacks functional groups on the surface to adsorb metal ions [92].

The MX-SAG biopolymer composite microspheres possess a unique spherical structure that entraps Hg^{2+} ions and provides sufficient binding locations [54, 93–95]. Figure 5.6 shows the comparative performance of MX-SAG biopolymer composite and GO/SAG biopolymer composite on the adsorption efficiency of mercuric ions. Zhang et al. [96] also revealed the applicability of MXene-based green polymer composites to adsorb environmental toxicants like heavy metal ions, radionuclides, organic dyes and gas molecules and even its sensing capability [96–97].

The applicability of MXenes in supercapacitor applications was discussed by Qin et al. [16]. The MXene/biopolymer composite paper electrodes are obtained by self-assembly of negatively charged $Mo_{1.33}$ CMXene and amine cations functionalized lignin. The intercalation of L-DEA (lignin-diethanolamine), along with MXene, enhanced the interlayer spacing between MXene sheets and promotes improved charge transfer and the composite electrode possessed a maximum capacitance of 503.7Fg^{-1}. The amine-modified lignin-based biopolymer with MXene also exhibited excellent stability and rate performance. The Mo1.33C/L-DEA/exfoliated graphene/ruthenium oxide nanoparticle composite device provides an energy density of 51.9 and 27.8 $Whkg^{-1}$ at a voltage window of 1.35 V and at power densities of 338.4and 40095 Wkg^{-1} and hence can be utilized

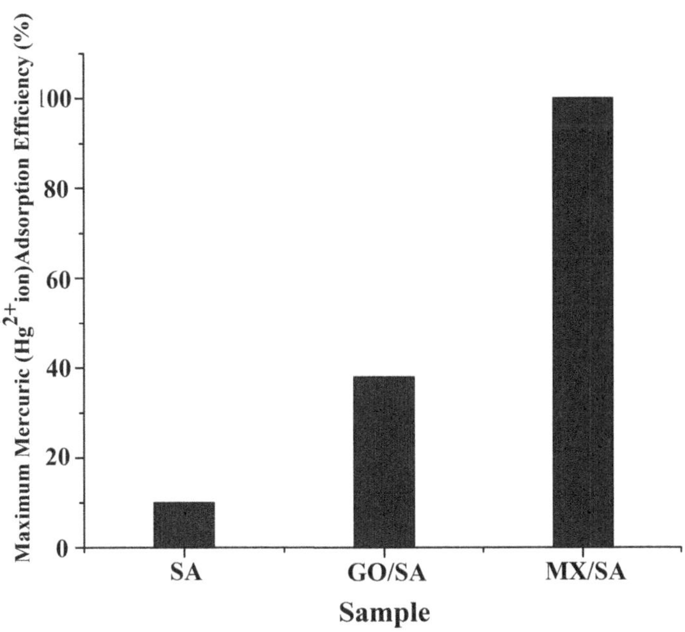

FIGURE 5.6 Adsorption efficiency of SAG biopolymer spheres along with MXene on mercuric (Hg^{2+}) ion removal in comparison with GO [54].

Source: Reprinted with permission from Elsevier.

to design high-potential supercapacitors with enhanced energy density [16]. The cellulose nanofiber/MXene (CNF/MXene) composite films could enhance the electrical conductivity by 621 S m^{-1}[53]. The reflection-absorption zigzag reflection mechanism of alternating multilayered CNF/MXene films demonstrated excellent EMI shielding effectiveness of approximately 40 dB. The authors [53] observed that owing to the impedance mismatch almost 90% of incident electromagnetic (EM) waves are immediately reflected. When EM waves pass through the MXene films, they hit holes, dipoles and electrons, causing massive polarization and losses at interfaces. These losses aid the high attenuation of EM-wave energy, causing EM-wave absorption by these films. This mechanism of EMI reflection and absorption of cellulose nanofiber/MXene films could contribute to EMI shielding performance but has relatively low electrical conductivity. This drawback could be circumvented by the presence of alternating multilayered CNF/MXene film that could raise the impedance mismatch of the specimen–air interface and thereby improve the reflection of incident electromagnetic waves at the interface. Furthermore, EM waves transmit from one MXene to the other MXene layers, and alternating layered structures cause a high-impedance mismatch between the cellulose nanofiber biopolymer and the MXene layer. This promotes zigzag reflection between the biopolymer layer and the MXene layer and results in a high-absorbing attenuation of EM waves and very little transmittance [76, 98–100]. This mechanism of MXene incorporated green nanocomposite accounts for a significant increase in EMI shielding effectiveness [53,101].

Another study [15] disclosed the significance of MXene-based green chitosan biopolymer composite as a conductive membrane enhancing OH^- conduction up to 172% owing to continuous ion conduction channels prevailing in the composite. This justifies the potential of MXene green nanocomposites in the development of anion exchange membranes for alkaline fuel cell applications [15]. The SPEEK/chitosan/$Ti_3C_2T_x$ composites demonstrate the enhanced proton conductivities and at incorporation of 10 wt% of MXene, conductivity enhanced by 81% and 146% under 120°C owing to the proton transfer pathways provided by sulfonated MXene intercalated in the polymer phase [71]. Wang et al. [102] proposed a green, flexible and efficient approach for the development of

3DMXene incorporated with PDMS matrix and these composites exhibited a high electrical conductivity of 5.5 S/cm. Even at low vol% of 2.5, MXene thermal conductivity enhancement was about 220% and decreased electrical resistance. The composite system has its relevance in the development of triboelectric nanogenerators with capacitor structures and could be applied to sensors and energy harvesters [33,37, 102–106]. Some other studies on MXene-based nanocomposites revealed their application in photocatalytic hydrogen production showing high catalytic activity [44,107].

5.6 CONCLUSION AND FUTURE DIRECTION

The objective of this chapter was to elucidate the most cost-effective, environmentally friendly, facile approach to synthesis strategies for MXene/biopolymer composites. A review of the methodology of preparation of green polymer/MXene nanocomposites, its morphological structure, various property enhancements demonstrating the relevance of MXene-based green nanocomposites in various sectors of materials engineering was presented. The percentage of enhancement of the properties of biopolymers, along with MXene, was reported, and the methods/alternative approaches for improving the properties of biopolymer were explicated. MXene/green biopolymer composites have received increasing attention owing to their wide range of compositions, enhanced conductivity and dispersibility. Although different synthesis routes and processing of MXene/biopolymer green nanocomposites have been reported in the past decade, there are still challenges and hurdles in its practical applications. Modulating the surface functional groups on MXenes is a major factor that produces active binding sites and improves the performance of MXenes once they are incorporated in a biopolymer. Based on the size of specific surface area, the porosity and the type of functional group performance of MXene green nanocomposites could be varied. The areas like electromagnetic shielding, antibacterial applications and electrochemical energy storage device development, among others, will continue to be the focus of research in the future.

REFERENCES

1. Jaya Prakash, N., Kandasubramanian, B. 2021. Nanocomposites of MXene for industrial applications. Journal of Alloys and Compounds, 862: 158547.
2. Riazi, H., Nemani, S.K., Grady, M.C., et al. 2021. Ti_3C_2 MXene–polymer nanocomposites and their applications. Journal of Materials Chemistry A, 9: 8051–8098.
3. Tian, W., Vahid Mohammadi, A., Reid, M.S., et al. 2019. Multifunctional nanocomposites with high strength and capacitance using 2D MXene and 1D nanocellulose. Advanced Materials, 31: 1902977.
4. Lipton, J., Weng, G.M., Alhabeb, M. 2019. Mechanically strong and electrically conductive multilayer MXene nanocomposites. Nanoscale, 11: 20295–20300.
5. Ren, X., Zhou, J., Qi, X. 2017. Few-layer black phosphorus nanosheets as electrocatalysts for highly efficient oxygen evolution reaction. Advanced Energy Materials, 7: 1700396.
6. Peng, L., Ye, S., Song, J. 2019. Solution-phase synthesis of few-layer hexagonal antimonene nanosheets via anisotropic growth. Angewandte Chemie International Edition, 58: 9891–9896.
7. Wang, X., He, J., Zhou, B. 2018. Bandgap-tunable preparation of smooth and large two-dimensional antimonene. Angewandte Chemie, 130: 8804–8809.
8. Chen, H., Yang, Z., Zhang, Z. 2019. Construction of a nanoporous highly crystalline hexagonal boron nitride from an amorphous precursor for catalytic dehydrogenation. Angewandte Chemie International Edition, 58: 10626–10630.
9. Manzeli, S., Ovchinnikov, D., Pasquier, D. 2017. 2D transition metal dichalcogenides. Nature Reviews Materials, 2: 1–15.
10. Lv, L., Yang, Z., Chen, K. 2019. 2D layered double hydroxides for oxygen evolution reaction: from fundamental design to application. Advanced Energy Materials, 9:1803358.
11. Zhao, S., Kang, W., Xue, J. 2015. MXene nanoribbons. Journal of Materials Chemistry C, 3: 879–888.
12. Mayerberger, E.A., Urbanek, O., McDaniel, R.M. 2017. Preparation and characterization of polymer-$Ti_3C_2T_x$ (MXene) composite nanofibers produced via electrospinning. Journal of Applied Polymer Science, 134: 45295.

13. Dong, X., Ding, B., Guo, H. 2018. Superlithiated polydopamine derivative for high-capacity and high-rate anode for lithium-ion batteries. ACS Applied Materials & Interfaces, 10: 38101–38108.
14. Li, T., Ding, B., Wang, J. 2020. Sandwich-structured ordered mesoporous polydopamine/MXene hybrids as high-performance anodes for Lithium-ion batteries. ACS Applied Materials & Interfaces, 12:14993–15001.
15. Wang, L., Shi, B. 2018. Hydroxide conduction enhancement of chitosan membranes by functionalized MXene. Materials, 11: 2335.
16. Qin, L., Tao, Q., Liu, L. 2020. Flexible solid-state asymmetric supercapacitors with enhanced performance enabled by free-standing MXene– biopolymer nanocomposites and hierarchical graphene–RuOx paper electrodes. Batteries & Supercaps, 3: 604–610.
17. Alnoor, H., Elsukova, A., Palisaitis, J. 2021. Exploring MXenes and their MAX phase precursors by electron microscopy. Materials Today Advances, 9:100123.
18. Naguib, M., Kurtoglu, M., Presser, V. 2011. Two-dimensional nanocrystals produced by exfoliation of Ti_3AlC_2. Advanced materials, 23: 4248–4253.
19. Barsoum, M.W. 2013. MAX phases: Properties of machinable ternary carbides and nitrides: John Wiley & Sons.
20. Eklund, P., Rosen, J., Persson, P.O.Å. 2017. Layered ternary Mn+1AXnphases and their 2D derivative MXene: An overview from a thin-film perspective. Journal of Physics D: Applied Physics, 50: 113001.
21. Yang, J., Naguib, M., Ghidiu, M., et al. 2016. Two-dimensional Nb-Based M4C3 solid solutions (MXenes). Journal of the American Ceramic Society, 99: 660–666.
22. Fu, L., Xia, W. 2021. MAX phases as nanolaminate materials: Chemical composition, microstructure, synthesis, properties, and applications. Advanced Engineering Materials, 23: 2001191.
23. Zang, X., Wang, J., Qin, Y., et al. 2020. Enhancing capacitance performance of Ti 3 C 2 T x MXene as electrode materials of supercapacitor: from controlled preparation to composite structure construction. Nano-Micro Letters, 12: 1–24.
24. Nyamdelger, S., Ochirkhuyag, T., Sangaa, D et al. 2020. First-principles prediction of a two-dimensional vanadium carbide (MXene) as the anode for lithium ion batteries. Physical Chemistry Chemical Physics, 22: 5807–5818.
25. Kim, H., Anasori, B., Gogotsi, Y., et al. 2017. Thermoelectric properties of two-dimensional molybdenum-based MXenes. Chemistry of Materials, 29: 6472–6479.
26. Ahmed, B., Anjum, D.H., Hedhili, M.N., et al. 2016. H 2 O 2 assisted room temperature oxidation of Ti 2 C MXene for Li-ion battery anodes. Nanoscale, 8: 7580–7587.
27. Zhan, X., Si, C., Zhou, J., et al. 2020. MXene and MXene-based composites: Synthesis, properties and environment-related applications. Nanoscale Horizons, 5:235–258.
28. Jimmy, J., Kandasubramanian, B. 2020. MXene functionalized polymer composites: Synthesis and applications. European Polymer Journal, 122: 109367.
29. George, S.M., Kandasubramanian, B. 2020. Advancements in MXene-polymer composites for various biomedical applications. Ceramics International, 46:8522–8535.
30. Carey, M.S. 2017. On the synthesis & characterization of Ti3C2Tx MXene polymer composites (Doctoral dissertation, Drexel University).
31. Zhang, T., Pan, L., Tang, H., et al. 2017. Synthesis of two-dimensional Ti3C2Tx MXene using HCl+ LiF etchant: Enhanced exfoliation and delamination. Journal of Alloys and Compounds, 695:818–826.
32. Huang, H., Dong, D., Li, W., et al. 2020. Synergistic effect of MXene on the flame retardancy and thermal degradation of intumescent flame retardant biodegradable poly (lactic acid) composites. Chinese Journal of Chemical Engineering, 28:1981–1993.
33. Ling, Z., Ren, C.E., Zhao, M.Q., et al. 2014. Flexible and conductive MXene films and nanocomposites with high capacitance. Proceedings of the National Academy of Sciences, 111:16676–16681.
34. Yi, Z., Yang, J., Liu, X., et al. 2020. Enhanced mechanical properties of poly (lactic acid) composites with ultrathin nanosheets of MXene modified by stearic acid. Journal of Applied Polymer Science, 137: 48621.
35. Hu, D., Huang, X., Li, S., et al. 2020. Flexible and durable cellulose/MXene nanocomposite paper for efficient electromagnetic interference shielding. Composites Science and Technology, 188:107995.
36. Li, H., Hou, Y., Wang, F., et al. 2017. Flexible all-solid-state supercapacitors with high volumetric capacitances boosted by solution processable MXene and electrochemically exfoliated graphene. Advanced Energy Materials.

37. Kannan, K., Sadasivuni, K.K., Abdullah, A.M., et al. 2020. Current trends in MXene-based nanomaterials for energy storage and conversion system: A mini review. Catalysts, 10:495.
38. Chen, X., Zhao, Y., Li, L., et al. 2021. MXene/polymer nanocomposites: preparation, properties, and applications. Polymer Reviews, 61: 80–115.
39. Szuplewska, A., Rozmysłowska-Wojciechowska, A., Poźniak, S., et al. 2019. Multilayered stable 2D nano-sheets of Ti 2 NT x MXene: synthesis, characterization, and anticancer activity. Journal of nanobiotechnology, 17: 1–14.
40. Zhang, H., Wang, L., Chen, Q., et al. 2016. Preparation, mechanical and anti-friction performance of MXene/polymer composites. Materials & Design, 92:682–689.
41. Riazi, H., Taghizadeh, G., Soroush, M., et al. 2021. MXene-based nanocomposite sensors. Acs Omega, 6:11103–11112.
42. Cao, Y., Li, M., Lu, J., et al. 2019. Bridging the academic and industrial metrics for next-generation practical batteries. Nature Nanotechnology, 14: 200–207.
43. Mayerberger, E.A., Street, R.M., McDaniel, R.M., et al. 2018. Antibacterial properties of electrospun Ti 3 C 2 T z (MXene)/chitosan nanofibers. RSC Advances, 8: 35386–35394.
44. Yin, J., Zhan, F., Jiao, T., et al. 2020. Facile preparation of self-assembled MXene@ Au@ CdS nanocomposite with enhanced photocatalytic hydrogen production activity. Science China Materials, 63: 2228–2238.
45. Huang, X., Wang, R., Jiao, T., et al. 2019. Facile preparation of hierarchical AgNP-loaded MXene/Fe3O4/polymer nanocomposites by electrospinning with enhanced catalytic performance for wastewater treatment. ACS Omega, 4: 1897–1906.
46. Gong, K., Zhou, K., Qian, X., et al. 2021. MXene as emerging nanofillers for high-performance polymer composites: A review. Composites Part B: Engineering, 108867.
47. Wang, D., Zhang, D., Li, P., et al. 2021. Electrospinning of flexible poly (vinyl alcohol)/MXene nanofiber-based humidity sensor self-powered by monolayer molybdenum diselenide piezoelectric nanogenerator. Nano-Micro Letters, 13: 1–13.
48. Habib, T., Patil, N., Zhao, X., et al. 2019. Heating of Ti 3 C 2 T x MXene/polymer composites in response to radio frequency fields. Scientific Reports, 9:1–7.
49. Wang, S., Shao, H.Q., Liu, Y., et al. 2021. Boosting piezoelectric response of PVDF-TrFE via MXene for self-powered linear pressure sensor. Composites Science and Technology, 202: 108600.
50. Abedi, A., Hasanzadeh, M., Tayebi, L. 2019. Conductive nanofibrous Chitosan/PEDOT: PSS tissue engineering scaffolds. Materials Chemistry and Physics, 237:121882.
51. Ghosal, K., Agatemor, C., Špitálsky, Z., et al. 2019. Electrospinning tissue engineering and wound dressing scaffolds from polymer-titanium dioxide nanocomposites. Chemical Engineering Journal, 358: 1262–1278.
52. Cao, W.T., Ma, C., Mao, D.S., et al. 2019. MXene-reinforced cellulose nanofibril inks for 3D-printed smart fibres and textiles. Advanced Functional Materials, 29:1905898.
53. Zhou, B., Zhang, Z., Li, Y., et al. 2020. Flexible, robust, and multifunctional electromagnetic interference shielding film with alternating cellulose nanofiber and MXene layers. ACS Applied Materials & Interfaces, 12:4895–4905.
54. Shahzad, A., Nawaz, M., Moztahida, M., et al. 2019. Ti3C2Tx MXene core-shell spheres for ultrahigh removal of mercuric ions. Chemical Engineering Journal, 368:400–408.
55. Yu, B., Tawiah, B., Wang, L.Q., et al. 2019. Interface decoration of exfoliated MXene ultra-thin nanosheets for fire and smoke suppressions of thermoplastic polyurethane elastomer. Journal of Hazardous Materials, 374: 110–119.
56. Lin, B., Yuen, A.C.Y., Li, A., et al. 2020. MXene/chitosan nanocoating for flexible polyurethane foam towards remarkable fire hazards reductions. Journal of Hazardous Materials, 381:120952.
57. Gao, Y., Wang, L., Li, Z. 2014. Preparation of MXene-Cu2O nanocomposite and effect on thermal decomposition of ammonium perchlorate. Solid State Sciences, 35: 62–65.
58. Yuan, B., Sun, Y., Chen, X., et al. 2018. Poorly-/well-dispersed graphene: Abnormal influence on flammability and fire behavior of intumescent flame retardant. Composites Part A: Applied Science and Manufacturing, 109:345–354.
59. Li, L., Liu, X., Wang, J., et al. 2019. New application of MXene in polymer composites toward remarkable anti-dripping performance for flame retardancy. Composites Part A: Applied Science and Manufacturing, 127: 105649.

60. Xue, Y., Feng, J., Huo, S., et al. 2020. Polyphosphoramide-intercalated MXene for simultaneously enhancing thermal stability, flame retardancy and mechanical properties of polylactide. Chemical Engineering Journal, 397:125336.
61. Zhou, Y., Lin, Y., Tawiah, B., et al. 2021. DOPO-decorated two-dimensional mxene nanosheets for flame-retardant, ultraviolet-protective, and reinforced polylactide composites. ACS Applied Materials & Interfaces.
62. Sheng, X., Li, S., Zhao, Y., et al. 2019. Synergistic effects of two-dimensional MXene and ammonium polyphosphate on enhancing the fire safety of polyvinyl alcohol composite aerogels. Polymers, 11: 1964.
63. Du, Z., Chen, K., Zhang, Y., et al. 2021. Engineering multilayered MXene/electrospun poly (lactic acid) membrane with increscent electromagnetic interference shielding (EMI) for integrated Joule heating and energy generating. Composites Communications:100770.
64. Liu, R., Li, W. 2018. High-thermal-stability and high-thermal-conductivity Ti3C2T x MXene/poly (vinyl alcohol)(PVA) composites. ACS Omega, 3: 2609–2617.
65. Xu, H., Yin, X., Li, X., et al. 2019. Lightweight Ti2CT x MXene/poly (vinyl alcohol) composite foams for electromagnetic wave shielding with absorption-dominated feature. ACS Applied Materials & Interfaces, 11: 10198–10207.
66. Chen, Z., Han, Y., Li, T., et al. 2018. Preparation and electrochemical performances of doped MXene/poly (3, 4-ethylenedioxythiophene) composites. Materials Letters, 220:305–308.
67. Rakhi, R.B., Ahmed, B., Anjum, D.et al. 2016. Direct chemical synthesis of MnO2 nanowhiskers on transition-metal carbide surfaces for supercapacitor applications. ACS Applied Materials & Interfaces, 8: 18806–18814.
68. Alhabeb, M., Maleski, K., Anasori, B., et al. 2017. Guidelines for synthesis and processing of two-dimensional titanium carbide (Ti3C2T x MXene). Chemistry of Materials, 29: 7633–7644.
69. Maleski, K., Ren, C.E., Zhao, M.Q., et al. 2018. Size-dependent physical and electrochemical properties of two-dimensional MXene flakes. ACS Applied Materials & Interfaces, 10:24491–24498.
70. Rafiee, M.A., Rafiee, J., Srivastava, I., et al. 2010. Fracture and fatigue in graphene nanocomposites. Small, 6:179–183.
71. Zhang, J., Liu, Y., Lv, Z., et al. 2017. Sulfonated Ti3C2Tx to construct proton transfer pathways in polymer electrolyte membrane for enhanced conduction. Solid State Ionics, 310:100–111.
72. Dikin, D.A., Stankovich, S., Zimney, E.J., et al. 2007. Preparation and characterization of graphene oxide paper. Nature, 448: 457–460.
73. Li, Z., Xu, J., O'Byrne, et al. 2012. Freestanding bucky paper with high strength from multi-wall carbon nanotubes. Materials Chemistry and Physics, 135: 921–927.
74. Ji, Z., Zhang, L., Xie, G., et al. 2020. Mechanical and tribological properties of nanocomposites incorporated with two-dimensional materials. Friction, 8: 813–846.
75. Li, N., Chen, Z., Ren, W., et al. 2012. Flexible graphene-based lithium ion batteries with ultrafast charge and discharge rates. Proceedings of the National Academy of Sciences, 109:17360–17365.
76. Cao, W.T., Chen, F.F., Zhu, Y.J., et al. 2018. Binary strengthening and toughening of MXene/cellulose nanofiber composite paper with nacre-inspired structure and superior electromagnetic interference shielding properties. Acs Nano, 12: 4583–4593.
77. Zhan, Z., Song, Q., Zhou, Z., et al. 2019. Ultrastrong and conductive MXene/cellulose nanofiber films enhanced by hierarchical nano-architecture and interfacial interaction for flexible electromagnetic interference shielding. Journal of Materials Chemistry C, 7: 9820–9829.
78. Mirkhani, S.A., Shayesteh Zeraati, A., Aliabadian, E., et al. 2019. High dielectric constant and low dielectric loss via poly (vinyl alcohol)/Ti3C2T x MXene nanocomposites. ACS Applied Materials & Interfaces, 11: 18599–18608.
79. Zhou, H., Zhang, R., Song, S., et al. 2018. Dopamine-assisted synthesis of MoS2 nanosheets on carbon nanotube for improved lithium and sodium storage properties. ACS Applied Energy Materials, 1: 5112–5118.
80. Gueon, D., Moon, J.H. 2019. Polydopamine-wrapped, silicon nanoparticle-impregnated macroporous CNT particles: Rational design of high-performance lithium-ion battery anodes. Chemical Communications, 55: 361–364.
81. Zhu, Y., Hu, A., Tang, Q., et al. 2018. Compact-nanobox engineering of transition metal oxides with enhanced initial coulombic efficiency for lithium-ion battery anodes. ACS Applied Materials & Interfaces, 10:8955–8964.

82. Liu, T., Kim, K.C., Lee, B., et al. 2017. Self-polymerized dopamine as an organic cathode for Li-and Na-ion batteries. Energy & Environmental Science, 10: 205–215.
83. Chen, J., Huang, Q., Huang, H., et al. 2020. Recent progress and advances in the environmental applications of MXene related materials. Nanoscale, 12:3574–3592.
84. Liao, H., Guo, X., Wan, P., et al. 2019. Conductive MXene nanocomposite organohydrogel for flexible, healable, low-temperature tolerant strain sensors. Advanced Functional Materials, 29: 1904507.
85. Feng, X., Yu, Z., Long, R., et al. 2020. Polydopamine intimate contacted two-dimensional/two-dimensional ultrathin nylon basement membrane supported RGO/PDA/MXene composite material for oil-water separation and dye removal. Separation and Purification Technology, 247: 116945.
86. Rasool, K., Mahmoud, K.A., Johnson, D.J., et al. 2017. Efficient antibacterial membrane based on two-dimensional Ti 3 C 2 T x (MXene) nanosheets. Scientific Reports, 7: 1–11.
87. Bitounis, D., Ali-Boucetta, H., Hong, B.H., et al. 2013. Prospects and challenges of graphene in biomedical applications. Advanced Materials, 25: 2258–2268.
88. Chung, C., Kim, Y.K., Shin, D., et al. 2013. Biomedical applications of graphene and graphene oxide. Accounts of Chemical Research, 46: 2211–2224.
89. Gurunathan, S., Han, J.W., Dayem, A.A., et al. 2012. Oxidative stress-mediated antibacterial activity of graphene oxide and reduced graphene oxide in Pseudomonas aeruginosa. International Journal of Nanomedicine, 7: 5901.
90. Rasool, K., Helal, M., Ali, A. 2016. Antibacterial activity of Ti3C2T x MXene. ACS Nano, 10:3674–3684.
91. Zhang, Y., Nayak, T.R., Hong, H., et al. 2012. Graphene: A versatile nanoplatform for biomedical applications. Nanoscale, 4: 3833–3842.
92. Hou, C., Jiao, T., Xing, R., et al. 2017. Preparation of TiO2 nanoparticles modified electrospun nanocomposite membranes toward efficient dye degradation for wastewater treatment. Journal of the Taiwan Institute of Chemical Engineers, 78: 118–126.
93. Manos, M.J., Petkov, V.G., Kanatzidis, M.G. 2009. H2xMnxSn3-xS6 (x= 0.11–0.25): A novel reusable sorbent for highly specific mercury capture under extreme pH Conditions. Advanced Functional Materials, 19:1087–1092.
94. Xue, H., Chen, Q., Jiang, F., et al. 2016. A regenerative metal–organic framework for reversible uptake of Cd (ii): from effective adsorption to *in situ* detection. Chemical Science, 7: 5983–5988.
95. Rasool, K., Lee, D.S. 2015. Characteristics, kinetics and thermodynamics of Congo Red biosorption by activated sulfidogenic sludge from an aqueous solution. International Journal of Environmental Science and Technology, 12: 571–580.
96. Zhang, Y., Wang, L., Zhang, N., et al. 2018. Adsorptive environmental applications of MXene nanomaterials: A review. RSC Advances, 8:19895–19905.
97. He, Y., Ma, L., Zhou, L. 2020. Preparation and application of bismuth/MXene nano-composite as electrochemical sensor for heavy metal ions detection. Nanomaterials, 10: 866.
98. Liu, R., Miao, M., Li, Y., et al. 2018. Ultrathin biomimetic polymeric Ti3C2T x MXene composite films for electromagnetic interference shielding. ACS Applied Materials & Interfaces, 10: 44787–44795.
99. Song, W.L., Cao, M.S., Lu, M.M. 2014. Flexible graphene/polymer composite films in sandwich structures for effective electromagnetic interference shielding. Carbon, 66: 67–76.
100. Shahzad, F., Alhabeb, M., Hatter, C.B., et al. 2016. Electromagnetic interference shielding with 2D transition metal carbides (MXenes). Science, 353: 1137–1140.
101. Li, X., Ran, F., Yang, F., et al. 2021. Advances in MXene films: Synthesis, assembly, and applications. Transactions of Tianjin University, 1–31.
102. Wang, D., Lin, Y., Hu, D., et al. 2020. Multifunctional 3D-MXene/PDMS nanocomposites for electrical, thermal and triboelectric applications. Composites Part A: Applied Science and Manufacturing, 130: 105754.
103. Ma, C., Ma, M.G., Si, C., et al. 2021. Flexible MXene-based composites for wearable devices. Advanced Functional Materials. 2009524.
104. He, W., Sohn, M., Ma, R., et al. 2020. Flexible single-electrode triboelectric nanogenerators with MXene/PDMS composite film for biomechanical motion sensors. Nano Energy, 78: 105383.
105. Wu, X., Han, B., Zhang, H.B., et al. 2020. Compressible, durable and conductive polydimethylsiloxane-coated MXene foams for high-performance electromagnetic interference shielding. Chemical Engineering Journal, 381:122622.

106. Song, D., Li, X., Li, X.P., et al. 2019. Hollow-structured MXene-PDMS composites as flexible, wearable and highly bendable sensors with wide working range. Journal of Colloid and Interface Science, 555: 751–758.
107. Yuan, W., Cheng, L., An, Y., et al. 2018. MXene nanofibers as highly active catalysts for hydrogen evolution reaction. ACS Sustainable Chemistry & Engineering, 6: 8976–8982.

6 Thermoset/MXene Nanocomposites

Benjamin Tawiah and Sarkodie Bismark

CONTENTS

6.1 INTRODUCTION

Polymers have become an important and ubiquitous part of human existence to the extent that one cannot imagine life without them. The application of polymers in recent years has expanded to hitherto unimagined areas thanks to the advancement in polymer science and engineering. One important area in polymer research and development spearheading the widespread application of polymers in many fields is polymer nanocomposites (PNCs; Aghamohammadi et al., 2021; Krishnamoorti and Vaia, 2007). PNCs have demonstrated distinctive physicochemical properties that are far distant from their components acting alone. The functionality of nanocomposites is fundamentally related to the efficient properties of the constituents, composition, structure, and interfacial interactions at the nanoscale level. The unique nanoscale properties of the individual components and the good interaction of the components as a result of their careful design results in excellent PNC products with wide engineering applications (Sliozberg et al., 2020). Among the various types of nanocomposites, thermoset/MXene nanocomposites have gained more attention recently because these nanocomposites exhibit unique and beneficial properties compared to other nanocomposites.

Thermoset/MXene nanocomposites are mostly a multicomponent solid products containing nanoscale material(s) (100nm or smaller in dimension) dispersed in a thermoset or thermoplastic matrix. These form a new group of nanocomposites possessing essential properties, such as high thermal stability, high mechanical performance, and excellent barrier properties. Currently, polymer nanocomposites are used in many industries, including vehicles, transportation, civil engineering, and tools used in sports and recreational activities (Gao et al., 2020).

DOI: 10.1201/9781003164975-6

With the quest to expand the usability of thermoset nanocomposites, the amelioration of functionality of polymer nanocomposites *via* the introduction of fillers has received keen interest recently (Aghamohammadi et al., 2021). Due to the important characteristics of inorganic nanomaterials, they find applications as fillers to enhance the performance and widen the applications of polymer nanocomposites. These fillers include fibers from textiles waste, wood, and MXene and other two-dimensional (2D) materials such as graphene, layered double hydroxide (LDH), molybdenum disulfide (MoS_2), hexagonal boron nitride (HBN), phosphorene and MXene. In recent years, the incorporation of MXene in polymer matrix has yielded numerous publications in many different areas of applications. MXene is a series of 2D compounds revealed to have a promising prospect for reinforcement of nanocomposites, due to their profitable mechanical and conductive behavior in heat and electricity. Stoichiometrically, MXene is defined as Mn+1XnTx, in which n = 1, 2, or 3, "M" is a transition metal "d", "X" is carbon and/or nitrogen, and the Tx is assignable to fluorine, oxygen, and/or hydroxyl terminations (F, O and/or OH) (Mishra et al., 2017). The fabrication strategies used to produce MXene from the MAX phase are responsible for the terminations. Experimental studies revealed that the M-A bonds are frail than M-X bonds in the MAX phase. As a result, the extraction of A elements from MAX phases by etchant leads to the formation of novel materials known as MXene (Alhabeb et al., 2017a; Kim et al., 2021). In addition, the development of essential ionic groups further makes MXene a Versatile nanomaterial (Come et al., 2015).

Resistance to elastic deformation, good conductivity to heat and electricity are among the veritable characteristic features of MAX phases. Also, the inherent functional groups of MXene, good mechanical properties, self-lubricating, and high thermal conductivity are considered to be the most preferable reinforcements for polymeric materials (Aghamohammadi et al., 2021). Moreover, the hydrophilic properties of MXene enable good interaction with polar polymeric constituents in composites, which is not commonly shown in other 2D nanomaterials. The hydrophilic properties of MXene is due to the surface terminating groups such as OH and –O–and F resulting from the chemical etching of Al. MXene has been widely used in research for the production of state-of-the-art energy materials, lubricant additives, and polymer-reinforced materials (Song et al., 2021b; Gao et al., 2020; Tontini et al., 2020; Ming et al., 2021; Verger et al., 2019; Sun et al., 2021; Jaya Prakash and Kandasubramanian, 2021; Zhang et al., 2016). Also, the physical and electrical functionalities of MXene have been established that the addition of low MXene loadings into polymer systems enhances the tribological and corrosion resistivity of nanocomposites.

Thermoset polymers have been mixed with MXene using many different means of nanocomposite preparation techniques to mutually benefit the final product (nanocomposites). Many studies on thermoset/MXene nanocomposites have been done with interesting results (Wang et al., 2019; Ji et al., 2020; Liu et al., 2020; Carey and Barsoum, 2021; Gao et al., 2020; Song et al., 2021b). Therefore, this chapter discusses thermoset polymer/MXene composites, the impact of MXene on thermoset polymers' mechanical and thermal behavior, fabrication strategies and the structural morphology, challenges, and applications of thermoset/MXene nanocomposites.

6.2 THERMOSET NANOCOMPOSITES

Thermosets are essentially polymers possessing the ability to crosslink their inter-and intramolecular chains to form a solid polymer network (Dodiuk, 2013). Unlike thermoplastic polymers, thermoset polymers network covalently crosslinks and become irreversible after manufacturing. This implies that thermoset polymers cannot be reshaped or remelted after fabrication. Upon heating or incorporating curing agents (such as amine hardeners) from the resin state to the final solid state, the basic polymeric constituents of thermosets undergo an irrevocable change in the chemical structure. The route through which the monomers are transformed into a solid, nonsoluble, and nonfusible matrix, which is mostly the last fabrication step of a polymerization reaction is called curing (Roller, 1986). Comparatively, the irreversible cross-linkage in thermosets grants good mechanical strength and chemical and heat resistance than thermoplastics. Additionally, thermosetting resins possess numerous benefits over thermoplastic resins regarding their processability, cost, dimensional stability, environmental stress cracking resistance, and scope for modification (Ratna, 2012). This makes thermoset polymer composites more preferable for protective coatings, aerospace composites, computer chip packaging, and structural applications.

Thermoset nanocomposites have the potency in advancing the utility of composites that surpass scientific discoveries in recent decades regarding materials employed in the production of nanocomposites. The nanostructure fabricated by a nano-phase in polymer matrix represents an indispensable booster to the structural properties of the polymer in recent years. As a result, the exploration of a huge variety of thermoset nanocomposites and broad knowledge of the fabrication methods and their properties have been studied extensively. Notably, what distinguishes thermoset nanocomposite from traditional composites is the extent of control in their production, processability, and functionality in a specific field of application that can be achieved nearly down to the atomic scale.

Also, in the field of advanced technology, thermoset nanocomposites are used as substitutes for wood and steel, especially in the aerospace and military sectors due to the multiple functionalities of the composites produced with the presence of thermosets reinforced with fillers like carbon fibers and glass. There are numerous types of thermosets possessing different properties, relatable to their structure and constituents. Thermoset polymers are classified into phenolic resins, amino resins (polyimides [dianhydride, diamine], and bismaleimides), polyester resins, silicon resins, epoxy resins, vinyl esters, and polyurethanes (Zaferani, 2018). Among these, epoxy has gained tremendous applications in many works due to its high performance.

Thermoset epoxy exhibits unique properties such as a low tendency to shrink, good compatibility, strong dielectric properties, long-term stability, relatively low cost, and so on. Epoxy resins possess many reactive groups, which determine the functionality of epoxy. The greater the number, the higher the functionality. Some of the commonly used epoxies include diglycidyl ether of bisphenol-A (DGEBA) and bisphenol-F (DGEBF), tetraglycidyl diamino diphenyl methane (TGDDM), and triglycidyl resin of p-aminophenol (TGAP), Figure 6.1. However, the most widely used are the DGEBA-based.

Curing as the final step of polymerization takes place by the reaction of reactive groups of epoxy and a primary diamine or an anhydride commonly referred to as hardeners.

In general, unsaturated polyesters compose of polymer units linked via ester groups are among the earliest categories of thermoset polymers widely employed in fiber-reinforced polymer composites. Polyester resins are viscous, even with some being solids having a low melting point and a low molecular weight. The unsaturated polyesters also possess C=C bonds in the main chain which permits

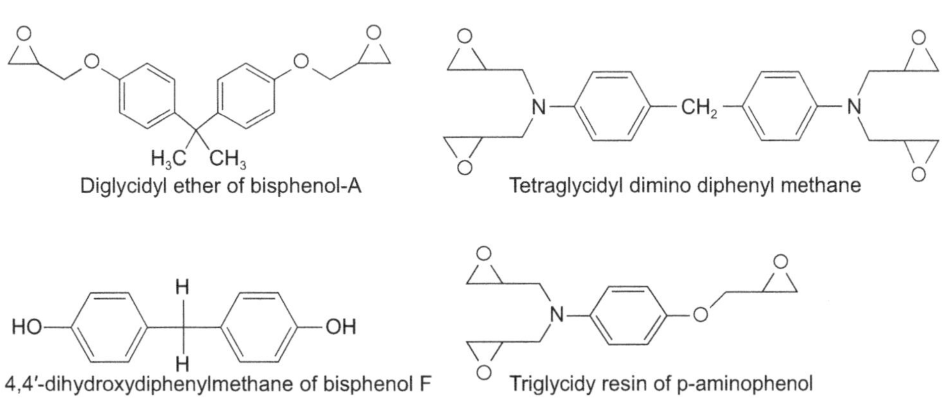

FIGURE 6.1 Chemical structure of some commonly used epoxies.

crosslinking reactions of polymer units. One characteristic advantage of unsaturated polyesters is their ability to react with the monomeric vinyl group due to the presence of C=C bonds, which facilitates the crosslinking of the chains. Primarily, styrene and α-methyl-styrene are employed as co-monomers to ensure the occurrence of required cross-linkages. Crosslinked polyester products, such as aminoplast or phenol-aldehyde resins, tend not to be particularly rigid (Miskolczi, 2013a). Thermoset polyester composites are typically fabricated by polycondensation reactions.

Aside from epoxy and unsaturated polyesters, vinyl ester resins are other types of thermoset polymers. These are a special type of polyester used to fabricate high-performance composites with discrete material properties at a low cost compared to numerous unsaturated polyesters. The molecular structure of vinyl esters are similar to polyesters but differs in the location of their reactive sites. Vinyl ester resins are primarily composed of mixtures of styrene and methacrylate epoxy. Unlike polyester, vinyl esters have fewer cross-linked bonds owing to the cross-linking of C=C double bonds at the end of the polymer chain (Miskolczi, 2013b). Styrene as a constituent of vinyl ester has one reactive vinyl group, but vinyl ester monomer has several reactive vinyl end groups. The reactive groups of vinyl ester provide crosslinking capacity and branching while styrene reveals only linear chain extension. The polymerization of vinyl ester units takes place by free radical chain growth (Chigwada et al., 2005). It has been fabricated with many different nanomaterials (such as nanoclay) to improve its performance and broaden its application. Vinyl ester finds application in fire retardancy, wind turbine manufacture, automotive industries, and constructions.

Another group of thermoset polymer system worth mentioning is those based on formaldehydes. Formaldehydes are typically adhesive resins with several merits in terms of cost, fast curing speeds, low curing temperatures, good panel performance, and, in most cases, colorless, which makes them ideal candidates for composites fabrication despite their perceived adverse health and environmental effects. The formaldehyde adhesive resins in this category include phenolic resin, melamine formaldehyde resin, and polyurethane (Martin et al., 2006; Singha and Thakur, 2009). The thermosetting formaldehyde resin based on urea is particularly produced at large scale globally, with production exceeding 5 million metric tons (t) annually due to its wide applications in plywood, fiberboard, and particleboards (Song et al., 2021a). The application of the various thermoset formaldehyde-based resins in composites fabrication has been explored. In a study conducted by Zheng and coworkers, graphene oxide–poly(urea–formaldehyde) was used to fabricate epoxy composite for corrosion protection in harsh environments, and excellent results were achieved (Zheng et al., 2018). Similarly, a thermally stable, lightweight, and biodegradable Sago-fiber particle board bonds were fabricated using urea formaldehyde to form a composite material meant for various applications (Chiang et al., 2016). Despite the wide applications of formaldehydes in composite fabrication, there are no sufficient data on the use of MXene with formaldehyde-based resins for composite fabrication.

In this decade, polymeric materials have found considerable applications due to their unique properties, such as low cost, low density, corrosion resistance, and ease of processing. For instance, polymeric coatings are mainly employed as protective coatings for metallic substrates to hinder the corrosion of the substrate. However, the low mechanical, electrical, energy storage capacity, and tribological properties of polymeric materials have restricted their broad utilization. Therefore, various approaches have been introduced for improving the mechanical, electrical, flame retardancy, tribological performance, and so on of thermoset polymer composites. A prominent route for ameliorating the performance of thermoset polymer composites is the incorporation of nanomaterials into their matrices. These materials help improve the properties of thermoset such as electrical, heat, magnetic, mechanical, wear properties, and others, hence increasing their level of utility in many other areas and widening their scope of application. Several nanomaterials in a form of 1D, 2D, and 3D have been used for elevating the properties of polymers. It has been demonstrated that the 2D nanomaterials, such as graphene, molybdenum disulfide, nanoclay, MXene, and others, are promising reinforcements for the polymeric matrices due to their discrete morphology and properties (Fu et al., 2019; Frigione and Lettieri, 2020).

Thermoset/carbon nanotube (CNT) nanocomposites have been successfully fabricated by resistive heating–assisted infiltration followed by curing of the polymer matrix resin (Kim et al., 2014).

The improved wetting and adhesion of the polymer resin to the CNT yielded a significant improvement of the thermoset/CNT nanocomposite, including enhanced mechanical properties (Kim et al., 2014). Also, CNTs have been incorporated in nanocomposites to increase thermal conductivity and thermal stability (Dong et al., 2018). Other usable fillers include fibers (carbon fibers, glass fibers, aramid fibers, etc.), starch, and chitosan (Matos Ruiz et al., 2000). These textiles waste in the form of waste or technical fibers are incorporated in thermoset to increases their mechanical properties as observed in a study by Zunjarrao Kamble and coworkers. Among all the thermosets, epoxies are one of the most versatile categories, with a wide range of application fields owing to their good reactivity, which enables them to bond well to fibers (Gibson, 2017).

In this era of sustainable production, the biodegradability of thermoset polymers has also been improved by the incorporation of biodegradable materials, such as cellulose and protein fibers, as reinforcing components of the matrix. In addition to their biodegradability, they combine good mechanical properties with low density (Matos Ruiz et al., 2000). New research development of thermoset nanocomposites in the area of flame retardancy is also offering significant advantages over conventional flame-retardant formulations and pure thermoset polymer composites. This can give rise to intrinsically new properties that are not displayed by the pure components. Such new properties typically originate from the change of the polymer nature in the surrounding of the filler. The polymer can be absorbed by the filler surface or trapped between fillers.

6.3 THERMOSET/MXene-BASED NANOCOMPOSITES

Despite the massive applications of thermosets, properties such as poor mechanical and electrical properties (poor flexibility, insulation to heat and electricity, poor wear properties, etc.) have hindered the growth of thermosets in certain applications particularly in areas where friction resistance is required, in energy storage, electromagnetic shielding, and vibration-resistant components in structures and gadgetry (Mgbemena et al., 2018; Paul and Robeson, 2008). One of the main reasons associated with the challenges is the difficulty in ameliorating the previously mentioned properties of epoxies by physical or chemical means. The incorporation of other materials is necessarily required to improve upon its qualities to expand its application. Thermoset epoxy resins are the most frequently used class of thermoset polymer, hence the recent increase in research on the preparation and application of thermoset/MXene nanocomposites.

Nanoparticles endow polymer composites with improved functionalities, such as conductivity, enhancement in catalytic activity for environmental remediation, and applications in detecting gases, and many more. Due to the discrete characteristics of MXene and their multiple variations, exploration into their usage as nanofillers has become a major focus since their discovery (Carey and Barsoum, 2021) Aside MXene, many different materials have been incorporated with thermoset to form nanocomposites. Among them are nanomaterials like graphene, CNTs, montmorillonite, and layered hydroxides. MXene is primarily formed by selectively extracting the A elements of the Ti_3AlC_2 MAX phase with transition metal elements via etching. Notably, MXene has received a broader application as a result of its unique characteristics, facile fabrication strategies, and its reproducibility at massive scale. It possesses good stability in heat, excellent electrical conductivity, and mechanical properties similar to graphene. Aside from the large surface area of the unique graphene-like nanomaterial MXene, it presents accessible surface hydroxyl, oxygen, and fluorine termination points than carbon, TiO_2, and other 2D nanomaterials. Owing to the physical barrier effect of its 2D nanosheets, MXene grants better retardancy and suppresses smoke when incorporated into polymers (Anasori et al., 2017; Hai et al., 2020). MXene has been incorporated in unsaturated polyester to mitigate peak heat release rate, enhance the suppression of smoke, and inhibit the release of carbon monoxide. These results reveal that the essential function of the chemical constituents of MXene impacts the catalytic attenuation effect of the thermosetting polymer nanocomposites during combustion. Undoubtedly, the chemical constituents of MXene have an influential role in determining the nature of the physical barrier effect, especially the flame-retardancy of MXene-based nanocomposites (Hai et al., 2020).

With regard to synthesis, MXene predominantly possesses reactive functional groups, useful in ensuring good reaction with other constituents during the fabrication of nanocomposites. Primarily, the magnitude of exfoliation of the MAX phases is chiefly dependent on determinants, namely, the variety of MAX phases, and the parameters (such as temperature, the length of time, and concentration) employed in the etching process. A hydrofluoric acid etchant is reported to be very efficient for the fabrication of MXene from MAX phases. Nevertheless, its high corrosiveness necessitated the usage of alternative etchants. The varieties of MAX phases usually employed for the fabrication of MXene by etching include Ti_3AlC_2 and Ti_2AlC. MXene prepared via wet chemical acidic etching is employed in two basic forms, that is, a single to a few flakes or in the form of multilayer MXene stacks (Carey and Barsoum, 2021; Verger et al., 2019). The synthesized MXene is incorporated in thermoset polymers via many different strategies of nanocomposite fabrication such as *in situ* polymerization, solvent mixing, hot press/injection molding, and others, which are further detailed in the next subtopic. Even though there exist more than 20 fabricated MXenes, such as Ti_3–xC_2Ty,V_2CTx, (Ti, V)$_3C_2$, and (Cr, V)$_3C_2$), Ti_3CN, Ti_4N_3, and Nb_3C_2Tx and many more (Zhao et al., 2019), as detailed by Babak Anasori et al.(Anasori et al., 2017), the most widely used MXene is the $Ti_3C_2T_x$. The$Ti_3C_2T_x$ was the first MXene discovered in 2011 (Gao et al., 2020; Sun et al., 2017). MXene has found application in energy storage (Ahmed et al., 2020; Anasori et al., 2017), catalysis (Ahmed et al., 2020; Zhao et al., 2019), sensors (Anasori et al., 2017; Li et al., 2021; Lee et al., 2019), flame retardance (Yu et al., 2021; Si et al., 2019), electronic magnetic interference shielding (Gao et al., 2020; Anasori et al., 2017; Hu et al., 2020), rechargeable batteries (Ming et al., 2021), fuel cells (Liu et al., 2018a), nanogenerators, and many others (Gao et al., 2020). As illustrated in Figure 6.2, the MXene suspension is added to the polymer solution and subsequently cured to

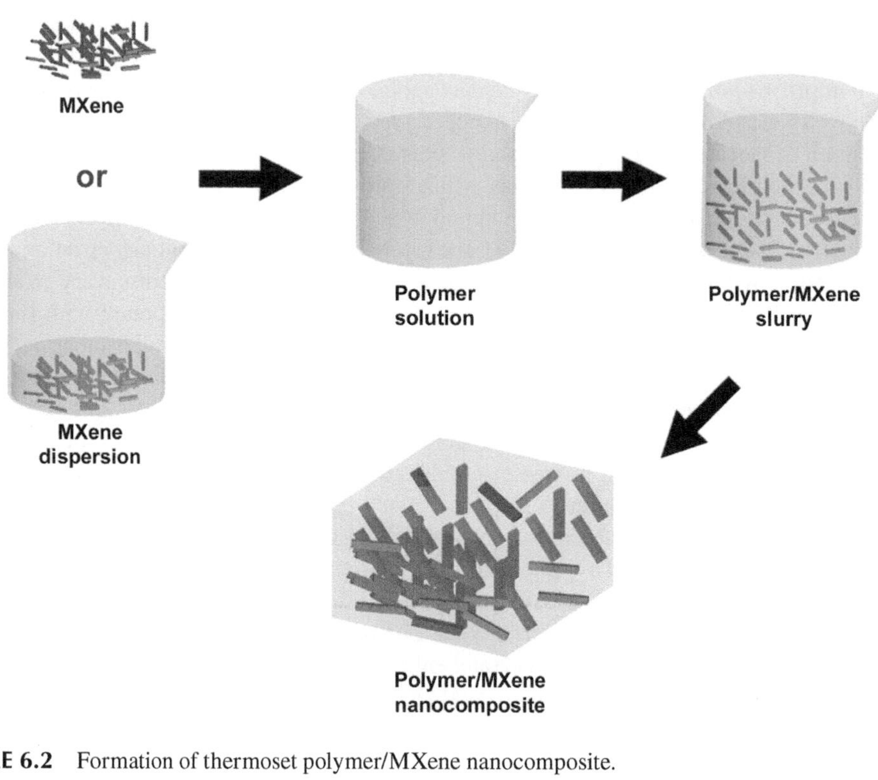

FIGURE 6.2 Formation of thermoset polymer/MXene nanocomposite.

TABLE 6.1
MXene, Application, and Properties

MXene	Field of Application	Properties
$Ti_3C_2T_x$ MXene-welded AgNW film	Aerospace and military applications	Electromagnetic interference shielding (Chen et al., 2020b)
$Ti_3C_2T_x$ modified membrane	Protection from microbes	High antimicrobial activity against *E. coli* and *B. subtilis* (Rasool et al., 2017)
$Ti_3C_2T_x$	Energy storage applications (supercapacitor)	High electrochemical performance of electrode (Fu et al., 2018; Li et al., 2017; Lukatskaya et al., 2013)
$Ti_3C_2T_x$ nanosheets	Transient electronic skin, intelligent robots, and human–machine sensors.	High flexible and sensitive piezo-resistive sensor, and excellent durability (Yan et al., 2020b)
$Ti_3C_2T_x$MXene@Pd colloidal nanoclusters paper film	Novel hydrogen sensor	Lightweight, flexible, and high sensitivity to hydrogen at room temperature at either flat or bent states (Zhu et al., 2020)

form a thermoset/MXene nanocomposite. The orientation of the MXene in the nanocomposite is influenced by the percentage of MXene fillers to polymer, solvent, and the fabrication method used. Some MXenes and their applications are outlined in Table 6.1.

6.4 THERMOSET/MXene NANOCOMPOSITES FABRICATION

Fabrication strategies form an influential factor in the outcome of products. They could affect the orientation, interaction, and shape of the nanocomposite. It is noteworthy that thermoset polymer resins are liquids with low viscosity, which grants easy processability during fabrication; hence, they do not require high pressure in the formation of the composite. The fabrication of thermoset/MXene nanocomposites can be done by many different routes, including *in situ* polymerization, solution mixing/blending, freeze-drying, hot press/injection molding, vacuum-assisted filtration, resin transfer molding, intercalation, template-assisted, and sonication-assisted fabrication, among others.

6.4.1 *In Situ* Polymerization Blending

In situ polymerization is an effective wet processing method of composite fabrication that enhances good dispersion of MXene in the thermosetting resin, solvent-based curing solution, and initiators. During synthesis, MXene fillers are initially dispersed in the monomer by subjecting them to physical stirring. Subsequently, the mixture is exposed to heat sonification and polymerized using the approach appropriate for a specific thermosetting resin. Aside from the preceding steps, MXene could also be dispersed in a solvent and added to the monomer solution before polymerization. To ensure a stable suspension, the solvent compatible with the resin is used for the dispersion of MXene before the initiation of polymerization. In some occurrences, interfacial agents could be added to enhance stability in the suspension. Most reports use solvents, namely, acetone or dimethylformamide during this process to accelerate dispersion. Also, studies on the *in situ* polymerizations of polypyrrole, polyaniline, polyacrylamide (PAM), and others in aqueous solutions using solvents, such as acetonitrile, have been disclosed. Figure 6.3 shows the process of fabricating nanocomposite by *in situ* polymerization. A good dispersion of MXene fillers with high interfacial strength in the interface of the MXene–polymer matrix is achievable with this method. However, this strategy is not well received for the fabrication of some particular nanocomposite materials, due to its demand for low elastomer viscosity before polymerization. In other words, low viscosity necessitates good interaction between constituents at their basic level. The strong interaction between thermoset and

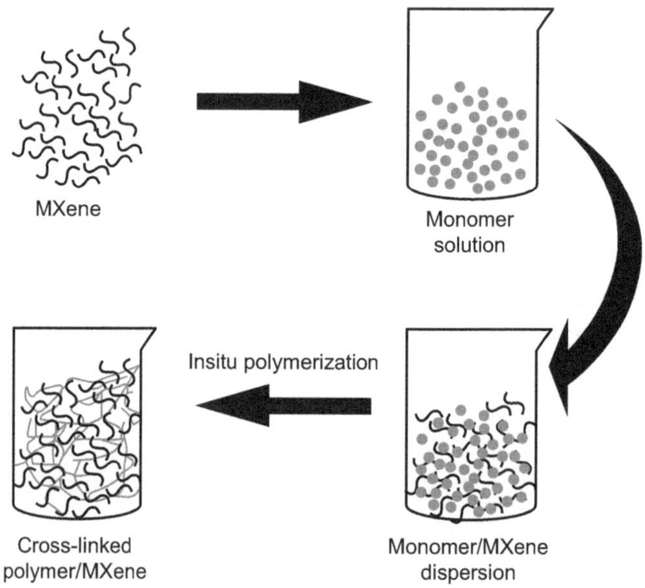

FIGURE 6.3 *In situ* polymerization process of thermoset/MXene nanocomposites fabrication.

MXene is beneficial in achieving the required functional properties. Particularly, this fabrication strategy is greatly efficient for the preparation of polymers, such as epoxides and polypropylene, among others, containing heterocyclic compounds or linear molecules (Ndayishimiye et al., 2020). These polymers have been integrated with MXene to prepare more functional nanocomposites to widen their application in many different fields.

Generally, direct mixing of the various constituents and intercalative polymerization of intercalation have been the most used synthesis methods in fabricating epoxy nanocomposites, owing to the fluidity of epoxy monomers. For *in situ* polymerization by intercalation, the epoxy molecular chains intercalate in between the layers of MXene, which consequentially leads to the increment in the distance between MXene layers and the stripping of layers after curing. Furthermore, the exfoliated nanosheets in the polymer matrix could contribute to the enhancement in surface area existing at the interface of the additive and the matrix. In other words, MXene nanosheets play a crucial role in attaining good interaction with the polymer in exfoliation via *in situ* intercalation.

A typical polymerization mechanism of DGEBA/MXene was prepared with the hardener acting as a catalyst initially resulting in ring opening and the breaking of C–O bonds in the monomer (Zhang et al., 2016). At the same time, an intermediate is formed, where the N–H links break and transfer H to O. Subsequently, the tertiary amine initiates the ring opening of the monomer to commence the reaction process. Upon the incorporation of MXene, its functional groups were attached to the epoxide groups (comparable to etherification reaction), alongside the proceeding ring opening. In other words, the MXene functioned as a mediator that promotes curing and the generation of a crosslinking between the numerous chains in the epoxy matrix.

6.4.2 Solution Mixing/Blending

A well-known and facile strategy for the fabrication of MXene/polymer nanocomposites is solution mixing/blending. Solution blending is mostly recommended for polar polymers. It constitutes the predominant technique used for composite fabrication due to the hydrophilic characteristic of MXene resulting from its large functional groups. In this fabrication process, MXene and polymer resin are separately dispersed, similar to the procedure used in *in situ* polymerization. The resin solution and MXene suspension are stirred together to ensure a good mixture of the two components. Alternatively, the MXene/polymer resin could be dispersed in a polar solvent that has good solubility for the polymer being processed. Upon stirring, a uniform viscous mixture is obtained. The final step in this method is the solvent evaporation process to remove the solvent. Thermoset polymer, such as polyurethane, have been successfully combined with MXene to produce nanocomposite using this method. The major demerits of this technique are as follows: weak mechanical properties, production of a large volume of environmental wastes resulting from the composites production process, and the challenge in drawing out solvents using evaporation. These challenges are considered impediments to the application of this technique in the fabrication of nanocomposites.

Aside from the production of MXene/polymer nanocomposites using hydrophilic polymers, namely, poly(diallyldimethylamonium chloride), poly(acrylamide), and polyethylene oxide, employing water as a solvent has been revealed. In addition to water, MXene/polymer nanocomposites are processable in other polar solvents. This is another area of study that remains unexplored, and an investigation of dispersion in a model polymer by this processing technique could make a significant contribution to the field of MXene/polymer nanocomposite research (Carey and Barsoum, 2021).

Another solvent-related method of preparation is ultrasonic mixing. In this technique, polymer and MXene are physically mixed, similar to the initial step of *in situ* polymerization. However, the ultrasonication process is applied for a while to achieve suitable dispersion of the MXene in the matrix, after which the composite is subjected to curing at the appropriate temperature. This method is especially suitable for epoxy/MXene-based composites. Another derivative of the solvent approach is the solution casting method, which involves separate dispersion of MXene and polymer materials in different solvents. The various dispersion is brought together by blending, followed by gradual removal of the solvent either by evaporation or freeze-drying approach (as shown in Figure 6.4), after which curing takes place at an appropriate temperature.

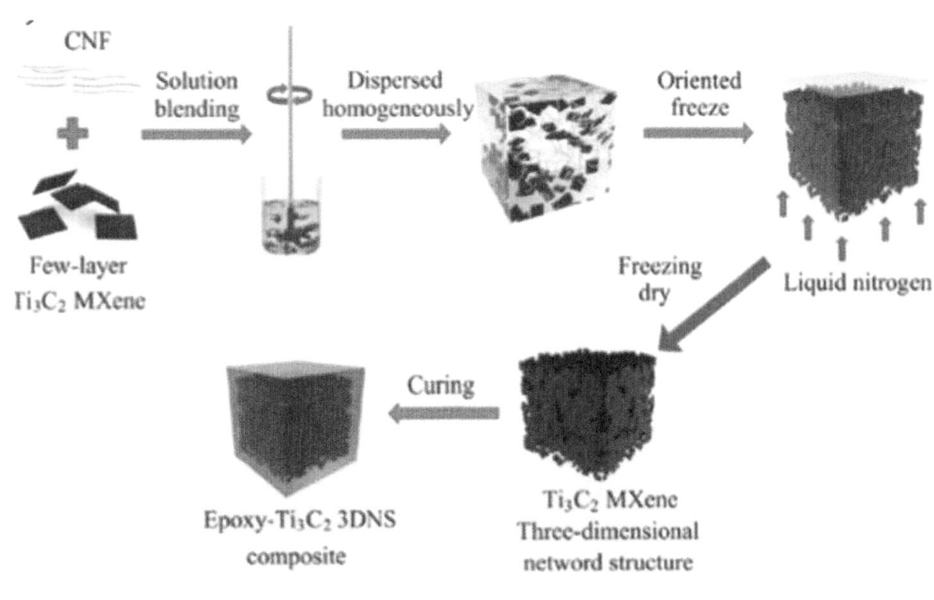

FIGURE 6.4 Freeze-drying of MXene/epoxy composite.

Source: Aghamohammadi et al. (2021).

6.4.3 Hot Press/Injection Molding

Hot press/injection molding is a facile technique used for fabricating MXene/polymers. This strategy is also referred to as melt blending. Unlike solution mixing, hot press/injection molding is mainly suitable for nonpolar polymers, which hinders its wider application. It is done by first melting the polymer, and subsequently adding the MXene to the polymer solution. Under a controlled condition, the dispersion of MXene and a melted polymer matrix are stirred to attain the desired viscosity. The composition is injected into a heated die within a period. It is then cooled down to obtain the solid nanocomposite. The extruder has been the main device used for mixing nanocomposite materials in this technique. Mostly, the mixing of MXene and polymer materials is done at high temperatures, specifically above the melting point of the polymer. This method has shown great potency in the production of nanocomposites in large quantities. Notably, the hot press/injection molding technique for nanocomposite fabrication finds several applications for hydrophobic polymers because hydrophobic polymers are not solution-or solvent-friendly (Jaya Prakash and Kandasubramanian, 2021).

Furthermore, due to the high bulk density of MXene (~4.26 g/cm^3), which is essential in easing the feeding process in this technique, MXene has significant advantages over graphene. Melt processing has been used to produce MXene/polymer nanocomposites with polymers, such as polyurethane and phthalonitrile (Derradji et al., 2019; Carey and Barsoum, 2021; Jaya Prakash and Kandasubramanian, 2021). With the quest to improve upon the compatibility of hydrophilic MXene and the nonpolar polymers, the surface of the MXene is subjected to modification. Compared to polymeric materials, MXene displays good stability at high degradation temperatures. Also, benefits such as the flexibility in readying the constituents, the noninvolvement of solvent, low-cost, and, above all, environmental-friendly.

6.4.4 Vacuum-Assisted Filtration

This technique is primarily used to dry nanocomposites by drawing out solvent from a dispersion. In this method, the driving force for filtration results from the application of suction at the area containing the filtrate as seen in Figure 6.5. It has been extensively used in the fabrication of nanocomposite films. Prior to suction, the desired quantity of MXene is added to a polymer

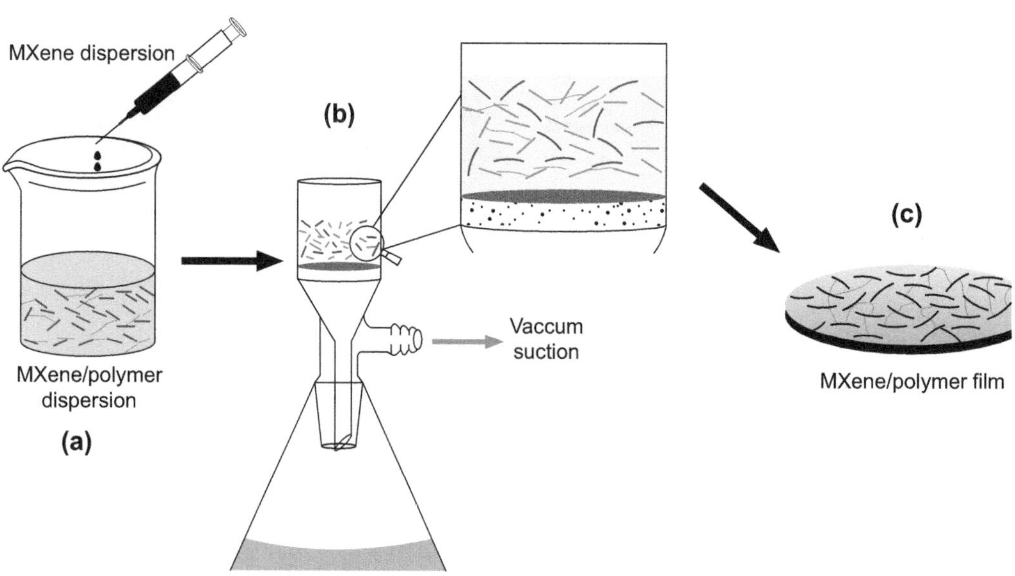

FIGURE 6.5 Vacuum-assisted filtration of MXene/polymer nanocomposite.

Source: Huang and Wu (2020).

aqueous solution, which might be containing a curing agent. The mixture could sometimes be diluted by deionized water, mixed for some minutes or hours, and/or sonicated to form uniform suspension before filtration with a membrane assisted by a vacuum pump to prepare MXene/polymer nanocomposite film. The final weight of the polymer used in the preparation of nanocomposite by this technique is most at times moderately lower than its nominal weight. The loss in weight is attributable to the loss of the polymer or some constituent in the polymer during the filtration. Subsequently, the film is heated in a vacuum oven to get rid of air pockets and then at an appropriate temperature and time (Mirkhani et al., 2019; Cao et al., 2018; Hu et al., 2020; Ji et al., 2020).

MXene/polymer film produced through regular vacuum-assisted filtration has the advantages of high conductivity, excellent mechanical performance, good surface wettability, and others. However, the production method is time and energy consuming. This limits the application of films prepared by vacuum-assisted filtration at a large scale (Wang et al., 2021; Wan et al., 2021; Chen et al., 2020a). A vacuum was used to dry a filter-coated $Ti_3C_2T_x$ dispersion by which a homogeneously dispersed and bubble-free polydimethylsiloxane solution was dropped on a polymeric filter. The decorated filter was sandwiched between two polyethylene terephthalate (PET) plates in its wet state and then cured at 100°C for 120 min (Hu et al., 2020).

6.4.5 Resin Transfer Molding

Resin transfer molding (RTM) is a flexible and economical substitute for autoclave processing used in the manufacture of intricate polymer composites that find application in structural constructions. It is a good method for producing dimensionally controlled parts, which are challenging to duplicate in series of productions. One of the main differences between this method and hot press/injection is the raw materials used. As the name implies, the low-viscous resin is used in RTM, while in hot press/injection molding, the polymer is melted to attain low viscosity. RTM is economically preferable to hot press/injection molding.

Two categories of RTM, namely, vacuum-assisted and pressure-assisted RTM methods, have been employed in the production of composites. In general, a resin is injected into the mold with the aid of a vacuum and/or pressure. The drawing of resin into the mold using vacuum rather than injection pressure is called vacuum-assisted RTM. The overriding merit of vacuum-assisted RTM is its prospect for lower cost in its setup and propensity to produce large-size composite parts. This process does not need a complex chamber or device for curing. Prior to curing a filtration of resin, a wet-state composite is done with a single-sided tool together with vacuum bag material. Before injection of resin/MXene suspension, it is of great importance to ensure that low viscosity is attained after stirring to facilitate the generation of the compact and less defective nanocomposite. To keep processing times to a minimum, the RTM is heated to accelerate the curing process and solidification of resin/MXene slurry. After curing, the mold cavity is opened, and the part removed. The MXene/polymer nanocomposites take the form of the mold and/or mandrel used during the fabrication. Although the advancement of RTM is still underway, it has been employed for decades in fabricating composites applied in the construction of commercial boats and high-performance composites. An illustration of RTM has been presented in Figure 6.6.

6.4.6 Template-Assisted Fabrication

Template-assisted fabrication method could also be produced thermoset/MXene nanocomposites. The template-assisted method enables the formation of porous architecture nanocomposite. This technique has been used for assembling 2D MXene into 3D porous macrostructures, so is a polymer. It is propounded to be more preferable in the formation because MXene has a high tendency to agglomerate and restack, which leads to the reduction in their specific surface area. This negatively affects the desirable properties of MXene. In light of that, the MXene flakes assemblage into 3D

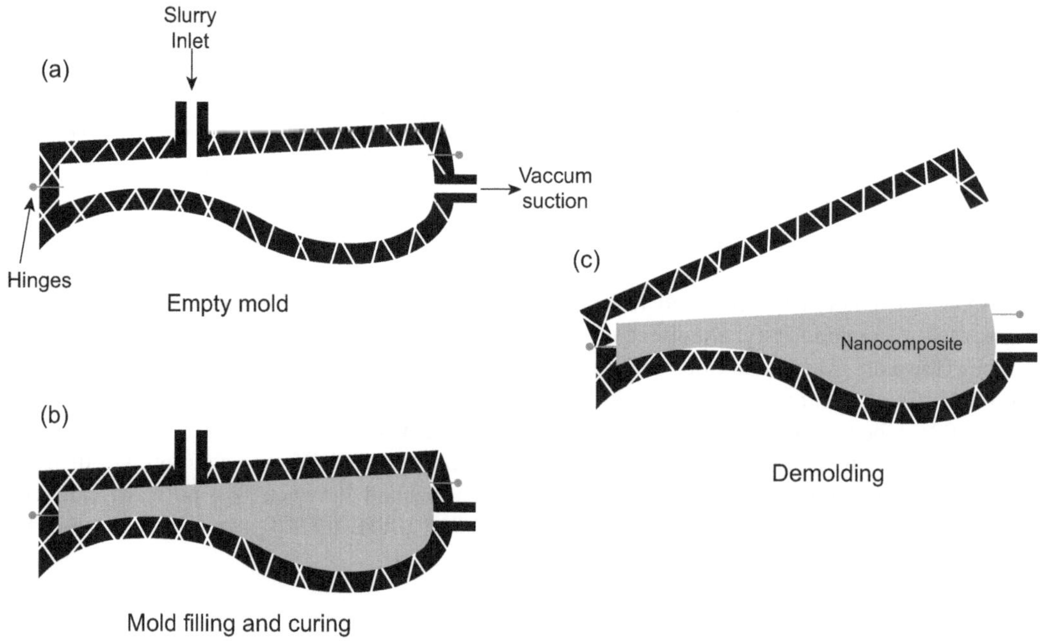

FIGURE 6.6 Graphical representation of the stages in RTM.

architecture is a good means of hindering stacking (Zhao et al., 2018). Primarily, the template-assisted method involves the employment of a sacrificial template to regulate the porosity and configuration of MXene/polymer in solvent. Some materials used as templates include salt, Al_2O_3, and others. These templates are leached out by dissolving in water or mostly by exposing the nanocomposite to heat treatment after the fabrication of the composites. Also, nanostructured materials in different shapes such as pillars, fibers, or spheres could be used as templates to create pores or architecture surfaces on the nanocomposite. Following the template-assisted fabrication technique, MXene flakes have been fabricated on the surface of the spherical sacrificial template to obtain a spherical hollow composite (Bu et al., 2020). The size and ratio of the sacrificial templates employed in templating the nanocomposites enact a critical role in tuning the dimensions of the nanocomposite's architecture.

The procedure for template-assisted fabrication is revealed in the following using some procedures used in some studies for the fabrication of nanocomposites. An MXene or a polymer could be templated before the incorporation of the two components. This implies that the thermoset polymer, in the form of solution could be incorporated after the preparation of the MXene architecture structure. Chen et al. used salt as a template in preparing epoxy/HBN–polyvinylidene difluoride nanocomposites, which yielded a relatively high conductivity to heat. Templating was done by first mixing the HBN–polyvinylidene difluoride with a salt followed by the dissolution of the polyvinylidene fluoride in acetone. The polyvinylidene fluoride in acetone served as an adhesive that raises the thinness in viscosity and hindered the precipitation of HBN–polyvinylidene difluoride. Upon the incorporation of a salt, insoluble in acetone, the HBN–polyvinylidene difluoride orients betwixt the salt particles, which leads to the formation of a continuous HBN–polyvinylidene difluoride complex after leaching out acetone and salt. The thermoset polymer solution was incorporated by vacuum-assisted impregnation to form the nanocomposite, during which the pores in the framework of HBN–poly vinylidene difluoride were occupied by epoxy without affecting its structure. This resulted in considerably good thermal conductivity of thermoset nanocomposites.

Another template-assisted fabrication technique used to obtain a 3D porous structure is freeze-casting. In this method, the architectural design structure of the end product is gorged from the shape of the mold employed during freeze-drying. This is to say, the shape of the mold determines the shape of the nanocomposites, because the nanocomposites take the form of the mold, hence the name freeze-casting. When employed effectively, not only the outer design but also the inner morphology, including the size and shape of the internal pores, can be tuned (Tontini et al., 2020). An ice-template method was used in preparing MXene and MXene/Ag aerogel before the fabrication of MXene-epoxy nanocomposite films. MXene and MXene/Ag were dispersed in poly(vinyl alcohol) aqueous solution, which was then poured into a cubic Teflon-coated mold glued on the surface of a copper pillar immersed in liquid nitrogen. After freezing the suspension, MXene/Ag aerogels were then prepared at the required conditions by freeze-drying at –55°C and 50 Pa for 2 days. They were subsequently annealed at 250°C for 2 h under Ar/H_2atmosphere. The template-assisted method has found few applications in fabricating thermoset nanocomposites; nevertheless, it is deemed appropriate for the fabrication of layer-to-layer film nanocomposites whereby the template materials are used to perforate a layer before the dispersion of MXene.

A cationic surfactant can be used to tune the hydrophilic character of MXene by the exchange of H^+ with the –OH groups to synthesize high internal-phase emulsions. The steadiness of the emulsions is highly dependent on the pH of the solutions. Furthermore, polymerization initiation of the binder remoted by temperature leads to the fabrication of a porous monolith with a cellular structure (Bian et al., 2018). Also, an electrospun polymer in the form of fibers or a 3D framework could be used as a template and immersed in an MXene dispersion, followed by drying in air or posttreatment to enhance the interaction of MXene on the polymer (Yuan et al., 2018; Estrany et al., 2016).

More so, a honeycomb Al_2O_3 structural-reduced graphene-assisted template was synthesized by high-temperature annealing and then freeze-dried using honeycombAl_2O_3 panels for MXene/epoxy composites. This method is suggested to provide regular structures, which could pave way for the regulation of design in terms of cell size, shape, and uniform dispersion of MXene in the matrix. The matrix was prepared according to the illustration in Figure 6.7, where the honeycomb

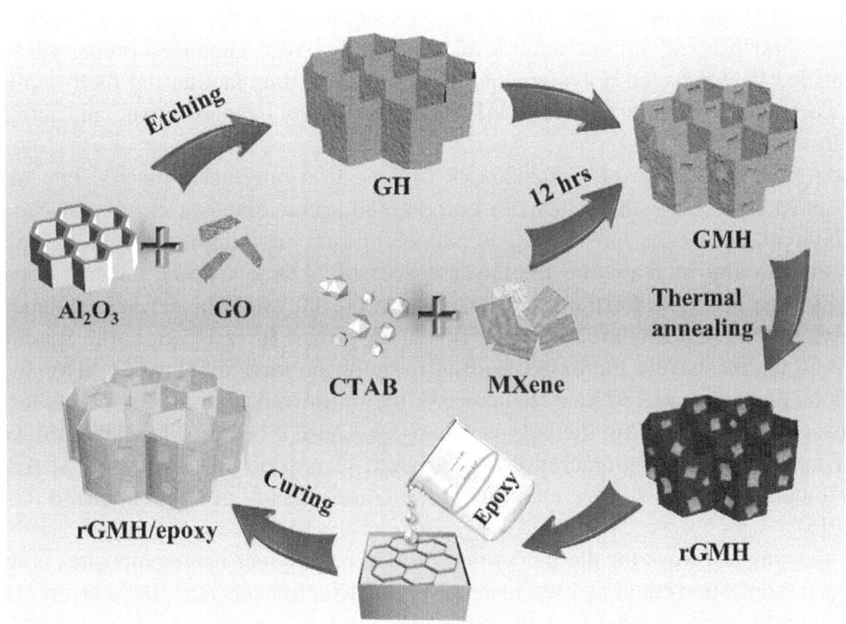

FIGURE 6.7 Honeycomb-assisted fabrication of rGMH/epoxy nanocomposites.

Source: Song et al. (2020).

structural reduced graphene oxide (rGO) was initially prepared with an Al_2O_3 honeycomb plate as a template. The MXene dispersion is mixed with quaternary ammonium surfactant to transform the cationic charge on the surface of MXene into negative charges. Subsequently, a honeycomb structural rGO-MXene with impeccable electrical conductivity, having good loadbearing is attained by MXene self-assembly on honeycomb structural rGO through electrostatic attraction and adsorption. After obtaining the honeycomb structural rGO-MXene, the mixtures of epoxy resin and curing agent are added and then cured using an appropriate time/duration to produce thermoset/epoxy nanocomposites (Song et al., 2020).

6.4.7 Ionic Intercalation and Sonication-Assisted Method

Prior to the discussion on thermoset/MXene intercalation, MXene has been mostly fabricated by ionic intercalation. Metal ions have been intercalated into MXene due to its high reactivity, chelating role, and ability to enhance the functional properties. Ionic intercalation of MXene has proved to be advantageous by reducing the pressure on less abundant lithium used in energy storage (Lei et al., 2020; Kajiyama et al., 2016). In addition, surface modification of MXene, such as Ti_3C_2Tx, with other nanoparticles (viz. sulfur and titanium dioxide) has ameliorated steadiness incapacity. Even though multivalent ionic species intercalated in MXene do have numerous interesting features, like higher charge-to-radii ratio, they are readily available than lithium and safer. However, drawbacks, such as low-charge efficiency, still exist. The challenges span from low suitability of electrolyte to unsatisfactory plating/stripping of the anode or excessive bondage between the MXene and the intercalated ions.

Even though MXene materials have displayed good properties for many applications in various fields, they are faced with serious challenges in achieving desirable flexible devices because of the high stiffness and relatively small lateral sizes. In light of that, flexibility could be introduced with polymers (Ma et al., 2016). Hence, thermoset/MXene nanocomposite, taking advantage of the flexible structure of polymer chains, molecular-level coupling between MXene and polymer molecules can be achieved by intercalating polymers into MXene flakes. In this way, the resultant MXene/polymer membranes could exhibit some level of flexibility with enhanced properties because of the synergistic effect between polymers and MXene flakes, thus facilitating their application for filtration, electromagnetic interference (EMI) shielding, energy storage, sensor, and actuators (Gao et al., 2020).

Depending on the degree of dispersion of MXene in a polymeric matrix, two main types of thermoset/MXene nanocomposites can be obtained: intercalated or exfoliated nanocomposites. The intercalation of thermoset/MXene nanocomposite in which the polymer chains diffuse between platelets and increase the interlayer spaces of MXene without causing separation of platelets has been revealed. Furthermore, exfoliation occurs when the stacked MXene structure is destroyed and the platelets are no longer bonded together. In exfoliation, the specific area of MXene could be extensively increased leading to more interactions between MXene and polymer solution. The exfoliated MXene structure is the optimized form of the 2D material which makes it easy to disperse within the polymer matrix because it is only a few layer thicker, hence, the reduced possibility of agglomeration. As a result, several properties such as stiffness, glass transition temperature, and storage modulus of the composite are greatly enhanced (Gauvin and Robert, 2015).

One of the simplest ways for the fabrication of MXene/polymer nanocomposites is the sonication approach. Sonication could be used to delaminate MXene sheets into single layers (Deshmukh et al., 2020). In this method, after initially stirring the mixture of polymer solution and MXene, the suspension is ultrasonicated for a period to achieve good distribution of the MXene. This method is usually employed in producing thermoset epoxy-based composites. Ultrasonication has been used together with other fabrication methods to improve the dispersion of MXene in an aqueous solution or polymer solution and for the intercalation of the polymer into MXene (Monastyreckis et al., 2020). It is mainly done by inserting the sonication rod in the suspension or placing a beaker containing the suspension in a sonication bath. The sonication-assisted method could be used to fabricate peroxide-decorated MXene, which needs no conventional initiator for polymerization (Riazi et al., 2021). MXene/polymer nanocomposite was fabricated by this method in which delamination and etching were achieved concurrently. It is suggested that cavitation bubbles, produced by sound waves generate hydrogen peroxide groups in the etching medium, which then reacts with the functional hydroxyl groups on the surface of MXene to generate peroxide groups. Through this process, the initiation of free-radical polymerization is feasible in the absence of a conventional initiator (Tao et al., 2019).

Aside from the extensively used vacuum-assisted filtration, casting, and hot press techniques, and so on, other fabrication strategies like layer-by-layer assembly, cold press, electrospinning, and electrochemical deposition have been used to produce 2D MXene/polymer nanocomposites with superior functional properties.

6.5 STRUCTURE AND MORPHOLOGY OF THERMOSET/MXene NANOCOMPOSITES

The structure and morphology of the individual components of the nanocomposite have some influence on the performance of the final product. As observed in the earlier sections, nanomaterials or nanocomposites could be structured by modification with templates, intercalation, exfoliation, and others. The morphological structure could also positively enhance the interaction of the components in a nanocomposite. The frequently used tools for examining the morphological structure of nanocomposites are X-ray diffraction (XRD) and scanning/transmission electron microscopy instruments. XRD could be used to determine the spaces between layers in structures using Bragg's law. For instance, a lower XRD diffraction angle is mostly indicative of a widened interlayer distance in the low-angle region, revealing the formation of an intercalated structure. On the other hand, the disappearance of peaks owing to the loss of structural registry of the layers indicates exfoliated component (MXene) or nanocomposite. Mostly, *in situ* polymerization leads to the formation of intercalated structures, whiles the exfoliation–adsorption process generates exfoliated nanostructures (Zhang et al., 2016), which could be confirmed by XRD diffraction peaks.

At low MXene fraction, the diffraction peaks related to MXene at 2^{θ}= 7.5° are weak and sometimes disappear, revealing that MXene layers have been exfoliated. Conversely, at a high fraction of MXene, the peak appears at 2^{θ}= 6.1°, indicating a large inter-distance layer that is most suitable for intercalation of thermoset polymer composites. Howbeit, nanocomposites containing a high load of MXene reveal a peak at 9.5°. This is evidenced that the MXene nanofillers are partially delaminated in the nanocomposites (Zhi et al., 2018). In the SEM surface structural micrograph (Figure 6.8), pure polymer showed a smooth and featureless fractured surface, but that of the MXene/polymer nanocomposite surfaces are typically rough, suggesting the intercalating of thermoset polymer molecular chains into the interlayers of MXene, which can serve as reinforcement for thermoset/MXene nanocomposites. The roughness of the nanocomposite increases as the ratio of MXene increases in the composition.

The morphology of MXene and thermoset epoxy/MXene composite reveals that the MXene develops a higher interlayer distance—typical of intercalated MXene prepared by etching (Zhang et al., 2016). In addition, the MXene Ti_3C_2Tx/epoxy nanocomposites display fractured and rough surfaces as compared to neat thermoset epoxy due to the intercalation of the polymeric molecular chains into Ti_2CTx layers as shown in Figure 6.8. Similar results have been reported in many studies related to thermoset/MXene nanocomposites (Aghamohammadi et al., 2021; Jaya Prakash and Kandasubramanian, 2021; Alhabeb et al., 2017b). The TEM micrographs give a clear view of the

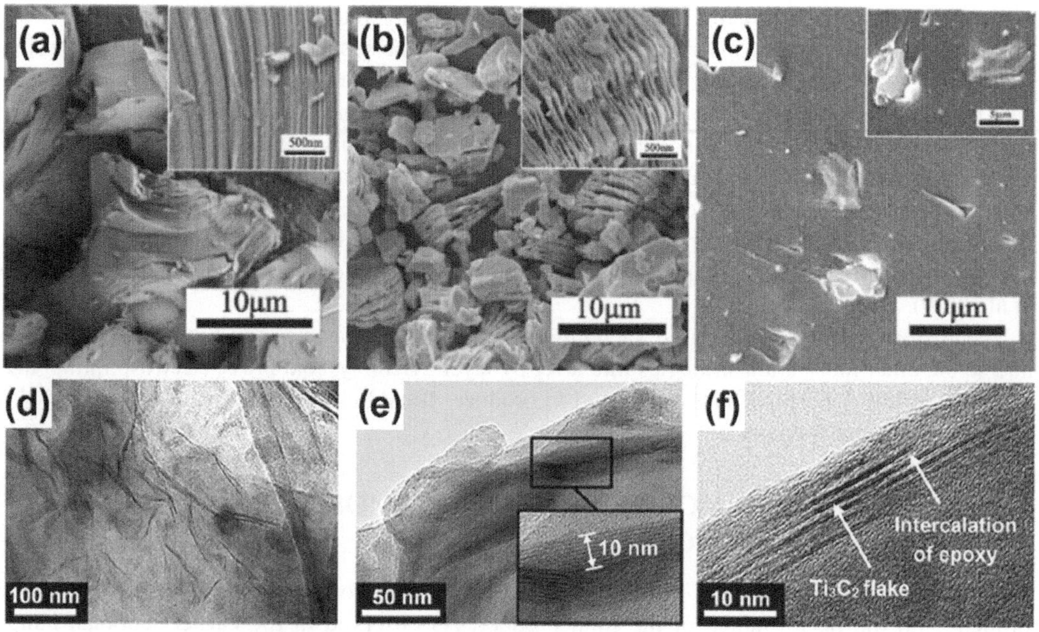

FIGURE 6.8 SEM and TEM images of MAX Ti_2AlC phase (a, d), MXene Ti_2CTx (b, e), MXene/epoxy (c, f) nanocomposite, respectively.

Source: Aghamohammadi et al. (2021).

intercalation of thermosetting epoxy polymer into the interlayers of MXene, thereby hindering the restacking of MXene, which results in good dispersion.

6.6 MECHANICS OF THERMOSET/MXene-BASED NANOCOMPOSITES

The composition of thermoset/MXene nanocomposite works hand in hand to present enhanced performance in many different areas. Understanding the interaction between polymer and MXene and their corresponding mechanisms in their application (wear resistance, electric shielding interference, and mechanical properties, etc.) is still premature. The mechanical properties of a nanocomposite can be enhanced by manipulating the crystallinity of the matrix. One of the major reasons that cause the enhancement in the performance in most fabricated thermoset/MXene nanocomposites is the existence of hydrogen bonding between OH groups of MXene and polymer. Strong interfacial interaction between Ti_3C_2Tx and the polymer matrix positively influences the surface hardness of the nanocomposite. Also, the magnitude of crosslinking in polymer could marginally increase at the interface of MXene, leading to the improvement of interaction between the polymer and MXene nanosheets. The unique layered morphology of Ti_3C_2Tx MXene enhances the mechanical properties of the polymer. Increasing the load of the MXene could result in more interaction at the MXene and polymer interface. This leads to the occurrence of more load transfer mechanisms to ensure wear resistance. It is noteworthy that excessive fraction of MXene could negatively affect the interfacial property and performance of the nanocomposite.

Furthermore, the amelioration in wear resistance of nanocomposite can also be attributed to van der Waal bonding. The mechanical and abrasion resistance of MXene/polymer nanocomposites could similarly be improved by the existence of van der Waals bonding, hydrogen bonding, and electrostatic forces at the interface of the polymeric material and the MXene layers as shown in

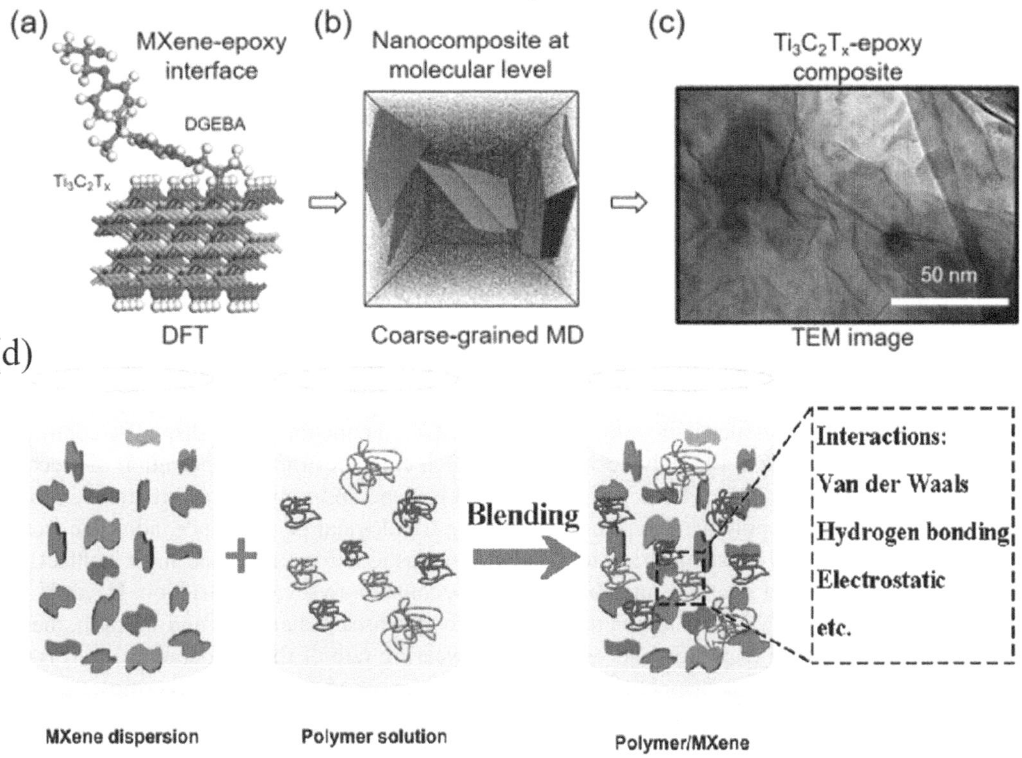

FIGURE 6.9 Interface bonding in MXene–epoxy composites: (a) Density function theory of DGEBA and MXene interface, (b) coarse-grained molecular dynamic simulation MXene–DGEBA composite interface, (c) TEM image of MXene–DGEBA interface, and (d) schematic illustration of MXene–DGEBA interface.

Source: Sliozberg et al. (2020) and Gao et al. (2020).

Figure 6.9d. In MXene/nonpolar polymers, a compatibilizer, such as maleic anhydride, is used to enhance hydrogen bonding in MXene/nonpolar polymer nanocomposites (Dong et al., 2016).

It has been established that the tensile strength of polymer nanocomposites can be enhanced by increasing the fraction of $Ti_3C_2T_x$ nanosheets in the matrix due to the physical hindrance of the polymer chains' mobility in the nanoconfinement structure. The stronger adhesion at Ti_3C_2Tx/polymer interface relates to the presence of multiple hydrogen bonds; as a result, the storage modulus and the elongation at break of nanocomposites get enhanced with a small percentage of $Ti_3C_2T_x$ nanosheets. The good distribution and exfoliation of the $Ti_3C_2T_x$ in the polymeric resin also account for the improvements in the mechanical properties of nanocomposites as the concentration of MXene increases. Also, the development of the confined nanoscale caused by the existence of numerous hydrogen bonds in the $Ti_3C_2T_x$ layers and polymer reduces the deformation and failure of the polymer chain to improve the strength of the nanocomposite. Generally, $Ti_3C_2T_x$ MXene/polymer nanocomposites display superior mechanical properties due to the good dispersion of MXene, the strong interaction, and the effective load transfer between the stiff MXene and the polymer matrix (Liu et al., 2021). Because of the versatile nature of MXene surface chemistry, organic addends that can form a good interconnection between MXene and polymers are used, resulting in the formation of reactions. The MXene acts as a bridge that holds polymer chains together. Besides the obvious effect of MXene on the mechanical properties of thermosets, their positive effects on heat conductivity can be also be assigned to strong bonding or compatibility in the nanocomposite between the MXene and polymer resulting from hydroxyl groups. Furthermore, the MXene has a high intrinsic thermal conductivity, which can elevate the thermal conductivity of polymer nanocomposites compared to most pristine polymers. In this function, the effects of MXene on thermal conductivity are more pronounced at relatively higher loading (Aghamohammadi et al., 2021). It has been established that MXene/polymer nanocomposite have superior advantages to 2D (graphene oxide/MoS_2/CNT)/polypropylene nanocomposites (Shi et al., 2019)

For thermal stability, a well-dispersed MXene in the nanocomposite could play a physical barrier role during thermal decomposition to improve the stability of the nanocomposite and concomitantly improve the mechanical properties. The transfer of heat is restrained due to the compatible interaction of the MXene layers with the polymeric material. In EMI shielding, the electromagnetic waves are reflected, scattered, and absorbed, leading to the lengthening of the paths of the electromagnetic waves, which results in electrical loss. In Ti_3C_2Tx MXene/natural rubber nanocomposite studies (Luo et al., 2019), larger absorption than reflection enhances the electromagnetic shielding effect of the MXene polymer nanocomposite. The $Ti_3C_2T_x$ nanocomposites display a continuous interactive filler network in the polymeric matrix, which enhances the transportation of electrons throughout the whole matrix. The negatively charged MXene and polymer ensure an even distribution of MXene at the polymer interfaces. Regarding tribological performance, adhesive wear is considered the principal wear mechanism. The severe plastic deformation and more contact area of the thermoset polymer coatings and counter-body become more severe in friction. Nevertheless, the incorporation of the MXene into the nanocomposite increases its hardness and strength, thereby restricting the distortion of plastic and reducing the wearing rate of the nanocomposite. It is proposed that the homogeneous dispersion and interconnection of MXene within thermoset polymer is the basis for the optimal efficiency of amino-functionalized $Ti_3C_2T_x$ in enhancing the wear and deformation resistance of nanocomposites, compared with MXene $Ti_3C_2T_x$ (Yan et al., 2020a). Generally, Young's modulus of MXene-thermoset epoxy composites is significant compared to the pristine epoxy because of the distribution of stress between the polymeric material and the MXene nanosheets. In terms of energy storage, the capacitance of nanocomposites such as MXene Ti_3C_2/polypyrrole membrane is about 30% higher than that of neat polyppyrole, with high stability in capacitance after thousands of charging/discharging cycles. In the electrochemical performance, the conductive polypyrrole chains were closely attached to the MXene layers, owing to the strong hydrogen bonds, which grants direct route for the transportation of electrons. Also, using electrochemical polymerization for polymer/MXene matrix, porous film structure could be achieved, due to the even dispersion of MXene flakes within the composite, which facilitates electrolyte permeation.

6.7 APPLICATION OF THERMOSET/MXene-BASED NANOCOMPOSITES

Due to the versatile and desirable properties of MXene in thermoset/MXene nanocomposites, it has remained the obvious option in multiple applications. The attractive qualities of thermoset/MXene nanocomposites such as good mechanical properties, electrical conductivity, and uniqueness in the structure make it an exceedingly favorable choice for high-performance composite applications. In addition to the previously mentioned properties, its chemical stability and ion inter-calatability make MXene more favorable for catalysis (electrocatalysis, photocatalysis), the detection of gas or changes in an environment (gas sensors, strain sensors, biosensors), biomedical (photothermal therapy, bio-imaging, bone regeneration, antimicrobial activity), energy storage (batteries, supercapacitors; Ji et al., 2020; Wei et al., 2020), photovoltaic devices, thermoelectric power generation), flame retardance, and EMI shielding applications (Wang et al., 2019; Liu et al., 2018b). With the increasing utilization of electronic equipment, the harm caused by electromagnetic radiation to human health and safety is ruinous. As a result, the use of MXene/polymer nanocomposites as EMI shielding materials has received great attention due to the excellent electromagnetic shielding performance of MXene in composites compared to most of the already existing materials (Wang et al., 2019). Aside from electromagnetic shielding, they are also used in other areas, such as opto/spintronic, for the processing of information, tribological, and medical applications. MXene also finds application in the manufacture of batteries and supercapacitors, providing a large energy density applicable for compact-volume uses in portable electronics and offering high power density and faster charging and discharging, which is ideal for mobile applications. Also, the existence of surface functional groups in large numbers and their biocompatibility can be tailored for biomedical applications. In MXene composites, the biocompatibility, high conductivity, and tunable surface properties of thermoset/MXene nanocomposites enhance their applicability in electrochemical biosensors (Xu et al., 2020). Coupled with stable mechanical properties and excellent surface characteristics, these materials are considered superior transducing materials for various sensors (such as bio-electrochemical, electrochemical, gas, strain, chemiluminescence, pressure, and photochemical sensors). Studies show that the low coefficient of friction, the possibility of super-lubricity effect through the uniaxial strains, and the low barrier for sliding make thermoset/MXene nanocomposites an excellent material for tribological applications.

6.8 CHALLENGES IN THERMOSET/MXene NANOCOMPOSITES: FABRICATION AND APPLICATIONS

Despite the extensive exploration of thermoset/MXene nanocomposites for various applications in recent years, challenges with environmental safety and of most polymers and advanced technologies for large exfoliation of MXene still exist. Hitherto, several resins employed in fabricating nanocomposites are obtained from petroleum sources. These resins are mostly not degradable and could exist for a number of years in an anaerobic environment. Undoubtedly, it is not worthwhile to have composites degrade during the application; nevertheless, they project a huge disadvantage to the environment after usage, owing to their nondegradability. Unlike the recyclability of thermoplastics, technology does not exist to reuse thermoset resins.

Although significant progress has been made with regards to MXene synthesis, most of the techniques do not ensure homogeneity in terminations across the surface of MXene layers. Theoretical studies have already shown that bare MXene shows better performance than surface-terminated ones in many applications, so it is recommendable to develop large-scale techniques to synthesize bare MXene (MAX) into refined MXene to enhance broader applications.

The issue of irrevocable restacking of MXene flakes during synthesis has been a major setback restricting further industrial exploitation and utilization for certain applications (Fu et al., 2018). The vacuum-assisted filtration that finds application in fabricating film nanocomposite materials is quite time-consuming and, therefore, not commercially sustainable. In other words, using this method to drain out solvent is inefficient, hence the need to improve on the fabrication method to reduce time and ensure high drying capacity (Chen et al., 2020a).RTM is considered to be an excellent strategy, but the issue with the cost of the tooling is a major setback for wide adoption. Also, process-induced defects such as microvoids, incomplete filling, a low degree of cure, and dry spots often hinder the broader application of nanocomposites fabricated by RTM in sophisticated or crucial applications in which high effectiveness and efficiency are required. Understanding the cause and the morphology of these defects, together with the mechanism and techniques for removing the nanocomposites is a vital step toward the development of quality RTM of nanocomposites.

The most used MXene (Ti_3C_2Tx) is prepared with hydrofluoric acid, which is a strong corrosive acid, which is harmful to human health and the environment. Additionally, the harsh etching aptness of hydrofluoric acid tends to cause defects on the surface of MXene layers. These defects come along with generated carbon balls on the surface of the MXene sheets, which is unfavorable for the application of MXene. This necessitates the quest to undertake studies to obtain a more effective and harm-free technique to etch MAX-phase TiAlC into MXene $Ti_3C_2T_x$ (Song et al., 2021b).

6.9 CONCLUSION

Generally, the remarkable attributes MXene has led to a great improvement in the functional properties of not only thermoset nanocomposites but also, other polymeric composites, even at low loadings. The functional groups of MXene and the generation of hydrogen bonding improve the dispersion of MXene within matrices and ensure good interfacial interaction which influences the mechanical properties of polymer nanocomposites. Also, the layered morphology and the self-lubricating films on the surface of polymers, spearing from their interaction with MXene enhances the tribological properties of polymeric composites. Many different fabrication strategies such as *in situ* polymerization blending, solution mixing, injection molding, vacuum-assisted filtration, template-assisted fabrication, and RTM are the cardinal techniques for preparing functionally novel nanocomposites. New emerging methods for the construction of architecture nanocomposites are underway to improve the properties of thermoset/MXene nanocomposites. Thermoset/MXene nanocomposites have found application in many areas even though most of them are still in their early stage of development. The proper dispersion of the MXene into the polymers, strong hydrogen bonds, and van der Waals bonding, which forms the major anchor in sustaining the nanocomposites, improved their functionalities by heat and electrical conduction, EMI shielding, and so on. Undoubtedly, the interfacial interaction greatly impacts the required outcome of the composite and is one of the huge drawbacks in the fabrication of quality thermoset/MXene nanocomposites for wider applications, especially for MXene/nonpolar polymers. Consequently, research focusing on enhancing properties in MXene/nonpolar polymer nanocomposites can be achieved by surface modification with a solid comprehension of the interaction between the nanofillers and the polymeric matrix.

REFERENCES

AGHAMOHAMMADI, H., AMOUSA, N. & ESLAMI-FARSANI, R. 2021. Recent advances in developing the MXene/polymer nanocomposites with multiple properties: A review study. *Synthetic Metals*, 273, 116695.

AHMED, B., GHAZALY, A. E. & ROSEN, J. 2020. i-MXenes for energy storage and catalysis. *Advanced Functional Materials*, 30, 2000894.

ALHABEB, M., MALESKI, K., ANASORI, B., LELYUKH, P., CLARK, L., SIN, S. & GOGOTSI, Y. 2017a. Guidelines for synthesis and processing of two-dimensional titanium carbide (Ti3C2T x MXene). *Chemistry of Materials*, 29, 7633–7644.

ALHABEB, M., MALESKI, K., ANASORI, B., LELYUKH, P., CLARK, L., SIN, S. & GOGOTSI, Y. 2017b. Guidelines for Synthesis and Processing of Two-Dimensional Titanium Carbide (Ti3C2Tx MXene). *Chemistry of Materials*, 29, 7633–7644.

ANASORI, B., LUKATSKAYA, M. R. & GOGOTSI, Y. 2017. 2D metal carbides and nitrides (MXenes) for energy storage. *Nature Reviews Materials*, 2, 16098.

BIAN, R., LIN, R., WANG, G., LU, G., ZHI, W., XIANG, S., WANG, T., CLEGG, P. S., CAI, D. & HUANG, W. 2018. 3D assembly of Ti3C2-MXene directed by water/oil interfaces. *Nanoscale*, 10, 3621–3625.

BU, F., ZAGHO, M. M., IBRAHIM, Y., MA, B., ELZATAHRY, A. & ZHAO, D. 2020. Porous MXenes: Synthesis, structures, and applications. *Nano Today*, 30, 100803.

CAO, W.-T., CHEN, F.-F., ZHU, Y.-J., ZHANG, Y.-G., JIANG, Y.-Y., MA, M.-G. & CHEN, F. 2018. Binary strengthening and toughening of MXene/cellulose nanofiber composite paper with nacre-inspired structure and superior electromagnetic interference shielding properties. *ACS Nano*, 12, 4583–4593.

CAREY, M. & BARSOUM, M. W. 2021. MXene polymer nanocomposites: A review. *Materials Today Advances*, 9, 100120.

CHEN, H., WEN, Y., QI, Y., ZHAO, Q., QU, L. & LI, C. 2020a. Pristine titanium carbide MXene films with environmentally stable conductivity and superior mechanical strength. *Advanced Functional Materials*, 30, 1906996.

CHEN, W., LIU, L.-X., ZHANG, H.-B. & YU, Z.-Z. 2020b. Flexible, transparent, and conductive Ti3C2Tx MXene–silver nanowire films with smart acoustic sensitivity for high-performance electromagnetic interference shielding. *ACS Nano*, 14, 16643–16653.

CHIANG, T. C., HAMDAN, S. & OSMAN, M. S. 2016. Urea formaldehyde composites reinforced with sago fibres analysis by FTIR, TGA, and DSC. *Advances in Materials Science and Engineering*, 2016, 5954636.

CHIGWADA, G., JASH, P., JIANG, D. D. & WILKIE, C. A. 2005. Fire retardancy of vinyl ester nanocomposites: Synergy with phosphorus-based fire retardants. *Polymer Degradation and Stability*, 89, 85–100.

COME, J., BLACK, J. M., LUKATSKAYA, M. R., NAGUIB, M., BEIDAGHI, M., RONDINONE, A. J., KALININ, S. V., WESOLOWSKI, D. J., GOGOTSI, Y. & BALKE, N. 2015. Controlling the actuation properties of MXene paper electrodes upon cation intercalation. *Nano Energy*, 17, 27–35.

DERRADJI, M., TRACHE, D., HENNICHE, A., ZEGAOUI, A., MEDJAHED, A., TARCHOUN, A. F. & BELGACEMI, R. 2019. On the preparation and properties investigations of highly performant MXene (Ti3C2(OH)2) nanosheets-reinforced phthalonitrile nanocomposites. *Advanced Composites Letters*, 28, 2633366X19890621.

DESHMUKH, K., KOVÁŘÍK, T. & KHADHEER PASHA, S. K. 2020. State of the art recent progress in two dimensional MXenes based gas sensors and biosensors: A comprehensive review. *Coordination Chemistry Reviews*, 424, 213514.

DODIUK, H. 2013. *Handbook of thermoset plastics*. William Andrew.

DONG, M., LI, Q., LIU, H., LIU, C., WUJCIK, E. K., SHAO, Q., DING, T., MAI, X., SHEN, C. & GUO, Z. 2018. Thermoplastic polyurethane-carbon black nanocomposite coating: Fabrication and solid particle erosion resistance. *Polymer*, 158, 381–390.

DONG, Y., ZHANG, C., ZHAO, G., GUAN, Y., GAO, A. & SUN, W. 2016. Constitutive equation and processing maps of an Al–Mg–Si aluminum alloy: Determination and application in simulating extrusion process of complex profiles. *Materials & Design*, 92, 983–997.

ESTRANY, F., CALVET, A., DEL VALLE, L. J., PUIGGALÍ, J. & ALEMÁN, C. 2016. A multi-step template-assisted approach for the formation of conducting polymer nanotubes onto conducting polymer films. *Polymer Chemistry*, 7, 3540–3550.

FRIGIONE, M. & LETTIERI, M. 2020. Recent advances and trends of nanofilled/nanostructured epoxies. *Materials*, 13, 3415.

FU, Q., WANG, X., ZHANG, N., WEN, J., LI, L., GAO, H. & ZHANG, X. 2018. Self-assembled Ti3C2Tx/SCNT composite electrode with improved electrochemical performance for supercapacitor. *Journal of Colloid and Interface Science*, 511, 128–134.

FU, S., SUN, Z., HUANG, P., LI, Y. & HU, N. 2019. Some basic aspects of polymer nanocomposites: A critical review. *Nano Materials Science*, 1, 2–30.

GAO, L., LI, C., HUANG, W., MEI, S., LIN, H., OU, Q., ZHANG, Y., GUO, J., ZHANG, F., XU, S. & ZHANG, H. 2020. MXene/polymer membranes: synthesis, properties, and emerging applications. *Chemistry of Materials*, 32, 1703–1747.

GAUVIN, F. & ROBERT, M. 2015. Durability study of vinylester/silicate nanocomposites for civil engineering applications. *Polymer Degradation and Stability*, 121, 359–368.

GIBSON, G. 2017. Chapter 27—epoxy resins. In: GILBERT, M. (ed.) *Brydson's plastics materials (Eighth edition)*. Butterworth-Heinemann.

HAI, Y., JIANG, S., ZHOU, C., SUN, P., HUANG, Y. & NIU, S. 2020. Fire-safe unsaturated polyester resin nanocomposites based on MAX and MXene: A comparative investigation of their properties and mechanism of fire retardancy. *Dalton Transactions*, 49, 5803–5814.

HUANG, X. & WU, P. 2020. A small amount of delaminated Ti3C2 flakes to greatly enhance the thermal conductivity of boron nitride papers by assembling a well-designed interface. *Materials Chemistry Frontiers*, 4, 292–301.

HU, D., HUANG, X., LI, S. & JIANG, P. 2020. Flexible and durable cellulose/MXene nanocomposite paper for efficient electromagnetic interference shielding. *Composites Science and Technology*, 188, 107995.

JAYA PRAKASH, N. & KANDASUBRAMANIAN, B. 2021. Nanocomposites of MXene for industrial applications. *Journal of Alloys and Compounds*, 862, 158547.

JI, C., WANG, Y., YE, Z., TAN, L., MAO, D., ZHAO, W., ZENG, X., YAN, C., SUN, R., KANG, D. J., XU, J. & WONG, C.-P. 2020. Ice-templated MXene/Ag–epoxy nanocomposites as high-performance thermal management materials. *ACS Applied Materials & Interfaces*, 12, 24298–24307.

KAJIYAMA, S., SZABOVA, L., SODEYAMA, K., IINUMA, H., MORITA, R., GOTOH, K., TATEYAMA, Y., OKUBO, M. & YAMADA, A. 2016. Sodium-ion intercalation mechanism in MXene nanosheets. *ACS Nano*, 10, 3334–3341.

KIM, J.-W., SAUTI, G., SIOCHI, E. J., SMITH, J. G., WINCHESKI, R. A., CANO, R. J., CONNELL, J. W. & WISE, K. E. 2014. Toward high performance thermoset/carbon nanotube sheet nanocomposites via resistive heating assisted infiltration and cure. *ACS Applied Materials & Interfaces*, 6, 18832–18843.

KIM, Y.-J., KIM, S. J., SEO, D., CHAE, Y., ANAYEE, M., LEE, Y., GOGOTSI, Y., AHN, C. W. & JUNG, H.-T. 2021. Etching mechanism of monoatomic aluminum layers during MXene synthesis. *Chemistry of Materials*, 33(16), 6346–6355.

KRISHNAMOORTI, R. & VAIA, R. A. 2007. *Polymer nanocomposites*. Wiley Online Library.

LEE, E., VAHIDMOHAMMADI, A., YOON, Y. S., BEIDAGHI, M. & KIM, D.-J. 2019. Two-dimensional vanadium carbide MXene for gas sensors with ultrahigh sensitivity toward nonpolar gases. *ACS Sensors*, 4, 1603–1611.

LEI, Y.-J., YAN, Z.-C., LAI, W.-H., CHOU, S.-L., WANG, Y.-X., LIU, H.-K. & DOU, S.-X. 2020. Tailoring MXene-based materials for sodium-ion storage: Synthesis, mechanisms, and applications. *Electrochemical Energy Reviews*, 3, 766–792.

LI, D., LIU, G., ZHANG, Q., QU, M., FU, Y. Q., LIU, Q. & XIE, J. 2021. Virtual sensor array based on MXene for selective detections of VOCs. *Sensors and Actuators B: Chemical*, 331, 129414.

LI, J., YUAN, X., LIN, C., YANG, Y., XU, L., DU, X., XIE, J., LIN, J. & SUN, J. 2017. Achieving high pseudocapacitance of 2D titanium carbide (MXene) by cation intercalation and surface modification. *Advanced Energy Materials*, 7, 1602725.

LIU, D., WANG, R., CHANG, W., ZHANG, L., PENG, B., LI, H., LIU, S., YAN, M. & GUO, C. 2018a. Ti3C2 MXene as an excellent anode material for high-performance microbial fuel cells. *Journal of Materials Chemistry A*, 6, 20887–20895.

LIU, L., YING, G., WEN, D., ZHANG, K., HU, C., ZHENG, Y., ZHANG, C., WANG, X. & WANG, C. 2021. Aqueous solution-processed MXene (Ti3C2Tx) for non-hydrophilic epoxy resin-based composites with enhanced mechanical and physical properties. *Materials & Design*, 197, 109276.

LIU, R., MIAO, M., LI, Y., ZHANG, J., CAO, S. & FENG, X. 2018b. Ultrathin biomimetic polymeric Ti3C2Tx MXene composite films for electromagnetic interference shielding. *ACS Applied Materials & Interfaces*, 10, 44787–44795.

LIU, Z., WANG, W., TAN, J., LIU, J., ZHU, M., ZHU, B. & ZHANG, Q. 2020. Bioinspired ultra-thin polyurethane/MXene nacre-like nanocomposite films with synergistic mechanical properties for electromagnetic interference shielding. *Journal of Materials Chemistry C*, 8, 7170–7180.

LUKATSKAYA, M. R., MASHTALIR, O., REN, C. E., DALL'AGNESE, Y., ROZIER, P., TABERNA, P. L., NAGUIB, M., SIMON, P., BARSOUM, M. W. & GOGOTSI, Y. 2013. Cation intercalation and high volumetric capacitance of two-dimensional titanium carbide. *Science*, 341, 1502–1505.

LUO, J.-Q., ZHAO, S., ZHANG, H.-B., DENG, Z., LI, L. & YU, Z.-Z. 2019. Flexible, stretchable and electrically conductive MXene/natural rubber nanocomposite films for efficient electromagnetic interference shielding. *Composites Science and Technology*, 182, 107754.

MARTIN, C., RONDA, J. & CADIZ, V. 2006. Development of novel flame-retardant thermosets based on boron-modified phenol–formaldehyde resins. *Journal of Polymer Science Part A: Polymer Chemistry*, 44, 3503–3512.

MA, S., WEBSTER, D. C. & JABEEN, F. 2016. Hard and flexible, degradable thermosets from renewable bioresources with the assistance of water and ethanol. *Macromolecules*, 49, 3780–3788.

MATOS RUIZ, M., CAVAILLÉ, J. Y., DUFRESNE, A., GÉRARD, J. F. & GRAILLAT, C. 2000. Processing and characterization of new thermoset nanocomposites based on cellulose whiskers. *Composite Interfaces*, 7, 117–131.

MGBEMENA, C. O., LI, D., LIN, M.-F., LIDDEL, P. D., KATNAM, K. B., THAKUR, V. K. & NEZHAD, H. Y. 2018. Accelerated microwave curing of fibre-reinforced thermoset polymer composites for structural applications: A review of scientific challenges. *Composites Part A: Applied Science and Manufacturing*, 115, 88–103.

MING, F., LIANG, H., HUANG, G., BAYHAN, Z. & ALSHAREEF, H. N. 2021. MXenes for rechargeable batteries beyond the lithium-ion. *Advanced Materials*, 33, 2004039.

MIRKHANI, S. A., SHAYESTEH ZERAATI, A., ALIABADIAN, E., NAGUIB, M. & SUNDARARAJ, U. 2019. High dielectric constant and low dielectric loss via poly(vinyl alcohol)/Ti3C2Tx MXene nanocomposites. *ACS Applied Materials & Interfaces*, 11, 18599–18608.

MISHRA, A., SRIVASTAVA, P., CARRERAS, A., TANAKA, I., MIZUSEKI, H., LEE, K.-R. & SINGH, A. K. 2017. Atomistic origin of phase stability in oxygen-functionalized MXene: A comparative study. *The Journal of Physical Chemistry C*, 121, 18947–18953.

MISKOLCZI, N. 2013a. 3—Polyester resins as a matrix material in advanced fibre-reinforced polymer (FRP) composites. In: BAI, J. (ed.) *Advanced fibre-reinforced polymer (FRP) composites for structural applications*. Woodhead Publishing.

MISKOLCZI, N. 2013b. Polyester resins as a matrix material in advanced fibre-reinforced polymer (FRP) composites. In: *Advanced fibre-reinforced polymer (FRP) composites for structural applications*. Elsevier.

MONASTYRECKIS, G., MISHNAEVSKY, L., HATTER, C. B., ANISKEVICH, A., GOGOTSI, Y. & ZELENIAKIENE, D. 2020. Micromechanical modeling of MXene-polymer composites. *Carbon*, 162, 402–409.

NDAYISHIMIYE, A., GRADY, Z. A., TSUJI, K., WANG, K., BANG, S. H. & RANDALL, C. A. 2020. Thermosetting polymers in cold sintering: The fabrication of ZnO-polydimethylsiloxane composites. *Journal of the American Ceramic Society*, 103, 3039–3050.

PAUL, D. R. & ROBESON, L. M. 2008. Polymer nanotechnology: Nanocomposites. *Polymer*, 49, 3187–3204.

RASOOL, K., MAHMOUD, K. A., JOHNSON, D. J., HELAL, M., BERDIYOROV, G. R. & GOGOTSI, Y. 2017. Efficient antibacterial membrane based on two-dimensional Ti(3)C(2)T(x) (MXene) Nanosheets. *Sci Rep*, 7, 1598.

RATNA, D. 2012. 3—Thermal properties of thermosets. In: GUO, Q. (ed.) *Thermosets*. Woodhead Publishing.

RIAZI, H., NEMANI, S. K., GRADY, M. C., ANASORI, B. & SOROUSH, M. 2021. Ti3C2 MXene–polymer nanocomposites and their applications. *Journal of Materials Chemistry A*, 9, 8051–8098.

ROLLER, M. 1986. Rheology of curing thermosets: A review. *Polymer Engineering & Science*, 26, 432–440.

SHI, Y., LIU, C., LIU, L., FU, L., YU, B., LV, Y., YANG, F. & SONG, P. 2019. Strengthening, toughing and thermally stable ultra-thin MXene nanosheets/polypropylene nanocomposites via nanoconfinement. *Chemical Engineering Journal*, 378, 122267.

SI, J. Y., TAWIAH, B., SUN, W. L., LIN, B., WANG, C., YUEN, A. C. Y., YU, B., LI, A., YANG, W., LU, H. D., CHAN, Q. N. & YEOH, G. H. 2019. Functionalization of MXene nanosheets for polystyrene towards high thermal stability and flame retardant properties. *Polymers (Basel)*, 11.

SINGHA, A. S. & THAKUR, V. K. 2009. Study of mechanical properties of urea-formaldehyde thermosets reinforced by pine needle powder. *BioResources*, 4, 292–308.

SLIOZBERG, Y., ANDZELM, J., HATTER, C. B., ANASORI, B., GOGOTSI, Y. & HALL, A. 2020. Interface binding and mechanical properties of MXene-epoxy nanocomposites. *Composites Science and Technology*, 192, 108124.

SONG, J., CHEN, S., YI, X., ZHAO, X., ZHANG, J., LIU, X. & LIU, B. 2021a. Preparation and properties of the urea-formaldehyde res-in/reactive halloysite nanocomposites adhesive with low-formaldehyde emission and good water resistance. *Polymers*, 13, 2224.

SONG, P., LIU, B., QIU, H., SHI, X., CAO, D. & GU, J. 2021b. MXenes for polymer matrix electromagnetic interference shielding composites: A review. *Composites Communications*, 24, 100653.

SONG, P., QIU, H., WANG, L., LIU, X., ZHANG, Y., ZHANG, J., KONG, J. & GU, J. 2020. Honeycomb structural rGO-MXene/epoxy nanocomposites for superior electromagnetic interference shielding performance. *Sustainable Materials and Technologies*, 24, e00153.

SUN, Y., CHEN, D. & LIANG, Z. 2017. Two-dimensional MXenes for energy storage and conversion applications. *Materials Today Energy*, 5, 22–36.

SUN, Y., DING, R., HONG, S. Y., LEE, J., SEO, Y.-K., NAM, J.-D. & SUHR, J. 2021. MXene-xanthan nanocomposite films with layered microstructure for electromagnetic interference shielding and Joule heating. *Chemical Engineering Journal*, 410, 128348.

TAO, N., ZHANG, D., LI, X., LOU, D., SUN, X., WEI, C., LI, J., YANG, J. & LIU, Y.-N. 2019. Near-infrared light-responsive hydrogels via peroxide-decorated MXene-initiated polymerization. *Chemical Science*, 10, 10765–10771.

TONTINI, G., GREAVES, M., GHOSH, S., BAYRAM, V. & BARG, S. 2020. MXene-based 3D porous macrostructures for electrochemical energy storage. *Journal of Physics: Materials*, 3, 022001.

VERGER, L., NATU, V., CAREY, M. & BARSOUM, M. W. 2019. MXenes: An introduction of their synthesis, select properties, and applications. *Trends in Chemistry*, 1, 656–669.

WAN, Y.-J., RAJAVEL, K., LI, X.-M., WANG, X.-Y., LIAO, S.-Y., LIN, Z.-Q., ZHU, P.-L., SUN, R. & WONG, C.-P. 2021. Electromagnetic interference shielding of Ti3C2Tx MXene modified by ionic liquid for high chemical stability and excellent mechanical strength. *Chemical Engineering Journal*, 408, 127303.

WANG, J., KANG, H., MA, H., LIU, Y., XIE, Z., WANG, Y. & FAN, Z. 2021. Super-Fast Fabrication of MXene Film through a Combination of Ion Induced Gelation and Vacuum-Assisted Filtration. *Engineered Science*.

WANG, L., CHEN, L., SONG, P., LIANG, C., LU, Y., QIU, H., ZHANG, Y., KONG, J. & GU, J. 2019. Fabrication on the annealed Ti3C2Tx MXene/Epoxy nanocomposites for electromagnetic interference shielding application. *Composites Part B: Engineering*, 171**,** 111–118.

WEI, D., WU, W., ZHU, J., WANG, C., ZHAO, C. & WANG, L. 2020. A facile strategy of polypyrrole nanospheres grown on Ti3C2-MXene nanosheets as advanced supercapacitor electrodes. *Journal of Electroanalytical Chemistry*, 877**,** 114538.

XU, B., ZHI, C. & SHI, P. 2020. Latest advances in MXene biosensors. *JPhys Materials*, 3.

YAN, H., ZHANG, L., LI, H., FAN, X. & ZHU, M. 2020a. Towards high-performance additive of Ti3C2/ graphene hybrid with a novel wrapping structure in epoxy coating. *Carbon*, 157**,** 217–233.

YAN, J., MA, Y., LI, X., ZHANG, C., CAO, M., CHEN, W., LUO, S., ZHU, M. & GAO, Y. 2020b. Flexible and high-sensitivity piezoresistive sensor based on MXene composite with wrinkle structure. *Ceramics International*, 46**,** 23592–23598.

YU, B., YUEN, A. C. Y., XU, X., ZHANG, Z.-C., YANG, W., LU, H., FEI, B., YEOH, G. H., SONG, P. & WANG, H. 2021. Engineering MXene surface with POSS for reducing fire hazards of polystyrene with enhanced thermal stability. *Journal of Hazardous Materials*, 401**,** 123342.

YUAN, W., YANG, K., PENG, H., LI, F. & YIN, F. 2018. A flexible VOCs sensor based on a 3D Mxene framework with a high sensing performance. *Journal of Materials Chemistry A*, 6**,** 18116–18124.

ZAFERANI, S. H. 2018. 1—Introduction of polymer-based nanocomposites. In: JAWAID, M. & KHAN, M. M. (eds.) *Polymer-based nanocomposites for energy and environmental applications.* Woodhead Publishing.

ZHANG, H., WANG, L., ZHOU, A., SHEN, C., DAI, Y., LIU, F., CHEN, J., LI, P. & HU, Q. 2016. Effects of 2-D transition metal carbide Ti2CTx on properties of epoxy composites. *RSC Advances*, 6**,** 87341–87352.

ZHAO, D., CHEN, Z., YANG, W., LIU, S., ZHANG, X., YU, Y., CHEONG, W.-C., ZHENG, L., REN, F., YING, G., CAO, X., WANG, D., PENG, Q., WANG, G. & CHEN, C. 2019. MXene (Ti3C2) Vacancy-confined single-atom catalyst for efficient functionalization of CO2. *Journal of the American Chemical Society*, 141**,** 4086–4093.

ZHAO, D., CLITES, M., YING, G., KOTA, S., WANG, J., NATU, V., WANG, X., POMERANTSEVA, E., CAO, M. & BARSOUM, M. W. 2018. Alkali-induced crumpling of Ti3C2Tx (MXene) to form 3D porous networks for sodium ion storage. *Chemical Communications*, 54**,** 4533–4536.

ZHENG, H., GUO, M., SHAO, Y., WANG, Y., LIU, B. & MENG, G. 2018. Graphene oxide–poly(urea–formaldehyde) composites for corrosion protection of mild steel. *Corrosion Science*, 139**,** 1–12.

ZHI, W., XIANG, S., BIAN, R., LIN, R., WU, K., WANG, T. & CAI, D. 2018. Study of MXene-filled polyurethane nanocomposites prepared via an emulsion method. *Composites Science and Technology*, 168**,** 404–411.

ZHU, Z., LIU, C., JIANG, F., LIU, J., MA, X., LIU, P., XU, J., WANG, L. & HUANG, R. 2020. Flexible and lightweight Ti3C2Tx MXene@Pd colloidal nanoclusters paper film as novel H2 sensor. *Journal of Hazardous Materials*, 399**,** 123054.

7 Thermal and Crystallization Behavior of MXene/Polymer Nanocomposites

Sarkodie Bismark and Benjamin Tawiah

CONTENTS

7.1 INTRODUCTION

Most natural and synthetic polymers have semi-crystalline structures, which affect their mechanical, optical, and thermal properties. For this reason, it is imperative to comprehend the behavior of polymer crystallization upon exposure to different conditions. Polymer crystallization can be induced by factors such as compositions, thermal history, chemical structures, spatial confinements, and pressure. To understand the crystallization phenomenon of polymeric materials, theoretical and simulation is usually combined with systematic experimental investigations [1]. Research works related to the crystallization process of polymers are of great importance in their processing, due to the dependency of the resulting physical properties on the morphology formed and the extent of crystallization [2].

Thermal behavior is a major limiting factor in the processability and application of polymer nanocomposites. This property is again closely linked crystallization behavior of polymers and their composites [3, 4]. With the quest to reduce thermal degradation of nanocomposites, especially for the obvious advantages of flame retardancy, polymer crystallization behavior, a thorough understanding of the crystallization and the melting phenomenon of polymers is vital. The crystallization behavior of most semi-crystalline polymers is affected when other materials are introduced into their matrix for composites. More often, the specific thermal characteristics of the polymer composites, such as glass transition temperature, crystallization phenomenon, and melting behavior, are either enhanced or compromised. As a result, it is important to understudy the dynamics of the various materials used in polymers for the formation of composites and their mechanistic effects on the ensuing composites. Usually, nanofillers are incorporated into polymer systems to ameliorate their thermal stability and improve mechanical performance. Nanofillers such as carbon nanotubes, nanoclays, graphene nanosheets, and delaminated MXene have become a common strategy to

DOI: 10.1201/9781003164975-7

improve the functionality of the polymer for several applications; however, their effect on the crystallization behavior of polymers varies greatly. Some nanofillers are known to increase or decrease the glass transition temperature of polymer composites while others are well known for affecting the nucleation of polymer crystals after rising or reduction in the rate of crystallization [5]. Many of these improvements are attributed to the changes in the properties of the polymer in the environs of the particles due to the interfacial interaction between the polymer and the filler [6]. Besides, the changes in glass transition and the crystallization induced by the presence of nanofillers, the melting temperature (thermal stability) of the polymer composites can also be affected. Most fillers with conductive properties, easily transmit heat through polymers quickly, and reduce the melting temperature of the composites slightly, whereas fillers that absorbs heat increase the melting point of the composites. Composites prepared with nanofillers often exhibit two peak effects when the heating rate is maintained constantly at a slow rate.

Also, the synergistic effect of nanofillers on the crystallization behavior of polymer composites varies. For instance, the organo-montmorillonite/nano-Cu has been identified to improve the mechanical performance of polymer nanocomposites. Most nanofillers improve the melting temperature (thermal stability) and the crystallization behavior of composites [3].

Unlike nanoclay with Cu, the synergistic effect between MXene and other materials not only improves the thermal and mechanical properties but also provides self-lubricating functionalities with high thermal conductivity and electromagnetic shielding effect, which makes it a unique two-dimensional (2D) material for applications in various materials [7–11]. MXene experiences marginal weight loss due to decomposition of the organic addends even under extremely high temperatures [12]. In this regard, MXene/polymer composites have been prepared for various technical applications. The synergistic efficiency between MXene and organic polymer solutions is influenced by several factors, including the class of polymer, the type of MXene, the method of exfoliation, the interfacial interaction, and its distribution in the composite [4].

MXene–polymer composites undergo multiphase transitions when exposed to heat. Glass transition, enthalpy of chain relaxation, cold crystallization, and melting are some of the typical multiphase transitions during the thermal study of polymers. This property is often affected due to the presence of MXene in the matrix [13]. This chapter describes the methods used in elucidating the thermal and crystallization behavior of polymer–MXene nanocomposites, the various methods of evaluating these thermal transitions, and their effect on the ensuing nanocomposite.

7.2 METHODS OF EVALUATING THE THERMAL BEHAVIOR OF POLYMER NANOCOMPOSITES

Different methods exist for evaluating the thermal behavior of polymers and composites. However, the choice of a particular approach is determined by the specific thermal phenomenon one wants to observe. The most popular methods include the use of thermogravimetric analysis (TGA), differential scanning calorimetry (DSC), thermomechanical analysis (TMA), and dynamic mechanical analysis (DMA). The four main techniques are often complementary; however, sometimes, only a combination of the four techniques can provide a full insight into the sample understudies. Other thermal analysis techniques include micro/nanoscale local thermal analysis (MNLTA) and dielectric analysis (DEA). This section provides an introduction to the two most popular thermal analysis techniques, thus TGA and DSC. Detailed information about these thermal analysis techniques, including instrumentation, measurement procedure, instrument calibration, and applications, with the necessary theoretical explanations for all the techniques mentioned can be found in a book titled *Thermal Analysis of Polymers: Fundamentals and Applications*, edited by Joseph D. Menczel and R. Bruce Prime [14].

TGA measures thermal transitions in the physical and chemical transformations of materials as its been subjected to temperature with respect to time. Fluctuations in the mass of a material can be ascribed to many thermal transformations such as desorption due to the removal of adsorbed water, vaporization, absorption, sublimation, reduction, oxidation, and decomposition. These changes are observed as the mass of the material experiences changes under cumulative temperature swoops. TGA is, therefore, a versatile approach suitable for the examination of volatile/gaseous products that are lost during the endothermal degradation/reactions in thermosets, elastomers, thermoplastics, and composites when it is coupled with Fourier transform infrared (FTIR) spectroscopy. The technique provides insight into vicissitudes that occur in materials at specific temperatures, and the chemical breaking down occurring in real time in various controlled atmospheres (argon, oxygen, nitrogen, etc.). TG-FTIR is a useful technique for monitoring the evolution of gases under thermal degradation. Some of the machines are fitted with additional functionality that help give information about the crystalline nature of materials, although this may not be their fundamental use.

DSC is a useful thermal analysis tool that provides a worth of information about materials like polymers, nanomaterials, and food products. For DSC thermo-analytic studies, the variance in heat essential to increase the temperature of a specimen under study and the reference specimen are determined regarding temperature and time at a programmed heating rate. During this process, thermal changes characteristic to most polymeric materials such as glass transition, crystallinity, polymorphism, and eutectic changes are observed. In thermoset polymers, important criteria, such as curing and degree of cure, and other properties are also observed. Generally, in conducting the DSC measurement, the sample and reference (usually an empty calibrated DSC pan) are kept at the same temperature throughout the experiment duration at a constant supply of heat with the reference specimen possessing a distinct heat capacity over the range of temperatures to be measured and analyzed. Most often, pertinent phase transitions of the specimen such as glass transition melting point, and the decompositions temperature are obtained and analyzed.

Besides the thermal analysis techniques, studies related to crystallization behavior can be obtained using X-ray diffraction. The change in crystallinity of the constituents in a pristine polymer or nanocomposites could be detected by the diffraction peak shift and spacing. The test is mostly done at a specific scanning rate in a variable range using a diffractometer [15]. Most often, DSC is used to assess the crystallization and thermal properties of nanocomposites with different MXene content or composition. With DSC, the non-isothermal melt and cold crystallization and subsequent melting behavior have been investigated by differential scanning calorimetry [16].

Polarized light microscopy (PLM) is one of the methods used in polymer crystallography. PLM in contrast to other techniques provides a contrast-enhancing effect that improves image quality with birefringent materials. Compared to darkfield and brightfield illumination, phase contrast, differential interference contrast (DIC), Hoffman modulation contrast (HMC), and fluorescence, PLM has a higher degree of sensitivity that makes it a versatile tool for both quantitative and qualitative studies directed at a wide range of anisotropic samples including polymer nanocomposites. The quantitative aspect of polarized light microscopy is employed in crystallography than the qualitative facets. As a result, PLM is combined with *ex situ* or *in situ* fast scanning calorimetry can provide profound insight into crystal formation and nuclei evolution even for materials that undergo fast-crystallizing like polymer nanocomposites. The downside, however, is that the size of the crystals or crystal superstructures like spherulites ought to be in the micrometer range to be detectable. PLM is designed to observe and photograph samples that are visible primarily due to their optical anisotropic properties as shown in Figure 7.1.

Lately, fast scanning chip calorimetry (FSC) has been developed as a precious tool in polymer research with the ability to provide real-time evidence about the crystallization phenomenon, crystal restructuring, glass transition, and the melting phenomenon visually. The advantage of FSC is that it has a short response time in the order of microseconds that permits a fast transition from scanning to isothermal modes and conversely. Beyond the performance of FSC–atomic force microscopy (AFM) coupled device has been reported, where the AFM sample holder is substituted by the FSC chip sensor to enable recurrent annealing at well-defined temperatures and sometimes with the AFM images capturing the phase transitions from the same spot of the sample concurrently [18]. With this coupled device, valuable data on nucleation rate and crystal growth could be obtained,

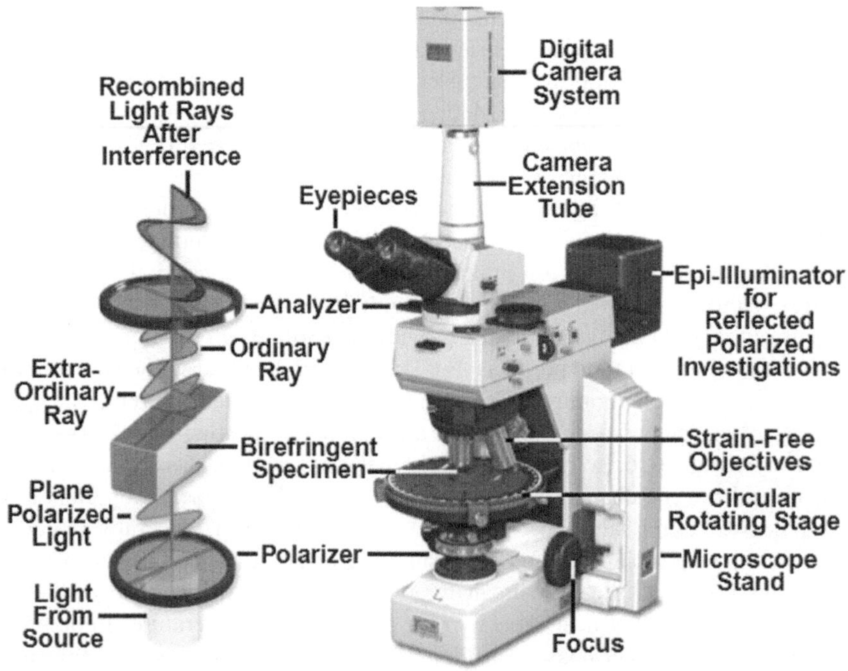

FIGURE 7.1 Electronic configuration of PLM

especially in situations in which the heterogeneous nucleation overcomes crystallization at a low supercooling rate of the melt, frequently resulting in spherulitic development of lamellar crystals in most semi-crystalline polymers.

7.3 THERMAL PROPERTIES OF MXene

The thermal properties of MXene are influenced by several factors such as the type of surface terminating groups, the method of synthesis, and the method of storage after synthesis. It is important to highlight the effect of surface terminating groups on the stability (thermal) of MXene and polymer composites. Also, the stability under storage and applications is very important because several studies have indicated that MXene degrades when exposed to moisture or exceptionally high temperatures [19–21]. In a study of Nb_2CTx MXene, it was found that the Nb adatoms located at the surface bonded to ambient oxygen when exposed and destabilized the system, resulting in the gradual degradation of the compound structure [21]. In a similar study, it was noticed that a Ti_3C_2Tx polyethylene terephthalate exposed to air for 70 hours presented a significant reduction (approximately 20%) in its initial conductivity and thermal properties [22]. To curtail the degradation associated with the storage of MXene, it has been suggested that MXene nanosheets should be stored in an environment free from oxygen with low light intensity. Also, a strong polar solvent such as dimethyl sulfoxide (DMSO) and n-methyl-2-pyrrolidinone (NMP) [23], or the application of solution filtration processes are used to produce MXene films, including a freeze-drying approach [23, 24]. MXene processed in this manner produces stable dispersions suitable for further processing into continuous films with adequate bonding to polymer matrices. The filtration processes ensure MXene nanosheets remain stacked, yet loosely packed, due to the effect of van der Waals force, and most of the time, in a wrinkled manner; hence, the morphological arrangement remains intact with little or no degradation internally and around the edges.

Regarding the method of synthesis and its effect on the thermal stability of the ensuing MXene nanosheet, it has been recommended that optimized etching procedures that reduce the adatoms on the surface of MXene be used to increase the stability and avert possible degradation [23]. The surface character of MXene (e.g., chemical addends and electrical properties) are undoubtedly leveraged as a foundation for optimized fabrication conditions. The type and elemental constituents of MXene influenced the thermal behavior. For example, the ABABAB arrangement (hexagonally closely packed stacking) of Mo atoms in Mo_2C in contrast to $Mo_3C_3T_x$ (ABCABC ordering [face-centered cubic stacking] of Mo atoms are deemed to grant high thermal stability of Mo_2CT_xthan $Mo_3C_2T_x$ [25]. MXene $Ti_3C_2T_x$ has been confirmed to possess high thermal stability in a nitrogen atmosphere from 0–950 °C [26]. Three insignificant degradation steps observed were assigned to the evaporation of free water held on the surface and between the layers of $Ti_3C_2T_x$ followed by the dissipation of bonding water and functional groups on the surface. The third stage was attributed to the oxidative decomposition of the titanium component. The breakdown temperature of the pure MXene $Ti_3C_2T_x$ itself was about 785 °C indicating better stability to heat [27].

In terms of thermal conductivity, the ion intercalated into the MAX phase might be metallic or semiconducting, or half-metallic. When MXene is functionalized, the electronic properties differ from metallic to semiconducting, contingent on the nature of the M, X, and T clusters. The majority of surface-terminated MXene, however, maintain their metallic properties whereas some undergo chemical changes into semiconductors due to a shift of the Fermi level. Some MXene that reveals metallic behavior includes Ti_2CF_2, $Ti_2C(OH)_2$, Ti_2NF_2, V_2CF_2, Cr_2NF_2, and Zr_2CF_2, while others, such as Cr_2CF_2, Ti_2CO_2, Zr_2CO_2, and Mo_2CF_2, are considered as semi-conducting [28, 29].

Typically, MXene thermal stability conducted in the argon atmosphere showed a three-stage decomposition attributed to many factors. The Ti_3C_2Tx mass loss of 0.38% from ambient temperature to 200 °C was attributed to the removal of adsorbed H_2O and hydrogen fluoride molecules whereas a mass loss of4.48% between 200–800 °C was attributed to OH group removal followed by and a 2.47% mass loss from 800 °C up to 1000 °C, attributed to the elimination of the fluorine atoms [30]. Similar weight loss was detected in other studies [31–33], which was later validated by thermogravimetric-coupled mass spectrometry (TG-MS) analysis [34].

To remove the organic addends and further improve the thermal properties, several studies have suggested annealing as a thermal treatment most effective for removing terminal groups (OH/F) since each can be detached in a precise temperature range [33]. The downside of the thermal treatment, however, is that the resultant material cannot be classified as pristine MXene but rather as an oxygenated one, except the heat treatment is done in a controlled environment without oxygen. In recent times, exposure of Ti_3C_2Tx at 750 °C under vacuum has been reported to remove the majority of the oxygen to obtain pristine MXene [35]. Practically, the evaluation of the thermal properties of MXene in air is required. In an oxygen atmosphere, the formation of rutile crystals occurs around 500 °C, and the MXene phase alteration occurs at considerably higher temperatures. Typically, a Mo-based MXene exhibited high thermal stability up to 530 °C in H_2/N_2 environment [36]. Also, the thermal properties of MXene have been studied extensively in a nitrogen atmosphere and results showed a much-improved thermal property. For instance, Zr_3C_2Tx was studied in N_2, and the results are shown in Figure 7.2. The MXene had excellent thermal stability with little mass loss. In this study, the initial loss in weight took place at 125 °C, which is attributable to the evaporation of absorbed water in the layers of the MXene sheets. Nevertheless, it has been revealed that the heat treatment processes have a significant impact on the structure and properties of MXene. Heat treatment at relatively high temperatures is liable to lead to the removal of functional groups containing –F or –OH/=O [3, 37]. The absence of these elements is beneficial in Li-ion batteries due to decreased Li storage capacity caused by the hydroxyl and the fluorine-terminated groups on the surface of Ti_3C_2 [3].

To evaluate the complete removal or otherwise of the surface terminating groups of MXene after annealing, an FTIR was by Wang et al., to verify the removal of surface addends, and the results showed that the intensity of peak at 3440 cm^{-1} reduced and the peak at 1400 cm^{-1} disappeared,

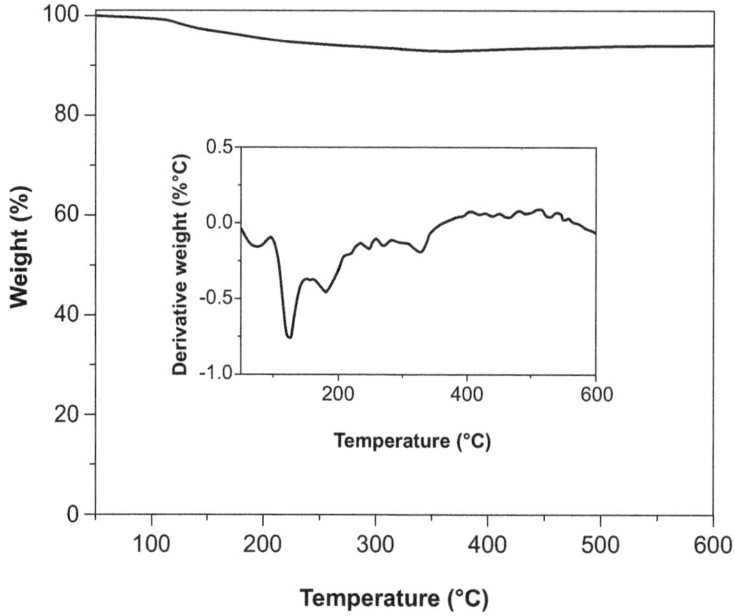

FIGURE 7.2 TGA and dTG curves (insert) of MXene powder.

indicating a decrease in the content of –F and –OH groups adhering to the surface of the annealed Ti_3C_2Tx nanosheets. However, the exposure of the MXene to thermal reduction could increase the ratio of C/O ratio—indicating that the thermal reduction had dissipated partial oxygen-containing groups of MXene. Also, XRD results revealed the shifting of the (002) peak of the annealed $Ti_3C_2T_x$ from 6.58° to 6.40°. Furthermore, the thermal reduction could be beneficial to the delamination of the $Ti_3C_2T_x$ and the increase in the corresponding layer spacing without necessarily exhibiting any by-product [38].

Furthermore, low-temperature conduction *ex situ* of the electronic properties of intercalated and de-intercalated MXene Ti_3CNT_x has established that annealing leads to the de-intercalation of MXene. Also, the intercalated MXene displays semiconductor-like behavior through the whole range of the Physical Property Measurement System (PPMS; from RT to 263 °C), while the annealed de-intercalated MXene displayed metallic behavior at about 150 °C and below. At this stage, the de-intercalated MXene Ti_3CNT_x presents a transition to the negative temperature dependence of resistance, analogous to the behavior of multilayer $Ti_3C_2T_x$ samples. They established that annealing Ti_3CNT_xbeyond 300 °C results in partial de-functionalization [29].

7.4 THERMAL BEHAVIOR OF MXene/POLYMER NANOCOMPOSITES

Due to the good abrasion resistance, processability, transparency, and mechanical performance of most polymers, they are widely used in many different areas of life, mostly in the form of nanocomposites in automobiles, medical devices, sporting goods, and decorations. However, factors, such as thermal conductivity and tensile strength, have been hindering their application in some fields [15]. Traditionally, the use of nanomaterials as fillers enhances the general properties of polymers such as epoxy. They are also used to increase the operational temperature window of polymer nanocomposite. A polymer such as epoxy has low thermal conductivity due to the indiscriminately disheveled chains that result in the scattering of the phonons [39]. MXene is one of the few best solutions as a reinforcement with excellent thermal stability and conductivity; hence, the addition of

MXene to polymers which are usually thermal insulators can improve the thermal properties. The addition of MXene as nanofillers for composites fabrications confines the mobility of the polymer chain, which directly influences the glass transition temperature and the subsequent crystallization of polymers [4]. Comparatively, MXene polymer–based hybrids exhibit better thermal stability even with lesser filler content than graphene-based nanocomposites. The excellent thermal conductivity of MXene reduces the thermal resistivity of the polymeric matrix resulting in a uniform and high conductivity, especially when the MXene load is sufficient to form a good network with the polymer chains [25]. As a result, MXene uniformly disperses within epoxy matrixes, and thus facilitates the intercalation of epoxy molecular chains into space between MXene layers as discussed in Chapter 6. Experimental studies have revealed the cost-effective MXene/polymer nanocomposites with improved thermal properties at low MXene loading, while in most occurrences, higher MXene loading negatively affects the thermal stability of the nanocomposites [4]. The thermal behavior of MXene/epoxy nanocomposites has been studied extensively due to the wide application of epoxy polymer nanocomposites [39].

It has been established that the interplay of PVA with MXene exhibits enhanced thermal stability in both nitrogen and oxygen atmospheres as shown in Figure 7.3[27]. The degradation of pure epoxy cured with presumed asbestos-containing material (PACM) occurs about 370°C. A minor enhancement was noticed at 5wt% MXene Ti_3CN filling, which extended the onset degradation to around 380°C. A further increase in MXene loading (in Ti_3CN-epoxy composites) resulted in low thermal stability and demonstrated a sharp drop in initial decomposition temperature. This phenomenon is attributed to Ti_3CN dispersion throughout the matrix. Noteworthy is that the increases in MXene gave considerably good stability at higher temperatures compared to the nanocomposites with lower MXene load. In other words, the degradation temperature of the nanocomposite reduces as the MXene filler content increases, arraying the positive impact of MXene in ameliorating the thermal stability of the polymer. The underlining explanation for the improved stability is that, in the presence of MXene nanofillers, the mobility of the polymeric chains is restricted, and thus, a higher temperature is required for the movement of the molecular chains. This occurrence is primarily related to the negatively charged MXene flakes having the potency to adsorb amine-based hardeners, which influences the structure of the polymer after the curing process [26].

Figure 7.3b shows a line graph of MXene content against temperature. The line indicates that little decomposition is experienced at temperatures between 300–380 °C when higher MXene content is included in a composite. The higher the MXene content, the lower the mass loss experienced within the temperature zone stated.

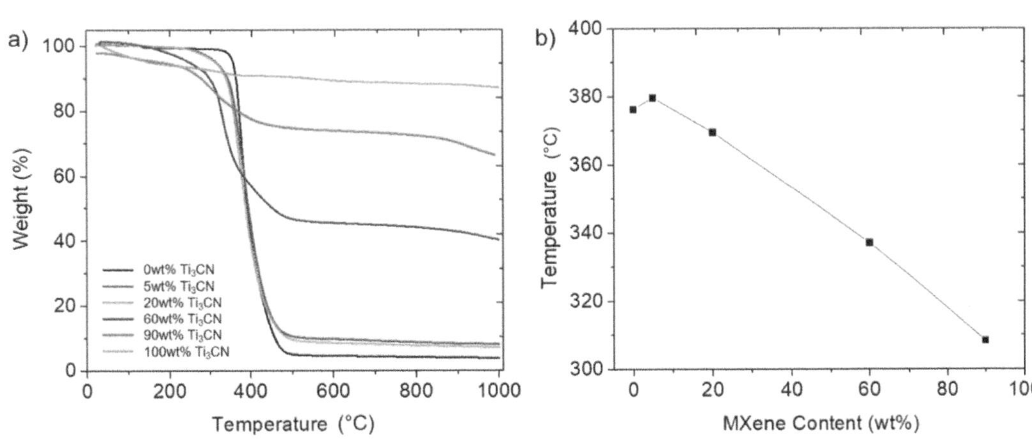

FIGURE 7.3 (a) TGA curves of MXene-epoxy composites with variable Ti_3CN filling, (b) diminution in onset degradation with increasing MXene filling.

Microstructural analysis of $Ti_3C_2T_x$/epoxy composite with varying MXene $Ti_3C_2T_x$ loading indicates that few-layer $Ti_3C_2T_x$ MXene disperses homogeneously in polymer matrixes than poorly delaminated MXenes with more stacked layers. MXene $Ti_3C_2T_x$ promotes thermal transfer between molecular chains and enhances the application of such nanocomposite for heat conduction. Also, the coefficient of thermal expansion (CTE) values decrease with increasing MXene in most epoxies due to its ability to restrain the mobility of the polymer chains [39]. Also, a high-temperature degradation of 1.74 vol% $Ti_3C_2T_x$MXene in epoxy nanocomposites was assigned to the proper delamination of the MXene nanosheets, resulting in improved barrier resistance and subsequent reduction in the evolution of organic volatiles due to the tortuous effect [40]. Similarly, it has been demonstrated that the initial decomposition temperature of polyurethane nanocomposite (with 0.5 wt.% polyethyleneglycol) improved significantly after the addition of MXene. Contrary to the result reported in most studies, the initial degradation temperature of MXene/polymer nanocomposites lowers as the MXene load increases but gets improved after 400 °C thanks to the high thermal conductivity, surface energy, and the uniform dispersion of MXene nanosheets, which enhances the thermal transfer in the matrix [15].

An analysis of the thermal conductivity of Ti_3C_2/poly(vinylidene fluoride) membranes with different MXene Ti_3C_2 percentages has established that low MXene Ti_3C_2 loading (<1.0 wt.%) grants significant elevation in the thermal conductivity while further increases in the MXene load leads to an excellent thermal conductivity. This was attributed to the varying phase structures in the nanofilms, which form sea-island structures on the incorporation of relatively low MXene Ti_3C_2. The incessant network formed at high MXene Ti_3C_2 loading leads to a rapid increase in conductivity. Other major reasons that facilitate the increase in the thermal conductivity are mostly as a result of the large surface area of MXene Ti_3C_2 nanoflakes and the development of hydrogen bonding between Ti_3C_2 and epoxy, which reduces the interfacial heat resistance, especially at higher MXene Ti_3C_2 loading [37].

Temperature-dependent and polarized laser power-dependent Raman tests were adopted to study the thermal conductivity of MXene Ti_3C_2/polyvinyl chloride (PVA) nanofilm and other polymers. It was realized that the thermal conductivity of the membrane was relatively lower than the pristine MXene Ti_3C_2 but significantly higher than the composite membrane containing iron, silicon dioxide, aluminum oxide, and stainless steel, respectively. Generally, the thermal properties and conductivity of MXene-filled polymer nanocomposites are leveraged by the percentage of filling, proper dispersal, phase structure, and the interfacial thermal resistance. In a typical study of the thermal degradation of MXene Ti_3C_2/epoxy [41], the temperature rises along with the increase in the concentration of MXene. However, similar to other studies alluded to in the preceding paragraphs, the thermal stability began to decline when the MXene Ti_3C_2 loading reached 5.44 wt.%. A similar occurrence was observed in MXene Ti_3C_2/linear low-density polyethylene (LLDPE) [3]. Usually, the improvement in the thermal stability of MXene Ti_3C_2 results from the formation of solid Ti–O bond between Ti_3C_2 and most polymeric matrixes, which decreases the thermal coefficient of the Eg 1 mode. It has been established that the Ti–O linkage is the most stable bond amid the functional groups of Ti–O, Ti–F, and Ti–OH, having respective bond strengths of 1.97 Å, 2.17 Å, and 2.19 Å [42].

More so, an experimental non-isothermal crystallization and theoretical calculation of the crystallization kinetics was studied using the Mo model, which adequately describes the non-isothermal crystallization kinetics of Ti_3C_2/LLDPE blends [3]. At 2.0 wt % Ti_3C_2 loading, the composite showed the highest crystallization rate; however, at 4.0 wt % Ti_3C_2 loading, the rate of crystallization becomes even slower crystallization than the neat LLDPE film. It was realized that LLDPE chains entangled on the surface of Ti_3C_2 and restricts movements and self-assembly instead of the filler serving as nucleating sites for easy crystallization. In certain polymers, for instance, MXene/nylon 6 nanocomposites, higher MXene loading increased the rate of crystallization and at the same time increases the melting temperature of the composite. Such enhancements in thermal stability have been linked to the enhanced barrier qualities due to the torturous path effect [40].

It is worth noting that the thermal decomposition and the crystallization behavior of MXene loading on different polymer nanocomposites vary. For poly(lactic acid) (PLA), the incorporation of 0.5% and 1% MXene intensifies the thermal stability of the nanocomposites significantly, whereas, at a higher MXene loading, the thermal stability gets reduced portentously. Hence, it is established that 1% MXene nanosheet filling is ideal for enhancing the thermal stability and improving the crystallization behavior of MXene/PLA composites. Further increases in the MXene content in the nanocomposites present a complicated result that is rarely reported because of the pretentious initial increase in thermal stability of PLA while displaying converse results as the heating progresses [43]. Therefore, the class or composition of polymer could be considered an essential factor in improving the interfacial interaction between the polymer and MXene filler toward high thermal properties.

7.5 THERMODYNAMICS OF CRYSTALLIZATION AND MELTING OF POLYMERS

Thermodynamics addresses the reason behind or the circumstance that leads to the crystallization of polymers, or their melt phenomenon. As already indicated, most polymers are semi-crystalline in nature, which implies that polymers are composed of amorphous and crystalline regions in a single polymer unit. Crystallization and the melting phenomenon are characteristically first-order phase evolutions between the amorphous and crystalline phases. An instantaneous change in the proportion or density of the composition is referred to as a first-order phase transition. This is complemented by the release of heat, also called the heat of transformation—which is mostly linked to the primary cause of changes in composition. The second-order phase transition occurs when the increasing distance of one end causes the transition of some physical quantity at the other end (equal to zero) to gradually increases from zero. In this case, the density and concentration of the composition in the nanocomposite vary incessantly, and the thermal effect created is not released or absorbed. The two phases are reversible in thermodynamic equilibrium at a certain temperature. The temperature is termed the equilibrium melting point of polymer crystals. At the polymer nanocomposite equilibrium melting point, there exists a free-energy change (zero), especially in the bulk system of polymers [1].

Among factors affecting the thermal behavior of polymers and nanocomposites, crystallization is one of the crucial factors from both a scientific and industrial standpoint, because the mechanical/thermal properties of polymers and their nanocomposites greatly depend on their crystallinity. Cold crystallization (which is discussed in the latter part of the chapter) occurs when polymers or polymer nanocomposites are quenched rapidly after melting. Also, crystallization endotherms are usually detected when polymers or polymer nanocomposites are slowly cooled after melting. The kinetics of the overall crystallization phenomenon is well ordered by nucleation mechanism and crystal growth rate. These two kinetics can be affected by many conditions including temperature, the type of fillers (size, orientation, interfacial properties, etc.). For instance, increasing the temperature facilitates the increment in crystallization in the crystal-growth sites of most polymers and nanocomposites, while there is a decrease in the crystallization at the nucleation-controlled sites. The degree of crystallinity can be estimated from the peak area of the curve obtained from a DSC machine or an XRD device.

In composite formation, the degree of crystallization can be affected by the addition of specific compounds. For instance, the addition of a plasticizer often results in an increase in the degree of crystallization by accelerating the spherulite growth rate, while a nucleation agent (which could be nanofillers like carbon nanotubes/fibers, graphene, or MXene) can also augment the rate of crystallization via cumulative nuclei density [44]. The MAX phases from which MXene flakes are obtained, are thermodynamically stable and display some metallic properties, such as high electrical and good mechanical properties, with some level of good thermal conductivity. Due to their versatility, they also have good stiffness, low density, and excellent corrosion resistance similar to ceramics. These unique characteristics of MAX phases endear them valuable for several industrial applications, including high-temperature ceramics, protective coatings, and electrical and electronic parts [25]

As indicated, the crystalline/amorphous structure of polymer materials gives specific properties. The melting point of crystalline polymers often occurs at elevated temperatures, and with the increasing temperature, it improves due to the changes in the crystallinity of the polymer. Also, alteration in the physical structure of the polymer from solid to a molten state is very rapid at increasing temperature. In polymers composed of a high percentage of amorphous structure, there is no abrupt change of physical structure like as observed in the crystalline polymer. When the temperature decreases from the molten to solid state, a gradual change in the physical appearance is revealed in amorphous polymers. The cooling down leads to the lowering of the thermal expansion of amorphous polymer, which happens abruptly due to decreasing temperature. In the level of crystallinity between these two temperatures, that is, the high temperature and low temperature, the thermal characteristics of polymers are revealed. Furthermore, above the melting point of most polymers, viscous characteristics are revealed; they show viscoelastic properties in the region of glass transition and melting temperature.

The formation of crystals in polymeric materials can occur during either the heating or the cooling cycle depending on the rate at which heat is supplied in the heating process or the rate at which the material is cooled down after heating. With the increasing crystalline structure, the melting temperature also increases. Regarding the thermal characteristics of polymers with high melting points, such as thermosets and elastomers, they often show amorphous characteristics by lowering the temperature from the liquid state. In addition, the decreasing temperature leads to the occurrence of crosslinks. These crosslinking restrains the crystallization of polymeric molecules and hinders the remelting and reliquefication of polymers owing to the structural cross-linkages. Given that the incorporation of nanofillers also influences the melting point of a polymer, an investigation into the structure and crystallization behavior of nanocomposites, such as MXene Ti_3C_2/polyethylene oxide films were done. Interestingly, in most nanocomposites, the rate of crystallization increases rapidly from the beginning and then decreases on the addition of varying MXene load. Similar to the thermal phenomenon, the fastest crystallization rate is usually observed at low concentrations of MXene. This occurrence is mostly attributed to the equilibrium between nucleation rate, which results in crystal formation, and the confinement effect created by the MXene nanofillers. Characteristically, low MXene loading hastens the crystallization rate by heterogeneous nucleation whereas higher MXene loading tends to slow down the crystallization rate because the nucleation rate gets reduced considerably due to an unyielding and limited confinement network.

7.6 CRYSTALLIZATION OF POLYMER NANOCOMPOSITES

Crystallization is a two-step route, that is, nucleation and growth, which changes state as a result of factors such as time and temperature. The absence of enough time or temperature does not allow the polymer to crystallize completely. Impurities in crystallites are liable to decrease the melting temperature. The crystallization of polymers and polymer nanocomposites can be categorized into three groups, namely,

a. crystallization during polymerization,
b. crystallization induced by orientation, and
c. crystallization under quiescent conditions.

The hallmark of polymerization crystallization is the development of macroscopically distinct crystals. In the course of single crystal formation, the original "monomer" crystals are conserved while the monomers developing into crystals link up into chains by solid-state polymerization. The terminal polymer crystal is attained upon the interfacial chemical interaction at the gas/solid or liquid/solid, which cannot be ascribed to changes in the physical state of the material as in the case of the

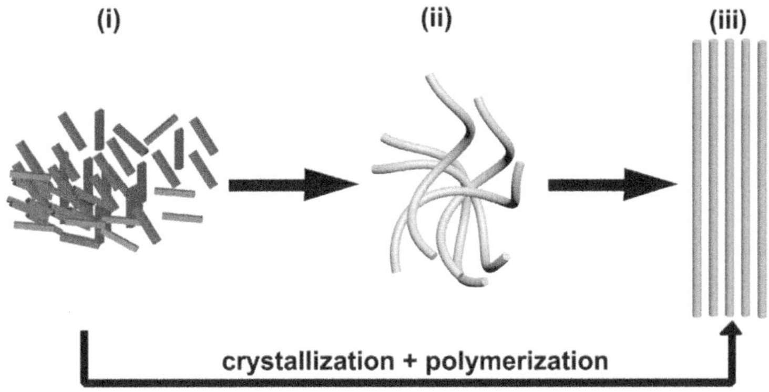

FIGURE 7.4 Polymerization subsequent to crystallization (i, ii) and crystallization in the course of polymerization (iii).

normal crystallization processes. The characteristics of the crystals formed via this method can be very interesting, because some have displayed good conduction ability for electricity along its axis even at relatively low temperatures. A brief description of such mechanisms is simultaneous polymerization and crystallization or successive polymerization and crystallization illustrated in Figure 7.4. In the concurrent polymerization and crystallization, the prime and ancillary linkages occur simultaneously. One of the advantages of simultaneous polymerization and crystallization of polymer is that it can occur on monomers in gas or melt state, beneficial to macromolecular crystallization that takes place on solely the melt or solution state. There is also the tendency that could cause the folding of chain crystals below the glass transition temperature of the ultimate polymer or polymer nanocomposite system.

For crystallization induced by orientation, the procedure can be termed as distending long polymer chains to yield fibrous crystals. This means of crystallization is the fundamental process identified in the development of fibers into smooth, thin, and lengthened chain morphology, which is challenging to achieve in the most seamless conditions. When the polymer is stretched, the polymeric chains tend to reorient from their initial configuration (mostly amorphous), which leads to a reduction in entropy (Figure 7.4). Nanofillers that exhibit a large surface area have a noticeable effect on the crystallization behavior of the polymer. The large surface area of the nanofillers and the movement in the polymeric chains near the particle surface could affect crystallization. During polymer processing (extrusion, injection molding, film blowing), non-isothermal crystallization conditions are involved; therefore, a thorough understanding of the factors that affect crystallization is essential for the optimization of the processing conditions and the fine-tuning of the properties of the end product. The conditions under which the crystallization of the polymer takes place affect numerous properties of nanocomposites such as the rate and the degree of crystallinity. Other known effects in the course of crystallization are heterogeneous nucleation, trans-crystallinity, and epitaxy. Also, the appearance of solid surfaces in the melt leads to the formation of heterogeneous nucleation. This nucleation is induced by the decreasing crucial enthalpy during nucleation at the melt and solid interactive sites. This indicates that the interaction between the polymer and the nanofillers affects the polymer morphology.

7.7 THE THERMAL PHENOMENON

7.7.1 Glass Transition Temperature

In the amorphous sections of the polymer, the molecules are considered to be in a frozen state at lower temperatures. In this state, the molecules can vibrate slightly but cannot experience significant mobility. The polymer is brittle, hard, and rigid compared to a crystalline solid with the molecular hysteria as a liquid with resemblance to glass, therefore the term *glassy state*. The glassy stature is analogous to a supercooled liquid in which the mobility of polymer molecules is in a frozen state. Upon heating, the polymer chains can waggle about each other and becomes soft, flexible, and lithe, just like rubber, hence the term *rubbery state*. The point (temperature) at which the glassy state transitions to a rubbery-like material is known as the glass transition temperature (*Tg*). This property affects the storage modulus of nanocomposites significantly. Usually, the glass transition happens somewhat in the amorphous regions of the polymer only. The crystalline region is largely intact during the glass transition period in most semi-crystalline polymers. When the mobility of the molecular chains takes place easily, the glassy state can transition into the rubbery state at a lower temperature. In case the movement of the molecular chains gets obstructed, then the glassy state becomes stable and, in so doing, restricts any breakage; however, it results in the immobility of the polymer chains at the lower temperature. The immobile molecules necessitate more energy to set them free and mobile, implying a rise in the *Tg*. The *Tg* is usually affected by such dynamics as molecular weight, measurement method, composition, plasticizer employed, and, more importantly, the heating and cooling rate.

The glass transition temperature (*Tg*) is an important parameter, that specifies the upper temperature limit for most applications. Also, changes in glass transition temperature figures have an essential connection with reinforcement phases and the interfacial relations between reinforcement materials such as MXene and polymeric resins [45]. In this regard, the effect of MXene on the crystallization behavior of epoxy nanocomposite synthesized via *in-situ* intercalation polymerization in which the epoxy molecular chains reacted with the hydroxyl groups on the surface of MXene Ti_2CT_x nanosheets to form a 3D crosslinked network structure during curing was studied [45]. The glass transition temperature increased gradually with a change in the concentration of MXene Ti_2CT_x. It was noticed that the integration of MXene Ti_2CT_x into epoxies increases the glass transition temperature due to the interfacial interaction and crosslinking effect between MXene Ti_2CT_x nanosheets and the molecular chains of epoxy. This implies that when the temperature increases,

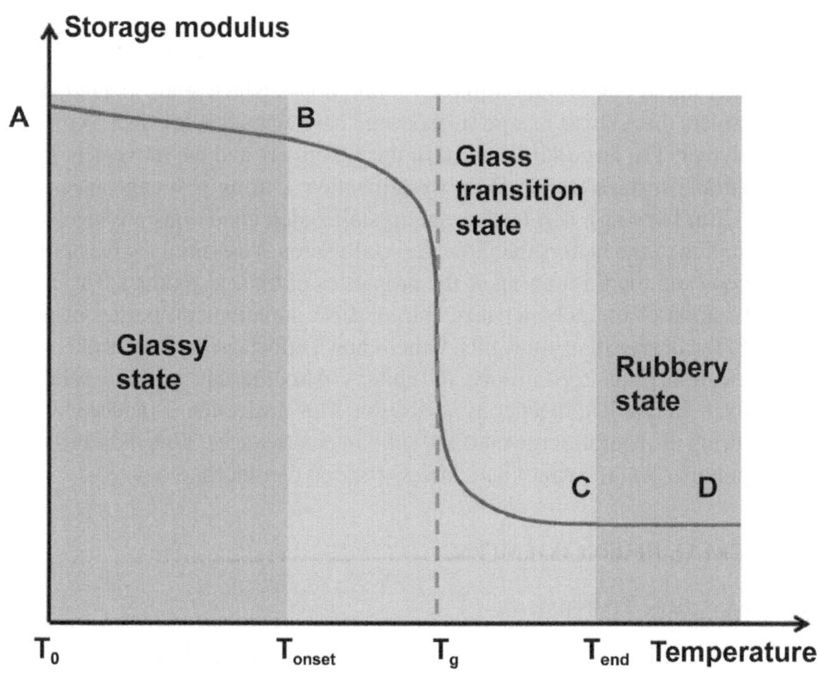

FIGURE 7.5 Graphical representation of temperature-dependent glass-transition process.

the surface of the stiff MXene Ti_2CT_x nanosheets forms an enhanced interfacial interaction with the epoxy molecules, which hinders the vibration of molecular chains and leads to an upsurge in the relaxation period [45].

A study on the glass transition temperature values of rGMH/epoxy nanocomposites also confirms a positive impact of MXene on nanocomposites as the mass load of MXene increases [47]. Epoxy resin was interwoven into 3D rGMH network to form the rGMH/epoxy before the incorporation of MXene. It was realized that when 3.3wt.% of MXene was loaded in the nanocomposite, the *Tg* of rGMH/epoxy nanocomposites increased to 150.9°C, which is 13.9°C higher than the value recorded on the pure epoxy resin (137.0°C). Upon the elevation in temperature, the surface of the stiff MXene interacts strongly with the epoxy molecular chains, thereby enhancing the delay time and obstructing the epoxy molecular chains vibration, thus increasing the *Tg* value, similar to the earlier reports [45]. The reason for such variation in *Tg* is due to the physical barrier effect of MXene, which limits the mobility of the epoxy molecular chains and reduces the thermal diffusion of rGMH/epoxy nanocomposites. This phenomenon also contributes to the delay of the thermal decomposition of rGMH/epoxy nanocomposites, which contributes to the improvement in the thermal stability of the ensued nanocomposites [47]. Also, the effect of MXene loading on the glass transition temperature of MXene/polymer nanofilm was studied by DMA, and the *Tg* values were obtained from the peak value of the loss factor from the DMA curves. There was a rise in the *Tg* of MXene Ti_3C_2/poly(vinylidene fluoride) membranes when the concentration of MXene Ti_3C_2 was increased. A similar phenomenon was observed in MXene Ti_3C_2/epoxy membrane by DSC. A gradual increase in the glass transition temperature of the nanocomposite was recorded with the increasing content of MXene Ti_3C_2 flakes in the composites. The elevation in the glass transition temperature was attributed to the restricted molecular motion and increased crosslinking density [4] [39]. Other studies have confirmed the effectiveness of MXene in improving the *Tg* of polymer nanocomposite due to the restriction in molecular motion. More important, it has been observed that the higher the degree of crosslinking, the greater the alteration in the dynamics of the polymer molecular chains. Also, the presence of functional groups on the surface of an MXene nanosheet improves the interfacial interaction between MXene and the polymer matrix, resulting in good and uniform dispersion of MXene within the matrix, which is beneficial for achieving a big shift in the *Tg* [37]. In totality, the incorporation of MXene into polymer produces better thermo-mechanical properties that perform better in a wide range of applications, such as separators, batteries, sensors and actuators, smart scaffolds, and filtration membranes. Table 7.1 outlines some of the established *Tg*s of polymer nanocomposites.

To give a comprehensive description of the melting temperature of a nanocomposite, a distinction between glass transition and melting temperature is necessary. The *Tg* is associated with changes that occur in the amorphous regions of the polymer, whereas the melting temperature is connected

TABLE 7.1
List of Polymer Nanocomposites and Their Corresponding Glass Transition Temperatures

Polymers	Glass Transition Temperature Range	Ref.
Pure epoxy	137 °C	[47]
rGO-epoxy nanocomposites	139.8 °C	[47]
rGO-(MXene)/epoxy nanocomposites	150.9 °C	[47]
Poly(vinylidene fluoride)	−32 °C	[37]
MXene/poly(vinylidene fluoride)	−25.5 °C	[37]
Polystyrene	90 °C	[48]
Nanogold/polystyrene	105 °C	[48]
Polymethyl methacrylate	40–140 °C	[49]

to the transitions in the crystalline region. Thermodynamically, the alterations are termed the first- and second-order transitions. *Tg* is the second-order transition, while the melting temperature is the first-order transition. Unlike melting temperature, the value of *Tg* is not unique because the glassy state is not always in equilibrium. All semi-crystalline polymers display both transitions corresponding to their crystalline and amorphous regions. Semi-crystalline polymers possess melting temperatures at which the well-ordered phase changes to disarrayed phases, while the amorphous sites soften in the glass transition temperature range. It is worth noting that melting point is absent in amorphous polymers, but all semi-crystalline polymers have *Tg*. Influential factors, such as the presence of aromatic groups, double bonds, and bulky or large side groups, could affect the state of polymer melting temperature. These outlined factors have the propensity to hinder the mobility of the polymer molecular chains, thereby limiting crystallization and affecting the melting temperature of nanocomposites. Furthermore, the branching of molecular chains in a polymer could also lead to a decrease in melting point. This negative effect caused by the branching is assignable to the defects created by the branches. Hence, the melting temperature is highly dependent on the previously mentioned factors.

The melting temperature of polypropylene-grafted MXene nanocomposites shows systematic changes in MXene loading for polypropylene-g nanocomposite. At a considerably lower $Ti_3C_2T_x$ loading, the melting temperature reduces while a higher $Ti_3C_2T_x$ loading above 0.5 wt.% results in increased melting temperature. The reason for the reduced nanocomposites melting temperature at low $Ti_3C_2T_x$ loading (below 0.5 wt.%) is due to the weak confinement of polypropylene-grafted chains in the MXene $Ti_3C_2T_x$/polymer nanocomposite, while the increase in temperature is due to higher $Ti_3C_2T_x$ loading is ascribed to the strong hydrogen bonding from the maleic anhydride-g-polypropylene and the –OH groups from the $Ti_3C_2T_x$, which restrains the motion of polypropylene-g chains [50]. It is also found that the addition of Ti_3C_2Tx nanocomposite into polypropylene-g matrix results in a slight improvement in the values of crystallinity. This phenomenon indicates that the slippage over H-bonds induced nanoconfinement and resulted in improved crystallinity. Furthermore, the crystallinity of polypropylene-g/$Ti_3C_2T_x$ nanocomposites decreases slowly with the increasing $Ti_3C_2T_x$ filling (that is when the load was more than 1 wt.%). It is worth mentioning that the crystallinity values are not consistent with the crystallinity values in the XRD analysis, due to the thermal capacity of the distorted polymeric molecules improving their intrinsic state or equilibrium mode after the relaxation process [50].

7.8 NON-ISOTHERMAL MELT AND COLD CRYSTALLIZATION OF MXene-BASED POLYMER COMPOSITES

Crystallization that occurs beyond the glass transition temperature when the polymer is heated in the glassy or amorphous state is referred to as "cold crystallization". This occurrence is often detected in supercooled phases of polymeric materials after the evolution in the thermodynamically stable crystalline phase [51]. Cold crystallization, also called post-crystallization, occurs above the *Tg* when the polymer chains gain adequate motion to arrange themselves into an ordered crystalline structure by chain folding. The cold crystallization phenomenon occurs between the *Tg* and the melting temperature. Cold crystallization is determined from the DSC exothermic peak. The importance of crystallization (cold crystallization) to industrial manufacturing of composites cannot be overemphasized, because the mechanical and optical performance and the long-term stability of final polymeric products in applications depends largely on this property. More so, the right processing temperature for most semi-crystalline polymer composites is determined by cold crystallization, which is necessary for product development and engineering.

The absence of crystallization during the cooling cycle reveals that the polymer has attained stable nuclei. In addition to the phenomenon of glass transition, cold crystallization also occurs when a polymer is composed of amorphous and crystallizable components. Thus, the exothermic cold crystallization attained during heating shows the reorganization of polymeric chains because the initially amorphous structure begins to crystallize. Cold crystallization of semi-crystalline plastics occurs at elevated temperatures and increases molecular mobility above the glass transition, even though some studies have revealed that it can take place at room temperature for rubber or thermoplastic polyurethane. Cold crystallization is a change in the physical structure that leads to an increase in the degree of crystallization and the lamella thickness, as well as to the perfection of the crystal structure. During cold crystallization, at the transition zone between the existing crystalline structures and the amorphous regions, new ordered structures (crystallites) grow. These newly formed crystals can be differentiated from the preexisting ones using their lower melting temperature.

Cold crystallization is mostly reported for macromolecules with pronounced structural mobility; however, cold crystallization of small molecules is possible but quite rare, hence the knowledge about the design guidelines for these molecules [51]. Cold crystallization gives a typical crystal morphologies with remarkable physical properties. In addition to the rare observation of cold crystallization in small molecules, some challenges remain with the clarification of the relationship between stable amorphous (glassy), other condensed states, and the development of the comprehensive molecular model concerning "cold crystallization" [51].

Melt crystallization is also known as crystallization from the melt (the melt is most correctly referred to as pure molten solid). Melt crystallization is the preferred separation technique for renewable materials. The fundamental development of the melt crystallization processes for renewable materials like acrylics provides good theoretical yields for the production of high-purity acrylic acid [52]. Despite the advantages of this process, melt crystallization has not been utilized most in polymer engineering. The inadequate understanding of the heat and mass transfer in melt crystallization and the difficulties in handling the solid transport and solid–liquid separation are under investigation to grant more understanding about melt crystallization. With the active research in this field of endeavor, there is a bright future in the fabrication and application of polymer nanocomposites.

The non-isothermal melt crystallization performance of PLA/MXene nanocomposites was investigated, and the cooling curves of the melt and the ensuing heating traces were thoroughly studied. It was observed that at a higher cooling rate, no exothermic peak is detected for pristine PLA due to its poor crystallization ability. On the contrary, the introduction of MXene (0.5 wt% loading) enhances the crystallization of PLA significantly. Upon further increases in MXene loading, the crystallization temperature (peak) remained virtually constant. However, as the cooling rate increases further, the single crystallization peak transitioned into a double peak. Further increases in the heating rate often result in changes from a double peak to a single peak. Isotactic polypropylene has been reported to behave in a similar way [53]. With such polymers, heterogeneous nucleation is dominantly reported a slower cooling rate due to the lower super-cooling rate, while at a higher cooling rate, heterogeneous nucleation is less likely; rather, homogeneous nucleation prevails. However, moderate cooling rates instigate the coexistence of heterogeneous and homogeneous nucleation and thus present double crystallization isotherms [53]. It is important to note that double-peak phenomenon in polymer composites is often associated with the presence of more than one crystal form occurring in the composite. It is worth stating that, the addition of MXene does not usually results in significant shift in crystallization temperature especially during cooling; however, the crystallization peak rather gets narrower [43]. This reveals the low crystallinity of the samples before cold crystallization. In certain polymers, cold crystallization peak is usually undetected, implying that the samples have a high degree of crystallinity after crystallizing from the melt due to melt recrystallization [43]

For instance, the crystallization thermograph for pure linear low-density polyethylene (LDPE) and MXene/LDPE nanocomposites at four diverse cooling rates was investigated. It was observed that the crystallization temperature moved to lower temperatures with a decrease in the crystallization enthalpy for 2 wt% MXene composite. The crystallization enthalpy decreased further with the increase in MXene loading in the composites as the cooling rate increases. For the nanocomposites crystallizing at a higher cooling rate, the production of the nuclei and the arrangement of chain segments were negatively affected due to insufficient time. In contrast, during a lower cooling rate, the nuclei activation occurred at a higher temperature and the array of chain segments remained in order. In other words, the higher cooling rate hinders the rate of crystallization, including nucleation and crystal growth. Moreover, when cooling rates reduce, the onset temperature of crystallization and the temperature of exothermic of the peak of MXene/LDPE nanocomposites increases significantly compared to that of pristine LDPE, which indicates that MXene is effective in enhancing the nucleation rate of polymers. However, the composites presented lower crystallization enthalpy at wider temperature ranges—indicating the presence of more lattice defects, poor movement of molecular chains, and low crystallinity [3].

7.9 EFFECT OF MXene ON THE CRYSTALLIZATION BEHAVIOR OF POLYMER NANOCOMPOSITES

Theoretically, crystallization is a two-step procedure involving (a) nucleation and (b) growth (crystal formation). The proportion of the nucleation rate to that of the growth rate is the most important determinant of the sequence of crystallization. The nucleation comprises both homogeneous and heterogeneous crystal formation. Due to the highly active surface of most nanofillers, the molecular chains of polymers gravitate to the surface of the nanofillers, serving as a heterogeneous nucleation site in improving the crystallinity of polymer molecules. From this perspective, it could be propounded that MXene (such as Ti_3C_2) acts as a nucleating agent in the nanocomposite in the slurry before fabrication or during solidification (drying/cooling) to provide a large surface to serve as nucleation centers to increase the rate of crystallization and the percentage of crystals in the molecules of the polymer. In other words, with a similar growth rate, there exist more nucleation sites which leads to large polymer crystallinity. With the increase in MXene load, the overall crystallization rate of polymer tends to increase [54]. The subtopics discussed earlier, buttressed with results in the literature, reveal more about the effect of MXene on the crystallinity of polymer [39, 50]. In a typical study of PVDF composites with 0–5 wt% MXene loading, an improvement in the PVDF thermal conductivity and the thermodynamic mechanical properties were obtained. Further study into the mechanism of crystallization revealed the existence of MXene fast-tracked PVDF crystallization at a low MXene filling, with the fastest crystallization occurring in the composite with 0.5 wt% MXene content [9].

In most cases, the existence of MXene $Ti_3C_2T_x$ hastens polymer crystallization at relatively low MXene loadings; however, the polymerization processes get impeded as MXene filling increases. The fastest crystallization rate has been reported mostly for nanocomposite having considerably low MXene loading due to its enhancement effect in nucleation rate, which results in faster crystal formation [5]. Nucleation is the process by which polymer molecules cluster together to make a larger structure of a definite critical size termed crystals. Thus, the MXene fillers act as nucleating agents, which aid the cluster formation of the polymer molecules. Nevertheless, as the content of MXene in the polymer matrix increases beyond a certain threshold, a phenomenon called nano confinement occurs, where the movement of polymer nuclei get limited (impeded) by the excess MXene fillers, hence reducing the rate of crystal formation. Generally, MXene increases the tensile strength and the elongation at the break of nanocomposites when the loading is minimally optimized. It is important to state that, the delaminated MXene provides more advantages in composites formation than its bulk character. Unlike the bulk MXene, delaminated MXene improves the mechanical and tribological properties of polymers because of the high level of dispersion and homogeneity besides the larger surface area. The most efficient ways to obtain a uniform dispersion of MXene nanofillers into a polymer matrix is through covalent functionalization, surface modification with functional molecules (surfactants), or with novel etchants. The reinforcement effect of MXene polymeric materials is usually attributed to the increase in interfacial interaction between the nanosheets and the neighboring polymer matrices. This interfacial interaction effect is not possible with bulk MXene, hence less usage for engineered polymer composites. The pie chart in Figure 7.6 shows the

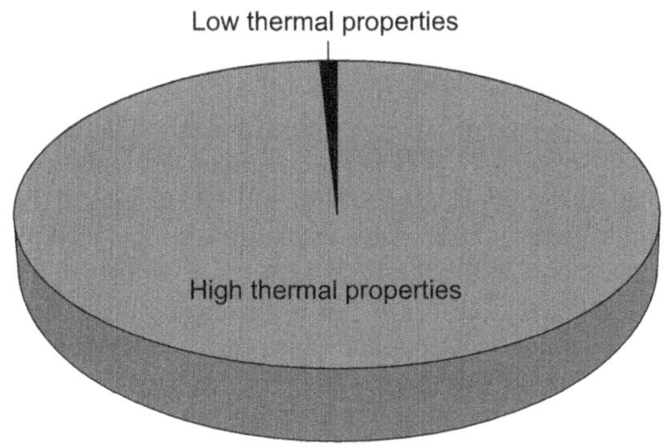

FIGURE 7.6 Graphical representation of the proportion of positive to negative impact of MXene on polymer nanocomposites' thermal properties.

proportion of literature that have reported significant improvement the thermal properties of polymer nanocomposite to that of studies that have reported adverse effect of MXene incorporation in polymer nanocomposites.

7.10 CONCLUSION

The thermal behavior and crystallization of polymer discussed in the chapter have revealed many different influential factors that combine to ensure good thermal properties of polymer nanocomposites. The addition of nanofiller such as MXene has a crucial impact on the thermal behavior and crystallinity of the polymer nanocomposites. Characterization techniques, such as TGA, DSC, and XRD, are primarily used for the evaluation of thermal properties. Three major processes of polymer crystallization and their behavior on the polymer crystals upon a change in conditions have been unveiled. It has been generally established that low MXene loading facilitates enhancement in crystallization, which is economically beneficial in the production and application of nanocomposite. The positive impact of MXene in polymer nanocomposite in enhancing the thermal and crystallinity of the polymer have been assigned to the strong interaction between the MXene and the monomeric component, dispersion, and bonds created between the nanofiller and polymer during the fabrication stage, which reduces the mobility of polymer chains. Factors such as *Tg* and the melting temperature of polymers have also been explicitly discussed. Conclusively, although the incorporation of MXene enhances the properties of polymer nanocomposites, excessive load is liable to reduce the required thermal properties of the nanocomposites.

REFERENCES

[1] W. Hu, L. Zha, Thermodynamics and kinetics of polymer crystallization, in: Polymer Morphology, John Wiley & Sons, Inc., 2016, pp. 242–258.

[2] P. Supaphol, J.E. Spruiell, Isothermal melt- and cold-crystallization kinetics and subsequent melting behavior in syndiotactic polypropylene: a differential scanning calorimetry study, Polymer, 42 (2001) 699–712.

[3] X. Cao, M. Wu, A. Zhou, Y. Wang, X. He, L. Wang, Non-isothermal crystallization and thermal degradation kinetics of MXene/linear low-density polyethylene nanocomposites, e-Polymers, 17 (2017) 373–381.

[4] L. Gao, C. Li, W. Huang, S. Mei, H. Lin, Q. Ou, Y. Zhang, J. Guo, F. Zhang, S. Xu, H. Zhang, MXene/polymer membranes: synthesis, properties, and emerging applications, Chemistry of Materials, 32 (2020) 1703–1747.

[5] Z. Huang, S. Wang, S. Kota, Q. Pan, M.W. Barsoum, C.Y. Li, Structure and crystallization behavior of poly(ethylene oxide)/Ti3C2Tx MXene nanocomposites, Polymer, 102 (2016) 119–126.

[6] F. Chen, A. Clough, B.M. Reinhard, M.W. Grinstaff, N. Jiang, T. Koga, O.K.C. Tsui, Glass transition temperature of polymer–nanoparticle composites: effect of polymer–particle interfacial energy, Macromolecules, 46 (2013) 4663–4669.

[7] G. Chigwada, P. Jash, D.D. Jiang, C.A. Wilkie, Fire retardancy of vinyl ester nanocomposites: synergy with phosphorus-based fire retardants, Polymer Degradation and Stability, 89 (2005) 85–100.

[8] Y. Hai, S. Jiang, C. Zhou, P. Sun, Y. Huang, S. Niu, Fire-safe unsaturated polyester resin nanocomposites based on MAX and MXene: a comparative investigation of their properties and mechanism of fire retardancy, Dalton Transactions, 49 (2020) 5803–5814.

[9] Y. Pan, L. Fu, Q. Zhou, Z. Wen, C.-T. Lin, J. Yu, W. Wang, H. Zhao, Flammability, thermal stability and mechanical properties of polyvinyl alcohol nanocomposites reinforced with delaminated Ti3C2Tx (MXene), Polymer Composites, 41 (2020) 210–218.

[10] J.Y. Si, B. Tawiah, W.L. Sun, B. Lin, C. Wang, A.C.Y. Yuen, B. Yu, A. Li, W. Yang, H.D. Lu, Q.N. Chan, G.H. Yeoh, Functionalization of MXene nanosheets for polystyrene towards high thermal stability and flame retardant properties, Polymers, 11 (2019).

[11] B. Yu, A.C.Y. Yuen, X. Xu, Z.-C. Zhang, W. Yang, H. Lu, B. Fei, G.H. Yeoh, P. Song, H. Wang, Engineering MXene surface with POSS for reducing fire hazards of polystyrene with enhanced thermal stability, Journal of Hazardous Materials, 401 (2021) 123342.

[12] K. Wang, Y. Zhou, W. Xu, D. Huang, Z. Wang, M. Hong, Fabrication and thermal stability of two-dimensional carbide Ti3C2 nanosheets, Ceramics International, 42 (2016) 8419–8424.

[13] L. As'habi, S.H. Jafari, H.A. Khonakdar, L. Häussler, U. Wagenknecht, G. Heinrich, Non-isothermal crystallization behavior of PLA/LLDPE/nanoclay hybrid: synergistic role of LLDPE and clay, Thermochimica Acta, 565 (2013) 102–113.

[14] J.D. Menczel, R.B. Prime, Thermal analysis of polymers—Fundamentals and applications, 2009, John Wiley & Sons, Inc., Publication.

[15] Q. Gao, M. Feng, E. Li, C. Liu, C. Shen, X. Liu, Mechanical, thermal, and rheological properties of Ti3C2Tx MXene/thermoplastic polyurethane nanocomposites, Macromolecular Materials and Engineering, 305 (2020) 2000343.

[16] Z. Ziaee, P. Supaphol, Non-isothermal melt- and cold-crystallization kinetics of poly(3-hydroxybutyrate), Polymer Testing, 25 (2006) 807–818.

[17] M.W.D. Philip, C. Robinson, Polarized light microscopy in: nikon microscopy U—The sources of Microscopy Education, 2021. https://www.microscopyu.com/techniques/polarized-light/polarized-light-microscopy.

[18] R. Zhang, E. Zhuravlev, R. Androsch, C. Schick, Visualization of polymer crystallization by *in situ* combination of atomic force microscopy and fast scanning calorimetry, Polymers, 11 (2019) 890.

[19] R.M. Ronchi, J.T. Arantes, S.F. Santos, Synthesis, structure, properties and applications of MXenes: current status and perspectives, Ceramics International, 45 (2019) 18167–18188.

[20] J. Wang, Y. Liu, Z. Cheng, Z. Xie, L. Yin, W. Wang, Y. Song, H. Zhang, Y. Wang, Z. Fan, Highly conductive MXene film actuator based on moisture gradients, Angewandte Chemie, 132 (2020) 14133–14137.

[21] R. Bian, S. Xiang, D. Cai, Fast treatment of MXene films with isocyanate to give enhanced stability, Chem Nano Mat, 6 (2020) 64–67.

[22] A. Lipatov, M. Alhabeb, M.R. Lukatskaya, A. Boson, Y. Gogotsi, A. Sinitskii, Effect of synthesis on quality, electronic properties and environmental stability of individual monolayer Ti3C2 MXene flakes, Advanced Electronic Materials, 2 (2016) 1600255.

[23] K. Maleski, V.N. Mochalin, Y. Gogotsi, Dispersions of two-dimensional titanium carbide MXene in organic solvents, Chemistry of Materials, 29 (2017) 1632–1640.

[24] L. Guo, Z. Zhang, M. Li, R. Kang, Y. Chen, G. Song, S.-T. Han, C.-T. Lin, N. Jiang, J. Yu, Extremely high thermal conductivity of carbon fiber/epoxy with synergistic effect of MXenes by freeze-drying, Composites Communications, 19 (2020) 134–141.

[25] K. Deshmukh, T. Kovářík, S.K. Khadheer Pasha, State of the art recent progress in two dimensional MXenes based gas sensors and biosensors: a comprehensive review, Coordination Chemistry Reviews, 424 (2020) 213514.

[26] C.B. Hatter, J. Shah, B. Anasori, Y. Gogotsi, Micromechanical response of two-dimensional transition metal carbonitride (MXene) reinforced epoxy composites, Composites Part B: Engineering, 182 (2020) 107603.

[27] R. Liu, W. Li, High-thermal-stability and high-thermal-conductivity Ti3C2Tx MXene/Poly(vinyl alcohol) (PVA) composites, ACS Omega, 3 (2018) 2609–2617.
[28] A. Champagne, J.-C. Charlier, Physical properties of 2D MXenes: from a theoretical perspective, Journal of Physics: Materials, 3 (2020) 032006.
[29] J.L. Hart, K. Hantanasirisakul, A.C. Lang, B. Anasori, D. Pinto, Y. Pivak, J.T. van Omme, S.J. May, Y. Gogotsi, M.L. Taheri, Control of MXenes' electronic properties through termination and intercalation, Nature Communications, 10 (2019) 522.
[30] M. Ghidiu, S. Kota, V. Drozd, M.W. Barsoum, Pressure-induced shear and interlayer expansion in Ti3C2 MXene in the presence of water, Science Advances, 4 (2018) eaao6850.
[31] O. Mashtalir, M.R. Lukatskaya, A.I. Kolesnikov, E. Raymundo-Pinero, M. Naguib, M. Barsoum, Y. Gogotsi, The effect of hydrazine intercalation on the structure and capacitance of 2D titanium carbide (MXene), Nanoscale, 8 (2016) 9128–9133.
[32] M. Seredych, C.E. Shuck, D. Pinto, M. Alhabeb, E. Precetti, G. Deysher, B. Anasori, N. Kurra, Y. Gogotsi, High-temperature behavior and surface chemistry of carbide MXenes studied by thermal analysis, Chemistry of Materials, 31 (2019) 3324–3332.
[33] H. Tang, Y. Yang, R. Wang, J. Sun, Improving the properties of 2D titanium carbide films by thermal treatment, Journal of Materials Chemistry C, 8 (2020) 6214–6220.
[34] J. Li, X. Yuan, C. Lin, Y. Yang, L. Xu, X. Du, J. Xie, J. Lin, J. Sun, Achieving high pseudocapacitance of 2D titanium carbide (MXene) by cation intercalation and surface modification, Advanced Energy Materials, 7 (2017) 1602725.
[35] P. Eklund, J. Rosen, P.O.Å. Persson, Layered ternary M n+ 1AX n phases and their 2D derivative MXene: an overview from a thin-film perspective, Journal of Physics D: Applied Physics, 50 (2017) 113001.
[36] H. Kim, B. Anasori, Y. Gogotsi, H.N. Alshareef, Thermoelectric properties of two-dimensional molybdenum-based MXenes, Chemistry of Materials, 29 (2017) 6472–6479.
[37] Y. Cao, Q. Deng, Z. Liu, D. Shen, T. Wang, Q. Huang, S. Du, N. Jiang, C.-T. Lin, J. Yu, Enhanced thermal properties of poly(vinylidene fluoride) composites with ultrathin nanosheets of MXene, RSC Advances, 7 (2017) 20494–20501.
[38] L. Wang, L. Chen, P. Song, C. Liang, Y. Lu, H. Qiu, Y. Zhang, J. Kong, J. Gu, Fabrication on the annealed Ti3C2Tx MXene/Epoxy nanocomposites for electromagnetic interference shielding application, Composites Part B: Engineering, 171 (2019) 111–118.
[39] H. Aghamohammadi, N. Amousa, R. Eslami-Farsani, Recent advances in developing the MXene/polymer nanocomposites with multiple properties: a review study, Synthetic Metals, 273 (2021) 116695.
[40] M.S. Carey, M. Sokol, G.R. Palmese, M.W. Barsoum, Water transport and thermomechanical properties of Ti3C2Tz MXene epoxy nanocomposites, ACS Applied Materials & Interfaces, 11 (2019) 39143–39149.
[41] Y. Zou, L. Fang, T. Chen, M. Sun, C. Lu, Z. Xu, Near-infrared light and solar light activated self-healing epoxy coating having enhanced properties using MXene flakes as multifunctional fillers, Polymers, 10 (2018) 474.
[42] Y. Shi, Z. Gui, B. Yu, R.K. Yuen, B. Wang, Y. Hu, Graphite-like carbon nitride and functionalized layered double hydroxide filled polypropylene-grafted maleic anhydride nanocomposites: comparison in flame retardancy, and thermal, mechanical and UV-shielding properties, Composites Part B: Engineering, 79 (2015) 277–284.
[43] Q. Zhao, B. Wang, C. Qin, Q. Li, C. Liu, C. Shen, Y. Wang, Nonisothermal melt and cold crystallization behaviors of biodegradable poly(lactic acid)/Ti3C2Tx MXene nanocomposites, Journal of Thermal Analysis and Calorimetry, 147.3 (2022) 2239–2251.
[44] L. Yu, H. Liu, K. Dean, L. Chen, Cold crystallization and postmelting crystallization of PLA plasticized by compressed carbon dioxide, Journal of Polymer Science Part B: Polymer Physics, 46 (2008) 2630–2636.
[45] H. Zhang, L. Wang, A. Zhou, C. Shen, Y. Dai, F. Liu, J. Chen, P. Li, Q. Hu, Effects of 2-D transition metal carbide Ti2CTx on properties of epoxy composites, RSC Advances, 6 (2016) 87341–87352.
[46] X. Xia, J. Li, J. Zhang, G.J. Weng, Uncovering the glass-transition temperature and temperature-dependent storage modulus of graphene-polymer nanocomposites through irreversible thermodynamic processes, International Journal of Engineering Science, 158 (2021) 103411.

[47] P. Song, H. Qiu, L. Wang, X. Liu, Y. Zhang, J. Zhang, J. Kong, J. Gu, Honeycomb structural rGO-MXene/epoxy nanocomposites for superior electromagnetic interference shielding performance, Sustainable Materials and Technologies, 24 (2020) e00153.

[48] S. Chandran, J.K. Basu, M.K. Mukhopadhyay, Variation in glass transition temperature of polymer nanocomposite films driven by morphological transitions, The Journal of chemical physics, 138 (2013) 014902.

[49] B.J. Ash, R.W. Siegel, L.S. Schadler, Glass-transition temperature behavior of alumina/PMMA nano-composites, Journal of Polymer Science Part B: Polymer Physics, 42 (2004) 4371–4383.

[50] Y. Shi, C. Liu, L. Liu, L. Fu, B. Yu, Y. Lv, F. Yang, P. Song, Strengthening, toughing and thermally stable ultra-thin MXene nanosheets/polypropylene nanocomposites via nanoconfinement, Chemical Engineering Journal, 378 (2019) 122267.

[51] Y. Tsujimoto, T. Sakurai, Y. Ono, S. Nagano, S. Seki, Cold crystallization of ferrocene-hinged π-conjugated molecule induced by the limited conformational freedom of ferrocene, The Journal of Physical Chemistry B, 123 (2019) 8325–8332.

[52] M. Le Page Mostefa, H. Muhr, E. Plasari, M. Fauconet, A purification route of bio-acrylic acid by melt crystallization respectful of environmental constraints, Powder Technology, 255 (2014) 98–102.

[53] D. Cavallo, L. Gardella, G.C. Alfonso, D. Mileva, R. Androsch, Effect of comonomer partitioning on the kinetics of mesophase formation in random copolymers of propene and higher α-olefins, Polymer, 53 (2012) 4429–4437.

[54] H. Zhang, L. Wang, Q. Chen, P. Li, A. Zhou, X. Cao, Q. Hu, Preparation, mechanical and anti-friction performance of MXene/polymer composites, Materials & Design, 92 (2016) 682–689.

8 Tribological Performance of MXene/Polymer Nanocomposites

Rodrigo Mantovani Ronchi, Hugo Gajardoni de Lemos, and Sydney Ferreira Santos

CONTENTS

8.1 INTRODUCTION

The term *tribology* was defined by (Jost 1966) as "The science and technology of interacting surfaces in relative motion and of associated subjects and practices". In his pioneer report, Jost highlighted that the reduction of wear, friction, and breakdowns associated with tribological failures could result in significant economic savings. He estimated that it could reach about 1% of the United Kingdom's gross national product (GNP), and over time, some estimation of economic savings has been published with values of 1.6% (Mate 2007) or even 4% of the GNP (Bhushan 2017). The name "tribology" is relatively new, but the tribology issues are part of everyday life throughout the history of humankind as can be exemplified by the production of fire by rubbing sticks, wheels for grinding cereals, writing with pencils, etc. At the time of Jost's report, the main interest of tribology was related to the reduction of wear and failure of moving mechanical parts of machines, which remain a relevant subject until now, but other ones came up over time, such as magnetic storage devices (e.g., wear at the head–disk interfaces in hard disk drives), and later the emerging field micro/nanoelectromechanical systems (MEMSs/NEMSs). Relevant tribology issues can be even inside us. Wear in joint prostheses is a very complex and important subject, and the enhancements in materials and design can extend the durability of these devices, with obvious improvements in the quality of life of patients. From the scientific point of view, tribology can be considered a multidisciplinary field that involves materials science and engineering, mechanical engineering, chemistry, physics, and so on. Thus, it demands a body of knowledge in a wide range of subjects such as the development of materials, materials processing and characterization, surface science, corrosion, mechanical design, hydrodynamics, heat transfer, colloidal suspensions and lubricants, and much more.

Today, polymers are used in several engineering applications due to their low unit cost, low weight, and easy processing. However, the use of polymers (in bulk) in tribological applications is very limited, because they usually do not achieve the needed requirements. Therefore, metallic and/or ceramic reinforcements are usually added to reduce their wear rates and optimize the coefficient of friction (COF), forming a polymer-matrix composite (PMC). For example, polytetrafluoroethylene (PTFE) is a self-lubricating fluorocarbon polymer that presents a very low COF but has a high

DOI: 10.1201/9781003164975-8

wear rate. In order to reduce its wear, it must be used in a composite, with the addition of ceramic particles or glass fibers, which increases its load-bearing capacity and maintains low levels of friction. Nowadays, PMCs are widely used in gears, bearings, artificial human joint bearing surfaces, tires, automobile brake pads, and others (Friedrich and Schlarb 2008).

However, the requirements for higher and tailored performances (*e.g.*, a miniaturized lightweight component with low wear rate at high pressures) make almost impossible the use of bare polymers or even traditional composites. One promising solution to this problem is the use of nanomaterials as reinforcements, since their lower dimensionality results in substantial properties enhancements when compared to traditional polymer composites. For example, different from traditional composites, nanocomposites can simultaneously enhance both strength and toughness (Friedrich and Schlarb 2008).

Several 2D nanomaterials such as graphene, boron nitride (BN), molybdenum disulfide (MoS_2) and, more recently, the transition metal carbides and nitrides (MXenes) have been investigated as promising alternatives for tribological applications (Yanbao Guo et al. 2021; Wyatt, Rosenkranz, and Anasori 2021). Regarding MXenes, these compounds are obtained from MAX phase (bulk) precursors through exfoliation and delamination (Alhabeb et al. 2017). MXenes have recently attracted much interest due to their unusual combination of high Young's modulus, thermal, and electrical conductivity with hydrophilic behavior, which provides better interaction with polymeric matrices, differently than graphene (Jing Chen et al. 2015; Naguib et al. 2016; Ling et al. 2014; Boota et al. 2017). Moreover, these nanocomposites can be fabricated through simple and low-cost methods (*e.g.*, solution mixing; Anasori et al. 2017; Zhang et al. 2018; Seyedin, Yanza, and Razal 2017) and are suitable to industrial techniques, such as electrospraying, emulsion (Zhi et al. 2018), extrusion printing, and wet spinning (Akuzum et al. 2018).

Studies of adding MXenes into nanocomposites are still in the beginning but already show great promises. Addition into ultra-high molecular weight polyurethane (UHMWPE) (Heng Zhang, Wang, Chen, et al. 2016) and epoxy (Heng Zhang, Wang, Zhou, et al. 2016) matrices showed an increase in wear resistance with a lower coefficient of friction compared to their pure matrices. For example, the addition of 2.0 wt% Ti_2CT_x in an epoxy matrix decreased its COF by nearly 65% (Heng Zhang, Wang, Zhou, et al. 2016).

In this chapter, the relevant aspects of MXenes and MXenes/polymer nanocomposites are reviewed. After this concise ***Introduction*** (Section 8.1), ***atomistic friction in MXenes*** is discussed in Section 8.2, where fundamental aspects involved in this phenomenon are depicted such as the influence of vacancies and temperature in friction. Furthermore, experimental setups used in atomic-scale friction measurements are briefly introduced. Section 8.3 reviews the use of ***MXenes as lubricant additives***, with emphasis on the production of stable colloidal suspensions, wear mechanisms in MXenes containing lubricants, and tribofilm formation. Section 8.4 is dedicated to ***MXenes/polymer nanocomposites***, in which the effect of MXenes on the friction coefficient and the wear behavior of these nanocomposites is discussed. The matrices and MXene compounds already used in these nanocomposites are reviewed, and the effect of Mxenes on mechanical properties is briefly discussed. Finally, in Section 8.5, the ***concluding remarks*** are presented, highlighting the trends and remaining open questions in this subject.

8.2 ATOMISTIC FRICTION IN MXenes

Tribology covers a wide range of dimensional scales with different dominant mechanisms at each scale. For example, the mechanical systems of automotive parts experience much lower contact stresses and pressures than the geological layers involved in earthquakes. The continuous miniaturization of mechanical components led to increasing attention regarding the tribological phenomena at the atomic scale. By way of illustration, the so-called micro/electromechanical systems (MEMSs), first reported in the 1980s, include sensors and micro-motors in the range of 100 nm to 1 μm using silicon-based materials. Since the surface area/volume ratio increases substantially with decreasing sizes, several surface-related effects arise and become critical to these devices, such as surface tension and adhesion. Thus, in order to tailor friction and wear at the micro/nanoscales, the atomic-related phenomena must be understood (Zhou et al. 2021; Guan et al. 2020). In this section, we focus on MXene nanofriction behavior, evaluated through both simulations and experimental studies.

Atomic-scale simulation methods are widely used in nanotribology, because they provide information about the atomic interactions between the layers. Moreover, since experimental frictional behavior is influenced by several factors (*e.g.*, normal load, temperature, layer number, humidity and others), the influence of only one variable is more clearly assessed by computational methods since it can be much more difficult to be isolated in the experiments.

Computational *ab initio* studies (H Zhang et al. 2017; Difan Zhang et al. 2017) have demonstrated some of MXenes' promising features to tribological applications, such as a low COF, a low barrier for interlayer sliding and the possibility of superlubricity effect through the uniaxial strains. From first principles calculations, (H Zhang et al. 2017) different oxygen-functionalized M_2CO_2 bilayers (M = Ti, Zr, Nb, Mo, Hf, Ta, and W) and the interlayer sliding resistances were explored. They found that larger lattice parameters and smaller interlayer distances induce higher sliding resistance. Furthermore, bare MXenes are much more difficult to slide because of their strong metallic bonds between the stacked layers.

In another work performed by first principles and classical molecular dynamics simulations (Difan Zhang et al. 2017), $Ti_3C_2O_2$ presented the lowest energy barrier (0.38 eV/nm^2) among Ti_2CO_2, $Ti_3C_2O_2$, and $Ti_4C_3O_2$. The COF was estimated between 0.24 and 0.27 at 10 K in these compounds, reaching coefficients lower than 0.04 at 298 K. Titanium and oxygen vacancies were found to increase friction (Figure 8.1.a) due to the increased surface roughness and attraction between the two layers. Finally, the authors evaluated different surface terminations for Ti_3C_2 compound, including –OH and –OCH_3, and found that both showed lower COFs, reaching 0.14 for –OCH_3 and even 0.10 for hydroxyl termination, as shown in Figure 8.1.b. This work highlights the influence of "n" layers (n = 2, 3, and 4), point defects and surface terminations into the COF aiding experimental observations. For example, it indicated that more controlled synthesis routes (*i.e.*, lower Ti and O vacancies) could optimize the lubricant properties for tribological applications.

Despite these studies, nanotribological simulations of MXenes still need to be explored. For instance, studies of fluorinated MXenes are still pendent, which is somewhat surprising since –F is the most common termination after hydrofluoric acid (HF) etching procedure (Hope et al. 2016;

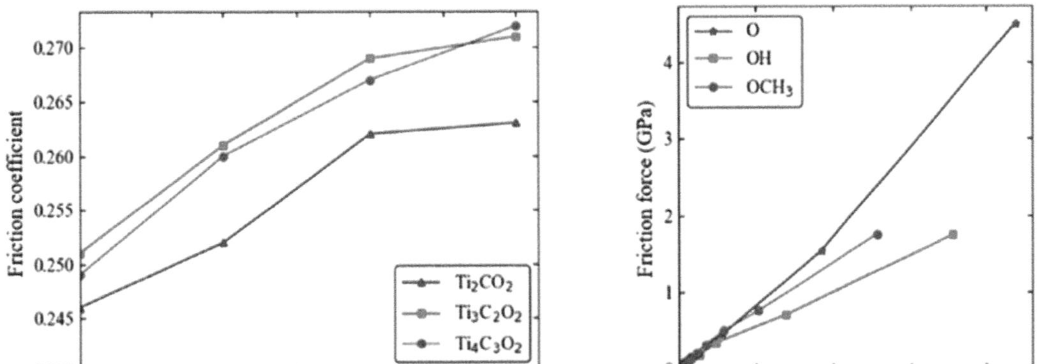

FIGURE 8.1 (a) Friction coefficient at different $Ti_{n+1}C_nO_2$ compounds (n = 1, 2, and 3) and the effect of oxygen vacancies; (b) effect of terminations on interlayer friction of $Ti_3C_2T_x$.

Source: Reproduced with permission (Difan Zhang et al. 2017). Copyright 2017, ACS Publications.

Naguib et al. 2011). Moreover, studies at larger systems evaluating mixed functional groups and/or number of layers, are still to be performed, in particular using molecular dynamics.

The predicted promising features of MXenes, such as low COF and low barrier for interlayer sliding, are very interesting for tribological applications and have attracted the attention of experimentalists. Atomic force microscopy (AFM) is the most common method used to probe friction at the atomic scale. In this equipment, surface forces, such as friction and adhesion, can be measured by the attractive or repulsive interactions between the surface atoms of the sample and those of the AFM tip (Kim and Kim 2009). A schematic diagram can be seen in Figure 8.2.a.

Experimental studies on MXenes identified the influence of pressure and temperature on the friction force (Guo et al. 2019; Guan et al. 2020). The measurements of friction and adhesion were performed by AFM analyses. The increased friction force under higher pressures was a direct result of the higher contact area between the tip and the surface while the decrease of friction with temperature was attributed to an increase in the MXene oxidation degree. Decreasing the relative humidity was also found to reduce MXenes' friction, because the lower number of water molecules adsorbed at the MXene surface decreases the interfacial adhesion and contact energy (Guan et al. 2020).

$Ti_3C_2T_x$ friction of a 6.5-nm $Ti_3C_2T_x$ nanosheet was approximately 35% of the one recorded on the silicon dioxide substrate (Rodriguez et al. 2021). This performance is inferior to graphene (5–15% of friction force; Vazirisereshk et al. 2019; Tripathi et al. 2018) but similar to MoS_2, which ranges from 15–40% (Acikgoz and Baykara 2020; Vazirisereshk et al. 2019). However, this result contrasts with a previous computational study that predicted an energy barrier for $Ti_3C_2O_2$ nearly 30% lower than MoS_2 (D. Zhang et al. 2017). Nonetheless, since there are some differences, such as the number of layers and the different functional groups, more systematic characterization and modeling must be performed.

The influence of the functional groups and different synthesis routes in nanotribological properties was evidenced through the comparison of $Ti_3C_2T_x$, fluorinated-$Ti_3C_2T_x$ and TMAOH-delaminated-$Ti_3C_2T_x$ compounds (Guan et al. 2020). Tetramethylammonium hydroxide (TMAOH) is a common intercalation agent used in MXenes synthesis procedures. It was shown that surface properties played a dominant role in both adhesion and friction. These authors reported that higher hydrophilicity increases adhesion and friction. Thus, the COF increased in the following order: $Ti_3C_2T_x$< fluorinated-$Ti_3C_2T_x$à TMAOH-delaminated-$Ti_3C_2T_x$.

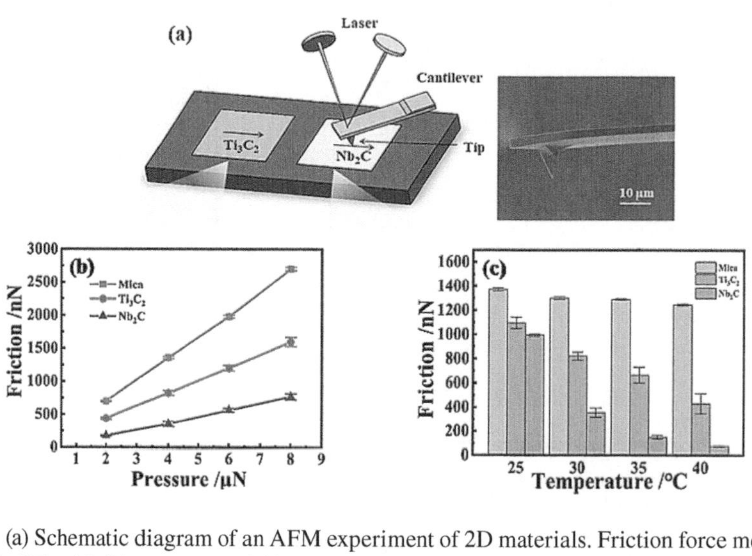

FIGURE 8.2 (a) Schematic diagram of an AFM experiment of 2D materials. Friction force measurements of $Ti_3C_2T_x$ and Nb_2CT_x with (b) pressure and (c) temperature.

Source: Adapted and reproduced with permission (Zhou et al. 2021). Copyright 2021, Elsevier.

Besides $Ti_3C_2T_x$, AFM was also used to evaluate the friction behavior of Nb_2CT_x with more than 20 layers (Fig. 8.2a). It was found that Nb_2CT_x presented less friction and adhesion than $Ti_3C_2T_x$ at all temperatures and/or pressures (Figures 8.2b and 8.2c). This behavior was attributed to the lower dipole moment of Nb_2CT_x, which induces a lower charge accumulation at the surface and a more uniform electron cloud distribution (Zhou et al. 2021).

The friction and adhesion of both materials increase with pressure (Figure 8.2b) due to the probe contact area growth, in agreement with previous works (Guo et al. 2019; Guan et al. 2020). With increasing temperatures (Figure 8.2c), the friction of both compounds decreased significantly. For example, Nb_2CT_x had its value decreased by nearly 92% at 40°C compared to 25°C. Researchers attributed this reduction to the lower polarization of the compound resulting from an increase in oxidation (Guo et al. 2019), faster atomic movements, and decreased strength of molecular interactions.

Table 8.1 summarizes some published data of atomic-scale friction of MXenes compared to other 2D nanomaterials. It can be observed that nanotribological investigations of MXenes are barely reported and need further investigations. Only $Ti_3C_2T_x$ and Nb_2CT_x were experimentally investigated, despite of more than 30 MXene compounds already synthesized, such as the newly discovered ordered in-plane and out-of-plane MXenes (Persson et al. 2018; Anasori et al. 2015). Moreover,

TABLE 8.1
Published Data of Atomic-Scale Friction of Mxenes Compared to Other 2D Nanomaterials

2D Material	Synthesis	Number of Layers (height)	Friction Measurements	Ref
$Ti_3C_2T_x$	Not informed	<8 (~6.5 nm)	F_{MXene}/F_{silica} = ~0.35	(Rodriguez et al. 2021)
$Ti_3C_2T_x$	HF	Monolayer (1.42 nm)	Friction factor 0.00657	(Guan et al. 2020)
Fluorinated $Ti_3C_2T_x$	HCl-LiF	Monolayer (1.46 nm)	Friction factor 0.00704	(Guan et al. 2020)
TMAOH- $Ti_3C_2T_x$	HF + TMAOH	Monolayer (1.60 nm)	Friction factor 0.02513	(Guan et al. 2020)
$Ti_3C_2T_x$	HF	>20 layers (~35 nm)	F_{MXene}/F_{mica} = ~0.58	(Zhou et al. 2021)
Nb_2CT_x	Not informed	>20 layers (~42.5 nm)	F_{MXene}/F_{mica} = ~0.25	(Zhou et al. 2021)
Graphene	Commercial chemically vapor deposited (CVD) graphene	Monolayer (~0.4 nm) Bilayer (~1.1 nm)	Normalized COF μ_{1LG}/μ_{silica} = 0.102; μ_{2LG}/μ_{silica} = 0.065	(Tripathi et al. 2018)
MoS2	mechanical exfoliation	Monolayer (~1.0 nm)	F_{MXene}/F_{silica} = ~0.4	(Acikgoz and Baykara 2020)
MoS2	CVD	Monolayer (~0.8 nm)	Friction force = ~ 3.95 nN μ_{1LG}/μ_{silica} = ~0.07	(Vazirisereshk et al. 2019)
Graphene	CVD	Monolayer (~0.4 nm)	Friction force = ~ 8.55 nN μ_{1LG}/μ_{silica} = ~0.151	(Vazirisereshk et al. 2019)
Graphene	CVD	7–9 layers (estimated by Raman)	$\mu_{Graphene}/\mu_{silica}$: ~0.08	(Berman et al. 2015)
Graphene Oxide (GO)	Commercial GO solution	Multilayer (not informed)	μ_{GO}/μ_{silica}: ~0.55	(Berman et al. 2015)
Graphene	mechanically exfoliated (Scotch tape method)	Bilayer (~0.6 nm) 3-layer (~0.9 nm) 4-layer (~1.2 nm) Bulk (>20 layers)	μ_{2LG}/μ_{1LG} = ~0.8 μ_{3LG}/μ_{1LG} = ~0.6 μ_{4LG}/μ_{1LG} = ~0,45 μ_{BULK}/μ_{1LG} = ~0,45	(Lee et al. 2010)
MoS2	mechanically exfoliated (Scotch tape method)	Bilayer 3-layer 4-layer Bulk (>20 layers)	μ_{2LG}/μ_{1LG} = ~0.9 μ_{3LG}/μ_{1LG} = ~0.7 μ_{4LG}/μ_{1LG} = ~0.6 μ_{BULK}/μ_{1LG} = ~0,3	(Lee et al. 2010)

(Continued)

TABLE 8.1
(Continued)

2D Material	Synthesis	Number of Layers (height)	Friction Measurements	Ref
h-BN	Mechanically exfoliated (Scotch tape method)	Bilayer 3-layer 4-layer Bulk (>20 layers)	μ_{2LG}/μ_{1LG} = ~0.85 μ_{3LG}/μ_{1LG} = ~0.65 μ_{4LG}/μ_{1LG} = ~0.6 μ_{BULK}/μ_{1LG} = ~0.3	(Lee et al. 2010)
Graphene	CVD	~4-layer (1.97 nm)	F ≈ 8 nN (cycle =200)	(Peng, Wang, and Zou 2015)
Graphene	Mechanically exfoliated (Scotch tape method)	~4-layer (1.90 nm)	F ≈ 2.5 nN(cycle =200)	(Peng, Wang, and Zou 2015)
Graphene Oxide	Modified Hummer's method	Bilayer (2.10 nm)	F ≈ 40 nN (cycle =200)	(Peng, Wang, and Zou 2015)
Reduced Graphene Oxide	chemical reduction of the GO	Bilayer (2.10 nm)	F≈ 35 nN (cycle =200)	(Peng, Wang, and Zou 2015)

as the topological structure and the interlayer forces are directly related to nano-friction phenomena, several parameters need to be systematically studied in order to understand their effects. Among others, it can be cited the number of layers and the influence of different synthesis procedures (the most common, HF [Naguib et al. 2011], HCl-LiF [Ghidiu et al. 2014], and MILD (Lipatov et al. 2016; Sang et al. 2016] etching), surface defects, and lateral sizes.

8.3 MXenes AS LUBRICANT ADDITIVES

Friction and wear critically affect a wide range of applications, resulting in massive waste of energy and loss of materials. The use of lubricants consists of one the most efficient methods to decrease these negative phenomena. The high-performance lubricants currently available usually consist of a base fluid and additives. The former comprises the main component of the lubricant by determining its primary characteristic. The additives, on the other hand, may tune the lubricant properties by providing a variety of features such as better antiwear capabilities, friction reduction, viscosity modification, antioxidant properties, or corrosion inhibitors.

Two-dimensional materials have been extensively studied as alternative additives for lubricants (Xiao and Liu 2017). These materials show layered structures with relatively low shear strength, which permits sliding among adjacent layers under shear force. They also exhibit high specific surface area that allows them to be adsorbed onto rubbing surfaces to form a protective tribofilm. The tribofilm can be defined as "a thin solid film created as result of sliding contact, which is adhered on its parent worn surface but has different chemical composition, structure and tribological behavior" (Luo 2013). It prevents direct contact between the rubbing surfaces, lowering the friction and wear. In addition, 2D materials can act by filling concave areas and gaps usually present in surfaces due to manufacturing limitations. The resultant smoother surface leads to a reduction of localized contact pressure and plastic deformation. Moreover, the 2D additives can affect the fluid viscosity and drag, which comprises a dominant factor in the load-bearing capacity and friction force of the lubricant. Figure 8.3 summarizes the main proposed mechanisms in the literature, which explains the role of 2D structures as additives in lubricants (Xiao and Liu 2017).

Graphene and its derivatives have shown promising results as additives for lubricants due to their enhanced mechanical properties and chemical stability (Berman, Erdemir, and Sumant 2014; Mungse, Kumar, and Khatri 2015). In addition, other 2D structures, such as transition metal dichalcogenides (Xu et al. 2015; Tang et al. 2014) and zirconium phosphates (Xiao et al. 2015), also have shown excellent lubricant properties. However, most of these 2D additives present disadvantages regarding their dispersibility into the lubricant, as they show a tendency to agglomerate and

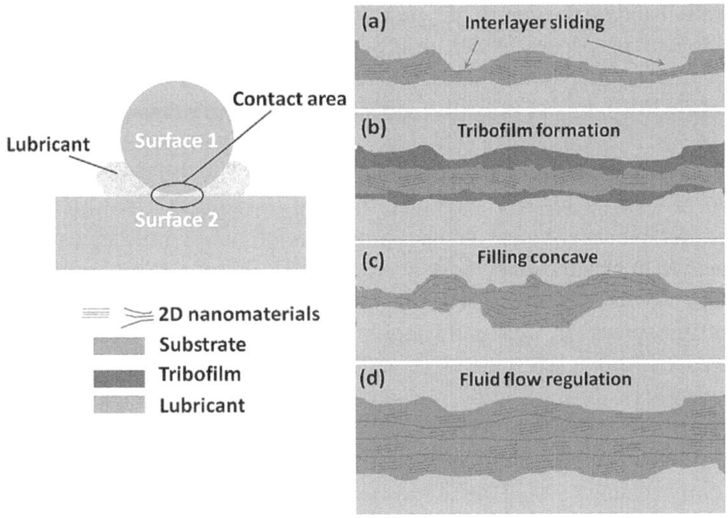

FIGURE 8.3 Proposed mechanisms of 2D nanosheets to enhance tribological properties: (a) sliding surfaces, (b) tribofilm formation, (c) filling the intrinsic gaps of surfaces, and (d) affecting the fluid viscosity.

Source: Reproduced with permission (Xiao and Liu 2017). Copyright 2017, Elsevier.

precipitate within a short period. This phenomenon results in negative effects on the shelf life and lubrication performance, which limit their widespread use as additives. Therefore, endeavors are searching for new materials that can attend to the desired tribological properties, chemical stability, and favored dispersibility in fluids.

Experimental and theoretical studies have evidenced the excellent mechanical and tribological properties of MXenes (Malaki and Varma 2020). Additionally, MXenes show good dispersion characteristics in a wide range of solvents due to the presence of varied functional groups, which could overcome the stability issues involving lubricant additives. Recent studies demonstrated the feasibility to obtain stable MXene solutions in nonpolar solvents. Carey et al. (2020) showed that MXenes become organophilic and form stable colloidal suspensions in nonpolar solvents, through the exchange of lithium cations between MXene multilayers after etching with di(hydrogenated tallow)benzyl methyl ammoniumchloride (DHT). Moreover, Lim et al. (2019) obtained stable nonpolar colloidal dispersions of MXenes chemically grafted with lipophilic octyltriethoxysilanes. Thereby, the progress in obtaining stable solutions of MXenes with different solvents has encouraged their use as additives in lubricants or reinforcing agents in composite materials.

Recent works have shown that even low content of these 2D materials may significantly enhance the antiwear and antifriction properties of fluids. (Yang et al. 2014) prepared Ti_3AlC_2 MAX phase by pressureless sintering followed by exfoliation using HF to obtain layered $Ti_3C_2T_x$. The authors showed that the addition of 1.0 wt% of the 2D material in paraffin base oil resulted in a decrease of the COF from 0.125 (pure oil) to 0.063 (1.0 wt% of MXene). The improvement on the antifriction property could be explained by the formation of a uniform lubricant film preventing direct contact between the mating surfaces. On the other hand, under higher concentrations of MXenes, an increase of the COF and a depreciation of antiwear property could be noticed due to the aggregation of the 2D material.

Similar results were obtained by Zhang et al. (2015), who studied the antifriction and antiwear properties of $Ti_3C_2(OH)_2$ nanosheets (10–20 nm of thickness) as additives for base oil lubricants. The authors found that under MXene concentrations below 1.0 wt%, a decrease of the COF could be obtained when compared to the oil without an additive. Through scanning electron microscopy (SEM) and energy dispersive X-ray spectroscopy (EDS), they proposed that the sliding friction of the $Ti_3C_2(OH)_2$ nanosheets and the presence of a stable and uniform tribofilm prevented a direct contact and decreased the shearing stress between the mating surfaces.

The concentration and dispersion of MXenes into the lubricant play a key factor for the enhancement of the tribological properties. It was proposed that under optimal conditions, the large surface area of the exfoliated $Ti_3C_2T_x$ contributes to change the fluid viscosity resulting in stable tribofilm formation (Liu et al. 2017). In addition, the high surface energy of $Ti_3C_2T_x$ nanosheets leads to their absorption into the surface valleys contributing to surface roughness depletion. On the other hand, under high additive concentrations, the colloidal system of $Ti_3C_2T_x$ nanosheets and base oil may become unstable, promoting additive agglomeration and consequent tribofilm damages.

In order to further enhance the tribological properties of $Ti_3C_2T_x$, strategies to incorporate other nanomaterials with MXene nanosheets have been explored. Xue et al. (2017) promoted the growth of TiO_2 nanoparticles onto $Ti_3C_2T_x$ by hydrothermal process. The $TiO_2/Ti_3C_2T_x$ hybrid nanocomposites showed an effective decrease in wear and friction when incorporated as additive in PAO 8 base oil. COF values decreased from 0.103 (pure oil) to 0.073 (1.0 wt% $TiO_2/Ti_3C_2T_x$) at 20N load condition. According to the authors, during sliding, desquamated TiO_2 nanoparticles can mend the scratched surfaces resulting in decrease of surface roughness. Moreover, the exfoliated $Ti_3C_2T_x$ nanosheets tend to be adsorbed onto the rubbing surfaces, preventing their direct contact through tribofilm formation. Similarly, Zhang et al. (2019) synthesized potassium titanate nanowires uniformly distributed onto $Ti_3C_2T_x$ surface. An evaluation of tribological performances of titanate-$Ti_3C_2T_x$-filled poly-α-olefin (PAO)8 oil, commonly applied in automotive and industrial gear oils, bearing lubricants, and so on, showed a decrease of COF and wear width by 4.9% and 22%, respectively, when compared to unmodified-$Ti_3C_2T_x$-filled oil. The superior performance of the titanate–$Ti_3C_2T_x$-filled oil could be attributed to the enlarged interlayer spacing of the nanocomposite nanosheets after intercalation treatment besides the enhanced stability and load capacity of the formed tribofilm.

Recently, environment-friendly water-based lubricants have attracted attention due to their high safety, flame retardant, and excellent cooling property. However, their low viscosity and relatively high surface tension when compared to oil-based lubricants limit their widespread use. The MXene functional groups (–O, –OH, and –F) provide high hydrophilic properties, making them a promising alternative as an additive for water-based lubricants. Chen and Zhao (2021) used an alternate pressure technique to obtain 2-nm $Ti_3C_2T_x$ nanosheets. Its thin lamellar structure resulted in high-stable MXene water dispersions, which contributed to the formation of a more continuous and effective lubricating film between the contact surfaces as shown by EDS analysis (Figures 8.4a and 8.4b). The MXene addition with an optimal concentration of 0.8 mg/ml decreased the COF and the wear rate by 34.74% and 45.58%, respectively, when compared to pure water-based lubricant (Figures 8.4c and 8.4d).

In order to further enhance the stability of $Ti_3C_2T_x$ nanosheets solution in water, (Yi et al. 2021) applied glycerol to a $Ti_3C_2T_x$ filled water-based lubricant. The $Ti_3C_2T_x$/glycerol/water system showed excellent dispersibility and stability rendering solutions without agglomeration for at least 60 days. $Ti_3C_2T_x$ on glycerol/water fluid lubricating performance was tested in a pin-on-disk tribometer with an Si_3N_4 ball and sapphire disk (Figure 8.5a). The $Ti_3C_2T_x$/glycerol/water system resulted in an impressive minimum COF value of 0.002. According to the authors, this triggered superlubricity was mainly attributed to a resultant mixed lubrication regime, which comprises the boundary and elastohydrodynamic lubrication (Figure 8.5b). For the former, tribochemical reactions between $Ti_3C_2T_x$ and Si_3N_4 lead to tribofilm formation primarily composed by colloidal silicon oxide and titanium oxide on the Si_3N_4 surface, whereas, at the sapphire surface, MXene nanosheets were physically adsorbed onto the surface wear tracks (Figures 8.5c–8.5e). These tribofilms, uniformly formed on both mating surfaces, contributed to prevent direct asperity contact between Si_3N_4 and sapphire. Besides, at the elastohydrodynamic lubrication, the strong affinity among the hydrophilic $Ti_3C_2T_x$ and water/glycerol molecules produced a hydration layer onto the nanosheets, which reduced the shear strength of the liquid film (Figure 8.5f).

Even though the MXenes family consists of a wide variety of compounds, to date, most of the work aiming at their use as lubricant additives were limited to the study of $Ti_3C_2T_x$. Recently, Cheng and Zhao (2022) showed that exfoliated Nb_2CT_x nanosheets could also be effectively used

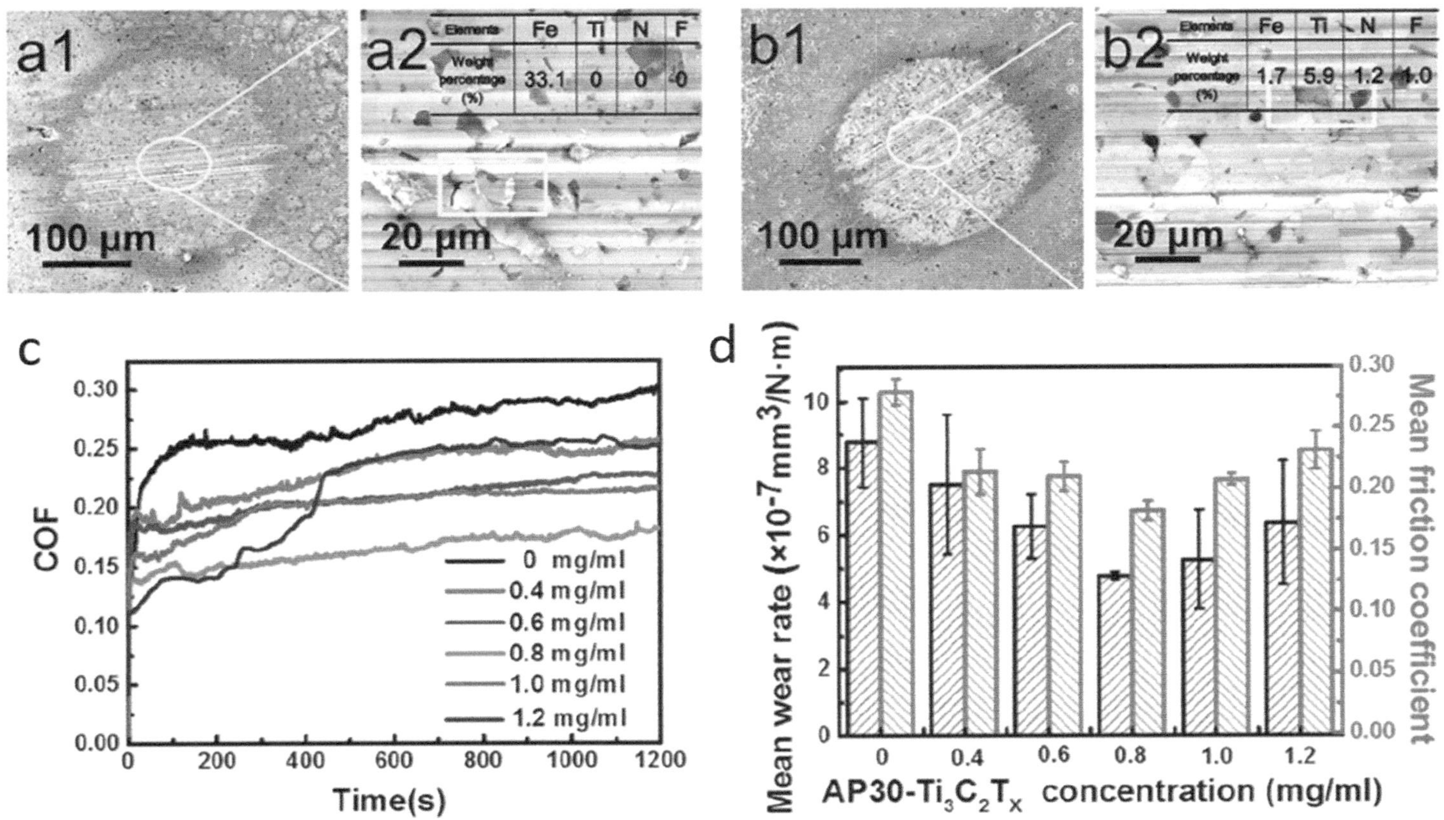

FIGURE 8.4 SEM and EDS analyses of wear scars on steel matching surfaces lubricated with pure water (a) and 0.8 mg/ml of $Ti_3C_2T_x$ (b), COF curves (c), and mean COF/mean wear rate (d) of $Ti_3C_2T_x$ filled water under different MXene concentrations.

Source: Reproduced with permission (Chen and Zhao 2021). Copyright 2020, Elsevier.

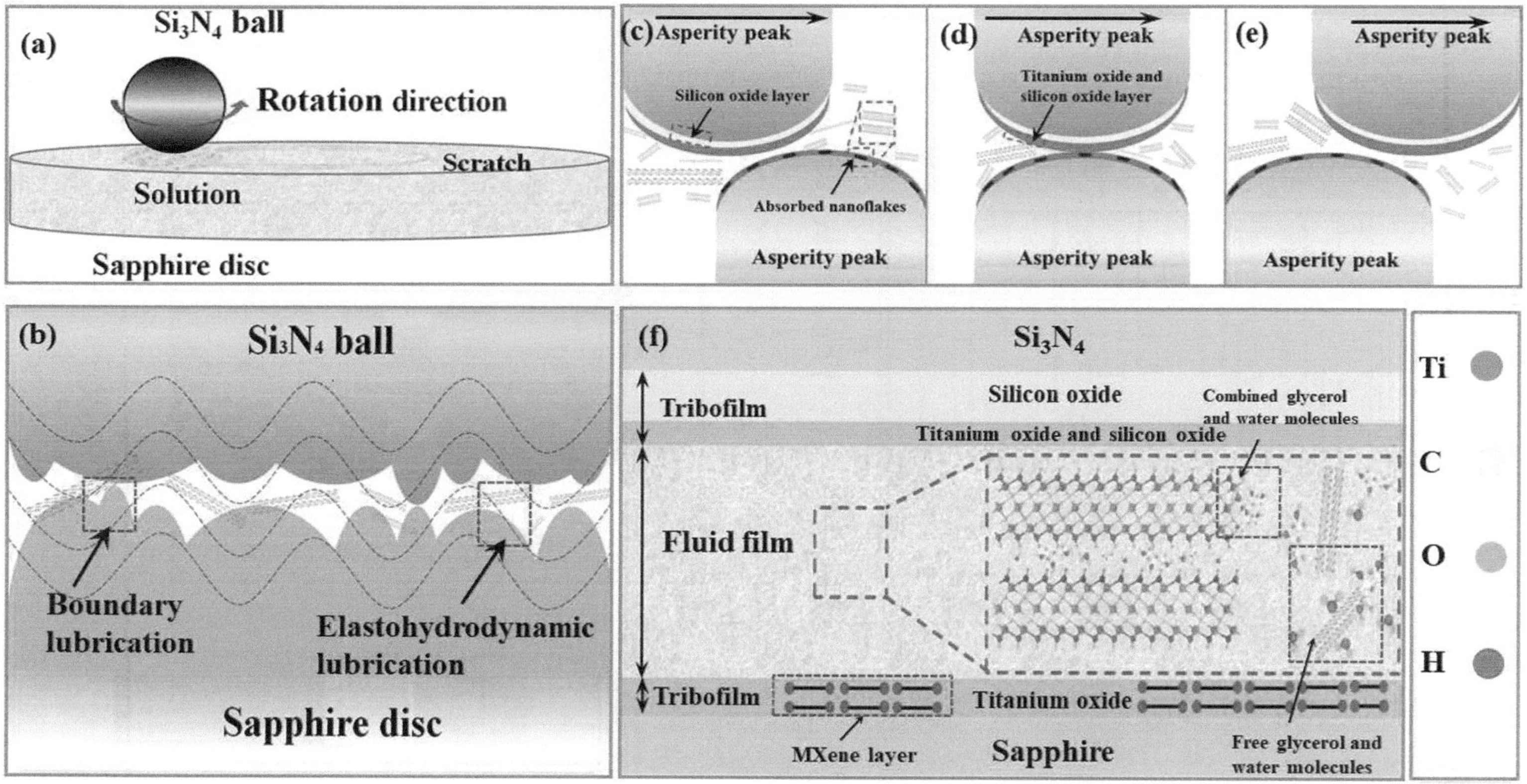

FIGURE 8.5 (a) Schematic representation of the pin-on-disk tribometer with Si_3N_4 ball and sapphire disk. (b) Schematic representation of the contact zone between Si_3N_4 and sapphire surfaces with the lubrication of the MXene-glycerol solution. (c–e) Boundary lubrication showing formed tribofilms on the Si_3N_4 and sapphire surface. (f) Schematic illustration of elastohydrodynamic lubrication showing the hydrated MXene–glycerol networks.

Source: Reproduced with permission (Yi et al. 2021). Copyright 2021, Elsevier.

as an additive for water-based lubricant. Tribological tests exhibited a COF reduction of about 81.2% for water-filled MXenes (0.5 mg/mL) when compared to pure water. The authors also studied the effect of Nb_2CT_x oxidation on the tribological performance. By oxidizing Nb_2CT_x sheets in solution, Nb_2O_5 nanoparticles and amorphous carbon could be formed onto MXenes sheets contributing to a further COF reduction of about 90.3% and a decrease of wear rate by 73.1% when compared with pure water. It is believed that the synergistic effect between Nb_2CT_x/Nb_2O_5 components strongly contributed to an enhanced antifriction and antiwear properties. The authors suggested that the Nb_2CT_x layers and carbon could be easily adsorbed on the rubbing surfaces, leading to a tribofilm formation, whereas Nb_2O_5 nanoparticles act by incoming and filling the worn tracks, reducing wear rate.

Table 8.2 summarizes the up-to-date published works involving MXenes as lubricant additives. A detailed inspection on these studies reveals that a favored dispersion and stability of

TABLE 8.2
Tribological Performance of MXene-Filled Lubricants

MXenes	Base Fluid	Rubbing Surfaces (test conditions)	Tribological Performance COF	Tribological Performance Wear	Ref.
$Ti_3C_2T_x$	Paraffin oil	440-C stainless steel ball (Ø4 mm) on 45# steel disc. 100 rpm, 15N	0.125 (base oil) 0.063 (1.0 wt%)	–	(Yang et al. 2014)
$Ti_3C_2(OH)_2$ (10–20 nm)	100SN oil	440-C stainless steel ball (Ø10 mm) on 45# steel disc. 200 rpm, 15N	0.092 (base oil) 0.075 (1.0 wt%)	*Width:* 220 μm (base oil) 200 μm (1.0 wt%)	(Zhang et al. 2015)
$Ti_3C_2T_x$	PAO8 oil	32100 steel ball (Ø9 mm) on steel disc. 240 rpm, 10N	0.126 (base oil) 0.114 (0.8 wt%)	*Volume:* 7.7% wear reduction (0.8 wt%)	(Liu et al. 2017)
TiO_2/ $Ti_3C_2T_x$	PAO8 oil	440-C stainless steel ball (Ø4 mm) on 45# steel disc. 150 rpm, 20N	0.103 (base oil) 0.073 (1.0 wt%)	*Width:* 280 μm (1.0 wt%)	(Xue et al. 2017)
Titanate-$Ti_3C_2T_x$	PAO8 oil	440-C stainless steel ball (Ø4 mm) on steel disc. 200 rpm, 5N	0.111 (base oil) 0.077 (1.0 wt%)	*Width:* 173 μm (base oil) 138 μm (1.0 wt%)	(Zhang et al. 2019)
$Ti_3C_2T_x$	PAO8 oil	440-C stainless steel ball (Ø4 mm) on steel disc. 200 rpm, 5N	0.111 (base oil) 0.081 (1.0 wt%)	*Width:* 173μm (base oil) 167μm (1.0 wt%)	(Zhang et al. 2019)
$Ti_3C_2T_x$	DI Water	SiC ball (Ø6 mm) on 316 steel disc. 2 Hz, 5N	0.275 (pure water) 0.180 (0.8 wt%)	*Rate:* $8.8 .10^{-7}$ $mm^3.N^{-1}.m^{-1}$ (base oil) $4.8 .10^{-7}$ $mm^3.N^{-1}.m^{-1}$ (0.8 wt%)	(Chen and Zhao 2021)
$Ti_3C_2T_x$	DI Water-Glycerol	Si_3N_4 ball (Ø4 mm) on saphire disc. 0–900 rpm, 3N	0.011 (base fluid) 0.002 (1.0 wt%)	*Width:* 278 μm (base fluid) 219 μm (1.0 wt%)	(Yi et al. 2021)
Nb_2C	DI Water	Al_2O_3 ball (Ø6 mm) on 316 steel disc. 5 Hz, 5N	0.44 (pure water) 0.08 (0.05 wt%)	*Rate:* $45 .10^{-7}$ $mm^3.N^{-1}.m^{-1}$ (bare) $18 .10^{-7}$ $mm^3.N^{-1}.m^{-1}$ (0.8 wt%)	(Cheng and Zhao 2022)
Nb_2C/ Nb_2O_5	DI Water	Al_2O_3 ball (Ø6 mm) on 316 steel disc. 5 Hz, 5N	0.44 (pure water) 0.04 (0.025 wt%)	*Rate:* $45 .10^{-7}$ $mm^3.N^{-1}.m^{-1}$ (bare) $12 .10^{-7}$ $mm^3.N^{-1}.m^{-1}$ (0.8 wt%)	(Cheng and Zhao 2022)
$Ti_3C_2T_x$	DI Water	440-C stainless steel ball (Ø3 mm) on 304 steel disc. 120 rpm, 3N	0.30 (pure water) 0.24 (5.0 wt%)	*Width:* 452μm (base oil) 236μm (5.0 wt%)	(Nguyen and Chung 2020)

MXenes in the lubricants play a key role in the enhancement of the tribological performance. Most of the published studies agree that a low concentration (usually 1.0 wt%) contributes to a better dispersion and prevents nanosheets agglomeration into the lubricants. In general, high degree of exfoliation usually results in a lower COF and wear rate (Malaki and Varma 2020). However, the effect of the nanosheet size should be further investigated. In addition, most of the works are limited to the study of $Ti_3C_2T_x$. MXenes comprise a wide family of 2D compounds for which many predicted structures have not yet been synthesized. Therefore, we can expect in the future to come across many studies focusing on other MXene compounds and their nanocomposites as additives for lubricants.

8.4 MXenes/POLYMER NANOCOMPOSITES

The use of MXenes to reduce friction and wear is not limited to additives in lubricants, as already discussed in Section 8.3. Some studies have evaluated MXene coatings and MXene-reinforced composites as wear-resistant and/or antifriction materials (Lian et al. 2018; Mai et al. 2019; Marian et al. 2020; Hu et al. 2020). However, despite these promising features, their use as reinforcements in composites remains significantly underexplored. For example, only UHMWPE (Zhang, Wang, Chen et al. 2016) and epoxy (Zhang, Wang, Zhou et al. 2016) polymeric matrices were screened. Therefore, there are several open questions and opportunities for scientific and technological investigations into this subject.

The first studies on the addition of $Ti_3C_2T_x$ into UHMWPE (Zhang, Wang, Chen, et al. 2016) and epoxy resins (Zhang, Wang, Zhou, et al. 2016) have shown similar behaviors: MXene addition enhanced wear resistance and decreased the COF compared with pure matrices. The COF was gradually decreased with MXene loading, reaching a reduction of 65% for 2.0 wt% of reinforcement. This decrease of the COF was associated with (1) the exfoliation and slipping of MXenes multilayer previously held only by weak van der Waals bonds, (2) tribofilm formation at the surface, and/or (3) an increase of mechanical properties and crystallinity degree. In the latter, MXenes nanosheets can act as nucleating substrates, increasing the crystallization rate.

The enhancement of wear resistance was also attributed to improvements of the nanocomposite mechanical properties, resulting in better plowing and scratching resistance and tribofilm formation, which promotes a lower contact area for plastic deformation between the composite and the counterbody (Zhang, Wang, Chen, et al. 2016; Zhang, Wang, Zhou, et al. 2016). However, the wear was evaluated mainly through optical and scanning electron microscopy, but no quantification was performed (*e.g.*, wear rate). In addition, since tribo-physical and chemical reactions during the tribofilm formation are very complex (Lihe Guo et al. 2021), more experimental investigations are necessary to allow a deeper understanding of the wear mechanisms involved.

Lihe Guo et al. (2021) performed more extended research on MXene/epoxy (EP) nanocomposites under ultra-low sulfur diesel (ULSD) oil lubrication. They found that the addition of 3 wt% $Ti_3C_2T_x$ decreased the wear rate by 79.6%, reaching 97% reduction when combined with Al_2O_3 nanoparticles. It is worth highlighting that 3.0 wt.% was not evaluated in the previous studies, which had maximum MXene loadings of 2.0 wt% (Zhang, Wang, Chen et al. 2016; Zhang, Wang, Zhou et al. 2016), and thus, their performance may be further enhanced if higher MXene concentration is used. In addition, for the first time, they reported a superlubricity effect of MXenes in the macroscale (COF lower than 0.01 after a 4.5-h sliding process), which was only previously predicted computationally (Zhang et al. 2017). From the results of Raman spectroscopy, Transmission electron microscopy (TEM), and EDS techniques, the authors proposed that, during the tribological tests, the epoxy molecules are fragmented and graphitized by heat and shear stress. These fragmented molecules, carbon material, and nanomaterials (MXene and Al_2O_3) are transferred to the counterpart steel surface and oriented parallel to the sliding direction due to shear stresses. Thus, the formed tribofilm (~100 nm) enables a high load-bearing capability and easys hearing, effectively separating the friction pairs and mitigating the oxidation of steel surface (Guo et al. 2021).

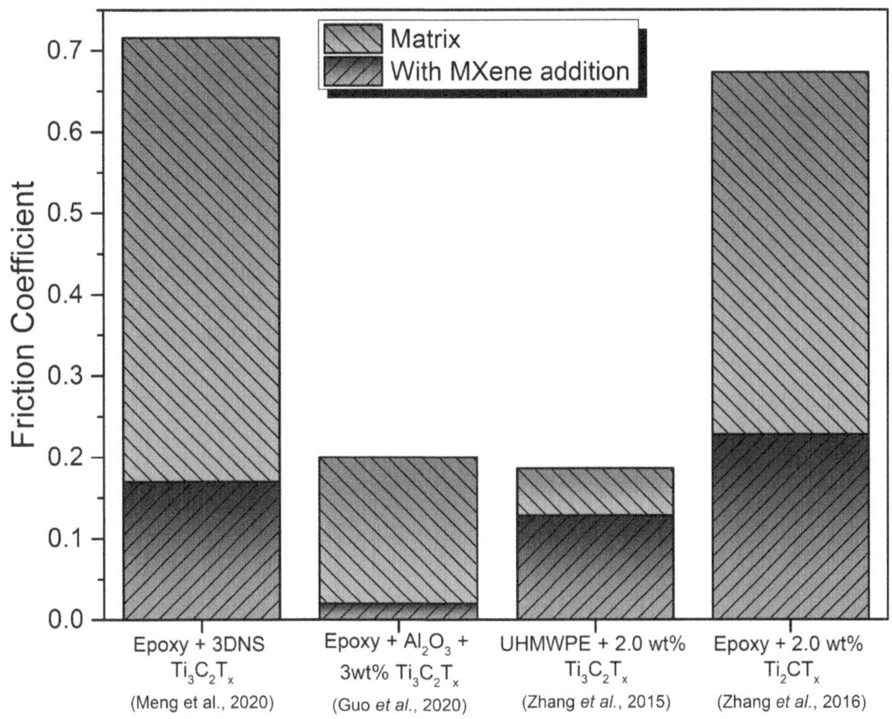

FIGURE 8.6 Comparison of friction coefficients with MXene addition in polymer-based composites.

Enhancements on the tribological response could also be seen in a freeze-dried nanocomposite of epoxy with a 3D $Ti_3C_2T_x$ structure with a modified carbon nanofiber (Meng et al. 2021). The wear rate decreased 67% than that of epoxy resin when 3 g of MXenes were used, while the COF decreased by 76% using only 1 g. A graphical summary showing the decrease of friction coefficient of all nanocomposites with MXenes discussed in this section is presented in Figure 8.6.

Although their outstanding potential as lubricants and reinforcements, as we presented, the research on this topic is quite scarce and needs to be further explored. First, apart from $Ti_3C_2T_x$, other MXenes should be evaluated since nearly 30 compounds were already synthesized. Several matrices of interest still need to be studied, such as poly ether ether ketone and PTFE (Friedrich and Schlarb 2008). Moreover, the influence of MXene synthesis and functional groups on their surfaces remains an open question, since their different termination may influence the matrix compatibility. Finally, the influence of different MXene-related features on tribological properties, such as lateral sizes and number of layers, still needs to be investigated.

8.5 CONCLUDING REMARKS

Among others, MXenes are 2D nanomaterials with promising applications in tribology, mostly as additives in lubricants and reinforcements in self-lubricating nanocomposites. Despite their bright prospect, the quantity of results from both experimental and simulations is still quite scarce. This scarcity is limited not only to applied research but also to fundamental ones. More information on the friction coefficient, Young modulus, thermal conductivity, and so on is still missing. This lack of information is especially true when we remember that the majority of these data were collected from the compound $Ti_3C_2T_x$ and several compositions are unexplored for tribological purposes. Likewise, when we consider the MXene nanocomposites, this lack of information is even more expressive. Different possible combinations of MXenes and matrices, different methods of exfoliation and delamination, large number of processing routes, surface functionalization, and many other significant variables make this subject almost inexhaustible. This scenario evidences the necessity of rationalization in the search for new MXene-based materials for tribological purposes. This rationalization may be performed using different approaches. One of them is to design and perform investigations bringing together computer simulation techniques (density functional theory (DFT), molecular dynamics, machine learning, etc.) and experiments (synthesis and characterization) in order to precisely determine and validate certain trends. These approaches will then provide the possibility to screen a large number of materials, their potential for tribological applications, coefficient of friction, supported bearing loads, wear mechanisms, and much more. In our perspective, some critical aspects hinder the development of MXene-based materials as potential wear-resistant materials and lubricants, which need to be addressed in the future. Some promising topics of research are the production of stable MXene colloidal solutions, allowing easier incorporation of MXenes in lubricant oils and polymeric matrices, development of new (self-lubricating) MXene nanocomposites and development of theoretical modeling to predict materials of interest.

REFERENCES

Acikgoz, Ogulcan, and Mehmet Z. Baykara. 2020. "Speed Dependence of Friction on Single-Layer and Bulk MoS2 Measured by Atomic Force Microscopy." *Applied Physics Letters* 116 (7). https://doi.org/10.1063/1.5142712.

Akuzum, Bilen, Kathleen Maleski, Babak Anasori, Pavel Lelyukh, Nicolas Javier Alvarez, E Caglan Kumbur, and Yury Gogotsi. 2018. "Rheological Characteristics of 2D Titanium Carbide (MXene) Dispersions: A Guide for Processing MXenes." *ACS Nano* 12 (3): 2685–2694. https://doi.org/10.1021/acsnano.7b08889.

Alhabeb, Mohamed, Kathleen Maleski, Babak Anasori, Pavel Lelyukh, Leah Clark, Saleesha Sin, and Yury Gogotsi. 2017. "Guidelines for Synthesis and Processing of Two-Dimensional Titanium Carbide (Ti3C2Tx MXene)." *Chemistry of Materials* 29 (18): 7633–7644. https://doi.org/10.1021/acs.chemmater.7b02847.

Anasori, Babak, Asia Sarycheva, Sara Buondonno, Zehang Zhou, Shu Yang, and Yury Gogotsi. 2017. "2D Metal Carbides (MXenes) in Fibers." *Materials Today* 20 (8): 481–482. https://doi.org/https://doi.org/10.1016/j.mattod.2017.08.001.

Anasori, Babak, Yu Xie, Majid Beidaghi, Jun Lu, Brian C Hosler, Lars Hultman, Paul R C Kent, Yury Gogotsi, and Michel W Barsoum. 2015. "Two-Dimensional, Ordered, Double Transition Metals Carbides (MXenes)." *ACS Nano* 9 (10): 9507–9516. https://doi.org/10.1021/acsnano.5b03591.

Berman, Diana, Ali Erdemir, and Anirudha V. Sumant. 2014. "Graphene: A New Emerging Lubricant." *Materials Today* 17 (1): 31–42. https://doi.org/https://doi.org/10.1016/j.mattod.2013.12.003.

Berman, Diana, Ali Erdemir, Alexander V. Zinovev, and Anirudha V. Sumant. 2015. "Nanoscale Friction Properties of Graphene and Graphene Oxide." *Diamond and Related Materials* 54 (1): 91–96. https://doi.org/10.1016/j.diamond.2014.10.012.

Bhushan, Bharat. 2017. "Introduction—Measurement Techniques and Applications." In *Nanotribology and Nanomechanics: An Introduction*, edited by Bharat Bhushan, 1–29. Cham: Springer International Publishing. https://doi.org/10.1007/978-3-319-51433-8_1.

Boota, Muhammad, Mariacecilia Pasini, Francesco Galeotti, William Porzio, Meng-Qiang Zhao, Joseph Halim, and Yury Gogotsi. 2017. "Interaction of Polar and Nonpolar Polyfluorenes with Layers of Two-Dimensional Titanium Carbide (MXene): Intercalation and Pseudocapacitance." *Chemistry of Materials* 29 (7): 2731–2738. https://doi.org/10.1021/acs.chemmater.6b03933.

Carey, Michael, Zachary Hinton, Varun Natu, Rahul Pai, Maxim Sokol, Nicolas J Alvarez, Vibha Kalra, and Michel W Barsoum. 2020. "Dispersion and Stabilization of Alkylated 2D MXene in Nonpolar Solvents and Their Pseudocapacitive Behavior." *Cell Reports Physical Science* 1 (4): 100042. https://doi.org/https://doi.org/10.1016/j.xcrp.2020.100042.

Chen, Jing, Ke Chen, Dingyi Tong, Youju Huang, Jiawei Zhang, Jianming Xue, Qing Huang, and Tao Chen. 2015. "CO2 and Temperature Dual Responsive 'Smart' MXene Phases." *Chem Comm* 51 (2): 314–317. https://doi.org/10.1039/C4CC07220K.

Chen, Junfeng, and Wenjie Zhao. 2021. "Simple Method for Preparing Nanometer Thick Ti3C2TX Sheets towards Highly Efficient Lubrication and Wear Resistance." *Tribology International* 153: 106598. https://doi.org/https://doi.org/10.1016/j.triboint.2020.106598.

Cheng, Hao, and Wenjie Zhao. 2022. "Regulating the Nb 2 C Nanosheets with Different Degrees of Oxidation in Water Lubricated Sliding toward an Excellent Tribological Performance." *Friction* 10: 398–410. https://doi.org/10.1007/s40544-020-0469-x.

Friedrich, Klaus, and Alois K Schlarb. 2008. "Tribology of Polymeric Nanocomposites." In *Tribology of Polymeric Nanocomposites: Friction and Wear of Bulk Materials and Coatings*, edited by Klaus Friedrich and Alois K Schlarb, 55:1–551. Tribology and Interface Engineering Series. Elsevier. https://doi.org/https://doi.org/10.1016/S1572-3364(13)70001-6.

Ghidiu, Michael, Maria R. Lukatskaya, Meng-Qiang Zhao, Yury Gogotsi, and Michel W Barsoum. 2014. "Conductive Two-Dimensional Titanium Carbide 'Clay' with High Volumetric Capacitance." *Nature* 516 (7529): 78–81. https://doi.org/10.1038/nature13970.

Guan, Yanxue, Miaomiao Zhang, Juan Qin, Xingxing Ma, Chen Li, and Jilin Tang. 2020. "Hydrophilicity-Dependent Distinct Frictional Behaviors of Different Modified MXene Nanosheets." *Journal of Physical Chemistry C* 124 (25): 13664–13671. https://doi.org/10.1021/acs.jpcc.0c01551.

Guo, Lihe, Yaoming Zhang, Ga Zhang, Qihua Wang, and Tingmei Wang. 2021. "MXene-Al2O3 Synergize to Reduce Friction and Wear on Epoxy-Steel Contacts Lubricated with Ultra-Low Sulfur Diesel." *Tribology International* 153: 106588. https://doi.org/https://doi.org/10.1016/j.triboint.2020.106588.

Guo, Yanbao, Xuanli Zhou, Kyungjun Lee, Hyun Chul Yoon, Quan Xu, and Deguo Wang. 2021. "Recent Development in Friction of 2D Materials: From Mechanisms to Applications." *Nanotechnology* 32 (31): 312002. https://doi.org/10.1088/1361-6528/abfa52.

Guo, Yanbao, Xuanli Zhou, Deguo Wang, Xiaqing Xu, and Quan Xu. 2019. "Nanomechanical Properties of Ti3C2 Mxene." *Langmuir* 35 (45): 14481–14485. https://doi.org/10.1021/acs.langmuir.9b02619.

Hope, Michael A., Alexander C. Forse, Kent J. Griffith, Maria R. Lukatskaya, Michael Ghidiu, Yury Gogotsi, and Clare P. Grey. 2016. "NMR Reveals the Surface Functionalisation of Ti3C2 MXene." *Physical Chemistry Chemical Physics* 18 (7): 5099–5102. https://doi.org/10.1039/C6CP00330C.

Hu, Jie, Shibo Li, Jing Zhang, Qiuying Chang, Wenbo Yu, and Yang Zhou. 2020. "Mechanical Properties and Frictional Resistance of Al Composites Reinforced with Ti3C2Tx MXene." *Chinese Chemical Letters* 31 (4): 996–999. https://doi.org/https://doi.org/10.1016/j.cclet.2019.09.004.

Jost, Peter. 1966. "Lubrication (Tribology)–A Report on the Present Position and Industry's Needs."In *Department of Education and Science*. London: HM Stationary Office.

Kim, Hyun Joon, and Dae Eun Kim. 2009. "Nano-Scale Friction: A Review." *International Journal of Precision Engineering and Manufacturing* 10 (2): 141–151. https://doi.org/10.1007/s12541-009-0039-7.

Lee, Changgu, Qunyang Li, William Kalb, Xin Zhou Liu, Helmuth Berger, Robert W. Carpick, and James Hone. 2010. "Frictional Characteristics of Atomically Thin Sheets." *Science* 328 (5974): 76–80. https://doi.org/10.1126/science.1184167.

Lian, Weiqi, Yongjin Mai, Cansen Liu, Liuyan Zhang, Shilin Li, and Xiaohua Jie. 2018. "Two-Dimensional Ti3C2 Coating as an Emerging Protective Solid-Lubricant for Tribology." *Ceramics International* 44 (16): 20154–20162. https://doi.org/https://doi.org/10.1016/j.ceramint.2018.07.309.

Lim, Sehyeong, Hyunsu Park, Jeewon Yang, Chaesu Kwak, and Joohyung Lee. 2019. "Stable Colloidal Dispersion of Octylated Ti3C2-MXenes in a Nonpolar Solvent." *Colloids and Surfaces A: Physicochemical and Engineering Aspects* 579: 123648. https://doi.org/https://doi.org/10.1016/j.colsurfa.2019.123648.

Ling, Zheng, Chang E Ren, Meng-Qiang Zhao, Jian Yang, James M Giammarco, Jieshan Qiu, Michel W Barsoum, and Yury Gogotsi. 2014. "Flexible and Conductive MXene Films and Nanocomposites with High Capacitance." *Proceedings of the National Academy of Sciences* 111 (47): 16676–16681. https://doi.org/10.1073/pnas.1414215111.

Lipatov, Alexey, Mohamed Alhabeb, Maria R Lukatskaya, Alex Boson, Yury Gogotsi, and Alexander Sinitskii. 2016. "Effect of Synthesis on Quality, Electronic Properties and Environmental Stability of Individual Monolayer Ti3C2 MXene Flakes." *Advanced Electronic Materials* 2 (12): 1600255. https://doi.org/https://doi.org/10.1002/aelm.201600255.

Liu, Yong, Xuefeng Zhang, Shangli Dong, Zhuyu Ye, and Yidan Wei. 2017. "Synthesis and Tribological Property of $Ti_3C_2T_x$ Nanosheets." *Journal of Materials Science* 52 (4): 2200–2209. https://doi.org/10.1007/s10853-016-0509-0.

Luo, Q. 2013. "Tribofilms in Solid Lubricants." In *Encyclopedia of Tribology*, edited by Q J Wang, Y W Chung, 3760–3767. Boston, MA: Springer. https://doi.org/10.1007/978-0-387-92897-5_1252.

Mai, Y J, Y G Li, S L Li, L Y Zhang, C S Liu, and X H Jie. 2019. "Self-Lubricating Ti3C2 Nanosheets/Copper Composite Coatings." *Journal of Alloys and Compounds* 770: 1–5. https://doi.org/https://doi.org/10.1016/j.jallcom.2018.08.100.

Malaki, Massoud, and Rajender S. Varma. 2020. "Mechanotribological Aspects of MXene-Reinforced Nanocomposites." *Advanced Materials* 32 (38): 1–20. https://doi.org/10.1002/adma.202003154.

Marian, Max, Gui Cheng Song, Bo Wang, Victor M Fuenzalida, Sebastian Krauß, Benoit Merle, Stephan Tremmel, Sandro Wartzack, Jinhong Yu, and Andreas Rosenkranz. 2020. "Effective Usage of 2D MXene Nanosheets as Solid Lubricant—Influence of Contact Pressure and Relative Humidity." *Applied Surface Science* 531: 147311. https://doi.org/https://doi.org/10.1016/j.apsusc.2020.147311.

Mate, C. Mathew. 2007. *Tribology on the Small Scale: A Bottom up Approach to Friction, Lubrication, and Wear.* New York: Oxford University Press. https://doi.org/10.1093/acprof:oso/9780198526780.001.0001.

Meng, Fanning, Zhenyu Zhang, Peili Gao, Ruiyang Kang, Yash Boyjoo, Jinhong Yu, and Tingting Liu. 2021. "Excellent Tribological Properties of Epoxy—Ti3C2 with Three-Dimensional Nanosheets Composites." *Friction* 9 (4): 734–746. https://doi.org/10.1007/s40544-020-0368-1.

Mungse, Harshal P., Niranjan Kumar, and Om P Khatri. 2015. "Synthesis, Dispersion and Lubrication Potential of Basal Plane Functionalized Alkylated Graphene Nanosheets." *RSC Advances* 5 (32): 25565–25571. https://doi.org/10.1039/C4RA16975A.

Naguib, Michael, Murat Kurtoglu, Volker Presser, Jun Lu, Junjie Niu, Min Heon, Lars Hultman, Yury Gogotsi, and Michel W Barsoum. 2011. "Two-Dimensional Nanocrystals Produced by Exfoliation of Ti3AlC2." *Advanced Materials* 23 (37): 4248–4253. https://doi.org/10.1002/adma.201102306.

Naguib, Michael, Tomonori Saito, Sophia Lai, Matthew S Rager, Tolga Aytug, M Parans Paranthaman, Meng-Qiang Zhao, and Yury Gogotsi. 2016. "Ti3C2Tx (MXene)–Polyacrylamide Nanocomposite Films." *RSC Advances* 6 (76): 72069–72073. https://doi.org/10.1039/C6RA10384G.

Nguyen, Huong Thi, and Koo-Hyun Chung. 2020. "Assessment of Tribological Properties of Ti3C2 as a Water-Based Lubricant Additive." *Materials* 13 (23). https://doi.org/10.3390/ma13235545.

Peng, Yitian, Zhuoqiong Wang, and Kun Zou. 2015. "Friction and Wear Properties of Different Types of Graphene Nanosheets as Effective Solid Lubricants." *Langmuir* 31 (28): 7782–7791. https://doi.org/10.1021/acs.langmuir.5b00422.

Persson, Ingemar, Ahmed el Ghazaly, Quanzheng Tao, Joseph Halim, Sankalp Kota, Vanya Darakchieva, Justinas Palisaitis, Michel W Barsoum, Johanna Rosen, and Per O Å Persson. 2018. "Tailoring Structure, Composition, and Energy Storage Properties of MXenes from Selective Etching of In-Plane, Chemically Ordered MAX Phases." *Small* 14 (17): 1703676. https://doi.org/https://doi.org/10.1002/smll.201703676.

Rodriguez, A., M. S. Jaman, O. Acikgoz, B. Wang, J. Yu, P. G. Grützmacher, A. Rosenkranz, and M. Z. Baykara. 2021. "The Potential of Ti3C2TX Nano-Sheets (MXenes) for Nanoscale Solid Lubrication Revealed by Friction Force Microscopy." *Applied Surface Science* 535 (August 2020). https://doi.org/10.1016/j.apsusc.2020.147664.

Sang, Xiahan, Yu Xie, Ming-Wei Lin, Mohamed Alhabeb, Katherine L Van Aken, Yury Gogotsi, Paul R C Kent, Kai Xiao, and Raymond R Unocic. 2016. "Atomic Defects in Monolayer Titanium Carbide (Ti3C2Tx) MXene." *ACS Nano* 10 (10): 9193–9200. https://doi.org/10.1021/acsnano.6b05240.

Seyedin, Shayan, Elliard Roswell S Yanza, and Joselito M Razal. 2017. "Knittable Energy Storing Fiber with High Volumetric Performance Made from Predominantly MXene Nanosheets." *J. Mater. Chem. A* 5 (46): 24076–24082. https://doi.org/10.1039/C7TA08355F.

Tang, Guogang, Jing Zhang, Changchao Liu, Du Zhang, Yuqi Wang, Hua Tang, and Changsheng Li. 2014. "Synthesis and Tribological Properties of Flower-like MoS2 Microspheres." *Ceramics International* 40 (8, Part A): 11575–11580. https://doi.org/https://doi.org/10.1016/j.ceramint.2014.03.115.

Tripathi, Manoj, Firas Awaja, Rafael A. Bizao, Stefano Signetti, Erica Iacob, Guido Paolicelli, Sergio Valeri, Alan Dalton, and Nicola Maria Pugno. 2018. "Friction and Adhesion of Different Structural Defects of Graphene." *ACS Applied Materials and Interfaces* 10 (51): 44614–44623. https://doi.org/10.1021/acsami.8b10294.

Vazirisereshk, Mohammad R., Han Ye, Zhijiang Ye, Alberto Otero-De-La-Roza, Meng Qiang Zhao, Zhaoli Gao, A. T. Charlie Johnson, Erin R. Johnson, Robert W. Carpick, and Ashlie Martini. 2019. "Origin of Nanoscale Friction Contrast between Supported Graphene, MoS2, and a Graphene/MoS2 Heterostructure." *Nano Letters* 19 (8): 5496–5505. https://doi.org/10.1021/acs.nanolett.9b02035.

Wyatt, Brian C., Andreas Rosenkranz, and Babak Anasori. 2021. "2D MXenes: Tunable Mechanical and Tribological Properties." *Advanced Materials* 33 (17): 2007973. https://doi.org/https://doi.org/10.1002/adma.202007973.

Xiao, Huaping, Wei Dai, Yuwei Kan, Abraham Clearfield, and Hong Liang. 2015. "Amine-Intercalated α-Zirconium Phosphates as Lubricant Additives." *Applied Surface Science* 329: 384–389. https://doi.org/https://doi.org/10.1016/j.apsusc.2014.12.061.

Xiao, Huaping, and Shuhai Liu. 2017. "2D Nanomaterials as Lubricant Additive: A Review." *Materials & Design* 135: 319–332. https://doi.org/https://doi.org/10.1016/j.matdes.2017.09.029.

Xu, Yufu, Yubin Peng, Karl D. Dearn, Xiaojing Zheng, Lulu Yao, and Xianguo Hu. 2015. "Synergistic Lubricating Behaviors of Graphene and MoS2 Dispersed in Esterified Bio-Oil for Steel/Steel Contact." *Wear* 342–343: 297–309. https://doi.org/https://doi.org/10.1016/j.wear.2015.09.011.

Xue, Maoquan, Zhiping Wang, Feng Yuan, Xianghua Zhang, Wei Wei, Hua Tang, and Changsheng Li. 2017. "Preparation of $TiO_2/Ti_3C_2T_x$ Hybrid Nanocomposites and Their Tribological Properties as Base Oil Lubricant Additives." *RSC Adv.* 7 (8): 4312–4319. https://doi.org/10.1039/C6RA27653A.

Yang, Jin, Beibei Chen, Haojie Song, Hua Tang, and Changsheng Li. 2014. "Synthesis, Characterization, and Tribological Properties of Two-Dimensional Ti3C2." *Crystal Research and Technology* 49 (11): 926–932. https://doi.org/10.1002/crat.201400268.

Yi, Shuang, Jinjin Li, Yanfei Liu, Xiangyu Ge, Jie Zhang, and Jianbin Luo. 2021. "In-Situ Formation of Tribofilm with Ti3C2Tx MXene Nanoflakes Triggers Macroscale Superlubricity." *Tribology International* 154 (July 2020): 106695. https://doi.org/10.1016/j.triboint.2020.106695.

Zhang, Chuanfang (John), Matthias P Kremer, Andrés Seral-Ascaso, Sang-Hoon Park, Niall McEvoy, Babak Anasori, Yury Gogotsi, and Valeria Nicolosi. 2018. "Stamping of Flexible, Coplanar Micro-Supercapacitors Using MXene Inks." *Advanced Functional Materials* 28 (9): 1705506. https://doi.org/https://doi.org/10.1002/adfm.201705506.

Zhang, Difan, Michael Ashton, Alireza Ostadhossein, Adri C.T. van Duin, Richard G Hennig, and Susan B. Sinnott. 2017. "Computational Study of Low Interlayer Friction in $Ti_{N+1}C_n$ (N=1, 2 and 3) MXene." *ACS Applied Materials & Interfaces* 9 (39): 34467–34479. https://doi.org/10.1021/acsami.7b09895.

Zhang, H, Z H Fu, D Legut, T C Germann, and R F Zhang. 2017. "Stacking Stability and Sliding Mechanism in Weakly Bonded 2D Transition Metal Carbides by van Der Waals Force." *RSC Advances* 7 (88): 55912–55919. https://doi.org/10.1039/C7RA11139H.

Zhang, Heng, Libo Wang, Qiang Chen, Ping Li, Aiguo Zhou, Xinxin Cao, and Qianku Hu. 2016. "Preparation, Mechanical and Anti-Friction Performance of MXene/Polymer Composites." *Materials & Design* 92: 682–689. https://doi.org/https://doi.org/10.1016/j.matdes.2015.12.084.

Zhang, Heng, Libo Wang, Aiguo Zhou, Changjie Shen, Yahui Dai, Fanfan Liu, Jinfeng Chen, Ping Li, and Qianku Hu. 2016. "Effects of 2-D Transition Metal Carbide Ti2CTx on Properties of Epoxy Composites."*RSC Advances* 6 (90): 87341–87352. https://doi.org/10.1039/C6RA14560D.

Zhang, Xianghua, Maoquan Xue, Xinghua Yang, Zhiping Wang, Guangsi Luo, Zhide Huang, Xiaoli Sui, and Changsheng Li. 2015. "Preparation and Tribological Properties of Ti3C2(OH)2 Nanosheets as Additives in Base Oil." *RSC Advances* 5 (4): 2762–27667. https://doi.org/10.1039/C4RA13800G.

Zhang, Xuefeng, Yu Guo, Yijia Li, Yong Liu, and Shangli Dong. 2019. "Preparation and Tribological Properties of Potassium Titanate-Ti3C2Tx Nanocomposites as Additives in Base Oil." *Chinese Chemical Letters* 30 (2): 502–504. https://doi.org/https://doi.org/10.1016/j.cclet.2018.07.007.

Zhi, Weiqiang, Shanglin Xiang, Renji Bian, Ruizhi Lin, Kaihua Wu, Tingwei Wang, and Dongyu Cai. 2018. "Study of MXene-Filled Polyurethane Nanocomposites Prepared via an Emulsion Method." *Composites Science and Technology* 168: 404–411. https://doi.org/https://doi.org/10.1016/j.compscitech.2018.10.026.

Zhou, Xuanli, Yanbao Guo, Deguo Wang, and Quan Xu. 2021. "Nano Friction and Adhesion Properties on Ti3C2 and Nb2C MXene Studied by AFM." *Tribology International* 153: 106646. https://doi.org/https://doi.org/10.1016/j.triboint.2020.106646.

9 Self-Healing Activity of MXene/Polymer Nanocomposites

Shikha Agarwal, Ayushi Sethiya, Divyani Gandhi, and Jay Soni

CONTENTS

9.1 INTRODUCTION

Presently, two-dimensional (2D) materials are one of the intense areas of research in the scientist community due to their large specific surface area, wide-band-gap modulation capacity, high catalytic activity, and so on, and these materials are used in disruptive applications such as medical appliances, catalysis, energy, optoelectronics, biological, and industrial applications.[1–2] These significant properties increase curiosity and thus lead to a novel field of research on 2D materials. Discovered in 2011, MXenes belong to an exclusive family of hydrophilic, conductive 2D nanomaterials.

The main objective of this chapter is to summarize the self-healing behavior of MXene-based nanocomposites. Herein, different nanocomposites of MXenes and their processing pathways, integration with polymer hosts, mechanism of action, and applications in different fields of science are deliberated. This chapter gives an outline of the diverse chemistries engaged in the synthesis of self-healable nanocomposites, with their pros and cons.

9.1.1 History of MXenes

The discovery of MXenes ($Ti_3C_2T_x$) took place in 2011 at Drexel University,[3] without any previous knowledge of the stability of this type of 2D material. Since that time, more than 30 compositions of MXenes have become available, and dozens more have been investigated by computational

DOI: 10.1201/9781003164975-9

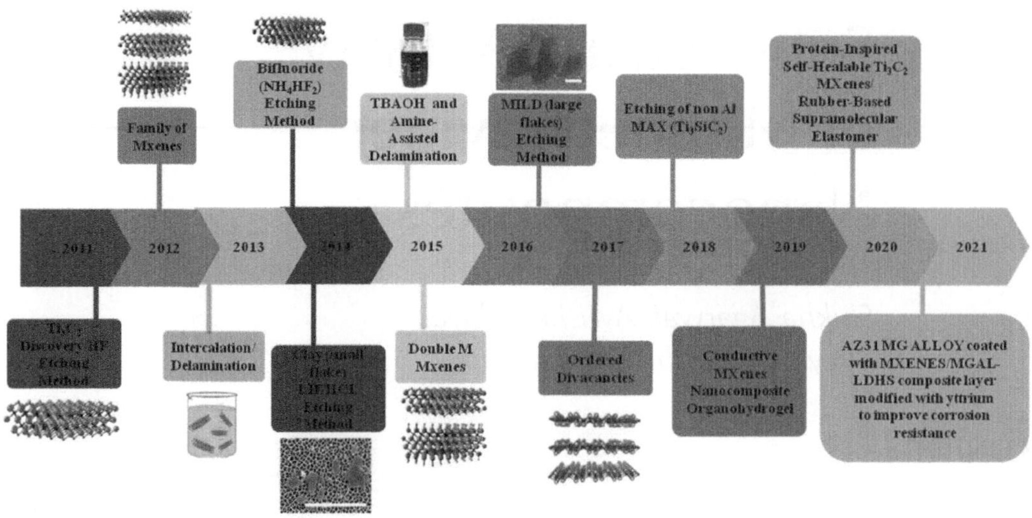

FIGURE 9.1 Development on MXenes in the past years.

Source: Modified from Reference no. 4.

techniques.[4] MXenes became famous due to fascinating characteristics that lead to their utility in several real-life applications.

MXenes are 2D sheets of transition metal carbonitrides, nitrides, and carbides that were prepared by $M_{n+1}AX_n$, or MAX phases, where 'M' is a transition metal, 'A' belongs to IV–V group elements, and 'X' denotes C or N. They are synthesized from MAX by particularly etching the 'A' layer.[3, 5–7] After etching, the MXene is best depicted like $M_{n+1}X_nT_z$, where T_z represents the functional groups that exchange during the etching procedure. T is supposed to maintain the charge neutrality in MXene. In general, T represents F, O, and OH.[8–11] They have an exclusive combination of characteristics due to the functionalized surfaces of MXenes that make them hydrophilic and prepared to bind with diverse classes. MXenes have a high value of (–) zeta-potential, high mechanical properties, and electrical conductivity. MXenes and their composites possess enormous applications in various fields, namely, acting as electrochemical capacitors, Li–ion, Na–ion capacitors, and batteries; on-chip energy storage; hybrid energy storage; electromagnetic interference (EMI) shielding; MXene inks for supercapacitors; binders for Si and carbon; electrocatalysis; textile industries; electrochromic devices; biomedical appliances; sensors, and many more.[4, 12–18] Recently, several review articles have been published on MXenes and their polymer nanocomposites,[19–20] membranes,[21] and their applications in wearable pressure sensors.[22] In 2021, Carey and colleauges[23] deliberated the various methods of synthesis of polymer nanocomposites and their properties. The development of MXenes in past years has been depicted in Figure 9.1.

9.1.2 History of Self-Healing Polymer Composites

Nature itself is an inspiration for scientific research. For example, skin acts as an efficient protective barrier for the human body and possesses self-healing capability. The conversion of these physiological characteristics into synthetic materials could unlock new prospects in numerous strategic fields from biomedical to robotics. Self-healing in compounds and material science is the capability of a material to refurbish its damages individually and keep its structural integrity. The demonstration of, quest for, and applications of self-healing polymers rose in the 21st century, motivated by the perceptive interfacial properties at the nanoscale and the progression of flexible devices. A

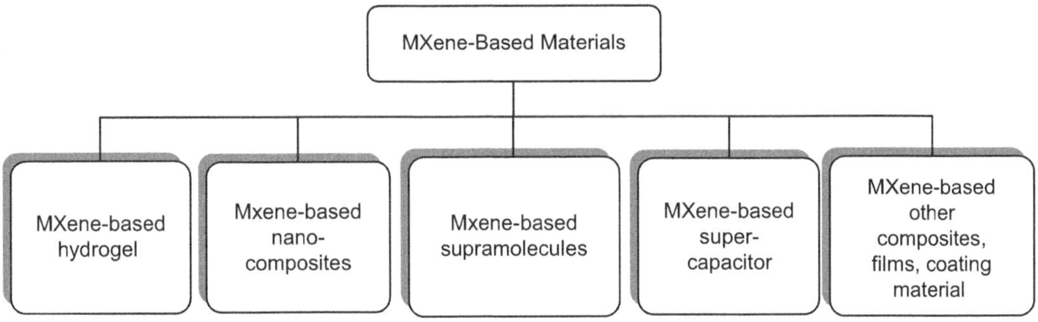

FIGURE 9.2 Classification of MXene-based composites.

swift investigation at the Web of Science demonstrated that approximately 360 papers were published from 1965 to 2000, and more than 5,000 papers were available since 2000 with the theme of "self-healing."[24–27]

Momentous development has been accomplished in the progress of materials that can sense and/or self-repair.[28–29] Supramolecules act as efficient composites for self-healing.[30–33] Polyborosiloxane (PBS) is a supramolecular polymer that demonstrates a fundamental self-healing property due to its dative bonds and is embedded with others to improve its characteristics.[34] The assimilation of self-healing and other functionalities in a strong and weightless material is still a subject of challenge. In the past years, diverse types of self-healing materials have been synthesized by employing different methodologies. The integration of appropriate functionalities with these materials stimulates a self-healing character.[35] MXene materials fulfilled these criteria as variations in stoichiometric compositions and surface functional groups allow the tailoring their physical and chemical properties. The key property of MXene is its self-healing nature. In this chapter, a detailed mechanistic approach to self-healing capacity of MXene and its composites has been deliberated. The chapter is divided in various subsections like MXene-based hydrogels, MXene-based nanocomposites, MXene-based supramolecules, MXene-based supercapacitors, and other compositions, with a detailed study about the synthesis and their self-healing behavior (Figure 9.2).

9.2 DESIGNING OF MXene-BASED COMPOSITIONS

In this chapter, efforts have been made to abridge the current advances in the self-healing behavior of MXene-based compositions. It has been instigated by reviewing diverse production pathways and their effects on the surface chemistry and structure of MXene-based composites. The arrangement, characteristics, constancy, and different types of layered MXenes are demonstrated over a period of time. Moreover, different types of MXene-based supercapacitors are abridged to emphasize the importance of MXene-based composites in making sensors, energy storage appliances, biomedical applications, and so on. Ultimately, challenges, limitations, and several aspects in this emerging area need to be explored for better development and progress in the self-healing behavior of MXenes.

9.2.1 Self-Healing Characteristics of MXene-Based Hydrogels

Currently, the self-healing characteristics of hydrogels based composites have provoked interest of researchers owing to their promising applications and potential to extend service life. Several strategies have been designed for the synthesis of hydrogels to balance self-healing efficiency and enhance high mechanical properties.[36–38] The hydrogels have been embedded with other 2D materials like graphene, MXene, carbon nanotubes, and others to boost their properties. The researchers are focusing on the designing of MXene-incorporated hydrogels to improve their self-healing characteristics as well as augment their industrial and medical applications.[39–41]

Zhang and colleagues[42] designed self-healable MXene-based polyvinyl alcohol (PVA) hydrogels with outstanding sensing abilities via a simple method through mixing of MXene nanosheets with commercially available hydrogel named "crystal clay" that include anti-dehydrating agents, water, PVA, and others. The obtained MXene-hydrogel exhibited notable stretchability (3400%). The addition of MXene nanosheets with negatively charged surfaces caused formation of secondary crosslinking as well as additional hydrogen bonding into the hydrogel system and, thus, facilitated the self-healing property. Furthermore, MXene hydrogels were also used in sensing materials to identify several bodily motions (facial expression, handwriting), monitoring vital signals with high accuracy and sensitivity.

Later on, in 2019, Zhang and coworkers[43] designed a highly efficient capacitive strain sensor by developing MXene/PVA-based hydrogel. The integration of PVA with MXene augmented the self-healability and conductivity of the hydrogel composite. The electrode displayed a high value of stretchability next to the break point (≈1200%) with spontaneous self-healing in just 0.15 s. The MXene/PVA hydrogel was synthesized as per the following method. Initially, 8 wt% PVA solutions was prepared using PVA powder having Mw ≈ 145000, by dissolving in 3 mL DI water and kept at 90°C for approximately 8 h. After this, the prepared solution was poured into 5 mg/mL MXene solution with stirring at room temperature for approximately 1.5 h. Then, a transparent sodium tetraborate solution was mixed in this solution to furnish the hydrogel. The addition of borax to this system caused crosslinking between the functionalities of PVA chains that formed the diol–diol bonds. Moreover, the functionalities of MXene having oxygen could make covalent bonds with borate ions, as well as with the PVA chains that trait to self-healability. The capacitive sensor based on MXene/PVA hydrogel demonstrated elevated linearity up to 200%, sensitivity up to 0.4, less hysteresis, fine mechanical strength, and durability.

In the same year, Zhang and coworkers[44] synthesized MXene hydrogels. Initially MXene was prepared using the "Yury mild method" and then acrylamide (15 wt%), lauryl methacrylate with concentration 1.5 mol% of the concentration of acrylamide and OP-10 (3 mol%), and an aqueous dispersion of MXene (1.3 mg/mL) were mixed. Nitrogen gas was passed to it, and the resultant mixture was stirred constantly for 50 min followed by the addition of potassium persulfate (KPS) having a concentration of 0.1 mol% acrylamide with 10 μL of N,N,N′,N′-tetramethylethylenediamine (TEMED). Furthermore, the solution was kept at room temperature for 3 days, and thus, hydrogels were prepared. Now N-isopropylacrylamide (NIPAM) (10 wt%), deionized H_2O, and N,N′-methylene diacrylamide (1 wt% of NIPAM) were added in a flask and stirred on ice bath for 1 h with a passing of nitrogen; subsequent to this, KPS having a concentration of 0.1 mol% NIPAM with 10 μL of TEMED were added. Thus, the nanocomposite double-network (NCDN)-based hydrogel, homogeneous and black in color, was achieved. Using this procedure, other hydrogels like double network (DN) hydrogel without MXene, hydrophobic polyacrylamide (HAPAM) hydrogel, graphene-based composite hydrogel, and pure poly(N-isopropyl acrylamide) hydrogels were synthesized. NCDN hydrogel having MXene demonstrated an elevated storage modulus, possibly due to the presence of crosslinking in hydrogels, therefore enhancing the mechanical strength of the hydrogel. The DN and NCDN hydrogels possessed self-healing characteristics due to H-bonding and hydrophobic interactions. A self-healed tensile test was used to determine the healing characteristics. The strip of NCDN hydrogel was cut in two parts; then, these two parts were placed together at room temperature. Three days later, the healed hydrogel was extended, and it did not break as depicted in Figures 9.3a and 9.3b. MXene embedded with hydrogels augmented the self-healing to 59.5% due to the hydrogen bond and crosslinking (Figure 9.3c and 9.3d).

A self-healing, conductive, antifreezing, and excellent mechanical strength, long-lasting moisture based on MXene nanocomposite organohydrogel (MNOH) was developed by Liao et al.[45] The MNOH was prepared by incorporating MXene nanosheets with a hydrogel polymer network prepared from polyacrylamide and PVA. Furthermore, there was a crosslinking interaction between

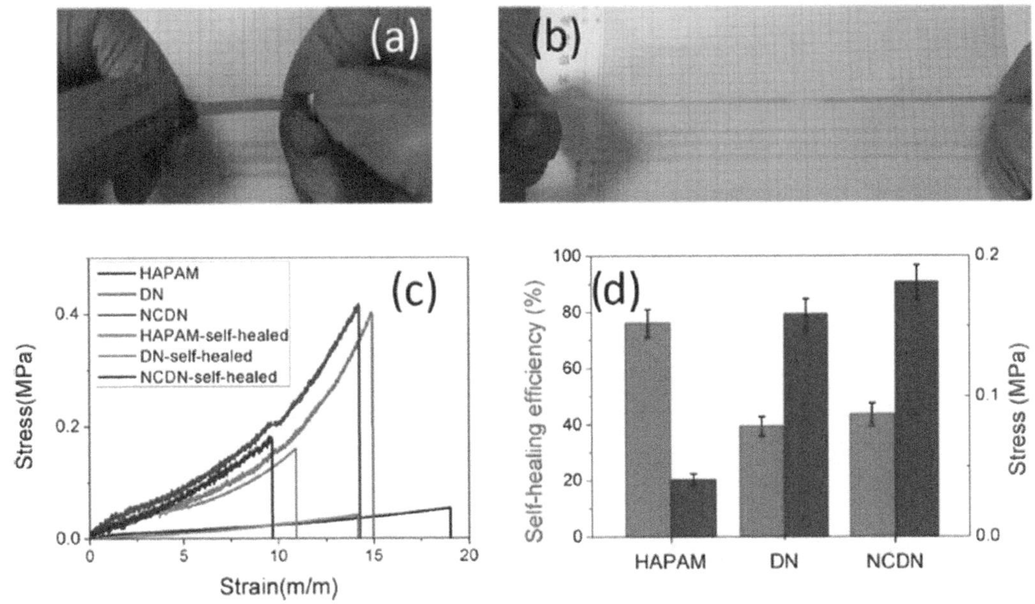

FIGURE 9.3 (a) Self-healing ability of NCDN hydrogel. (b) Stretching image of NCDN hydrogel. (c) Tensile curves. (d) Self-healing efficiency of the HAPAM, DN, NCDN-healed hydrogels (error bars, SD; n = 3).

Source: Reprinted with the consent from Reference no. [44] with Copyright {2019} American Chemical Society

the OH group and the tetrahydroxyl borate ions and a supramolecular interaction between the PVA, ethylene glycol, and MXene, resulting in an outstanding self-heal ability in the developed MNOH. A high number of hydrogen bonds formed between H_2O, and ethylene glycol molecules hindered the evaporation of water as well as prevented ice formation. The fabricated MNOH showed a strong antifreezing ability at very low temperatures (−40°C), moisture retention for approximately 8 days, healing ability, and higher mechanical strength. Furthermore, the fabricated hydrogels may have applications as wearable strain sensors to distinguish human activities with high-sensing sensitivity at extremely low temperatures (−40°C).

Wang et al.[46] synthesized MXene nanosheet self-assembled with polyacrylic acid (PAA) hydrogel having anti-aggregation and ultra-stretchable (nearly 1400%) properties. For the synthesis, TiO_2 nanoparticles were grown on the MXene surface. Furthermore, the reduced TiO_2@ MXene initiated rapid polymerization of monomer acryl amide (AA) without heating and induced crosslinked polymer chains to make hydrogel in very short time (in min). Moreover, the structural, mechanical, conductive and adhesive properties of synthesized hydrogel were modified by varying the content of TiO_2@MXene. The developed pathway minimized the aggregate of MXene nanosheets and thereby increased its stretching power. The synthesis of TiO_2@MXene-PAA hydrogel involved peeling off thick multilayer of (M-MXene) to form delaminated MXene (d-MXene) by ultrasonication followed by *in situ* development of TiO_2 NPs on d-MXene surface by solvo-thermal method (Figure 9.4). Further, an ultrasonic treatment was done to increase the interaction among AA and ammonium persulfate (APS) and maintain dispersion. Afterward, a redox reaction took place between the reduced TiO_2@MXene nanosheets with oxidant APS. There is decrease in decomposition activation energy (Ed) of APS and generated a large amount of SO_4–. Then, OH· radicals were generated by hydrolysis of SO_4– that initiated ultrafast polymerization of AA to PAA chain having a vine-like structure. The developed stable hydrogel showed outstanding mechanical properties, namely, stretchability up to 15 times the original length and recoverability. Furthermore, it did not rupture even after being subjected to 95% compression.

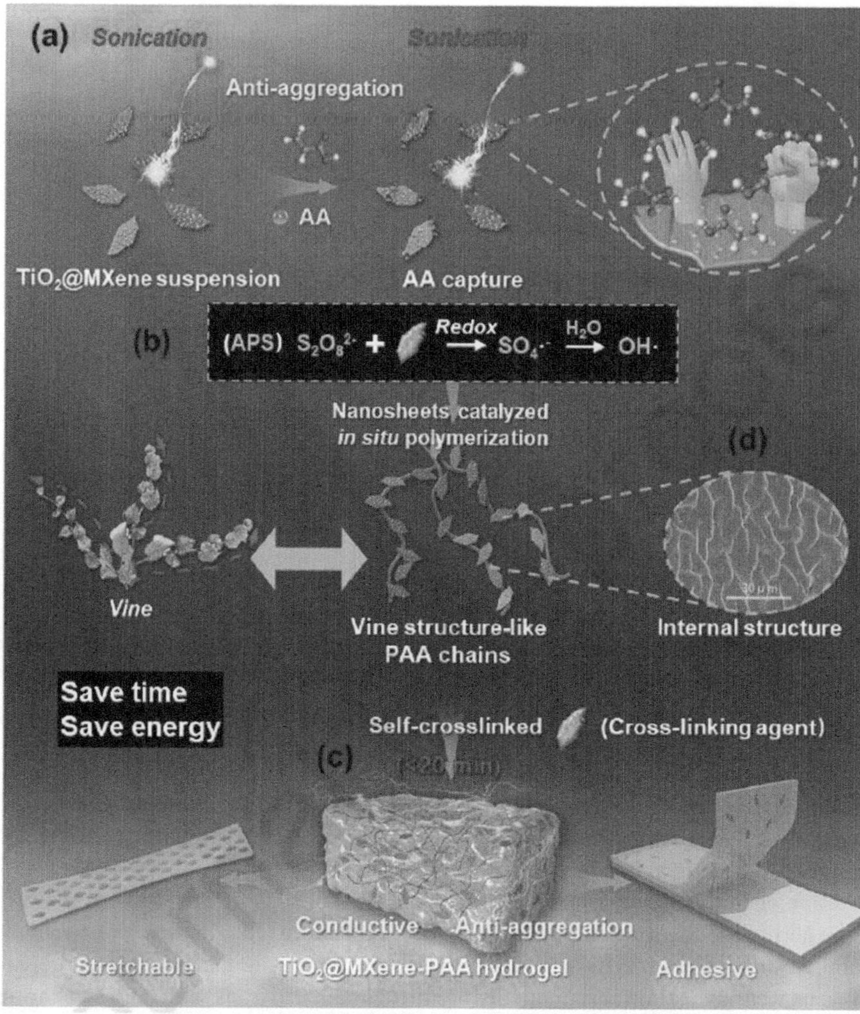

FIGURE 9.4 (a) The excitation of acrylamide monomers confined on the surface of TiO_2@MXene in presence of sonication.(b) The TiO_2@MXene catalyzed preparation of PAA.(c) Self-crosslinked and self-assembled characteristics of PAA with TiO_2@MXene.(d) Scanning electron microscopy images of synthesized hydrogel.

Source: Reprinted with the permission from Ref. [46] Copyright {2020} Elsevier.

MXene containing PVA/polyvinyl pyrrolidine (PVP) double-network hydrogels (MDN hydrogels) were synthesized by Lu et al.[47] for flexible strain sensors. The mechanical properties of MDN hydrogels were highly improved by the strong interaction between MXene and PVP and PVA matrix. The MDN hydrogels with 3 wt% PVP content and molecular weight 55,000 exhibited the best performance. The synthesized MDN hydrogels showed outstanding mechanical properties of puncture resistance, high stretchability (2400%), excellent stress tolerance, and cyclic stability. Moreover, the strain sensor based on the MDN hydrogel showed high sensitivity (10.75k/Pa), excellent durability, quick response (33.5 Ms), and low detection limit (0.87 Pa). The sensors could be used to detect human motions and embedded with certain devices to detect pressure and provide several applications in health diagnosis.

A superfast self-healable PVA/Bn/PEI/$Ti_3C_2T_x$ (PBPM-2)-based dual-network flexible hydrogels were prepared by PVA, and polyethylenimine (PEI) with Bn and further adding MXene to furnish the desired polymer network. Fabricated hydrogels PVA/Bn/PEI/MXene (PBPM-2) exhibited direction aware and fast self-healing ability with nearly 0.06 s self-healing time and rapid response performance (nearly 0.12 s). The outstanding self-healing ability was achieved by the formation of versatile chemical bonds, and supramolecular interaction between PEI, PVA, and MXene in the polymer network. Bn acted as a crosslinker between the functional groups OH of PVA and functional group NH_2 of PEI that formed several imine and borate bonds and induced self-healing property. The fabricated hydrogel exhibited a high direction recognizing ability that can be used for electronic skin.[48]

In general, MXene-based hydrogel did not fulfill the self-healing and mechanical necessities needed to be applied for soft electronics but due to the luminosity of intricate molecular interactions made by MXenes composites, stretchability, quick gelation characteristics, and self-healing properties in different polymer-based hydrogels are perhaps achievable. MXene-based hydrogels have become good candidates for stretchable electronics. In 2021, Gang ge et al.[49] synthesized MXene-based hydrogels, mixed $Ti_3C_2T_x$ nanosheets with a monomer, like acrylic acid with $(NH_4)_2S_2O_8$ and glycerol, to produce poly(acrylic acid)MXene hydrogels.

$Ti_3C_2T_x$ MXene acts as a crosslinker for activating fast gelation of a broad range of hydrogels (monomers to polymer-based precursors). The monomer precursors include AA (acrylic acid), DMA (N,N-dimethylacrylamide), AM (acrylamide), HEMA (hydroxyethyl methacrylate), NIPAM, ANI (aniline), and PEGDA (poly(ethylene glycol) diacrylate and the polymer based precursors include agar, gelatine, PVA, chitosan, and alginate. The mechanism of fast gelation involved easy generation of free radicals using $Ti_3C_2T_x$ MXene and molecular interaction between MXene and polymers. MXene, a versatile crosslinker, added high stretchability, adhesion, and self-healing properties to hydrogels in perspective of stretchable and wearable electronics. The data analysis showed that the induced gelation process of mixing MXene with polymer chains was slower compared to mixing with only monomers. The properties like threshold concentration and gelation of polymer-based MXene hydrogels mainly depends on characteristics of polymers. It is interesting to note that $PAAMXene_{1.43}$ hydrogels could uphold 530.81% healing strain with 147.86 kPa recovered. After healing for 6 h, the restored strains for PAA $MXene_{2.86}$ was 992.86% and $PAAMXene_{4.29}$ was achieved to be 1503.35%, respectively. The $PAAMXene_{4.29}$ was used as a reference to evaluate the self-healing property (Figure 9.5a–g). Both $PAAMXene_{2.83}$ and $PAAMXene_{4.29}$ hydrogels displayed good self-healing capability even after 15 days' storage. (Note: The subscript in MXene represents the concentration of MXene in hydrogel.)

Yu and coworkers[50] designed a two-step procedure to synthesize hydrogel-based flexible electrode by crosslinked PAA/chitosan/Ti_3C_2Tx (PCT). The exclusive 2D-layered-type structure of MXene Ti_3C_2Tx facilitated the hydrogel electrode to endure substantial alteration in pressure without breaking it and has tremendous tensile property. The electrode demonstrated an outstanding tensile power up to 0.083 MPa. This give research and expansion approach for nontoxic, binder-free, and easily portable storage devices.

A multifunctional conductive epidermal sensor based on stretchable, biocompatible, healable hydrogel was developed by Li and coauthors.[51] The developed hydrogel was coated with MXene network with (PAA) and amorphous $CaCO_3$. The developed epidermal sensors can be applied for the identification of human movements with a quick response time (20 ms). Moreover, the fabricated sensors could be degraded in the solution of phosphate buffer saline that caused no pollution to the environment.

Ai *et al.*[52] fabricated self-healing MXene nanocomposite hydrogels (MNH) using MXene and PVA as basic framework crosslinked by diol–borate ester bonding. The fabricated MNH exhibited high self-healing performance and outstanding tensile and fatigue resistance. They also showed 92% self-healing ability, enhance salination capacity (51 mg/g), good mechanical property, electrochemical healing property (95.8%) and excellent cyclic stability. The developed materials could self-heal damaged interfaces with applications in capacitive deionization.

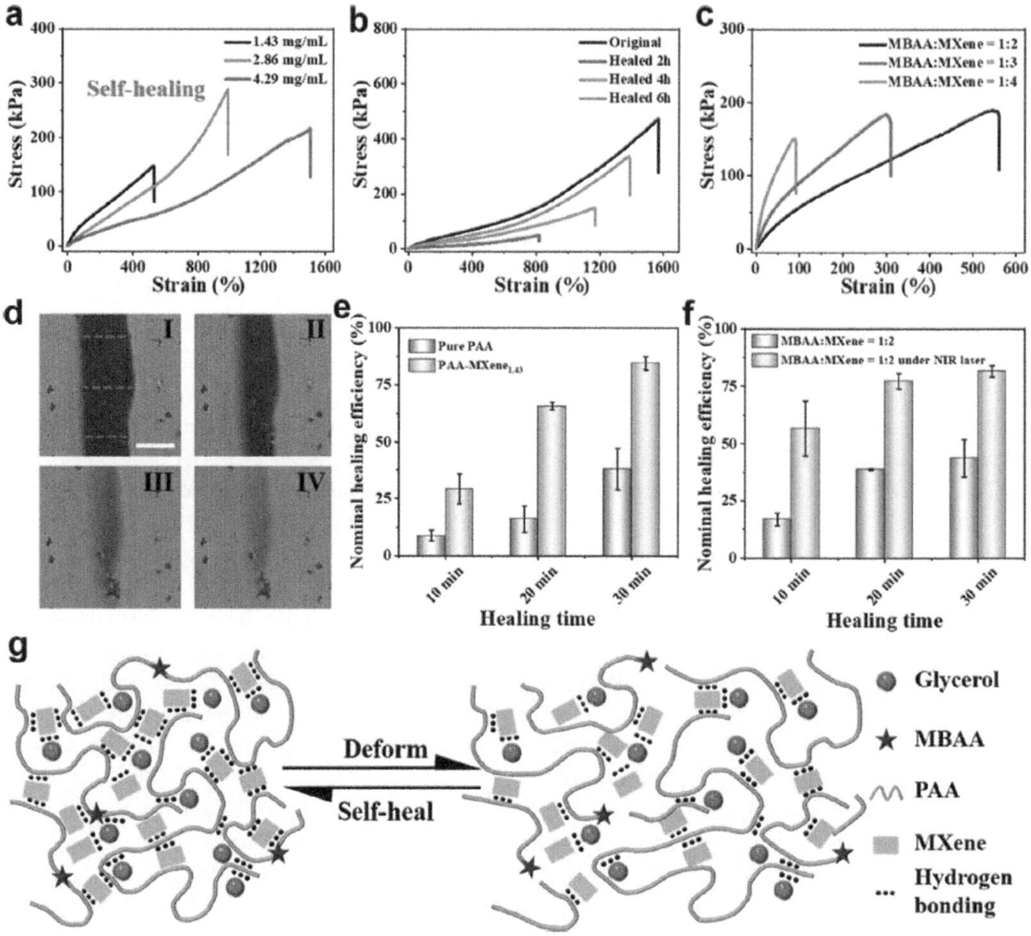

FIGURE 9.5 (a) The pictorial representation of the self-healing property of hydrogels at different MXene concentrations. (b) Stress–strain arc for PAAM-Xene$_{4.29}$ hydrogels in different time intervals. (c) N,N'-methylene-bisacrylamide (MBAA) concentration effect on self-healing characteristics of hydrogels. (d) The images by the optical microscope of the cut area after healing for subsequent 0 (I), 10 (II), 20 (III), and 30 (IV) min. (e) Interpretation of healing power of PAA (red) and PAAMXene$_{1.43}$ (green) hydrogels. (f) Near-infrared laser–induced self-healing power. (g) Mechanism of self-healing.

Source: Reprinted with the consent from Ref. [49] with copyright {2020} American Chemical Society

9.2.2 Self-Healing Characteristics of MXene-Based Nanocomposites/Nanosheets

The designing of wearable and translucent materials with self-healing characteristics, which may recuperate their functionality even after mechanical demolishment is essential. In lieu of this, Fan and coworkers[53] synthesized hybrid sheets of Ag-Np@MXene with polyurethane (PU). The incorporation of MXene with Ag nanoparticles gave rise to high photo-thermal conversion efficiency because of the grouping in plasmonic characteristics of Ag-NPs in combination with the thermal conductivity of the MXene. The synthesis of Ag-NP@MXene hybrid is done as per the previously reported literature.[54–56]

Wu *et al.*[57] synthesized an adhesive, conductive, moist, and healable MXene-based nanocomposite hydrogel using MXene nanosheets, phenylboronic acid grafted with sodium alginate (Alg-PBA), dopamine-grafted sodium alginate (Alg-DA), and polyacrylamide in the a presence of mixture of

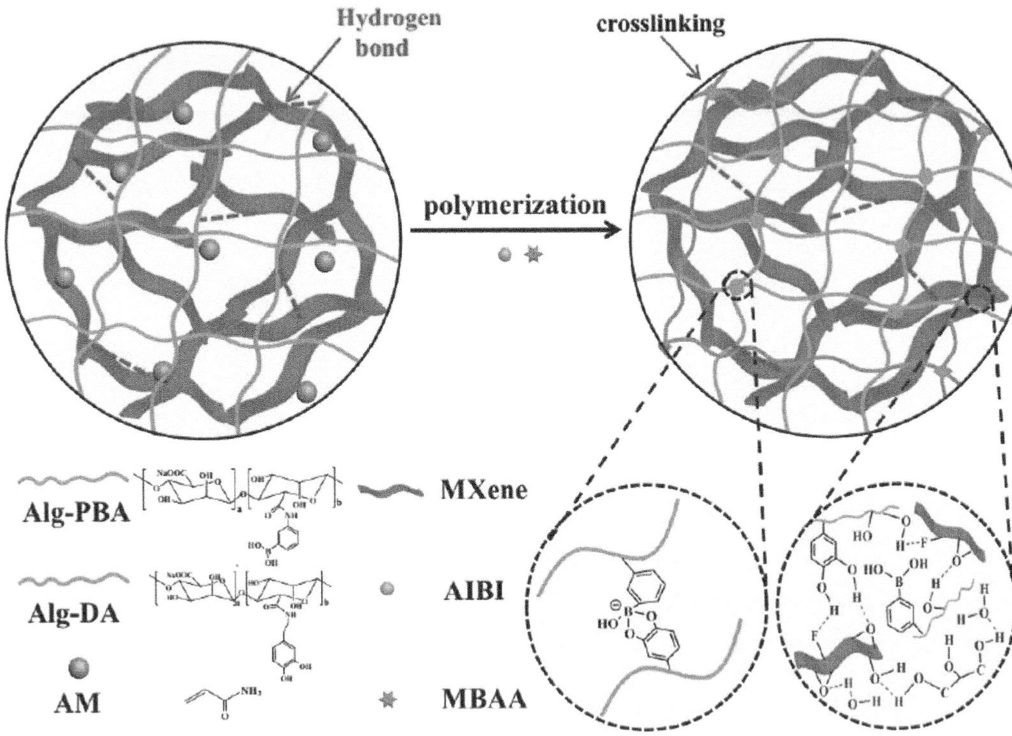

SCHEME 9.1 Graphical presentation of fabrication of MXene nanocomposite organo-hydrogel.

Source: Reprinted with the consent from Reference no. [57] Copyright {2020} Royal Society of Chemistry

water and glycerol. First, MXenes were synthesized using LiF/HCl method. In parallel to this, acrylamide was mixed to mixture of glycerol and water. Now, Alg-PBA and Alg-DA were mixed to this solution and stirred continuously. Later on, upon complete dissolution, the MXene nanosheets solution and Tris-HCl were added, with subsequent addition of N,N'-methylene-bisacrylamide and 2,2'-azobis[2-(2-imidazolin-2-yl)propane] dihydrochloride. The resultant mixture was stirred at 60°C for 2 h to obtain the organo-hydrogels. The schematic presentation of synthesis procedure is depicted in (Scheme 9.1). Polyacryl amide was integrated with the organo-hydrogels to improve the mechanical characteristics of organo-hydrogels. The mass ratio of water and glycerol affected the mechanical and other properties of composites. With increase in glycerol ratio, the conductivity of organo-hydrogels diminished whereas the other properties, like self-adhesiveness and the tensile strain, were augmented. These MXene nanocomposite organo-hydrogels exhibited a self-healing capability attributable to PBA catechol bonds as shown in Figure 9.6. When their healability was tested by cutting the sheet into two parts and then were put back together, they healed at room temperature after 12 h. The healed organo-hydrogels can be stretched up to 71.0% with the healing efficiency (88.75%). The self-adhesive property is attributed to OH groups of glycerol and catechol group of PDA, and due to these miscellaneous activities, these composites have been used in wearable epidermal sensors.

Zheng and coworkers[58] synthesized a multifunctional hydrogel scaffold to cure infection-impaired skin. Herein, multifunctional hydrogel scaffold CeO_2@MXene nanocomposites were synthesized *via* merging the 2D antibacterial MXenes ($Ti_3C_2T_x$) and antioxidant CeO_2. These multifunctional hydrogels have a plethora of properties like anti-inflammatory, self-healable, antibacterial, conductive bioactivities, antioxidative, hemostatic capacity, and tissue adhesiveness. Here, the

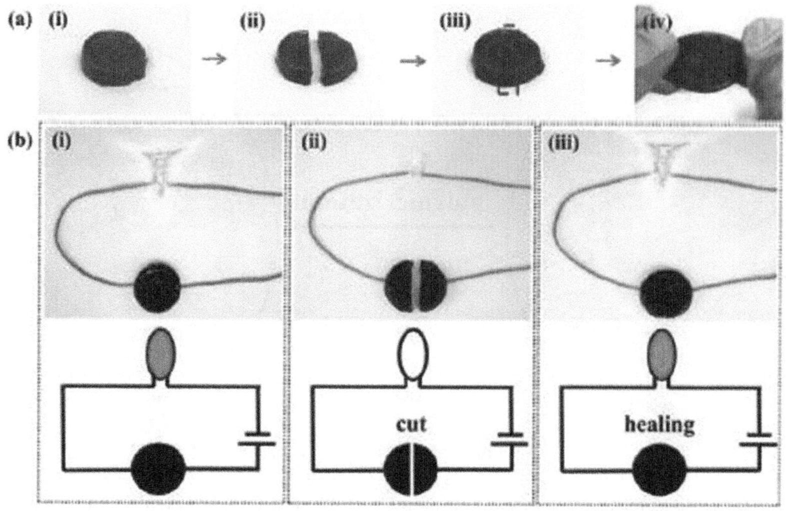

FIGURE 9.6 Images of self-healing process for MXene nanocomposite organo-hydrogels.

Source: Reprinted with the consent from Reference no. [57] Copyright {2020} Royal Society of Chemistry

authors established the imperative role of multi-functional CeO_2@MXene in various applications like multidrug-resistant infection therapy, collagen deposition, promoting fibroblasts proliferation, and granulation tissue formation, helpful in wound healing.

MXene-based composites have gained much scientific interest in electronic sensors, batteries, catalysis, purification, biomedical applications, and EMI shielding. On the other hand, their anti-corrosion property is hardly investigated due to oxidizing nature of MXene nanosheets. Zhao and coworkers[59] synthesized antioxidant Ti_3C_2 MXene nanosheets *via* noncovalent functionalization of MXene with an ionic liquid (IL) and used them as anticorrosion coating. The resultant IL@MXene nanosheets were employed as barrier provoked by waterborne epoxy (WBE) to improve the anticorrosion activity. The interpretation of anticorrosion activity of steels coated with IL@MXene–WBE displayed greater corrosion inhibitory activity than WBE. Furthermore, the healing characteristics of the IL@MXene–WBE coating were established by the SVET. This study gives an approach to devise the synthesis of MXene-based composites for anticorrosion coatings.

9.2.3 Self-Healing Characteristics of MXene-Based Supramolecules

Zhang and coworkers[60] deliberated and synthesized conductive silicone-based polymer composite. MXenes and amino polydimethylsiloxane (PDMS) were customized through an esterification reaction and Schiff base reactions of tiny bio-organic molecules. MXene was tailored by the reaction between the functional groups COOH and OH of D-asparagine, which was extracted from *Liliaceae* plants. In the meantime, 3,4-dihydroxybenzaldehyde was embedded in PDMS macromolecules. The A-MXenes@D-PDMS has self-healing ability at room temperature due to the presence of crucial interactions, namely, H-bonding and imine interaction. These crucial interactions also improve the tensile characteristics and healing efficiency. The customized MXenes are evenly dispersed, bestowing the newly fabricated composites with good electrical conductivity.

The self-healing characteristic was demonstrated using a stereomicroscope.

It was estimated by using a formula:

$$\textit{Healing}\ \text{Efficiency} = S_{\text{healable}}/S_{\text{org}} \times 100\%.$$

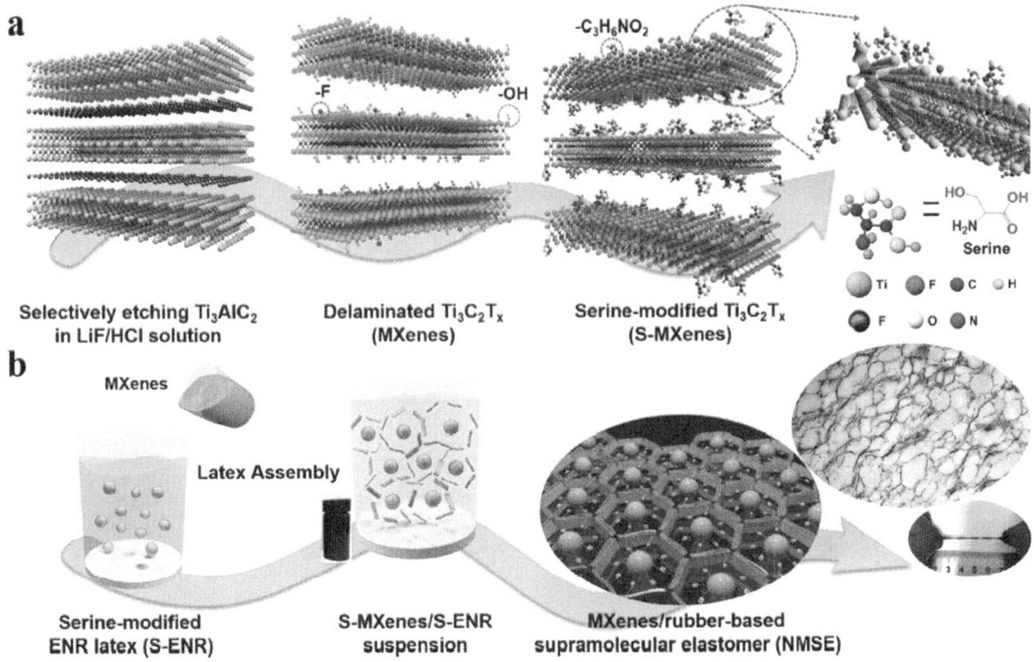

SCHEME 9.2 Synthesis of MXene-based elastomer *via* the latex assembly method.

Source: Reprinted with the consent from Reference no. [61] Copyright {2020} American Chemical Society.

S_{org} and $S_{healable}$ were related to the tensile capability of the original sample and the healable one, respectively.

The designed composites having 10 wt% of A-MXenes exhibited stretching capacity up to 81% and mechanical power (1.81MPa). Tensile characteristics and electrical conductivity can be regained to 98.4% and 97.6%, respectively, after regeneration. The healable composites can also precisely perceive petite human motion.

Supramolecular-based hydrogen bond interactions in a polymer composite are an efficient approach for bestowing designed composites with self-healing characteristics. In 2020, Guo and coworkers[61] devised an efficient, healable nanostructured Ti_3C_2MXenes@rubber (NMSE) for intelligent sensing. For this, nanoflakes of MXenes were customized *via* amino acid serine by the condensation reaction of OH group of MXene and COOH group of serine. The H-bonding amid the surface of customized MXenes and elastomer helped recuperate its mechanical and electrical performance at room temperature and provided outstanding self-healing properties (Scheme 9.2).

The as-synthesized supramolecular elastomers offered concurrently improved mechanical power and higher healing ability without external stimulus (Figure 9.7). The designed NMSE gave desirable toughness (12.34 MJ/m^3) and tremendous healing characteristics of nearly 100% just at room temperature. The sensors that are based on NMSE have high gauge factor (107.43), fast responding time (50 ms), and low strain detection limit (0.1%) can accurately determine subtle human activity like speech, heartbeat, facial expression, and pulse, as well as moisture variation, even after cut/healing procedure.

9.2.4 MXene-Based Supercapacitors and Their Self-Healing Characteristics

Micro-supercapacitors (MSCs) are one of the reliable energy storage devices for powering nano/micro devices. Flexible MXene-based MSCs are currently in demand due to their low-cost

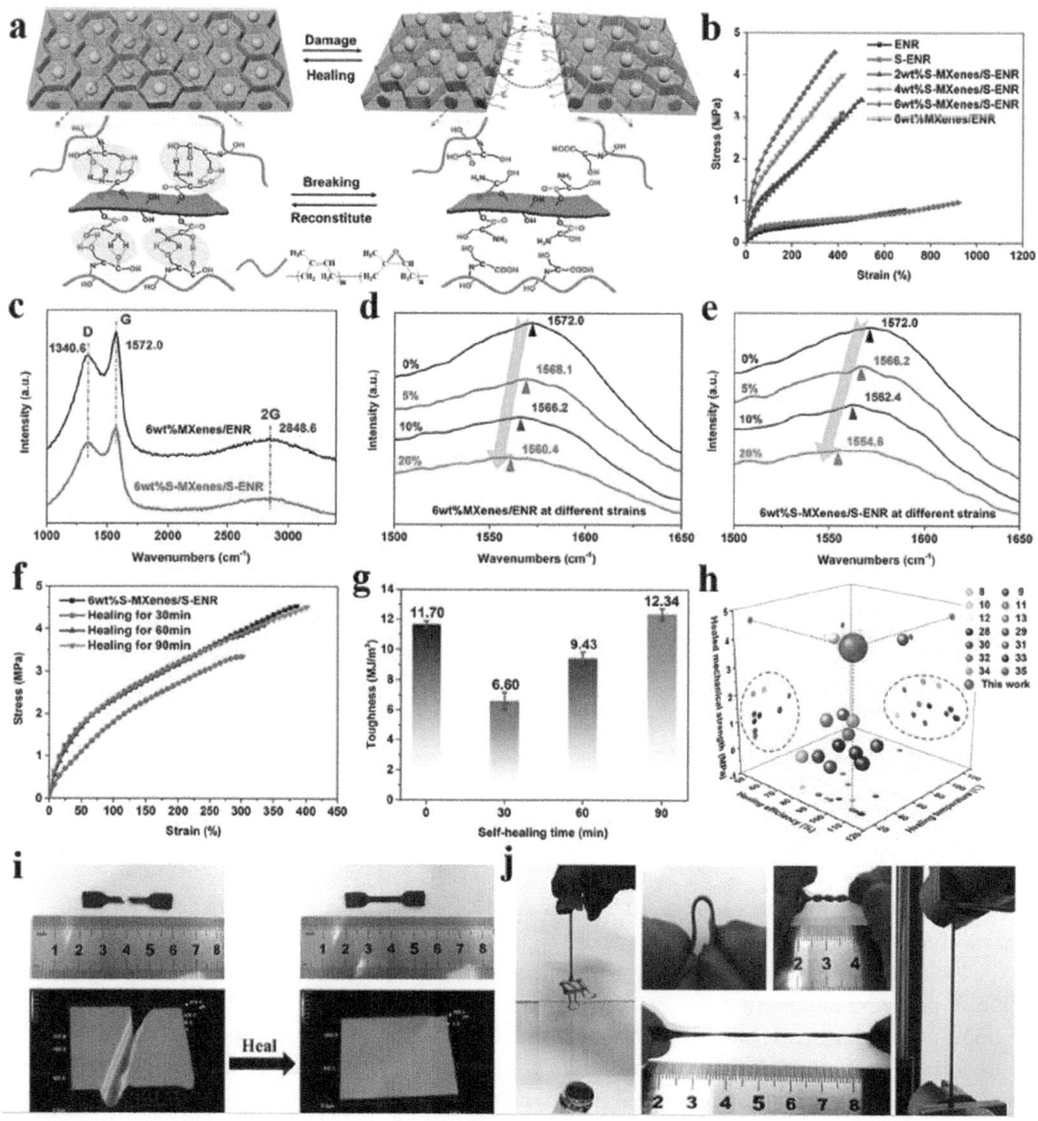

FIGURE 9.7 Self-healing characteristics of flexible NMSE: (a) Dynamic rupture and reformation of supramolecular H-bond in NMSE. (b) Stress–strain curves. (c) Raman spectrum. (d) Representation of blueshift of the G band at different strains in MXenes/ENR. (e) Blueshift for SMXenes/S-ENR samples.(f) Tensile curves for S-MXenes (6 wt%) at dissimilar times. (g) Tensile curve for S-MXenes at different times. (h) Evaluation of healed mechanical characteristics amid diverse healable elastomers. (i) Microscopic image of rough and healed samples. (j) Different images of a healed sample showing loading weight, twisting, stretching, and bending.

Source: Reprinted with the consent from Reference no.[61] with Copyright {2020} from American Chemical Society.

fabrication process and cycling stability. Several studies have been done in the field of flexible MXene-based MSCs, but some limitations need to be overcome:

- These MSCs are certainly smashed by internal and external deformation.
- MXene's conductivity corresponds to its nanosheet size.

In order to combat these limitations, a new kind of self-healable MSCs, dependent on the size of the MXene, was reported using spray-coating and laser-engraving methods.[62] Large and small MXene-based nanosheets (L-MXene, S-MXene) were synthesized by changing conditions of mechanical exfoliation. This approach involved fabrication of films based on MXene by spraying the solution of L-MXene and S-MXene onto cellulose paper with the help of an airbrush and further dried under vacuum for 3 h at 50°C. Then, the interdigital patterns of MXene were engraved by laser-engraving method. Furthermore, an H_2SO_4-PVA gel electrolyte was dropped on the interdigital patterns for assembling MXene-based MSCs. Finally, they were wrapped by carboxylated PU to convert them into self-healable MXene-based MSCs. PU is liable for the self-healing. The prepared micro-supercapacitors based on MXene showed good flexibility, excellent stability (>87.5%), and high areal capacitance (upto 73.6 mF/cm^2). The synthesized MSCs also showed self-healable performance, and the capacitance was retained (>80%) even after a fifth healing.

Yue *et al.*[63] designed a self-healable 3D micro-supercapacitors (MSCs) using ($Ti_3C_2T_x$)-@ graphene-(reduced GO) composites as an electrode material and enclosed with gel electrolyte composed of PVA-H_2SO_4 and carboxylated PU as self-healing materials. Simple freeze-drying and laser-cutting methods were used to synthesize MXene–rGO composite aerogel-based electrode. The carboxylated PU bears a high interfacial H-bond that helped in the outstanding self-healing properties, with 81.7% after a fifth healing. The fabricated MSCs showed a high exterior-area capacitance (34.6 mF/cm^2), with a scan rate (1mV/s) and excellent cycling performance.

9.2.5 MXene-Based Films, Electrodes, Other Compositions and Their Self-Healing Behavior

A library of healable, waterborne polyurethane (ADWPU) films was developed by combining PU with Ti_3C_2Tx MXene mechanically.[64] The ADWPU@ MXene emulsions were synthesized from castor oil, that is, an eco-friendly biomass polyol. 2-aminophenyl sulfide, the chain extender gave a reversible versatile network that led to outstanding self-healing power. The developed MXene-based films showed excellent EMI shielding (51.37 dB) and displayed negotiable alteration in shielding efficiency even after bending and stretching 200 times. Furthermore, Zhao and coworkers[65] designed an eco-friendly, safe zinc–ion micro-battery (ZIMB) having an anode made up of MXene-TiS_2 (de)intercalation and a cathode based on carbon nanotubes-vanadium dioxide (MWCNTs-VO_2), with $ZnSO_4$-PAM hydrogel as the electrolytic solution and PU acting like a defensive shield incorporated. The ZIMB displayed an adequate electrochemical capacity of 40.8 μAh cm^{-2}, a maximum energy density up to 1.2 mW cm^{-2}, and a maximum power density up to 32.5 μWh cm^{-2}. Moreover, ZIMB possessed thermo-stability, outstanding flexibility, and high self-healability. This research gives a viable approach to develop superior, highly consistent, and steady micro energy-storage devices for integrated electronics and wearable devices.

Wu and coworkers[66] designed a complex structure, by incorporating MXenes with magnesium–aluminum-layered double hydroxides (MXenes@Mg-Al-LDHs) with $Y(OH)_3$. The synthesis of these composites involved smart coating on Mg alloy AZ31 through the following steps:

- The multilayer MXene was converted to small-layer MXenes.
- $Y(NO_3)_3$ was added to a solution of small-layer MXenes *via* a single-step hydrothermal treatment.
- Due to the electronegative nature of MXenes, it could proficiently take up cations from the reaction mixture and endorse growth of Mg–A–LDHs and $Y(OH)_3$ on its surface.
- As per the corrosion analysis, the results specified that coating of yttrium have a splendid anticorrosion property and self-healing ability. Their self-healing behavior assists anticorrosion activity.

Cai and coauthors[67] designed a composite heterostructure by *in situ* assembly of $Ti_3C_2T_x$ @MgAl–LDH *via* electrostatic interaction. The interface amid the MXene $Ti_3C_2T_x$ and MgAl–LDH is examined by DFT theory. This $Ti_3C_2T_x$@MgAl–LDH accomplished tremendous compatibility as well as dispersibility in presence of epoxy resin. The synthesized $Ti_3C_2T_x$@MgAl–LDH/epoxy coating (C-MXene@LDH) demonstrated acceptable shear and wear defense with a definite self-healing property, because of the excellent dispersibility, the corrosion inhibitor discharge, and the barrier effect of $Ti_3C_2T_x$@MgAl–LDH. This study is a huge implication for MXene-based heterostructure analogs to acquire an enormous pathway approach for the anticorrosion and antiwear applications.

Zou *et al.*[68] synthesized flakes that are light-induced healable having an epoxy coating on MXene. This healable coating has reversible crosslinking network between the maleimide groups and the furan groups from oligomers by Diels–Alder reaction. As the temperature became high, the bonding between the multi-furan and the multi-maleimide broke, which produced a movable phase that streamed and filled up cracks. Healing is done at low temperatures. MXene was prepared by dissolving 10 g of Ti_3AlC_2 in 100 mL HF solution (49%) at optimum temperature of 30°C for a period of 7days. The solution was centrifuged and washing with solvents like water and ethanol. The ppt. was treated with 50 mL of tetrapropylammonium hydroxide, with stirring for 3 days at room temperature. Then this solution was centrifuged several times and washed with water and ethanol to remove excess of tetrapropylammonium hydroxide. In subsequent to this, oligomers were synthesized using DGEBA and FA was liquefied with DMF. This mixture was stirred for 6 h at 100°C. BMI was mixed with this solution. A definite quantity of delaminated MXene was dispersed with DMF (1 mL) under sonication for 1 h and stirred briskly. Leveling agents can also be mixed as needed. In this way, MXenes coated with epoxy were synthesized. With addition of the MXene, gel content (G) slightly increased and swelling ratio (Q) reduced. The addition of MXene improves not only the thermal stability but also the mechanical and anticorrosion properties. Temperature affects the self-healing behavior. The hole with a width of 10 μm in MX-0 started shutting after exposure of temperature 150°Cfor 5 s and then steadily vanished in 30 s.

For the easier understanding of the readers, a comparison table for self-healing efficiency, self-healing time of some MXene-based compounds has been appended in Table 9.1.

TABLE 9.1
Self-Healing Efficiency and Self-Healing Time of Some MXene-Based Compounds

S.No.	MXene-Based Compounds	Self-Healing Efficiency	Self-Healing Time	Reference
1.	MXene hydrophobically associated/polyacrylamide/ poly(N-isopropylacryl amide)	59.5%	72 h	[44]
2.	MXene nanocomposite organohydrogel synthesized from ethylene glycol	>85%	12 h	[45]
3.	AgNP@MXene-PU composites	97%	5 min of photo-thermal healing	[53]
4.	PAA-MXene hydrogels	52.21% with 818.81% strain	2 h	[49]
1.		74.68% with 1171.70% strain	4 h	
1.		88.29% with 1385.03% strain	6 h	
5.	MXene-Based Nanocomposite Synthesized by dopamine grafted sodium alginate, phenylboronic acid Grafted sodium alginate and polyacrylamide	88.75%	12 h	[57]

9.3 APPLICATIONS OF SELF-HEALABLE MXene-BASED MATERIALS

MXenes have an exclusive permutation of properties, together with the mechanical properties and high electrical conductivity of carbides/nitrides of the transition metal. The surface functionalization, which makes MXene-based materials hydrophilic, provokes their binding with other composites to enhance their properties and applications. The miscellaneous applications of MXenes in different fields of science are depicted in Figure 9.8.

Transparent and skin-mountable devices are extremely preferred for future applications of electronic devices. Designing of wearable, lightweight, easily portable, and transparent materials that have healable properties and maintain their original functionality even after mechanical damage, is inevitable. MXenes cover several areas of the basic sciences, including organic chemistry, synthesis, biomedical, structure designing, wearable sensors, energy storage, membranes, catalysis, and optics. This diversity of fields depict a foremost discovery and research in MXene uses in comparison to other 2D materials. MXene-based composites are used for electronic and photonic devices,[69] supercapacitors,[70]*in vitro* perspiration analysis,[71] flexible strain and pressure sensors,[47] drug delivery, diagnostic imaging, and biosensors,[72–74] among others.

Although MXenes possess a plethora of applications, their use is restricted. In order to increase their applications, self-healing property has been introduced in MXenes by embedding the self-healing polymer composites to it. This integration not only enhances the properties of MXenes but also provokes their applications. Due to the introduction of self-healing characteristics, MXene and its composites, hydrogels, films, and so on have been designed to cover enormous applications like corrosion coating, fire protection, e-skin, wearable sensors, and many more. MXenes embedded with hydrogels and supramolecules have been used to design sensors that can identify body motions[42, 45, 61] and be used to synthesize wearable sensors[60]and micro-supercapacitors.[50, 62–63] The designed AgNP@MXene hybrids are efficient molecular heaters, energy transformers, and photon captors. They are used to design wearable devices.[55] MXene nanocomposites have been used as epidermal sensors,[57]e-skin and infection-impaired skin multimodal therapy,[58] EMI shielding,[64] anti-corrosion coating,[59, 67, 75] electrodes,[65] MXene/chitosan-coated cotton fabric for intelligent fire protection,[76]and so on.

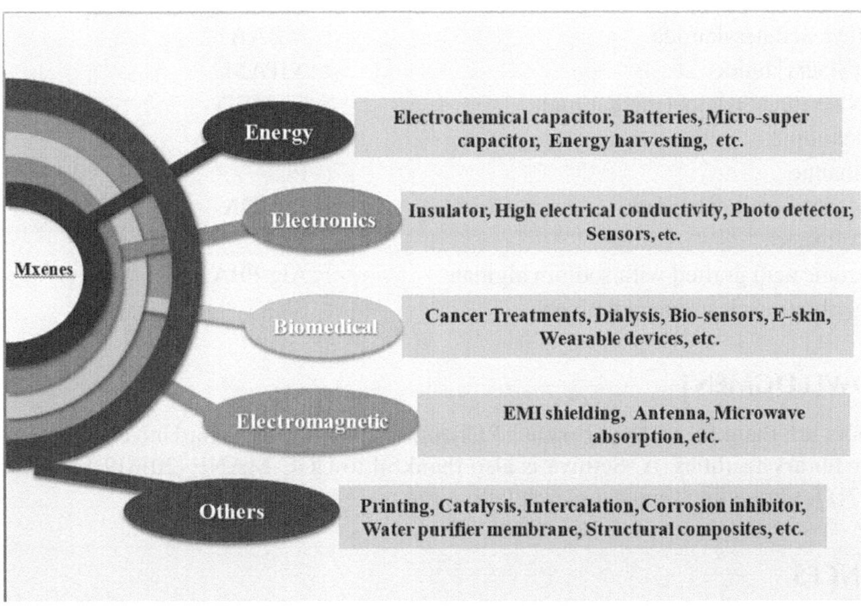

FIGURE 9.8 Miscellaneous applications of MXene-based compounds.

9.4 CONCLUSION

In this chapter, the current advancements and success in self-healable MXene-based materials have been discussed. The self-healing property was enhanced by integrating the MXene with other polymers/supramolecules/hydrogels. There has been an enormous rise in MXene-based self-healable materials over the past decades. These materials work as potential candidates in extensive applications, namely, electromagnetic interference shielding, electrocatalysis, the purification of water, and wearable devices, catalysis, biomedical applications, and others. MXene polymer–based compositions are noticeably promising for an extensive range of applications, still this field of research is in its infancy, Furthermore, the chapter investigates the emerging potential of the self-healing property of MXene-based composites up to date. We suppose that this work will stimulate more research interest in MXene-based compounds in other fields of chemistry and a quick focus on scalable production of these materials is indeed required.

ABBREVIATIONS

Acrylic acid	:	AA
Acrylamide	:	AM
Aniline	:	ANI
2, 2'-azobis[2-(2-imidazolin-2-yl)propane] dihydrochioride	:	AIBI
Amino polydimethylsiloxane	:	PDMS
Bis(4-maleimidopheny)methane	:	BMI
Diglycidylether of bisphenol A	:	DGEBA
Dopamine grafted sodium alginate	:	Alg-DA
Furfurylamine	:	FA
4-Formyl benzoboric acid	:	Bn
Hydroxyethyl methacrylate	:	HEMA
Lauryl methacrylate (96%)	:	LMA
Methyl methacrylate	:	MMA
N,N-dimethylacrylamide	:	DMA
N,N′-methylenediacrylamide	:	MBAA
N-isopropylacrylamide	:	NIPAM
N,N,N′,N′-tetramethylethylenediamine	:	TEMED
Polyethylenimine	:	PEI
Polyureathanne	:	PU
Poly(ethylene glycol) diacrylate	:	PEGDA
Polyborosiloxane	:	PBS
Phenylboronic acid grafted with sodium alginate	:	Alg-PBA
Scanning vibrating electrode technique	:	SVET

ACKNOWLEDGMENT

The authors are thankful to Department of Chemistry, M. L. Sukhadia University, for providing necessary library facilities. A. Sethiya is also thankful to UGC-MANF (201819-MANF-2018-19-RAJ-91971) for providing senior research fellowship.

REFERENCES

1. Huo, C., Yan, Z., Song, X., and Zeng, H. 2015. 2D materials via liquid exfoliation: A review on fabrication and applications. *Sci. Bull.* 60: 1994–2008. DOI: 10.1007/s11434-015- 0936-3.

2. Zhu, J., Ha, E., Zhao, G., Zhou, Y., Huang, D., Yue, G., Hu, L., Sun, N., Wang, Y., Lee, L. Y. S, Xu, C., Wong, K.Y., Astruc, D., and Zhao, P. 2017. Recent advance in MXenes: A promising 2D material for catalysis, sensor and chemical adsorption. *Coord. Chem. Rev.* 352: 306–327. DOI: 10.1016/j.ccr. 2017.09.012.
3. Naguib, M., Kurtoglu, M., Presser, V., Lu, J., Niu, J., Heon, M., Hultman, L., Gogotsi, Y., and Barsoum, M.W. 2011. Two-dimensional nanocrystals produced by exfoliation of Ti_3AlC_2. Adv. *Mater.* 23: 4248–4253.
4. Gogotsi, Y., and Anasori, B. 2019. The rise of MXenes. *ACS Nano* 13(8): 8491–8494.
5. Ghidiu, M., Lukatskaya, M.R., Zhao, M.-Q., Gogotsi, Y., and Barsoum, M.W. 2014. Conductive two-dimensional titanium carbide 'clay' with high volumetric capacitance. *Nature* 516: 78–81.
6. Halim, J., Lukatskaya, M.R., Cook, K.M., Lu, J., Smith, C.R., Näslund L.Å., May, S.J., Hultman, L., Gogotsi, Y., Eklund, P., and Barsoum, M.W. 2014. Transparent conductive two-dimensional titanium carbide epitaxial thin films. *Chem. Mater.* 26: 2374–2381.
7. Bae, S., Kang, Y.-G., Khazaei, M. Ohno, K., Kim, Y.-H. Han, M.J., and Chang, K.J., Raebiger, H. 2021. Electronic and magnetic properties of carbide MXenes the role of electron correlations. *Mater. Today Adv.* 9: 100118.
8. Jolly, S., Paranthaman, M.P., and Naguib, M. 2021. Synthesis of $Ti_3C_2T_z$ MXene from low-cost and environmentally friendly precursors. *Mater. Today Adv.* 10: 100139.
9. Hope, M.A., Forse, A.C., Griffith, K.J., Lukatskaya, M.R., Ghidiu, M., Gogotsi, Y., and Grey, C.P. 2016. MR reveals the surface functionalisation of Ti_3C_2 MXene. *Phys. Chem. Chem. Phys.* 18: 5099–5102.
10. Ghidiu, M., Halim, J., Kota, S., Bish, D., Gogotsi, Y., and Barsoum, M.W. 2016. Ion-exchange and cation solvation reactions in Ti_3C_2MXene. *Chem. Mater.* 28: 3507–3514.
11. Verger, L., Natu, V., Ghidiu, M., Barsoum, M.W. 2019. Effect of cationic exchange on the hydration and swelling behavior of $Ti_3C_2T_z$ MXenes. *J. Phys. Chem. C* 123: 20044–20050.
12. Li, J.; Wang, H., and Xiao, X. 2020. Intercalation in two-dimensional transition metal carbides and nitrides (MXenes) toward electrochemical capacitor and beyond. *Energy Environ. Mater.* 3: 306–322.
13. Ahmed, A., Hossain, M.M., Adak, B., and Mukhopadhyay, S. 2020. Recent advances in 2D MXene integrated smart-textile interfaces for multifunctional applications. *Chem. Mater.* 32(24): 10296–10320.
14. Oliveira, F.M., and Gusmão, R. 2020. Recent advances in the electromagnetic interference shielding of 2D materials beyond graphene. *ACS Appl. Electron. Mater.* 2(10): 3048–3071.
15. Xiong, D., Li, X., Bai, Z., and Lu, S. 2018. Recent advances in layered $Ti_3C_2T_x$ MXene for electrochemical energy storage. *Nano Micro. Small* 14: 1703419.
16. Nan, J., Guo, X., Xiao, J., Li, X., Chen, W., Wu, W., Liu, H., Wang, Y., Wu, M., and Wang, G. 2019. Nanoengineering of 2D MXene-based materials for energy storage applications. *Nano small* 17: 1902085.
17. Li, R., Ma, X., Li, J., Cao, J., Gao, H., Li, T., Zhang, X., Wang, L., Zhang, Q., Wang, G., Hou, C., Li, Y., Palacios, T., Lin, Y., Wang, H., and Ling, X. 2021. Flexible and high-performance electrochromic devices enabled by self-assembled 2D TiO_2/MXene heterostructures. *Nat. Commun.* 12: 1587. https://doi.org/10.1038/s41467-021-21852-7.
18. Michael, J., Qifeng, Z., and Danling, W. 2019. Titanium carbide MXene: Synthesis, electrical and optical properties and their applications in sensors and energy storage devices. *Nanomater. Nanotechno.* 9: 1–9. DOI: 10.1177/1847980418824470
19. Jimmy, J., and Kandasubramanian, B. 2019. MXene functionalized polymer composites: Synthesis and applications, *Eur. Polym. J.* 122: 109367.
20. Chen, X., Zhao, Y., Li, L., Wang, Y., Wang, J., Xiong, J., Du, S., Zhang, P., Shi, X., and Yu, J. 2021. MXene/polymer nanocomposites: Preparation, properties, and applications. *Polym. Rev.* 61: 80–115. DOI: 10.1080/15583724.2020.1729179.
21. Gao, L., Li, C., Huang, W., Mei, S., Lin, H., Ou, Q., Zhang, Y., Guo, J., Zhang, F., Xu, S., and Zhang, H. 2020. MXene/polymer membranes: Synthesis, properties, and emerging applications. *Chem. Mater.* 32: 1703–1747.
22. Lei, D., Liu, N., Su, T., Wang, L., Su, J., Zhang, Z., and Gao, Y. 2020. Research progress of MXenes-based wearable pressure sensors. *APL Mater.* 8: 110702. DOI: 10.1063/5.0026984.
23. Carey, M., and Barsoum, M.W. 2021. MXene polymer nanocomposites: A review. *Mater. Today Adv.* 9: 100120.
24. Zwaag, V.D. 2007. *Self-Healing Materials: An Alternative Approach to 20th Centuries of Material Science*, Springer.

25. Wu, D.Y., Meure, S., and Solomon, D. 2008. Self-healing polymeric materials: A review of recent developments. *Prog. Polym. Sci.* 33: 479–522.
26. Ghosh, S.K.2009. *Self-Healing Materials: Fundamentals, Design Strategies, and Applications*, Wiley-VCH. ISBN: 978-3-527-31829-2.
27. Zhai, L., Narkar, A., and Ahn, K. 2019. Self-healing polymers with nanomaterials and nanostructures. *Nano Today* 30: 100826.
28. Toohey, K.S., Sottos, N.R., Lewis, J.A., Moore J.S. and White, S.R. 2007. Self-healing materials with microvascular networks. *Nat. Mater.* 6: 581–585.
29. Tee, B.C-K., Wang, C., Allen, R., and Bao, Z. 2012. An electrically and mechanically self-healing composite with pressure- and flexion-sensitive properties for electronic skin applications. *Nat. Nanotechnol.* 7(12): 825–832.
30. Cordier, P., Tournilhac, F., Soulie- Ziakovic, C., and Leibler, L. 2008. Self-healing and thermo reversible rubber from supramolecular assembly. *Nature*451: 977–980.
31. Burnworth, M., Tang, L., Kumpfer, J.R. Duncan, A.J., Beyer F. L., Fiore G.L., S.J. Rowan, and Weder, C. 2011. Optically healable supramolecular polymers. *Nature* 472: 334–337.
32. VanGemert, G.M.L., Peeters, J.W., Söntjens, S., Janssen, H. M., Anton, W., and Bosman, A.W. 2011. Self-healing supramolecular polymers in action. *Macromol. Chem. Phys.* 213: 234–242.
33. Zhang, A., Yang, L., Lin, Y., Yan, L., Lu, H., and Wang, L. 2013. Self-healing supramolecular elastomers based on the multi-hydrogen bonding of low-molecular polydimethylsiloxanes: Synthesis and characterization. *J. Appl. Polym. Sci.* 129: 2435–2442.
34. Wu, T., and Chen, B. Synthesis of multiwalled carbon nanotube-reinforced polyborosiloxane nanocomposites with mechanically adaptive and self-healing capabilities for flexible conductors. 2016. *ACS Appl. Mater. Interfaces* 8(36): 24071–24078.
35. Thakur, V.K., and Kessler, M.R. 2015. Self-healing polymer nanocomposite materials: A review. *Polymer.* 69: 369–383.
36. Wu, Q., Wei, J., Xu, B., Liu, X., Wang, H., Wang, W., Wang, Q., and Liu, W. 2017. A robust, highly stretchable supramolecular polymer conductive hydrogel with self-healability and thermo-processability. *Sci. Rep.* 7: 41566.
37. Talebian, S., Mehrali, M., Taebnia, N., Pennisi, C.P., Kadumudi, F.B., Foroughi, J., Hasany, M., Nikkhah, M., Akbari, M., Orive, G., and Pirouz, A.D. 2019. Self-healing hydrogels: The next paradigm shift in tissue engineering? *Adv. Sci.* 6(16): 1801664.
38. Zhang, L., Tian, M., and Wu, J. 2016. *Hydrogels with Self-Healing Attribute*, Intech Open. DOI: 10.5772/64138
39. Zhang, J., Chen, H., Zhao, M., Liu, G., and Wu, J. 2020. 2D Nanomaterials for tissue engineering application. *Nano Res.* 13: 2019–2034.
40. Yang, L., Chen, W., Yu, Q., and Liu, B. 2021. Mass production of two-dimensional materials beyond graphene and their applications. *Nano Res.* 14(6): 1583–1597.
41. Zhang, Y.Z., El-Demellawi, J.K., Jiang, Q., Ge, G., Liang, H., Lee, K., Dong, X. 2020. Alshareef, H.N. MXene hydrogels: Fundamentals and applications. *Chem. Soc. Rev.* 49: 7229–7251.
42. Zhang, Y-Z., Lee, K.H., Anjum, D.H., Sougrat, R., Jiang, Q., Kim, H., Alshareef, H. N. 2018. MXenes stretch hydrogel sensor performance to new limits. *Sci. Adv.* 4: eaat0098
43. Zhang, J., Wan, L., Yang Gao, Y., Fang, X., Lu, T., Pan, L., and Xuan, F. 2019. Highly stretchable and self-healable MXene/polyvinyl alcohol hydrogel electrode for wearable capacitive electronic skin. *Adv. Electron. Mater.* 5: 1900285.
44. Zhang, Y., Chen, K., Li, Y., Lan, J., Yan, B., Shi, L., and Ran, R. 2019. High-strength, self-healable, temperature-sensitive, MXene- containing composite hydrogel as a smart compression sensor. *ACS Appl. Mater. Interfaces* 11: 47350–47357.
45. Liao, H., Guo, X., Wan, P., and Yu, G. 2019. Conductive MXene nanocomposite organohydrogel for flexible, healable, low-temperature tolerant strain sensors. *Adv. Funct. Mater.* 29(39):1904507.
46. Wang, Q., Pan, X., Lin, C., Gao, H., Cao, S., Ni, Y., and Ma, X. 2020. Modified $Ti_3C_2T_X$ (MXene) nanosheet-catalyzed self-assembled, anti-aggregated, ultra-stretchable, conductive hydrogels for wearable bioelectronics. *Chem. Eng. J.* 401: 126129. https://doi.org/10.1016/j.cej.2020.126129
47. Lu, Y., Qu, X., Zhao, W., Ren, Y., Si, W., Wang, W., Wang, Q., Huang, W., and Dong, X. 2020. Highly stretchable, elastic, and sensitive MXene-based hydrogel for flexible strain and pressure sensors. *Research*, 2020, Article ID 2038560, https://doi.org/10.34133/2020/2038560

48. Peng, W., Han, L., Huang, H., Xuan, X., Pan, G., Wan, L., Lu, T., Xu, M., and Pan, L. 2020. Direction-aware and ultrafast self-healing dual network hydrogel for flexible electronic skin strain sensor. *J. Mater. Chem. A* 8: 26109–26118. DOI: 10.1039/D0TA08987G.
49. Ge, G., Zhang, Y.-Z., Zhang, W., Yuan, W., El-Demellawi, J.K., Zhang, P., Fabrizio, E.D., Dong, X., and Alshareef, H.N. 2021. $Ti_3C_2T_x$ MXene-activated fast gelation of stretchable and self-healing hydrogels: a molecular approach. *ACS Nano* 15(2): 2698–2706.
50. Yu, T., Lei, X., Zhou, Y., Chen, H. 2021. $Ti_3C_2T_x$ MXenes reinforced PAA/CS hydrogels with self-healing function as flexible supercapacitor electrodes. *Polym. Adv. Technol.* 32(8): 3167–3179. DOI: 10.1002/pat.5329.
51. Li, X., He, L., Li, Y., Chao, M., Li, M., Wan, P., Zhang, L. 2021. Healable, degradable, and conductive MXene nanocomposite hydrogel for multifunctional epidermal sensors. *ACS Nano* 15(4): 7765–7773.
52. Aia, J., Lia, J., Lia, K., Yu. F., and Macde, B. 2021. Highly flexible, self-healable and conductive poly(vinyl alcohol)/$Ti_3C_2T_x$ MXene film and it's application in capacitive de-ionization. *Chem. Eng. J.* 408: 127256. https://doi.org/10.1016/j.cej.2020.127256.
53. Fan, X., Ding, Y., Liu, Y., Liang, J., and Chen, Y. 2019. Plasmonic $Ti_3C_2T_x$ MXene enables highly efficient photothermal conversion for healable and transparent wearable device. *ACS Nano* 13(7): 8124–8134. https://doi.org/10.1021/acsnano.9b03161.
54. Pandey, R. P., Rasool, K., Madhavan, V. E., Aïssa, B., Gogotsi, Y., and Mahmoud, K. A. 2018. Ultrahigh-flux and fouling-resistant membranes based on layered silver/MXene ($Ti_3C_2T_x$) nanosheets. *J. Mater. Chem. A* 6: 3522–3533.
55. Satheeshkumar, E., Makaryan, T., Melikyan, A., Minassian, H., Gogotsi, Y., and Yoshimura, M. 2016. One-step solution processing of Ag, Au and Pd@MXene hybrids for SERS. *Sci. Rep.* 6: 32049.
56. Han, X.-W., Meng, X.-Z., Zhang, J., Wang, J.-X., Huang, H.-F., Zeng, X.-F., and Chen, J.-F. 2016. Ultrafast synthesis of silver nanoparticle decorated graphene oxide by a rotating packed bed reactor. *Ind. Eng. Chem. Res.* 55: 11622–11630.
57. Wu, X., Liao, H., Ma, D., Chao, M., Wang, Y., Jia, X., Wan, P., and Zhang, L. 2020. A wearable, self-adhesive, long-lastingly moist and healable epidermal sensor assembled from conductive MXene nanocomposites. *J. Mater. Chem. C* 8: 1788–1795.
58. Zheng, H., Wang, S., Cheng, F., He, X., Liu, Z., Wang, W., Zhou, L., and Zhang, Q. 2021. Bioactive anti-inflammatory, antibacterial, conductive multifunctional scaffold based on MXene@CeO_2 nanocomposites for infection-impaired skin multimodal therapy. *Chem. Eng. J.* 424: 130148. https://doi.org/10.1016/j.cej.2021.130148,
59. Zhao, H., Ding, J., Zhou, M., and Yu, H. 2021. Air-stable titanium carbide MXene nanosheets for corrosion protection. *ACS Appl. Nano Mater.* 4(3): 3075–3086.
60. Zhang, K., Sun, J., Song, J., Gao, C., Wang, Z., Song, C., Wu, Y., and Liu, Y. 2020. Self-healing Ti_3C_2 MXenes/PDMS supramolecular elastomer based on small biomolecules modification for wearable sensor. *ACS Appl. Mater. Interfaces* 12(40): 45306–45314.
61. Guo, Q., Zhang, X., Zhao, F., Song, Q., Su, G., Tan, Y., Tao, Q., Zhou, T., Yu, Y., Zhou, Z., and Lu, C. 2020. Protein-inspired self-healable Ti_3C_2 MXenes/rubber-based supramolecular elastomer for intelligent sensing. *ACS Nano* 14(3): 2788–2797.
62. Li, X., Ma, Y., Shen, P., Zhang, C., Yan, J., Xia, Y., Luo, S., and Gao, Y. 2020. Highly-healable micro-supercapacitors with size-dependent 2D MXene. *Chem. Electro. Chem.* 7(3): 821–829.
63. Yue, Y., Liu, N., Ma, Y., Wang, S., Liu, W., Luo, C., Zhang, H., Cheng, F., Rao, J., Hu, X., Su, J., and Gao, Y. 2018. Highly self-healable 3D microsupercapacitor with MXene–graphene composite aerogel. *ACS Nano* 12: 4224–4232.
64. Lu, J., Zhang, Y., Tao, Y., Wang, B., Cheng, W., Jie, G., Song, L., and Hu, Y. 2021. Self-healable castor oil-based waterborne polyurethane/MXene film with outstanding electromagnetic interference shielding effectiveness and excellent shape memory performance. *J. Colloid Interface Sci.* 588: 164–174.
65. Zhao, B., Wang, S., Yu, Q., Wang, Q., Wang, M., Ni, T., Ruan, L., and Zeng, W. 2021. A flexible, heat-resistant and self-healable "rocking-chair" zinc ion microbattery based on MXene-TiS_2 (de)intercalation anode. *J. Power Sources* 504: 230076.
66. Wu, Y., Wu, L., Yao, W., Jiang, B., Wu, J., Chen, Y., Chen, X., Zhan, Q., Zhang, G., and Pan, F. 2021. Improved corrosion resistance of AZ31 Mg alloy coated with MXenes/MgAl-LDHs composite layer modified with yttrium. *Electrochimica Acta* 374: 137913

67. Cai, M., Fan, X., Yan, H., Li, Y., Song, S., Li, W., Li, H., Lu, Z., and Zhu, M.H. 2021. *In situ* assemble $Ti_3C_2T_x$ MXene@MgAl-LDH heterostructure towards anticorrosion and anti-wear application. *Chem. Eng. J.* 419: 130050.
68. Zou, Y., Fang, L., Chen, T., Sun, M., Lu, C., and Xu, Z. 2018. Near-infrared light and solar light activated self-healing epoxy coating having enhanced properties using MXene flakes as multifunctional fillers. *Polymers* 10(5): 474.
69. Kim, H., and Alshareef, H.N. 2020. MXetronics: MXene-enabled electronic and photonic devices. *ACS Materials Lett.* 2: 55–70.
70. Lee, H., Le, T.A., Tran, N.Q., and Hong, Y. 2019. Intertwined titanium carbide MXene within 3D tangled polypyrrole nanowires matrix for enhanced supercapacitor performances. *Chem. Eur. J.* 25(4): 1037–1043.
71. Lei, Y., Zhao, W., Zhang, Y., Jiang, Q., He, J.-H., Baeumner, A.J., Wolfbeis, O. S., Wang, Z. L., Salama, K.N., and Alshareef, H.N. 2019. A MXene-based wearable biosensor system for high-performance in vitro perspiration analysis. *Small* 15: 1901190.
72. George, S.M., and Kandasubramanian, B. 2020. Advancements in MXene-polymer composites for various biomedical applications. *Ceram* 46: 8522–8535.
73. Yang, X., Feng, M., Xia, J., Zhang, F., and Wang, Z. 2020. An electrochemical biosensor based on AuNPs/Ti3C2 MXene three-dimensional nanocomposite for microRNA-155 detection by exonuclease III-aided cascade target recycling. *J. Electroanal. Chem.* 878: 114669.
74. Chen, K., Chen, Y., Deng, Q., Jeong, S.-H., Jang, T.-S., Du, S., Kim, H.-E., Huang, Q., and Han, C.-M. 2018. Strong and biocompatible poly(lactic acid) membrane enhanced by $Ti_3C_2T_z$ (MXene) nanosheets for guided bone regeneration. *Mater. Lett.* 229: 114–117.
75. Yan, H., Li, W., Li, H., Fan, X., Zhu M. 2019. Ti_3C_2 MXene nanosheets toward high-performance corrosion inhibitor for epoxy coating. *Prog. Org. Coat.* 135: 156–167.
76. Wang, B., Lai, X., Li, H., Jiang, C., Gao, J., and Zeng, X. 2021. Multifunctional MXene/chitosan-coated cotton fabric for intelligent fire protection. *ACS Appl. Mater. Interfaces* 13(19): 23020–23029.

10 Microwave Absorption and EMI Shielding Performance of MXene-Filled Polymer Nanocomposites

Bhavya Bhadran, Avinash R. Pai, and Sabu Thomas

CONTENTS

10.1 INTRODUCTION

Polymers are revolutionary materials that can be synthesized, processed, and designed to meet a broad range of applications. The science and technology of these fascinating materials are running their course in new directions. Very recently, a self-healing polymer based on aminopropylmethacrylamide that can mature, restore, and strengthen using carbon fixation was designed by a group of researchers.[1] In other words, polymers are synthetic materials that are capable of mimicking the properties of nature through proper modifications. With their lightweight nature, self-healing capabilities, and high thermal stability, polymers can satisfy the fast-growing technological demands and stringent requirements of space missions and stealth technologies, among others, to a greater extent. But the main drawback is that they possess very low tensile strength and are inferior to some properties like thermal and chemical stability, conductivity, and so on, which limit their application. So the search for high-performance polymers ended up in nanoparticles because of their outstanding mechanical, thermal, electrical, magnetic, optical, rheological, flame retardant, and biomedical properties.[2,3,4,5,6,7] Also, the size, shape, and geometry of nanoparticles play a crucial part in deciding the overall properties of a composite.

DOI: 10.1201/9781003164975-10

Among the 1D (one-dimensional), 2D, and 3D nanoparticles, the intriguing physicochemical properties of layered 2D nanoparticles, such as graphene, boron nitride, nanoclays, and others, have received significant interest because of their high aspect ratio, surface area, conductivity, and so on. The latest addition to this 2D family is the MXene– transition metal carbides and carbonitrides. They are fabricated by selective etching of "A" layers from MAX phases ($M_{n+1}AX_n$), where M describes early transition metals, A constitutes group III or IV elements, and X corresponds to carbon, nitrogen, or both and n = 1, 2, or 3 (Figure 10.2a). The general formula used to represent MXenes is $M_{n+1}X_nT_x$, where T_x symbolizes the surface functionalities like hydroxyl group (OH), O, halogens like fluorine or chlorine, and so on. The first type of MXene ($Ti_3C_2T_x$) was discovered in 2011 by Yuri Gogotsi, and it was named after graphene.

The high metallic conductivities (~6000–8000 S/cm) and hydrophilic nature make them superior to other 2D materials like graphene. The large negative–zeta potential value of MXenes enables them to form stable colloidal dispersions in water. MXenes are of two types—the first type belongs to out-of-plane ordered double transition-metal MXenes ($M_2'M''X_{2A}$) and the second is in-plane double transition-metal $(M'_{2/3}M''_{1/3})_2X$ type.[8] More than 30 compositions of MXenes are known to date.[9,10] Its inherent electrical conductivity, active functionalized surfaces, dielectric and magnetic properties, and the like make them tunable to a myriad of applications, comprising energy storage and conversion, hydrogen storage, catalysis, water purification, fabrication of biosensors, optical and transparent conductive electrodes, piezoelectric materials, and others. The importance of this 2D material in electronic applications can be clearly understood from the bar chart indicating publication statistics (data collected from Web of Science) given in Figure 10.1. According to the data collected, there has been a huge increase in the number of research outputs related to applications in the sphere of electronics since 2013. It should be noted that although a lot of research is going on regarding MXenes, the work based on radiation shielding and microwave absorption properties of MXenes is still in its infancy stage.

One of the major concerns now faced by humankind is exposure to high-energy electromagnetic radiations arising from electronic devices as a result of advanced technologies. The forthcoming 5G telecommunication system, the growing popularity of portable and wearable devices,

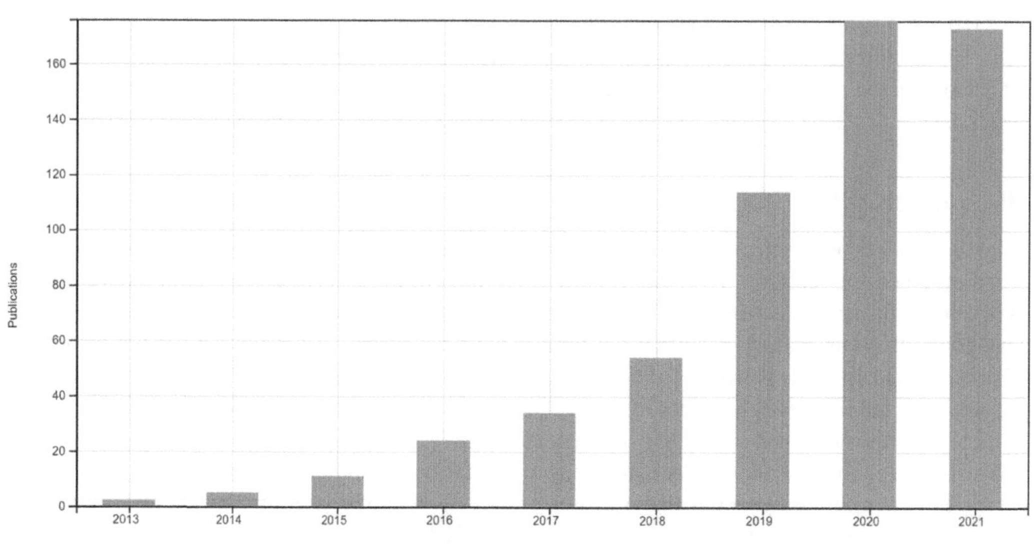

FIGURE 10.1 Number of publications in MXenes in the field of electronics versus year.

Source: Data collected from Web of Science.

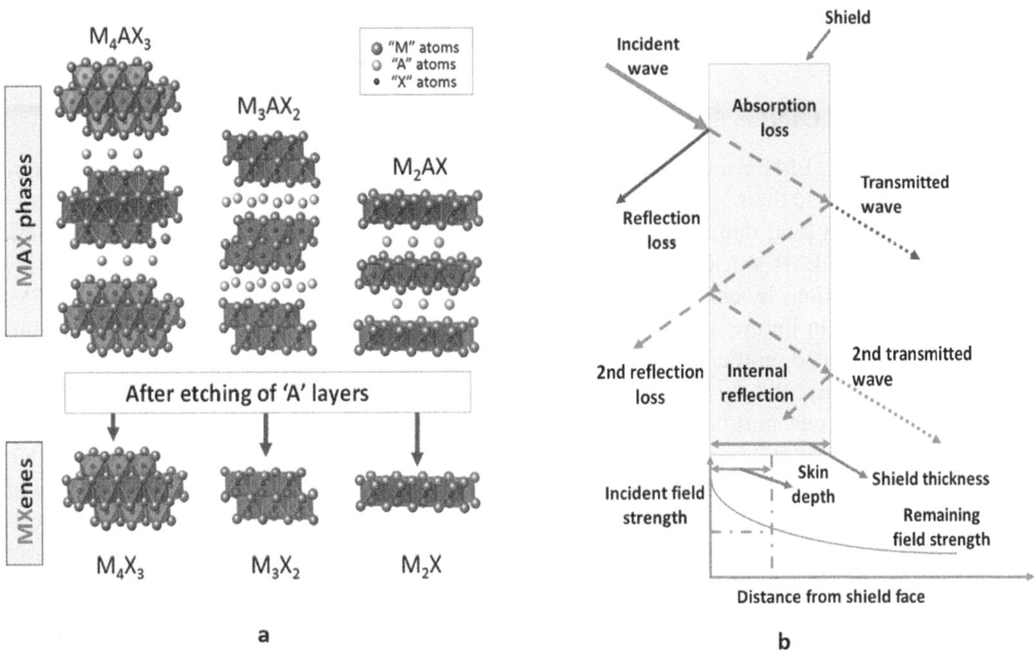

FIGURE 10.2 (a) MAX phase to MXene by selective etching of A layer. (b) Mechanism of EMI shielding.

and the like generate harmful electromagnetic pollution, which adversely affects human health as well as the operation of other communication and electronic devices. Thus, electromagnetic interference (EMI) is a perilous problem that arises from electrostatic discharge (ESD), electronic gadgets, lightning, solar flares, and so on. The only way to reduce the effect of these radiations is to fabricate lightweight, EMI shielding, and microwave-absorbing materials that can operate at a broader bandwidth. Materials with strong conductivity, better magnetic permeability, and dielectric constant are used for EMI shielding purposes. A great deal of effort has been put to diminish EMI using metals, dielectric/magnetic materials, and so on. But the lightweight, flexible, and corrosion-resistant nature of polymers make them more attractive for manufacturing shielding components. Polymer nanocomposites with conducting fillers provide effective shielding by both absorption and reflection methods by enhancing the δ value or skin-depth value (the distance that electromagnetic radiation can penetrate at a certain high frequency). Thus, conducting nanomaterials like graphene, silver nanowires, carbon nanoribbons, carbon nanofibers, carbon nanotubes (CNTs), and the like are incorporated into polymer matrices to develop high-performance shielding polymer nanocomposites. EMI shielding devices find application in various domains like telecommunication, military (in radars), aerospace, biomedical science, etc.[11]

Today, soft, flexible, and stretchable electronic devices are gaining significant attention as they constitute the next generation of smart devices. The ability to bend, compress, and twist makes them suitable for a myriad of applications (wearable electronics, soft robotics, etc.) that cannot be accomplished using conventional rigid, planar devices. Polydimethylsiloxanes (PDMS), polyurethanes (PU), liquid crystalline elastomers (LCE), styrene butadiene rubber (SBR), and the like are some of the stretchable elastomers used as soft substrates for manufacturing electronic devices.[12] The flexible nature of PDMS, along with excellent mechanical properties, thermal stability, oxidation resistance, and low glass transition temperature (T_g), makes it a potential candidate for fabricating materials that can meet high-end requirements.[13] Literature reports suggest that block copolymers based on polyimide and polydimethylsiloxanes can also be used for low-earth-orbit applications.[14]

LCEs are considered to be excellent soft substrates for innumerable applications in the areas like soft robotics, electronics, deployable devices, and so on.[15]

10.2 EMI SHIELDING

As discussed earlier, EMI is an unwanted electromagnetic signal that disturbs electronic and communication devices and their performance. They sometimes cause sudden power fluctuations, which ultimately results in total damage to the devices. The process of blocking these electromagnetic radiations is called EMI shielding.[16] Thus, the shielding materials function as a barrier against these harmful radiations by either absorbing, reflecting, or subduing them. The mechanism of EMI shielding[17] is given in Figure 10.2b. It is stated that while the mechanism of absorption shielding is completely grounded on the presence of electric and magnetic dipoles, reflection is governed by electrical conductivity, and the subduing or multiple internal reflections of the radiations can be due to irregularities, porous, multilayered structures, and so on in the shield. If the shielding material is electrically conductive, it contains mobile charge carriers in the form of electrons or holes. When an electromagnetic wave strikes the surface of the shield, these charge carriers come in contact with the wave and cause an opposing electromagnetic field due to the redistribution of electrons or holes in the shield. This opposing electromagnetic field formed within the shield cancels the external electromagnetic field and thus protects the device from harmful radiation. If the shielding material displays better dielectric constant and magnetic permeability, they contain electric and magnetic dipoles that block the passage of external magnetic field through the shield and thus act as a protective barrier. Sometimes, induced eddy currents in the shield also increase the shielding capacity of the material. Shielding due to multiple internal reflections is predominantly due to the scattering of electromagnetic waves from the cavities or porous structures or multilayers and the like present in the shield. The shielding efficiency is calculated as a decrease in the amount of incident power on moving through the shield. Most commonly shielding materials were made of dielectric, conducting, or magnetic materials. Due to their high conductivity, metals are often preferred for making EMI shielding materials, but their density, low flexibility, and corrosive nature limit their application in fabricating lightweight structures. These limitations can be solved using polymer-based nanocomposites because of their high flexibility, mechanical strength, tailor-made properties, and lightweight nature. An ideal EMI shielding device is supposed to show minimum reflection with maximum absorption of EM waves, which is still considered the most challenging task.

10.2.1 Mechanism of EMI Shielding

In the case of polymer-based nanocomposites, the shielding mechanism relies on the amount as well as proper dispersion of nanoparticles in the polymer matrices. Reflection or absorption of electromagnetic radiations depends on whether the nanoparticle loading is above or below the percolation threshold value. To be more precise, it is the value at which the electrical conductivity of a material or polymer displays a steep increase at a particular volume fraction of the conductive nanoparticle. It is well known that polymer nanocomposites with high electrical conductivity display superior shielding efficiency. But most commonly, a higher amount of conductive fillers must be loaded into polymer matrices to achieve high electrical conductivity and thus better shielding efficiency. That is, the percolation threshold value will be very high. As a result, the weight of the composite increases, which compromises its flexibility as well as processability and application. So polymer nanocomposites with a low percolation threshold value are considered to be more significant to fabricate suitable lightweight, flexible, and stretchable EMI shielding devices.

MXenes, the flagship of 2D nanomaterials, in this context, show extremely low percolation threshold values when compared to graphene, carbon nanotubes, and others. Due to its inherent electrical conductivity, MXene is considered as the future material for applications like 5G, artificial intelligence, soft robotics, and more.

10.2.1.1 Shielding Efficiency

Shielding efficiency is expressed in decibels (dB), and it analyses how effectively a sample material blocks the electromagnetic energy of a definite frequency when sweeping through it. Different interactions arising when electromagnetic radiation passes through a material are depicted in Figure 10.2b. When these radiations strike on the surface of a sample material, some amount of the incident power (P_I) gets reflected (P_R), whereas a certain amount of it gets absorbed and released in form of energy, while some portion of the power gets transmitted (P_T) across the surface of the material. Thus absorbance, reflection, and multiple internal reflections are the major three criteria that determine the shielding efficiency of a material. Shielding efficiency (SE) is the function of the logarithmic ratio of the incident and transmitted power (P). SE can be calculated in terms of incident and transmitted electric (E) fields and magnetic (H) fields also.

$$SE_T = 10\log\frac{P_I}{P_T} = 20\log\frac{E_I}{E_T} = 20\log\frac{H_I}{H_T}$$

The overall EMI shielding effectiveness (SE_T) of a material can be calculated by employing the equation

$$SE_T = SE_R + SE_A + SE_M,$$

where SE_R, SE_A, SE_M represents shielding effectiveness due to reflection, absorption, and multiple internal reflections, respectively. When SE > 10 dB, shielding efficiency due to multiple internal reflections is not counted. R and T are the coefficients of reflection and transmission, respectively.

10.2.1.2 Absorption Loss

When an electromagnetic wave passes through a medium, its amplitude decreases exponentially. This is due to the fact that the currents generated in the medium give rise to ohmic losses and result in the heating of the sample. In this case, E_1 and H_1 can be calculated by employing the equation

$$E_1 = E_o e^{-t/\delta} \text{ and } H_1 = H_o e^{-t/\delta}.$$ [18]

Thus, skin depth can be defined as the distance needed by the wave to be attenuated to or the depth at which the field strength falls to 1/e or 37% of the incident value. So SE_A can be calculated as

$$SE_A\left(dB\right) = 20\frac{t}{\delta}loge = 8.68\frac{t}{\delta},$$

Where t gives the thickness of the shielding material in millimeters; f represents the frequency in megahertz, μ expresses the relative permeability, and σ gives the conductivity relative to the metal Cu. The δ value, which is otherwise called skin depth in the preceding equation, can be calculated as

$$\delta = \sqrt{2/\sigma\omega\mu'}.$$

10.2.1.3 Reflection Loss

The reflection loss is associated with the relative discrepancy between the incident wave and the surface impedance of the shield. The magnitude of the reflection loss can be analyzed by using the following equations:

$$R_E = K_1 10\log\left(\frac{\sigma}{f^3 r^2 \mu}\right)$$

$$R_H = K_2 10\log\left(\frac{fr^2\sigma}{\mu}\right)$$

$$R_P = K_3 10\log\left(\frac{f\mu}{\sigma}\right),$$

where R_E (dB) represents the reflection losses for the electric field; R_H, for the magnetic field; and R_P, for the plane-wave field. The relative conductivity with respect to the metal Cu is denoted by the term σ. The terms f, μ, and r in the preceding equations symbolize frequency in Hertz, relative permeability with respect to free space, and the distance from the source to the shielding material (m), respectively.

10.2.1.4 Multiple Reflections

The SE due to multiple internal reflections can be positive or negative (usually it is found to be negative), and it can be neglected when SE_A is greater than 10 decibels. It is taken into account only when the frequency is very low ($< \approx 20$ kHz) and the metals used were thin. SE_M can be measured by employing the equation

$$SE_M = -20\log\,(1 - e^{-2t/\delta}).$$

The relative permittivity and permeability of the materials and their conductivity, as well as the polarization loss of fillers, size, shape, and morphology of the nanoparticles, the thickness of the sample, and others, are some of the crucial factors that affect the EMI SE of a material.

10.2.2 Measurement Techniques

EMI SE can be measured using network analyzer instruments. They are of two types, namely, scalar network analyzers (SNAs) and vector network analyzers (VNAs). The SNAs measure only signal amplitudes whereas the VNAs measure signal magnitude along with different phases. So VNAs are used to analyze shielding efficiency under far-field conditions. VNA consists of two ports, namely, S1 and S2. S1 and S2 indicate the incident and transmitted waves as complex scattering S parameters, that is, S_{11} or S_{22} and S_{21} or S_{12}, respectively. S_{11} represents the forward reflection coefficient, S_{22} gives an account of the reverse reflection coefficient, S_{12} gives the forward transmission coefficient, and S_{21} represents the backward transmission coefficient. Numerous methods like short circuit line (SCL), the National Institute of Standards and Technology (NIST) iterative, new non-iterative, delta function, transmission line theory, and others are used to calculate complex permittivity and permeability values by employing these scattering parameters. Among them, the Weir and Nicolson method[19,20] is widely accepted for calculating the permeability and permittivity in the frequency range 50 MHz to 18 GHz.

Reflection and transmission coefficients (S parameters) are determined by introducing the material to be tested in a specimen holder associated with the waveguide flanges of the VNA (Figure 10.3). In the rectangular waveguide measurements, the sample size varies with the frequency range. For example, 22.8 ×10 mm^2 is employed for X band (8.2–12.4 GHz) and 15.8 ×7.9 mm^2 is used for P band (12.4–18 GHz).

10.3 DESIGNING A SHIELDING MATERIAL

An ideal radiation shielding material must be designed in such a way that it should possess an appropriate structural geometry, and must exhibit better magnetic permeability (μ), electrical conductivity (σ), and dielectric permittivity (ε). The major concern is that shield must be conducting in nature

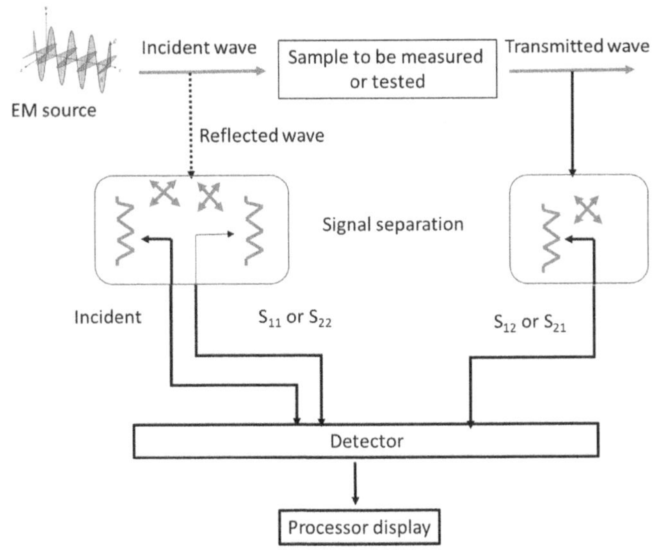

FIGURE 10.3 Internal block diagram of a spectrum analyzer.

since the first step in the EMI shielding mechanism is the reflection of incident waves when it strikes the surface of the shield. So the material to be used for shielding must contain mobile charge carriers like electrons or holes that can interact with the incident radiations to create ohmic losses in the material. The second important factor to be taken into account is that it must be magnetic in nature; that is, it must contain electric or magnetic dipoles to absorb (absorption mechanism) the incoming electromagnetic radiations. If the shielding material is moderately conducting in nature, then the dielectric permittivity has been taken into consideration since such materials exhibit dynamic dielectric and/or magnetic loss when electromagnetic waves strike on their surface.

Most commonly metals were used as EMI shielding materials, but because they are prone to corrosion and due to their heavy nature, they cannot be employed for certain applications. The importance of polymers comes at this juncture. They are lightweight in nature, corrosion-resistant, and able to exhibit multifunctional properties when properly tailored. Since shielding is associated with high electrical conductivity, the polymers should be conductive in nature. In order to make the polymers conductive, conductive fillers like CNTs, carbon nanofibers, graphene, carbon black, and others are introduced into the polymers by adopting different processing techniques like melt mixing, spin coating, solvent casting, *in situ* polymerization, and the like. A combination of fillers or hybrid fillers can also be used to make the polymers conductive. Coating the nanoparticles with metals, the addition of metal oxide nanoparticles like iron oxide, among others, are also alternative methods adopted to enhance the conductivity of polymers.

Intrinsically conducting polymers (ICPs) with microwave non-transparency, inherent conductivity, and flexibility are good alternatives to make better EMI shielding devices.[21,22,23] The mixing of ICPs with conductive fillers has been proved to show better conductivity and shielding efficiency when compared to nonconducting polymers with conductive fillers. Moreover, ICPs are easily processable and can be compatibilized with other polymers according to the nature of the application. Designing polymer nanocomposites with complex architectures like foams, multilayers, hybrid multilayers or Salisbury screens, honeycomb structures, and others is also considered a good strategy to fabricate radiation shielding devices[24,25] By using this strategic approach, the weight of devices, as well as the reflection loss, can be reduced. They can act as better microwave absorbers also. In the case of foam-like structures, the influence of air inside the cavities helps in reducing the reflection loss to a great extent, thereby decreasing the permittivity values.

10.4 MXene-BASED POLYMER COMPOSITES FOR EMI SHIELDING AND MICROWAVE ABSORPTION

According to literature reports, 2D materials with strong microwave attenuation capacity can act as good radiation absorbers. Thus, MXene, with its excellent conductivity, diverse surface chemistry, and better Young's modulus, can be used to fabricate better radiation shielding materials. A comprehensive literature review reveals that only 7% of works were communicated on microwave attenuation and radiation shielding studies in MXenes.[26] Out of 28 types of MXenes, only Ti- and Mo-based MXenes (Ti_2CT_x, $Ti_3C_2T_x$, $Mo_2TiC_2T_x$, $Mo_2Ti_2C_3T_x$) were utilized for EMI studies hitherto. Among them, most of the works were based on $Ti_3C_2T_x$. This may be because of the high inherent conductivity and semiconducting nature of $Ti_3C_2T_x$ and $Mo_2TiC_2T_x$, respectively. The active surface functionalities, strong M-A bonds in MAX phases, etching conditions, polarization loss, and so on are some of the challenges that have to be addressed in fabricating MXene-based shielding materials. MXenes surface can be engineered with graphene and magnetite nanoparticles to get better EMI shielding properties. Very recently, a smart EMI shielding actuation system based on Janus MXene/grapheneoxide film was developed by Li et al. using a simple vacuum filtration method.[27] According to them, the developed flexible Janus film shows an electromagnetic shielding efficiency of 40 dB to 55 dB, with the thickness of MXene layers ranging from 1.5 to 4.5 micrometers. They had developed a smart actuator system based on this Janus film to regulate the transmission of microwave signals in which the mechanical grippers were employed for the lifting and releasing of polymer foam (Figures 10.4a and 10.4b). The

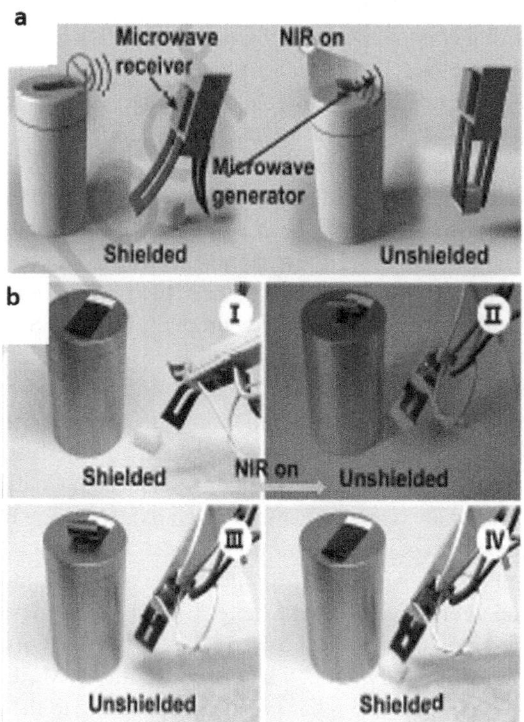

FIGURE 10.4 (a) Figure shows Stimuli-responsive electromagnetic interference shielding material made of a smart Janus film actuator. (b) Flexible Janus film used to regulate the lifting and releasing of a polymer foam.

Source: Reprinted from Smart MXene-Based Janus films with multi-responsive actuation capability and high electromagnetic interference shielding performances, Vol number 175, Lulu Li, Sai Zhao, Xin-Jie Luo, Hao-Bin Zhang, Zhong-Zhen Yu. Copyright (2021), Reproduced with permission from Elsevier.

interfacial morphology also plays a pivotal role in deciding the overall properties of the system, In particular, nanocomposites with a characteristic interconnected network of fillers in the matrix can provide fascinating properties.

10.4.1 MXene-Based Aerogels, Foams, and Hierarchical Architectures for EMI Shielding Applications

Recently, mechanically robust, low-density polymer composites, namely, aerogels, have gained significant interest because of their exceptional properties. Aerogels are highly porous, lightweight solid materials with a continuous 3D network structure processed by replacing the solvent in a gel with air without disrupting the network structure of the gel body.[28] They can act as excellent thermal insulators and can provide better shielding efficiency as per the end requirement.[29] Their low density, high surface area, and porosity make them suitable for fabricating materials for space missions like spacesuits, dust particle absorbers, and the like.[30,31] MXene aerogels are a new class of materials with a 3D network structure lighter than air.[32] The literature reports indicate that 3D MXene architectures are good alternatives for electrochemical energy storage and conversion applications.[33] High mechanical strength, electrical and thermal conductivity,[34] and adsorption capacity make them potential candidates for making radiation shielding devices, fuel cells, supercapacitors, and more.

Recently, MXene/reduced Graphene oxide (rGO) aerogel was used to fabricate piezoresistive sensors for wearable device applications.[35] According to Bian et al.[36] foam-like structures can absorb more electromagnetic waves due to impedance matching, multiple reflections, and scattering phenomena occurring within the pores. Based on this, Wang et al. synthesized a lightweight 3D porous MXene/rGO aerogel via a hydrothermal-assisted freeze-drying method with minimum reflection loss (31.2 dB at 8.2 GHz) for broadband microwave absorption applications.[37] Liu et al. synthesized a hydrophobic, freestanding, flexible foam made of $Ti_3C_2T_x$ and compared its electromagnetic shielding efficiency with respect to the corresponding MXene film.[38] They found that hydrophobic foam exhibited a much higher shielding efficiency of 70 dB when compared to MXene film (53 dB). According to them, the significant enhancement in the shielding efficiency of the foam is due to the presence of a porous network. The freeze-drying method was adopted to prepare a conductive, fire-resistant $Ti_3C_2T_x$ aerogel based on cellulose nanofiber and ammonium polyphosphate. The aerogel material also displayed a better shielding efficiency of 55 decibels. The shielding mechanism, in this case, was found to be absorption-driven.[39]

According to Jiang et al.,[40] conducting polymer composites consisting of segregated networks, double percolation structures, layer-by-layer assembly, foam-like structures or aerogels, and others can increase the electromagnetic absorption to a greater extent. As discussed earlier, foam-like structures exhibit absorption-dominated shielding attenuation. The pores in the foam help to absorb a major amount of radiation falling on its surface and thereby act as better shielding materials. Recently, Liu et al. synthesized a super-elastic, lightweight, highly thermal resistant, and conductive (4.0 Sm^{-1}), MXene aerogel with polyimide as the matrix, which accounts for almost 80% of reversible compressibility and excellent microwave absorption property.[41] Wang et al. synthesized highly conductive cellulose nanofiber– and carbon foam–impregnated 3D MXene ($Ti_3C_2T_x$)/epoxy nanocomposites with an EMI shielding efficiency of 74 dB and 46 dB, respectively. According to them, the absorption factor contributes more to the shielding mechanism due to the highly conductive porous networks.[42,43] A study conducted by Zhao et al. stated that MXene/rGO aerogels can also be used to fabricate epoxy nanocomposites with high electrical conductivity and better EMI shielding effectiveness (>50 dB).[44] Similarly, Luo et al. prepared a natural rubber (NR) latex–$Ti_3C_2T_x$MXene composite with an interconnected interfacial structure, exhibiting good electrical conductivity and a better EMI shielding effectiveness of 53 dB (Figures 10.5a and 10.5b) in the X band.[45] According to them, the composite displayed better shielding efficiency due to the formation of a continuous conductive network as a result of the MXene nanoparticles and NR latex particles. The increase in the shielding efficiency with respect to the increase in the amount of MXene is due to the inherent

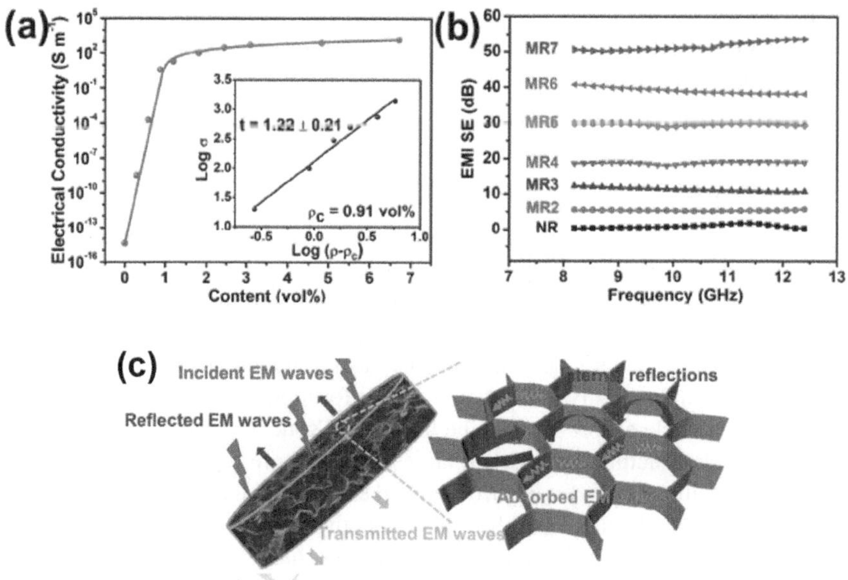

FIGURE 10.5 (a) Electrical conductivity of MXene-NR composites. (b) SE of MXene–NR composites at various filler loadings. (c) Shielding mechanism in a continuous MXene network.

Source: Reprinted from Flexible, stretchable and electrically conductive MXene/NR nanocomposite films for efficient electromagnetic interference shielding, Vol number 182, Jia-Qi Luo, Sai Zhao, Hao-Bin Zhang, Zhiming Deng, LuluLi, Zhong-Zhen Yu. Copyright (2021), Reproduced with permission from Elsevier.

electrical conductivity of $Ti_3C_2T_x$, the presence of surface functionalities, and the formation of an interconnected porous network between the latex particle and MXene. Here the mechanism was found to be shielding due to absorption.

Literature reports propose that adopting several strategies like designing unique hierarchical structures can improve the shielding effectiveness of the composite. A study conducted by Liang et al. reports that MXene aerogel/porous biocarbon composites with "mortar–brick" microstructures can function as excellent EMI shielding materials.[46] Likewise, Liu et al. developed a multifunctional nacre-mimetic composite of $Ti_3C_2T_x$MXene/polyurethane with better mechanical properties and radiation shielding efficiency.[47] Recently, Lin et al. illustrated an ion-diffusion-induced gelation method for fabricating scalable MXene foams with desired shapes and properties for terahertz shielding purposes.[48] Wu et al. fabricated a compressible, ultra-high EMI shielding MXene/sodium alginate aerogel coated with PDMS by employing directional-freezing accompanied with freeze-drying method (Figure 10.6). The foam exhibited excellent conductivity with a shielding efficiency of 53.9 dB (X-band) at 6.1 wt% of $Ti_3C_2T_x$. The polydimethylsiloxane-coated MXene foam, when used as an EMI shielding gasket, continued to display an efficiency of 48.2 decibels even after going through 500 compression-release cycles.[49]

The layer-by-layer assembly of hybrid nanoparticle polymer systems with core layer and skin layers can be used to fabricate material with better EMI shielding, electrical insulation, and high thermal conductivity.[50] Hence, Weng et al. prepared a semitransparent composite made of MXene–CNT with polyvinyl alcohol and polystyrene sulfonate (PSS) using layer-by-layer assembly architecture with excellent shielding efficiency and conductivity.[51] Hu et al.[52] prepared an aramid nanofiber $Ti_3C_2T_x$ composite that can withstand extreme environmental conditions using layer-by-layer self-assembly. The composite showed better mechanical properties and can function as a sensor as well as a shielding device. The specific SE was reported as 28,190 dB cm^2/g at 40 wt.% of $Ti_3C_2T_x$. The thickness of the composite was 9 μm.

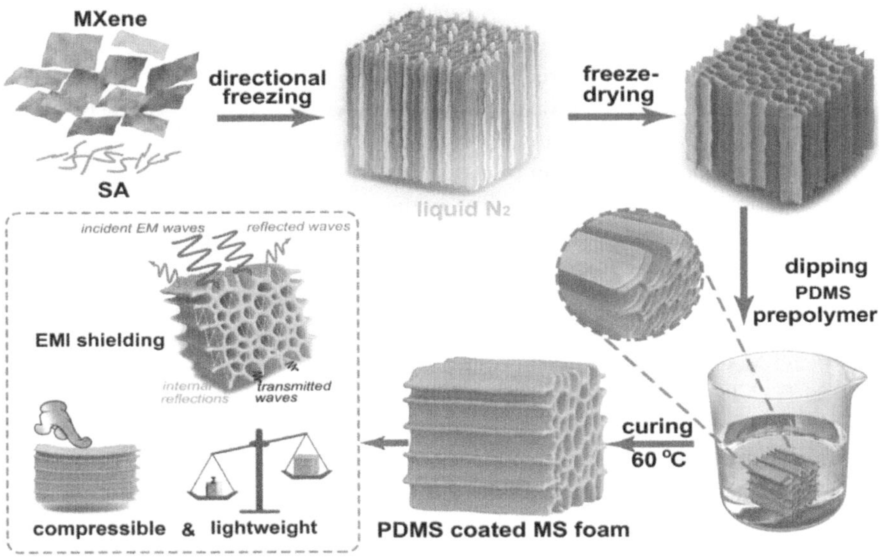

FIGURE 10.6 Compressible and conductive PDMS-coated MXene foams from MXene–sodium alginate aerogels.

Source: Reprinted from Compressible, durable and conductive polydimethylsiloxane-coated MXene foams for high-performance electromagnetic interference shielding, Vol number 381, Xinyu Wu, Bingyong Han, Hao-Bin Zhang, Xi Xie, Tingxiang Tu, Yu Zhang, Yang Dai, Rui Yang, Zhong-Zhen Yu. Copyright (2021), Reproduced with permission from Elsevier.

Very recently, a novel self-healing EMI shielding sponge based on MXene/melamine/PU was fabricated by Ma et al.[53] They took porous melamine sponge and impregnated it with MXene nanoparticles through a dip-coating method so that the MXene will get deposited into the sponge skeleton and can function as shielding capsules covering the porous structure to form membranes. PU provides the self-healing ability to the whole structure. During the shielding process, multiple internal reflections happen at the interfaces of these porewalls, making the MXene/melamine/PU sponge an excellent radiation shielding material. The spongy material exhibited a maximum shielding efficiency of 90.49 dB at only 0.82 vol% of MXene.

Over the last few years, additive manufacturing (AM) has grabbed much attention because it can be used to prepare complex microstructures, which are considered to be difficult to accomplish with usual polymeric processing methods like injection molding, extrusion molding, foaming, and the like. The major advantage of additive manufacturing is that it can be used for the development of scalable batch-size productions. Among the different AM techniques, extrusion-based 3D printing or direct ink writing is a cost-effective and facile route to design 3D structures or miniaturized prototypic models for high-performance applications. Moreover, it can be used to design honeycomb-like microstructures based on 2D functional nanomaterials[54] having diverse surface chemistry to fabricate flexible and stretchable electronic devices.[55],[56],[57] Recently, Orangi et al. reported the fabrication of high-energy-density micro-supercapacitors based on additive-free $Ti_3C_2T_x$MXene ink using 3D printing technology.[58] Literature reports also account for the fact that ordered hierarchical microstructures can provide better mechanical stability to the materials since they can withstand the load due to effective strain transfer.

10.4.2 Thermoplastic MXene Nanocomposites

Thermoplastic nanocomposites constitute a major part of the industry and play an important role in the designing and manufacturing of radiation shielding materials because of their lightweight and flexible nature. Recent research work by Sun et al.[59] describes the synthesis of a polystyrene/MXene nanocomposite with excellent EMI SE. They synthesized $Ti_3C_2T_x$ from Ti_3AlC_2 using LiF and HCl as etching agents and prepared a corresponding MXene polystyrene nanocomposite by electrostatic self-assembly of $Ti_3C_2T_x$ on polystyrene microspheres accompanied by high-pressure compression molding. As a result, a highly conductive network is formed between the MXene nanoparticles and polymer networks, imparting a better shielding property (62 dB) at a very low MXene concentration of 1.90 vol%. According to them, the composite displayed better shielding properties due to the formation of a highly compact conductive network (1081 S/m) with $Ti_3C_2T_x$ at the interfaces of polystyrene microspheres, making the shielding mechanism more absorption oriented.

When compared to graphene, MXenes are hydrophilic in nature, and as a result, it is very difficult to fabricate MXene polyolefin composites. By taking account of this, Xu et al.[60] prepared a polypropylene $Ti_3C_2T_x$ nanocomposite with an EMI SE greater than 60 dB in the X-band at a very low percolation threshold of 0.027 vol% of the filler. To fabricate this composite, they first activated the surface of polypropylene using oxygen plasma treatment, then with polyethyleneimine in order to make it hydrophilic, and then dip-coated it with $Ti_3C_2T_x$, followed by vacuum-assisted compression molding. Here also, the mechanism of shielding was found to be more absorption governed because of the development of a highly conductive interconnected network of MXene nanosheets in the pololefin matrix.

Gao et al.[61] prepared thermoplastic polyurethane films based on multilayered $Ti_3C_2T_x$ with high anisotropic thermal conductivity, electrical conductivity, and EMI SE (50.7 dB at 28.6 wt% of MXene) using a different approach, called the layer-by-layer spraying technique. According to them, the better shielding efficiency is due to reflection, multiple internal reflections as well as absorption of radiations between the multilayers of MXene. As the composite shows high anisotropic thermal conductivity, the authors also claim that it can act as excellent Joule heaters, which is considered to be very significant in the field of flexible and wearable devices.

Polyvinylidene fluoride (PVDF) is a thermoplastic polymer with excellent dielectric properties. MXene-incorporated PVDF nanocomposites also exhibit superior shielding properties. The presence of a conductive network along with better heat dissipation ability can improve the SE of the devices to a greater extent. Recent work by Rajavel et al.[62] points out that frequent exposure to radiation causes heating in shielding materials, which will lead to its deterioration, and therefore, the devices must be designed in such a way that they must display high thermal conductivity also. So they fabricated a PVDF–MXene nanocomposite film with good shielding properties (approximately 48 dB at 22.5 vol% of $Ti_3C_2T_x$) and thermal conductivity (0.767 W/m/k). The shielding property increased with an increase in $Ti_3C_2T_x$ owing to the development of a strong interfacial conductive network between MXene and PVDF. The optimum threshold value of the filler was found to be 6.7 vol%. Similarly, Li et al.[63] fabricated a PVDF nanocomposite film based on $Ti_3C_2T_x$ with a hierarchical brick-and-mortar assembly using a blade-coating method. The specific shielding efficiency of the composite was found to be 19504.8 dB cm^2/g (thickness of the film, $t = 17\ \mu m$). They also state that the nanocomposite can withstand extreme environments like annealing, corrosion, deformation, and so on by keeping the SE floating within 5%. In order to achieve better shielding parameters at low ultra-low filler content, double-percolated systems can be applied. For example, MXene-incorporated PVDF and PS can be used to construct a double-percolated system with layered or random structure through the solution-casting or hot-pressing techniques.[64] The solution-cast layered double-percolated structure portrayed a better SE of 55 dB at 12 wt% of MXene when compared to the random double-percolated structure. This is because of the fact that layered PS/PVDF/$Ti_3C_2T_x$ structures can reflect most of the radiation striking it due to their better electrical conductivity, the presence of a conductive network, multiple internal reflections between the layers, the presence of polar bonds, and more.

The synergistic effect of MXene with its 2D analogue graphene in polymers can also impart better shielding efficiency to a material. Recent research work by Vu et al.[65] incorporates both MXene and reduced graphene oxide in polymethylmethacrylate (PMMA) beads to form a 3D structure using hot compression method. According to them, the shielding efficiency and electrical, as well as thermal, conductivity can be tuned to desired values by varying the ratio of the fillers in the matrix. At 2 vol% of the hybrid filler concentration, the PMMA nanocomposite exhibited a shielding efficiency of 28–61 dB.

Wang et al.[66] fabricated an electromagnetic shielding material based on polypyrrole (PPy) decorated MXene/poly(ethylene terephthalate) textile, which is highly hydrophobic, conductive (1000 S/m), and joule heating in nature. The shielding efficiency of the silicon-coated hydrophobic textile material was found to be 90.3 dB. These materials can be used to design flexible, smart, and artificially intelligent wearable materials.

10.4.3 Thermosetting Plastic MXene Nanocomposites

Thermosetting polymers are mainly used for high-performance applications requiring stringent conditions. Thus, they can be used as an excellent base material to fabricate radiation shielding devices that can withstand extreme environments like high temperatures, corrosion, and the like. The thermoplastic nanocomposites can be based on epoxy resins, phenolic resins, polyesters, and so on. Owing to their low viscosity and ease of processing bisphenol epoxies are most widely used to fabricate nanocomposites. Wang et al.[67] prepared an epoxy nanocomposite (bisphenol F) based on few-layered thermally annealed MXene ($Ti_3C_2T_x$) with an EMI SE of41 dB. According to them, the annealed nanocomposite displayed better electrical conductivity when compared to non-annealed MXene/epoxy systems. For non-annealed systems, the electrical conductivity was 38 S/m at 15 wt% of MXene, but for annealed systems, the conductivity exhibited a sharp increase of 105 S/m with the same amount of MXene concentration. The increase in conductivity and shielding parameters can be attributed to the formation of interfaces and dipoles in annealed MXene, which further enhanced multiple reflections and electron transport, among other aspects, through the matrix. The shielding mechanism was found to be absorption-dominated. The composite also displayed better Young's modulus and hardness.

A similar study by Song et al.[68] reports the fabrication of an epoxy nanocomposite using $Ti_3C_2T_x$/reduced graphene oxide hybrid with a honeycomb structure for EMI shielding purposes. According to them, though MXene with a 3D porous structure can provide good radiation shielding properties, it is a laborious task to control the size, shape, and distribution of cells. So they first designed graphene oxide with a honeycomb structure using Al_2O_3 as a honeycomb template. Then $Ti_3C_2T_x$–cetyltrimethyl ammonium bromide solution was impregnated into the structure through electrostatic self-assembly followed by thermal annealing to prepare reduced graphene-oxide/$Ti_3C_2T_x$ honeycomb structure. It is then mixed with epoxy resin and curing agent to get the corresponding nanocomposite with better electrical conductivity and EMI SE of 387 S/m and 55 dB, respectively, at 1.2 wt% reduced graphene oxide and 3.3 wt% $Ti_3C_2T_x$ hybrid filler ratio. The shielding mechanism was absorption-governed. They also stated that the honeycomb structured composites displayed a fivefold increase in the EMI shielding value when compared to those without a honeycomb structure with the same amount of filler loading.

Wang et al.[69] also fabricated an epoxy nanocomposite with significant absorption governed radiation shielding properties based on a 3D $Ti_3C_2T_x$/C hybrid foam. They have adopted a sol-gel method accompanied by freeze-drying and thermal reduction to prepare the hybrid carbon foam/MXene. A mixture of resorcinol and HCHO (formaldehyde) was used as the carbon source. The corresponding bisphenol F epoxy nanocomposites were fabricated using a vacuum-assisted impregnation method using diethyl methyl benzene diamine as a curing agent. Owing to its highly conductive 3D network the composite exhibited an excellent shielding efficiency of about 46 decibels in the X-band, which was reported to be 4.8 times higher than the neat epoxy system.

10.4.4 Elastomer MXene Nanocomposites

Elastomers are considered the workhorse of the industry, and owing to their flexible and stretchable nature, they have attained a significant position in the field of electronics with the upcoming of miniaturization technology, smart wearable devices, artificial robotics, sensors, and more. A recent work by Wang et al.[70] investigated the influence of different hierarchical architectures on the radiation shielding and tensile properties of NR/MXene nanocomposite. They have prepared NR Latex/MXene composite films based on both a honeycomb structure and a brick-and-mortar structure and concluded that those with a honeycomb structure exhibited better mechanical properties and EMI shielding (39.5 MPa and 63.5 dB, respectively) with respect to the brick-and-mortar structure (57 dB of SE with 28.5 MPa tensile strength). The enhanced shielding property of honeycomb structured latex/MXene film is due to the presence of isolated units which can also act as polarization areas. When the radiation strikes the surface of the film, these units help in reflection, absorption, and multiple scattering of the radiations.

Yang et al.[71] fabricated a NR latex composite based on hydrophobic MXene $Ti_3C_2T_x$ exhibiting excellent flame retardancy, mechanical properties, and SE. The film displayed an SE of 47.8 dB with 34 MPa tensile strength. The material proved to possess excellent durability and shielding efficiency (32.8 dB) when subjected to multiple bending cycles even after storing in water for 2weeks.

Lu et al.[72] prepared a nanocomposite based on ethylene propylene diene rubber (EPDM) and $Ti_3C_2T_x$ with better electrical and thermal conductivity of106 S/m and 1.57 W/m K, respectively. They also investigated the shielding effectiveness of the sample at 6 wt% of MXene in the X- band and Ku-band. The percolation threshold reported was 2.7 wt%, thickness, t = 0.3 mm, and the shielding values obtained were 48 dB and 52 dB, respectively, in the corresponding bands. The mechanism of shielding was more absorption-oriented. It is reported that 3 wt% of MXenes is enough to fabricate shielding materials for commercial applications in the industry.

10.4.5 Intrinsically Conducting Polymer–MXene Nanocomposites

As mentioned earlier, ICPs can function as excellent shielding devices owing to their microwave non-transparency, inherent conductivity, and the like. Recently, a lightweight MXene–polyaniline ferromagnetic nanomaterial featuring high conductivity and a shielding efficiency of21–23 dB was reported by Kumar et al.[73] They also prove that the material possesses excellent microwave absorption throughout the X-band since the shielding parameters owing to absorption, as well as reflection, are constant. The layered 2D structure of MXenes is considered a boon since it extends the pathway of radiations as a result of multiple scattering inside the layers. Tong et al.[74] fabricated a multilayered polypyrrole MXene nanocomposite with better radiation absorption properties. On incorporating 25 wt% of polypyrrole/$Ti_3C_2T_x$ in paraffin, the material displayed a reflection loss of −49.2 dB at 8.5 GHz. According to them, the presence of dipoles, multilayered structure, a strong hydrogen bonding interaction between N-H groups of polypyrrole and terminal atoms (O/F) present in the MXene were responsible for better shielding properties. Recently Bora et al.[75] fabricated poly(3,4-ethylenedioxythiophene)/poly(styrene sulfonate) (PEDOT:PSS)/$Ti_3C_2T_x$ nanocomposite with an SE of 41 dB at 50-μL loading of MXene. The robust interfacial bonding between the PEDOT:PSS and $Ti_3C_2T_x$ nanosheets, high electrical conductivity, etc. were responsible for the absorption governed shielding mechanism. Similarly, Jia et al.[76] fabricated a compressible PDMS/$Ti_3C_2T_x$ polymer foam bead composite foams with very low reflection loss for electromagnetic shielding purpose. To prepare the composite, they first synthesized polypropylene foam beads and then decorated it with polyaniline followed by dip-coating of MXene. The corresponding foam was made compressible by incorporating PDMS. The composite exhibited an EMI SE of approximately 23.5–39.8 dB with a very low concentration of $Ti_3C_2T_x$.

10.5 CONCLUSION

Although a lot of stoichiometric compositions of MXenes are known, most of the research works based on shielding and attenuation properties were concentrated only on $Ti_3C_2T_x$. So there are a lot of research opportunities are available in unexplored MXenes. Also, the exact mechanism of shielding in MXene-based composites still requires further clarification. Moreover, only limited studies are conducted in MXene soft elastomeric aerogels developed using extrusion-based 3D printing technology for EMI shielding studies, as 3D printed structures of different microstructures are very difficult to achieve using conventional processing methods. The main challenge in fabricating these materials is to introduce a continuous conductive network that helps in the absorption of radiations. The concept of developing conductive inks for EMI shielding is also not widely explored yet. The concept of fabricating a suitable device to store and convert the radiations for useful purposes is also in its infancy stage. The behavior of the material regarding mechanical, rheological, and other physical and chemical properties is yet to be investigated. So there are a lot of areas and factors that still remain unexplored in the field of microwave absorption and EMI shielding when MXene-based polymer nanocomposites are taken into consideration.

NOTES

1 Kwak, S.Y., Giraldo, J.P., Lew, T.T.S., Wong, M.H., Liu, P., Yang, Y.J., Koman, V.B., McGee, M.K., Olsen, B.D. and Strano, M.S., 2018. Polymethacrylamide and carbon composites that grow, strengthen, and self-repair using ambient carbon dioxide fixation. *Advanced Materials*, 30(46), p. 1804037.

2 Thomas, T., Kanoth, B.P., Nijas, C.M., Joy, P.A., Joseph, J.M., Kuthirummal, N. and Thachil, E.T., 2015. Preparation and characterization of flexible ferromagnetic nanocomposites for microwave applications. *Materials Science and Engineering: B*, 200, pp. 40–49.

3 George, N., Chandra, J., Mathiazhagan, A. and Joseph, R., 2015. High performance natural rubber composites with conductive segregated network of multiwalled carbon nanotubes. *Composites Science and Technology*, 116, pp. 33–40.

4 Mathew, T.V. and Kuriakose, S., 2013. Photochemical and antimicrobial properties of silver nanoparticle-encapsulated chitosan functionalized with photoactive groups. *Materials Science and Engineering: C*, 33(7), pp. 4409–4415.

5 George, S.M., Thomas, S., Jose, A.J. and Parameswaranpillai, J., 2016. Rheological behavior of MWCNTs/modified epoxy-DDM system. *Polymers Research Journal*, 10(4), p. 187.

6 Jin, X., Wang, J., Dai, L., Liu, X., Li, L., Yang, Y., Cao, Y., Wang, W., Wu, H. and Guo, S., 2020. Flame-retardant poly (vinyl alcohol)/MXene multilayered films with outstanding electromagnetic interference shielding and thermal conductive performances. *Chemical Engineering Journal*, 380, p. 122475.

7 Philip, P., Jose, T., Prakash, J. and Cherian, S.K., 2021. Surface plasmon resonance-enhanced bathochromic-shifted photoluminescent properties of pure and structurally modified electrospun poly (methyl methacrylate)(PMMA) nanofibers incorporated with green-synthesized silver nanoparticles. *Journal of Electronic Materials*, pp. 1–16.

8 Khazaei, M., Mishra, A., Venkataramanan, N.S., Singh, A.K. and Yunoki, S., 2019. Recent advances in MXenes: From fundamentals to applications. *Current Opinion in Solid State and Materials Science*, 23(3), pp. 164–178.

9 Gogotsi, Y. and Anasori, B., 2019. The rise of MXenes. *ACS Nano*, 13(8), pp. 8491–8494.

10 Alhabeb, M., Maleski, K., Anasori, B., Lelyukh, P., Clark, L., Sin, S. and Gogotsi, Y., 2017. Guidelines for synthesis and processing of two-dimensional titanium carbide ($Ti_3C_2T_x$MXene). *Chemistry of Materials*, 29(18), pp. 7633–7644.

11 Thomassin, J.M., Jerome, C., Pardoen, T., Bailly, C., Huynen, I. and Detrembleur, C., 2013. Polymer/carbon based composites as electromagnetic interference (EMI) shielding materials. *Materials Science and Engineering: R: Reports*, 74(7), pp. 211–232.

12 Abraham, J., Xavier, P., Bose, S., George, S.C., Kalarikkal, N. and Thomas, S., 2017. Investigation into dielectric behaviour and electromagnetic interference shielding effectiveness of conducting styrene butadiene rubber composites containing ionic liquid modified MWCNT. *Polymer*, 112, pp. 102–115.

13 Planes, M., Le Coz, C., Soum, A., Carlotti, S., Rejsek-Riba, V., Lewandowski, S., Remaury, S. and Solé, S., 2016. Polydimethylsiloxane/additive systems for thermal and ultraviolet stability in geostationary environment. *Journal of Spacecraft and Rockets*, 53(6), pp. 1128–1133.
14 Gurr, P.A., Scofield, J.M., Kim, J., Fu, Q., Kentish, S.E. and Qiao, G.G., 2014. Polyimide polydimethylsiloxane triblock copolymers for thin film composite gas separation membranes. *Journal of Polymer Science Part A: Polymer Chemistry*, 52(23), pp. 3372–3382.
15 Davidson, E.C., Kotikian, A., Li, S., Aizenberg, J. and Lewis, J.A., 2020. 3D printable and reconfigurable liquid crystal elastomers with light-induced shape memory via dynamic bond exchange. *Advanced Materials*, 32(1), p. 1905682.
16 Shukla, V., 2019. Review of electromagnetic interference shielding materials fabricated by iron ingredients. *Nanoscale Advances*, 1(5), pp. 1640–1671.
17 Mishra, R.K., Thomas, M.G., Abraham, J., Joseph, K. and Thomas, S., 2018. Electromagnetic interference shielding materials for aerospace application: A state of the art. *Advanced Materials for Electromagnetic Shielding: Fundamentals, Properties, and Applications*, pp. 327–365.
18 Singh, A.P., Mishra, M. and Dhawan, S.K., 2015. Conducting multiphase magnetic nanocomposites for microwave shielding application. *Nanomagnetism*, 10, pp. 246–277.
19 Weir, W.B., 1974. Automatic measurement of complex dielectric constant and permeability at microwave frequencies. *Proceedings of the IEEE*, 62(1), pp. 33–36.
20 Nicolson, A.M. and Ross, G.F., 1970. Measurement of the intrinsic properties of materials by time-domain techniques. *IEEE Transactions on Instrumentation and Measurement*, 19(4), pp. 377–382.
21 Gopakumar, D.A., Pai, A.R., Pottathara, Y.B., Pasquini, D., Carlos de Morais, L., Luke, M., Kalarikkal, N., Grohens, Y. and Thomas, S., 2018. Cellulose nanofiber-based polyaniline flexible papers as sustainable microwave absorbers in the X-band. *ACS Applied Materials & Interfaces*, 10(23), pp. 20032–20043.
22 Gopakumar, D.A., Pai, A.R., Pottathara, Y.B., Pasquini, D., de Morais, L.C., Khalil HPS, A., Nzihou, A. and Thomas, S., 2021. Flexible papers derived from polypyrrole deposited cellulose nanofibers for enhanced electromagnetic interference shielding in gigahertz frequencies. *Journal of Applied Polymer Science*, 138(16), p. 50262.
23 Pai, A.R., Paoloni, C. and Thomas, S., 2021. Nanocellulose-based sustainable microwave absorbers to stifle electromagnetic pollution. In *Nanocellulose Based Composites for Electronics* (pp. 237–258). Elsevier.
24 Yang, Y., Gupta, M.C., Dudley, K.L. and Lawrence, R.W., 2005. Novel carbon nanotube– polystyrene foam composites for electromagnetic interference shielding. *Nano Letters*, 5(11), pp. 2131–2134.
25 Yuen, S.M., Ma, C.C.M., Chuang, C.Y., Yu, K.C., Wu, S.Y., Yang, C.C. and Wei, M.H., 2008. Effect of processing method on the shielding effectiveness of electromagnetic interference of MWCNT/PMMA composites. *Composites Science and Technology*, 68(3–4), pp. 963–968.
26 Cao, M.S., Cai, Y.Z., He, P., Shu, J.C., Cao, W.Q. and Yuan, J., 2019. 2D MXenes: electromagnetic property for microwave absorption and electromagnetic interference shielding. *Chemical Engineering Journal*, 359, pp. 1265–1302.
27 Li, L., Zhao, S., Luo, X.J., Zhang, H.B. and Yu, Z.Z., 2021. Smart MXene-based janus films with multiresponsive actuation capability and high electromagnetic interference shielding performances. *Carbon*, 175, pp. 594–602.
28 Kistler, S.S., 1931. Coherent expanded aerogels and jellies. *Nature*, 127(3211), pp. 741–741.
29 Pai, A.R., Binumol, T., Gopakumar, D.A., Pasquini, D., Seantier, B., Kalarikkal, N. and Thomas, S., 2020. Ultra-fast heat dissipating aerogels derived from polyaniline anchored cellulose nanofibers as sustainable microwave absorbers. *Carbohydrate Polymers*, 246, p. 116663.
30 Crowell, C., Reynolds, C., Stutts, A. and Taylor, H., 2015. *Aerogel Fabrics in Advanced Space Suit Applications. Journal of Undergraduate Materials Research,* 5, pp. 31–37.
31 Jones, S.M. and Sakamoto, J., 2011. Applications of Aerogels in Space Exploration. In *Aerogels Handbook* (pp. 721–746). Springer, New York, NY.
32 Bian, R., He, G., Zhi, W., Xiang, S., Wang, T. and Cai, D., 2019. Ultralight MXene-based aerogels with high electromagnetic interference shielding performance. *Journal of Materials Chemistry C*, 7(3), pp. 474–478.
33 Li, K., Liang, M., Wang, H., Wang, X., Huang, Y., Coelho, J., Pinilla, S., Zhang, Y., Qi, F., Nicolosi, V. and Xu, Y., 2020. 3D MXene architectures for efficient energy storage and conversion. *Advanced Functional Materials*, 30(47), p. 2000842.

34 Wang, Z., Cheng, Z., Fang, C., Hou, X. and Xie, L., 2020. Recent advances in MXenes composites for electromagnetic interference shielding and microwave absorption. *Composites Part A: Applied Science and Manufacturing*, 136, p. 105956.

35 Ma, Y., Yue, Y., Zhang, H., Cheng, F., Zhao, W., Rao, J., Luo, S., Wang, J., Jiang, X., Liu, Z. and Liu, N., 2018.3D synergistical MXene/reduced graphene oxide aerogel for a piezoresistive sensor. *Acs Nano*, 12(4), pp. 3209–3216.

36 Bian, R., He, G., Zhi, W., Xiang, S., Wang, T. and Cai, D., 2019. Ultralight MXene-based aerogels with high electromagnetic interference shielding performance. *Journal of Materials Chemistry C*, 7(3), pp. 474–478.

37 Wang, L., Liu, H., Lv, X., Cui, G. and Gu, G., 2020. Facile synthesis 3D porous MXene$Ti_3C_2T_x$@ RGO composite aerogel with excellent dielectric loss and electromagnetic wave absorption. *Journal of Alloys and Compounds*, 828, p. 154251.

38 Liu, J., Zhang, H.B., Sun, R., Liu, Y., Liu, Z., Zhou, A. and Yu, Z.Z., 2017. Hydrophobic, flexible, and lightweight MXene foams for high-performance electromagnetic-interference shielding. *Advanced Materials*, 29(38), p. 1702367.

39 Zhang, Y., Yu, J., Lu, J., Zhu, C. and Qi, D., 2021. Facile construction of 2D MXene ($Ti_3C_2T_x$) based aerogels with effective fire-resistance and electromagnetic interference shielding performance. *Journal of Alloys and Compounds*, 870, p. 159442.

40 Jiang, D., Murugadoss, V., Wang, Y., Lin, J., Ding, T., Wang, Z., Shao, Q., Wang, C., Liu, H., Lu, N. and Wei, R., 2019. Electromagnetic interference shielding polymers and nanocomposites-a review. *Polymer Reviews*, 59(2), pp. 280–337.

41 Liu, J., Zhang, H.B., Xie, X., Yang, R., Liu, Z., Liu, Y. and Yu, Z.Z., 2018. Multifunctional, superelastic, and lightweight MXene/polyimide aerogels. *Small*, 14(45), p. 1802479.

42 Wang, L., Song, P., Lin, C.T., Kong, J. and Gu, J., 2020. 3D shapeable, superior electrically conductive cellulose nanofibers/$Ti_3C_2T_x$MXene aerogels/epoxy nanocomposites for promising EMI shielding. *Research*, 2020.

43 Wang, L., Qiu, H., Song, P., Zhang, Y., Lu, Y., Liang, C., Kong, J., Chen, L. and Gu, J., 2019. 3D $Ti_3C_2T_x$MXene/C hybrid foam/epoxy nanocomposites with superior electromagnetic interference shielding performances and robust mechanical properties. *Composites Part A: Applied Science and Manufacturing*, 123, pp. 293–300.

44 Zhao, S., Zhang, H.B., Luo, J.Q., Wang, Q.W., Xu, B., Hong, S. and Yu, Z.Z., 2018. Highly electrically conductive three-dimensional $Ti_3C_2T_x$MXene/reduced graphene oxide hybrid aerogels with excellent electromagnetic interference shielding performances. *Acs Nano*, 12(11), pp. 11193–11202.

45 Luo, J.Q., Zhao, S., Zhang, H.B., Deng, Z., Li, L. and Yu, Z.Z., 2019. Flexible, stretchable and electrically conductive MXene/natural rubber nanocomposite films for efficient electromagnetic interference shielding. *Composites Science and Technology*, 182, p. 107754.

46 Liang, C., Qiu, H., Song, P., Shi, X., Kong, J. and Gu, J., 2020. Ultra-light MXene aerogel/wood-derived porous carbon composites with wall-like "mortar/brick" structures for electromagnetic interference shielding. *Science Bulletin*, 65(8), pp. 616–622.

47 Liu, Z., Wang, W., Tan, J., Liu, J., Zhu, M., Zhu, B. and Zhang, Q., 2020. Bioinspired ultra-thin polyurethane/MXene nacre-like nanocomposite films with synergistic mechanical properties for electromagnetic interference shielding. *Journal of Materials Chemistry C*, 8(21), pp. 7170–7180.

48 Lin, Z., Liu, J., Peng, W., Zhu, Y., Zhao, Y., Jiang, K., Peng, M. and Tan, Y., 2020. Highly stable 3D $Ti_3C_2T_x$MXene-based foam architectures toward high-performance terahertz radiation shielding. *ACS Nano*, 14(2), pp. 2109–2117.

49 Wu, X., Han, B., Zhang, H.B., Xie, X., Tu, T., Zhang, Y., Dai, Y., Yang, R. and Yu, Z.Z., 2020. Compressible, durable and conductive polydimethylsiloxane-coated MXene foams for high-performance electromagnetic interference shielding. *Chemical Engineering Journal*, 381, p. 122622.

50 Zhang, X., Zhang, X., Yang, M., Yang, S., Wu, H., Guo, S. and Wang, Y., 2016. Ordered multilayer film of (graphene oxide/polymer and boron nitride/polymer) nanocomposites: An ideal EMI shielding material with excellent electrical insulation and high thermal conductivity. *Composites Science and Technology*, 136, pp. 104–110.

51 Weng, G.M., Li, J., Alhabeb, M., Karpovich, C., Wang, H., Lipton, J., Maleski, K., Kong, J., Shaulsky, E., Elimelech, M. and Gogotsi, Y., 2018. Layer-by-layer assembly of cross-functional semi-transparent MXene-carbon nanotubes composite films for next-generation electromagnetic interference shielding. *Advanced Functional Materials*, 28(44), p. 1803360.

52 Hu, D., Wang, S., Zhang, C., Yi, P., Jiang, P. and Huang, X., 2021. Ultrathin MXene-aramid nanofiber electromagnetic interference shielding films with tactile sensing ability withstanding harsh temperatures. *Nano Research*, pp. 1–9.

53 Ma, W., Cai, W., Chen, W., Liu, P., Wang, J. and Liu, Z., 2021. A novel structural design of shielding capsule to prepare high-performance and self-healing MXene-based sponge for ultra-efficient electromagnetic interference shielding. *Chemical Engineering Journal*, p. 130729.

54 Wang, Z., Gao, W., Zhang, Q., Zheng, K., Xu, J., Xu, W., Shang, E., Jiang, J., Zhang, J. and Liu, Y., 2018. 3D-printed graphene/polydimethylsiloxane composites for stretchable and strain-insensitive temperature sensors. *ACS Applied Materials & Interfaces*, 11(1), pp. 1344–1352.

55 Yang, W., Yang, J., Byun, J.J., Moissinac, F.P., Xu, J., Haigh, S.J., Domingos, M., Bissett, M.A., Dryfe, R.A. and Barg, S., 2019. 3D printing of freestanding MXene architectures for current-collector-free supercapacitors. *Advanced Materials*, 31(37), p. 1902725.

56 Zhang, Y.Z., Wang, Y., Jiang, Q., El-Demellawi, J.K., Kim, H. and Alshareef, H.N., 2020. MXene printing and patterned coating for device applications. *Advanced Materials*, 32(21), p. 1908486.

57 Hassan, K., Nine, M.J., Tung, T.T., Stanley, N., Yap, P.L., Rastin, H., Yu, L. and Losic, D., 2020. Functional inks and extrusion-based 3D printing of 2D materials: A review of current research and applications. *Nanoscale*, 12(37), pp. 19007–19042.

58 Orangi, J., Hamade, F., Davis, V.A. and Beidaghi, M., 2019.3D printing of additive-free 2D $Ti_3C_2T_x$ (MXene) ink for fabrication of micro-supercapacitors with ultra-high energy densities. *ACS nano*, 14(1), pp. 640–650.

59 Sun, R., Zhang, H.B., Liu, J., Xie, X., Yang, R., Li, Y., Hong, S. and Yu, Z.Z., 2017. Highly conductive transition metal carbide/carbonitride (MXene)@ polystyrene nanocomposites fabricated by electrostatic assembly for highly efficient electromagnetic interference shielding. *Advanced Functional Materials*, 27(45), p. 1702807.

60 Xu, M.K., Liu, J., Zhang, H.B., Zhang, Y., Wu, X., Deng, Z. and Yu, Z.Z., 2021. Electrically conductive Ti3C2T x MXene/polypropylene nanocomposites with an ultralow percolation threshold for efficient electromagnetic interference shielding. *Industrial & Engineering Chemistry Research*, 60(11), pp. 4342–4350.

61 Gao, Q., Pan, Y., Zheng, G., Liu, C., Shen, C. and Liu, X., 2021. Flexible multilayered MXene/thermoplastic polyurethane films with excellent electromagnetic interference shielding, thermal conductivity, and management performances. *Advanced Composites and Hybrid Materials*, 4(2), pp. 274–285.

62 Rajavel, K., Luo, S., Wan, Y., Yu, X., Hu, Y., Zhu, P., Sun, R. and Wong, C., 2020. 2D $Ti_3C_2T_x$MXene/polyvinylidene fluoride (PVDF) nanocomposites for attenuation of electromagnetic radiation with excellent heat dissipation. *Composites Part A: Applied Science and Manufacturing*, 129, p. 105693.

63 Li, Y., Zhou, B., Shen, Y., He, C., Wang, B., Liu, C., Feng, Y. and Shen, C., 2021. Scalable manufacturing of flexible, durable $Ti_3C_2T_x$MXene/Polyvinylidene fluoride film for multifunctional electromagnetic interference shielding and electro/photo-thermal conversion applications. *Composites Part B: Engineering*, 217, p. 108902.

64 Wang, J., Yang, K., Wang, H. and Li, H., 2021. A new strategy for high-performance electromagnetic interference shielding by designing a layered double-percolated structure in PS/PVDF/MXene composites. *European Polymer Journal*, 151, p. 110450.

65 Vu, M.C., Mani, D., Kim, J.B., Jeong, T.H., Park, S., Murali, G., In, I., Won, J.C., Losic, D., Lim, C.S. and Kim, S.R., 2021. Hybrid shell of MXene and reduced graphene oxide assembled on PMMA bead core towards tunable thermoconductive and EMI shielding nanocomposites. *Composites Part A: Applied Science and Manufacturing*, 149, p. 106574.

66 Wang, Q.W., Zhang, H.B., Liu, J., Zhao, S., Xie, X., Liu, L., Yang, R., Koratkar, N. and Yu, Z.Z., 2019. Multifunctional and water-resistant MXene-decorated polyester textiles with outstanding electromagnetic interference shielding and joule heating performances. *Advanced Functional Materials*, 29(7), p. 1806819.

67 Wang, L., Chen, L., Song, P., Liang, C., Lu, Y., Qiu, H., Zhang, Y., Kong, J. and Gu, J., 2019. Fabrication on the annealed $Ti_3C_2T_x$MXene/Epoxy nanocomposites for electromagnetic interference shielding application. *Composites Part B: Engineering*, 171, pp. 111–118.

68 Song, P., Qiu, H., Wang, L., Liu, X., Zhang, Y., Zhang, J., Kong, J. and Gu, J., 2020. Honeycomb structural rGO-MXene/epoxy nanocomposites for superior electromagnetic interference shielding performance. *Sustainable Materials and Technologies*, 24, p.e00153.

69 Wang, Lei, Hua Qiu, Ping Song, Yali Zhang, Yuanjin Lu, Chaobo Liang, Jie Kong, Lixin Chen, and Junwei Gu., 2019. 3D $Ti_3C_2T_x$MXene/C hybrid foam/epoxy nanocomposites with superior electromagnetic

interference shielding performances and robust mechanical properties. *Composites Part A: Applied Science and Manufacturing*, 123: 293–300.

70 Wang, Y., Liu, R., Zhang, J., Miao, M. and Feng, X., 2021. Vulcanization of $Ti_3C_2T_x$MXene/natural rubber composite films for enhanced electromagnetic interference shielding. *Applied Surface Science*, 546, p. 149143.

71 Yang, W., Liu, J.J., Wang, L.L., Wang, W., Yuen, A.C.Y., Peng, S., Yu, B., Lu, H.D., Yeoh, G.H. and Wang, C.H., 2020. Multifunctional MXene/natural rubber composite films with exceptional flexibility and durability. *Composites Part B: Engineering*, 188, p. 107875.

72 Lu, S., Li, B., Ma, K., Wang, S., Liu, X., Ma, Z., Lin, L., Zhou, G. and Zhang, D., 2020. Flexible MXene/EPDM rubber with excellent thermal conductivity and electromagnetic interference performance. *Applied Physics A*, 126(7), pp. 1–12.

73 Kumar, S., Kumar, P., Singh, N. and Verma, V., 2019. Steady microwave absorption behavior of two-dimensional metal carbide MXene and polyaniline composite in X-band. *Journal of Magnetism and Magnetic Materials*, 488, p. 165364.

74 Tong, Y., He, M., Zhou, Y., Zhong, X., Fan, L., Huang, T., Liao, Q. and Wang, Y., 2018. Hybridizing polypyrrole chains with laminated and two-dimensional $Ti_3C_2T_x$ toward high-performance electromagnetic wave absorption. *Applied Surface Science*, 434, pp. 283–293.

75 Bora, P.J., Anil, A.G., Ramamurthy, P.C. and Tan, D.Q., 2020.MXene interlayered crosslinked conducting polymer film for highly specific absorption and electromagnetic interference shielding. *Materials Advances*, 1(2), pp. 177–183.

76 Jia, X., Shen, B., Zhang, L. and Zheng, W., 2021. Construction of compressible Polymer/MXene composite foams for high-performance absorption-dominated electromagnetic shielding with ultra-low reflectivity. *Carbon*, 173, pp. 932–940.

11 MXene–Polymer Nanocomposites for Gas- and Vapor-Sensing Applications

Vishal Chaudhary

CONTENTS

DOI: 10.1201/9781003164975-11

11.1 INTRODUCTION

Rapid industrial and technological advancements have not only contributed to development but also contributes to various global ecological issues, including global warming, air pollution, ozone depletion, water and land pollution, and scarcity of drinking water. These are caused due to emission of various environmental contaminants, including poisonous gases and vapors, sewage, the release of harmful chemicals, the production of nondegradable waste and electronic waste, and various ecological misbalances. Among all, the primary concern is the release of air contaminants in the form of gases, including ammonia, oxides of sulfur, oxides of nitrates, and oxides of carbon, along with vapors (alcohol, hydrazine, acetone), and other solid pollutants such as particulate matter (Chaudhary et al., 2016, 2017, 2022c). Hence, monitoring and controlling the release of these pollutants are required to preserve the ecosystem (Bai et al., 2007). Monitoring the release of air contaminants at each emission site is done through gas and vapor sensors.

11.2 GAS AND VAPOR SENSORS

A sensor system is an interface device similar to the human nervous system that converts a change in any of its properties in the presence of analyte into electric signals, which can be further detected using appropriate electronic circuitry (Bai et al., 2007; Baker et al., 2017). Generally, gas sensors consist of a material, which changes its properties, such as electrical, optical, acoustical, or mechanical, in the presence of air contaminants (Bai et al., 2007; Chaudhary et al., 2022b). The variation in any property of the material interacting with air contaminants is called its sensing parameter. The variation in any property of material interacting with air contaminants is called its sensing parameter. Various sensing parameters reported in the literature include electrical resistance, capacitance, reflectance, absorbance, dielectric properties, capacitance, resonant frequency, thermal and electrical conductivity, and refractive index (Bai et al., 2007). The variation in sensing parameters in the presence of a specific analyte has been ascribed to various physical and chemical phenomena, including adsorption, absorption, swelling, and chemical reactions (Anichini et al., 2018). Sensors are classified based on sensing parameters, including electrochemical, spectroscopic, solid-state, calorimetric, and mass-sensitive sensors (Bai et al., 2007). Most of the reported gas and vapor sensors are solid-state-type chemiresistor, in which a change in resistance in the presence of analyte is taken as a sensing parameter (Anichini et al., 2018; Bai et al., 2007; Chaudhary et al. 2022d). Chemiresistor are consists of a sensing layer, electrode system, substrate and electronic circuitry as shown in Figure 11.1. Various substrates include flexible and nonflexible substrates

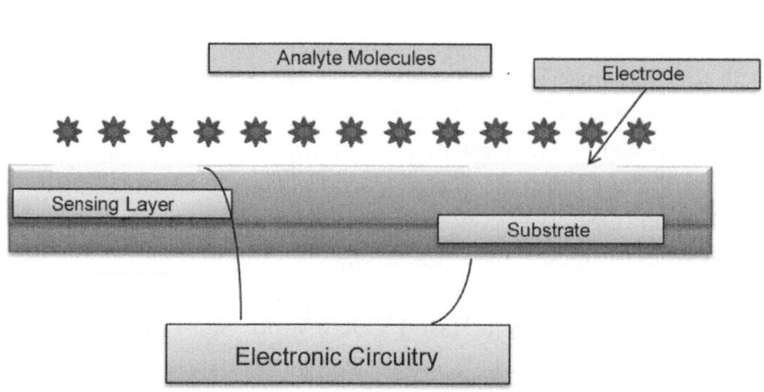

FIGURE 11.1 Schematic of a chemiresistor.

such as glass substrate, indium tin–oxide (ITO) substrate, polyindole (PI), polyethylene terephthalate (PET), and graphite papers (Bai et al., 2007). The electrode system can be simply two parallel electrodes type or interdigitated-type electrodes to record a maximum change in sensing parameter. Mainly highly conducting material contact electrodes such as gold are used to avoid any alteration or noise in obtained sensing signal due to the Schottky effect (Bai et al., 2007). The electronic circuitry of sensors is majorly based on the Wheatstone bridge principle. Change in resistance of one of its arms leads to current flow, creating an alarming or detection signal. Since the analyte detecting performance of a sensor depends on its sensing layer material, the researcher's primary concern is dedicated to developing sensing layer material for enhancing its various sensing characteristics.

11.3 SENSING CHARACTERISTICS

Various sensing characteristics essential for the commercial development of a sensor include three Ss (selectivity, sensitivity, stability) along with five essential Rs (room temperature operation, range, reproducibility, response and recovery time, repeatability).

11.3.1 Three Essential Ss

Sensitivity or sensing response is the prime sensing characteristic that determines the efficiency of a sensor. Generally, it is defined as the percentage change or degree of change in sensing parameter in the presence of the analyte (Chaudhary et al., 2015a, 2016). It depends on various topological and morphological properties, including effective morphology, dimensions, surface area, roughness, porosity, chemical properties including chemical structure, constituent components, and physical properties including charge carrier pathway, band gap, and conductivity (Bai et al., 2007; Baker et al., 2017; Anichini et al., 2018). Thus, enhancing the sensitivity through altering these parameters is the prime focus of the research community.

Degree in change (S) of sensing parameter (R) is given by

$$S = \Delta R/R_a. \ldots., \qquad (11.1)$$

where ΔR is the change in the resistance of the sensor in the presence of analyte (R_a–R_g), with R_g is the value of the sensing parameter in ambient conditions and R_a is the value of the sensing parameter in the presence of analyte.

Percentage change in sensing response

$$S(\%) = (\Delta R/R_a) \times 100 \ldots.. \qquad (11.2)$$

A typical sensing signal with few characteristics is shown in Figure 11.2.

The **selectivity** of a sensor is defined as the degree of detection of a particular analyte in the presence or absence of other analytes (Chaudhary et al., 2017, 2022a; Singh et al., 2022). A sensor detecting many analytes may be regarded as a 'versatile sensor', but at the same time, it is challenging to differentiate the detected analyte. Thus, selectivity towards a particular analyte in the presence of other analytes is a significant concern of a gas sensor. Since the sensing phenomenon is the adsorption process majorly, the selectivity is attributed to the adsorption energies of various analytes over specific surfaces and the presence of specific reactive functional groups in sensing material towards a particular analyte (Chaudhary et al., 2015b, 2017). A selectivity test is done by measuring the sensing response of sensors in different analytes and comparing them.

The **stability** of a sensor is required for its commercial development. A developed sensor must be highly stable under different environmental conditions, such as a change in humidity, pressure, and temperature. A stability test includes recording sensing parameters in varying humidity and temperature conditions (Bai et al., 2007).

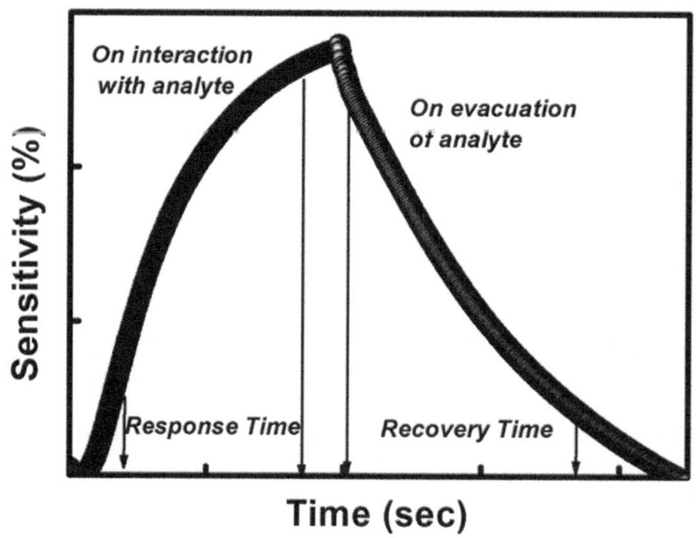

FIGURE 11.2 Sensing signal with sensitivity, response, and recovery.

Thus, the three Ss are the basic requirements of a sensor for its commercial development.

11.3.2 Five Desirable Rs

The **room temperature operation** of a sensor is essential for its economic and ecological development (Chaudhary et al., 2017; Bai et al., 2007; Singh et al., 2022). A sensor that can be operated at room temperature excludes micro-heating assemblages that reduce its production cost and energy requirement for its operation (Chaudhary et al., 2015c, 2022b). It also averts the merge of interparticle grain boundaries of sensing material, which increases its lifetime and performance.

The **ranges of detection** include both the upper range (high concentration of analyte up to which it detects analyte linearly and efficiently) and lower range (minimum concentration of analyte that can be detected by designed sensor). Linear detection range is considered a criterion for the commercial development of a sensor. Linear detection range test is done by linearly fitting the sensing response of sensor at various concentrations of analyte and measuring its regression value (Bai et al., 2007; Chaudhary et al., 2016). For the perfect linear range, the regression value must be equal to unity, which is not experienced during measurements.

The **response and recovery time** are defined as the 90% of elevation curve and the recovery curve of the sensing signal (Chaudhary et al., 2015a). The response time depends on the time and degree of interaction of the analyte with sensing material. However, recovery time depends on the degree and time of desorption of analyte molecule from sensing material. A suitable sensor must possess a short response and recovery time, which means it must detect the analyte rapidly and recover fast to its original state after the removal of the analyte.

A sensor must reproduce its results repeatedly and can be reproducible without any influence of any external environmental factor for its commercial development (Bai et al., 2007). A onetime high response or a onetime possible fabrication of a sensor is of no use. Thus, a sensor must be **repeatable and reproducible**. These tests are done by measuring the sensing response of the sensor for many cycles of analyte exposure (Chaudhary et al., 2017).

All the five Rs, along with the three Ss, together set the criterion for a suitable sensor and its commercial development. Apart from these characteristics, flexibility (Anichini et al., 2018) and self-driven sensors using nanogenerators (Wu et al., 2020) are also additional features for a sensor.

Flexibility can be achieved by using flexible substrates such as PI or PET or self-sustainable films (Alrammouz et al, 2018). The flexibility test measures the sensing response by bending the sensor film at different angles and for multiple folds (Alrammouz et al, 2018). The majority of developed sensing materials reported in the literature are characterized based on these essential sensing characteristics.

11.4 SENSING MATERIALS

The central part of a sensing device that plays a vital function in converting a chemical or physical signal into a detectable electrical signal is its sensing material. Nanotechnology plays a vital role in designing sensing materials with enhanced sensing characteristics (Bai et al., 2007). Since the interaction of the analyte with sensing materials is a surface phenomenon, high effective surface area (due to large surface-to-volume ratio) of nanomaterials with enhanced porosity, chemical and physical properties make them a promising candidate in sensing technology (Khan et al., 2020). An increase in surface area surges the probability of analyte interaction with the material, which results in enhanced sensing characteristics (Khan et al., 2020). Due to the abundance of natural materials, various nanomaterials have been used to detect different analytes (Khan et al., 2020). Out of them, two-dimensional (2D) materials, including graphene and its derivatives, borophene, 2D organic polymers, black phosphorus, molybdenum disulfide, and MXenes, due to their high surface-to-volume ratio, versatile surface chemistries, and room temperature response, are widely researched for gas-and vapor-sensing studies (Alrammouz et al, 2018; Bai et al., 2007; Anichini et al., 2018; Khan et al., 2020). Out of them,2D materials, including graphene and its derivatives, brophene, 2D organic polymers, black phosphorus, molybdenum disulfide, and MXenes, due to their high surface-to-volume ratio, versatile surface chemistries, and room temperature response are widely researched for gas-and vapor-sensing studies (Anichini et al., 2018; Wang et al., 2021a; Sheth et al., 2022; Chaudhary et al., 2022a). On the contrary, MXenes area family of 2D transition metal nitrides and carbides, which exhibited inordinate sensing properties due to their large effective surface area, good hydrophilicity, excellent electrical conductivity, high mechanical stability, biocompatibility, and abundant surface-terminated groups (–OH, –O, or –F) (Riazi et al., 2021). However, the use of MXene-based sensors has been limited to mostly reducing analytes such as ammonia (NH_3) due to high conductivity with the moderate sensing response (Riazi et al., 2021; Ho et al., 2021). Due to the high absorption energy between MXene and analyte, the recovery time is too significant or total recovery is not achieved, limiting its commercial development (Wang et al., 2021b). For commercial-scale development, MXene-based sensors must be produced in a large quantity to keep their conductivity high, and they must possess high mechanical endurance and flexibility for machine processability (Riazi et al., 2021; Zamhuri et al., 2021). For detection and monitoring of oxidizing analytes with efficient sensitivity, its conductivity needs to be controlled. Thus, controlling the electrical and mechanical properties of MXene is a challenge for sensor development, which is addressed by making its nanocomposites with polymers (Riazi et al., 2021; Carey et al., 2021). Nanostructured polymers such as polyaniline (PANI), polypyrrole (PPy), polyvinyl alcohol (PVA), and polystyrene (PS) have shown significant sensing response with enhanced room temperature–sensing characteristic toward various reducing and oxidizing analytes (Wong et al., 2020; Nazemi et al, 2019). They are easy to fabricate, cost-effective, environmental- and user-friendly, and energy-efficient. It has also been noticed that MXene-polymer composites have shown tremendous performance in energy storage applications due to multi-interactions, including hydrogen bonding, electrostatic interactions, or van der Waals forces (Gao et al., 2020a; Chaudhary et al., 2022b). Thus, to incorporate the merits of both types of materials, 2D hybrid MXene–polymer nanocomposites have been evaluated for gas-sensing performance (Zhan et al., 2020; Chen et al., 2021). However, there is dedicated literature on other inorganic-polymer nanocomposites for gas/vapor-sensing monitoring (Aarya et al., 2020). Most reports on inorganic-polymer nanocomposites demonstrate enhanced gas-sensing performance due to the formation of heterojunctions among the precursors. Nevertheless, the use of MXene as one of the precursors in nanocomposites makes it more unique and more efficient for gas/volatile organic compounds (VOC) detection. MXene-polymer hybrid systems show unique hetero-interfacial effects, which surges the synergistic effect, interfacial charge transfer, and contact with target analyte molecules due to the presence of abundant surface functionalities (Chen et al., 2021; Zhan et al., 2020). The inclusion of polymers into MXene can further protect its oxidation if they possess a more considerable affinity toward MXene (Chen et al., 2021). These MXene polymers have been reported to be fabricated using various ex situ and *in situ* approaches.

11.5 FABRICATION OF MXene–POLYMER-BASED GAS AND VAPOR SENSORS

Nanocomposites are materials possessing at least one of the dimensions of a parent component in a nanometric scale (<100nm) (Chaudhary et al., 2021a, 2015c). Nanocomposites can be classified as into three categories based on their morphology including phase-separated systems, intercalated systems, and exfoliated systems (Fu et al., 2019). MXene–polymer nanocomposites can be fabricated using an ex situ approach, such as mixing two parents after separate synthesis, or by *in situ* approaches, such as synthesizing both parents together or one parent in the presence of others (Kausar et al., 2021; Carey et al., 2021). These techniques include solvent blending, *in situ* polymerization, surface etching, melt blending, and RIR-MAPLE deposition (Chen et al., 2021; Zhan et al., 2020).

In general procedure, synthesis of MXene involves three steps etching (removing A element from its MAX phase through etchant including hydrofluoric acid, hydrochloric acid, lithium fluoride or their combination), washing (reestablishing pH to neutral value through decantation, centrifugation or dispersion in fresh water), and delamination (obtaining single layer MXene by ultrasonication, shaking or adding intercalants; Ho et al., 2021). It is followed by an *in situ* or ex situ technique to obtain nanocomposite (Carey et al., 2021). Furthermore, a sensing film is made on a flexible substrate (PET or PI) or nonflexible (glass or ITO-coated glass slides) using various techniques such as spin coating, electrospinning, and wet spinning (Gao et al., 2020b). Furthermore, two conductive electrodes have been deposited over sensing film for further analyzing the sensing performance (Wang et al., 2021b).

11.5.1 Ex Situ Approaches

In the ex situ approach, both the parent material (MXene and polymer) are synthesized separately and then dissolved in the appropriate solvent (Carey et al., 2021; Zhan et al., 2020). The sensing layer is further fabricated using MXene–polymer mixture using appropriate processing techniques such as electrospinning or wet spinning over the desired substrate or of self-standing nature or drop-casting (Carey et al., 2021). A nanocomposite-sensing layer of water-soluble polymer and MXene, which is hydrophilic, can be fabricated by mixing them in water to form an aqueous solution (Carey et al., 2021; Riazi et al., 2021). The formed layer is further dried in vacuum oven at a suitable temperature dependent on the thermal stability of the parent constituents, specifically polymer (Riazi et al., 2021).

11.5.2 *In Situ* Approaches

These synthesis techniques include synthesizing one parent constituent in the presence of another or synthesizing both the parent together (Carey et al., 2021). The most commonly used *in situ* technique is *in situ* polymerization, in which the polymer is polymerized in the presence of pre-synthesized MXene (Gao et al., 2020a). Furthermore, the sensing layer of nanocomposite is formed using suitable techniques such as dip coating, inkjet printing, or drop-casting (Carey et al., 2021). It is followed by drying the formed sensing layer or film at a specific temperature as per parent constituents (Carey et al., 2021).

Furthermore, a specific type of conductive electrode system, such as interdigitated electrodes or parallel electrodes, is deposited over the sensing layer for a sensing measurement using an appropriate technique such as inkjet printing, thermal deposition, spin coating, or dip casting (Zhan et al., 2020; Riazi et al., 2021).

Various fabrication techniques have been used to design a sensor to detect a particular analyte, discussed in the subsequent section. The chapter is further divided in terms of reports present in the literature for the detection of various analytes.

11.6 SUITABILITY OF MXene–POLYMER NANOCOMPOSITES STRUCTURES FOR ANALYTE SENSING

The inclusion of polymer into MXene prevents stacking of its layers or sheets (Carey et al., 2021), as shown in Figure 11.3. Due to the etching and delamination process, various functional groups including –OH, –O, and –F are present on MXene sheets, which binds polymers via electrostatic interactions resulting in the formation of the lamellar structure of MXene sheets with polymer chains between them (Carey et al., 2021). Such a lamellar structure with polymer chains possesses a high effective surface area with large adsorption sites for analyte molecules, resulting in enhanced sensing characteristics.

11.7 SENSING MEASUREMENTS OF MXene–POLYMER NANOCOMPOSITE-BASED SENSORS

Sensing measurements for gas sensors are generally recorded in a chemiresistive mode. A typical sensing apparatus consists of a sensing chamber, with the provision to insert a definite concentration of analyte molecules; a measuring device, such as digital multimeter, which records change in sensing parameter; computer interfacing with computer through software, such as Lab View; and temperature and humidity controller system (Chaudhary et al., 2015c, 2021b). Sensing measurements are done either in static mode or in dynamic mode. To test the sensitivity, the sensor is placed in an air-tight sensing chamber and connected to a measuring device, such as digital multimeter to detect the change in sensing parameter. In static mode, a definite amount of analyte in parts per million (ppm) is inserted into the sensing chamber, and the altered sensing parameter of the sensor is measured after its saturation (Bai et al., 2007). In dynamic mode, continuous flow of definite amount of analyte is made through sensing chamber, and respective change is measured (Chaudhary et al., 2016). Furthermore, all sensing characteristics are obtained using the appropriate formulation, as discussed in the coming sections.

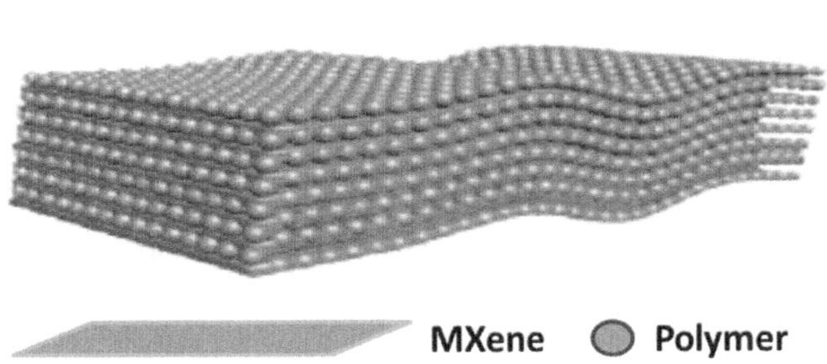

FIGURE 11.3 Structure of MXene–polymer nanocomposites.

11.8 MONITORING OF ANALYTES THROUGH MXene–POLYMER NANOCOMPOSITES

MXene is generalized by a chemical formula $M_{n+1}X_nT_x$, where M denotes an early transition metal, X denotes carbides or nitrides (with n = 1, 2 or 3) and T denotes various functional groups such as –O, –F and –OH (Gogotsi et al., 2019). Thus, based on the type of MXene (majorly based on the value of 'n'), MXene–polymer nanocomposites can be further classified. Thus, we can generalize a formula for MXene–polymer composites as $M_{n+1}X_nT_{x-P}$, where P stands for a type of polymer or combination of polymers. There are specific reports on MXene-based gas sensors with polymer substrate (Chen et al., 2020; Aghaei et al., n.d.; Lee et al., 2019; Hasan et al., 2021), which is entirely different from MXene–polymer nanocomposites. Since, in the case of nanocomposites, all the precursors are an integral part of the hybrid. However, MXene deposited over polymer substrate does not include polymer as one of the precursors and is not termed a nanocomposite. In this chapter, we have restricted our discussion to MXene–polymer nanocomposite-based gas sensors. In this section, we discuss each type of analyte one after one detected through various MXene–polymer-based nanocomposites in detail, including their synthesis route, sensor fabrication technique, type of sensor, sensing performance, and sensing phenomena with their merits and demerits in view of literature. We also discuss the importance of monitoring specific analytes for human welfare and environmental balance.

11.9 MXene–POLYMER NANOCOMPOSITE-BASED AMMONIA SENSORS

Ammonia gas is extensively used as raw materials in various chemical industries including manufacturing units of plastics, nitric acid, textiles, pesticides and explosives, refrigeration industries, petrochemical industries, and manufacturing industries of gallium nitrides and silicon nitrides in electronic industries (Chaudhary et al., 2015a, 2021b). The exposure limit for humans to ammonia is 25 ppm for 8 hours and 35 ppm for 10 minutes as per Occupational Safety and Health Administration (Timmer et al., 2005). Beyond these prescribed limits, ammonia is hazardous to human respiratory, olfactory, nervous, dermal, and excretory systems (Timmer et al., 2005). Exceeding the ammonium salt in the human body above 200–500 mg/kg of body weight may cause kidney failure, lung edema, and nervous disorder (Chaudhary et al., 2015a; Timmer et al., 2005). Thus, the monitoring of ammonia is critically required and done through ammonia sensors. There are various reports in the literature for the detection of ammonia using various sensing materials. Since ammonia is a reducing analyte, materials with high conductivity are best suited for its detection due to their significant possibility of oxidation in the presence of ammonia (Timmer et al., 2005). There are reports on the detection of ammonia through MXene and polymers separately in the literature (Aarya et al., 2020; Insausti et al., 2020; Ho et al., 2021). However, the inclusion of both materials in a single sensing layer has been reported to enhance their sensing response through many folds due to their merits (Chen et al., 2021; Riazi et al., 2021). MXenes, due to their 2D structure, possess a high surface-to-volume ratio, which amplifies the sensing signal due to more interaction between the sensing layer and ammonia molecules. In contrast, polymer provides stability and selectivity (Riazi et al., 2021; Chaudhary et al., 2021b). Various MXene–polymer nanocomposites have been further discussed in the succeeding subsection.

11.9.1 M_3X_2-BASED MXENE–POLYMER COMPOSITES ($M_3X_2T_x$-PS)

Various studies on density functional theory (DFT) calculations (Xiao et al., 2016; Kim et al., 2018; Hajian et al., 2018) had predicted that MXenes hold specific selectivity toward ammonia due to surficial terminated groups on them. Thus, they have been considered apt for the detection of ammonia. Titanium carbide (Ti3C2Tx) was the first M3X2-based MXene, explored for its gas-sensing properties by Kim et al. (2018) and Lee et al. (2017). This report was followed by few reports on ammonia and other vapor detection through M_3X_2-based MXenes (Kim et al., 2018).

However, it has been noticed that the sensing responses of M_3X_2-based MXenes are poor toward ammonia (Kim et al., 2018; Zhao et al., 2019). Thus, various researchers combined these MXenes with polymers including polyaniline (PANI) and Poly 3,4-ethylenedioxythiophene (PEDOT) to form M_3X_2-based MXene–polymer composite and investigated them for ammonia-sensing characteristics (Jin et al., 2020; Riazi et al., 2021; Li et al., 2020; Aghaei et al., n.d.). Jin et al. (2020) have reported the use of poly 3,4-ethylene dioxythiophene: poly (4-styrene sulfonate)/titanium carbide (PEDOT:PSS/$Ti_3C_2T_x$) composite for enhanced ammonia sensing performance. Li et al. (2020) reported flexible ammonia sensor based on polyaniline (PANI)/$Ti_3C_2T_x$ for agriculture ammonia volatilization monitoring. The two reports were similar in detecting ammonia at different ppm levels to analyze various ammonia sensing characteristics. However, Li et al. (2020) used PANI-M_3X_2-based MXene sensor to measure ammonia volatilization in agricultural applications through specially designed simulations.

11.9.1.1 Fabrication of $M_3X_2T_x$-P-Based Ammonia Sensors

$M_3X_2T_x$-P nanocomposites are generally fabricated by *in situ* approach (Riazi et al., 2021), in which desired polymer is polymerized in the presence of pre-synthesized $M_3X_2T_x$. Jin et al. (2020) reported the synthesis of PEDOT:PSS/$Ti_3C_2T_x$ nanocomposites using *in situ* polymerization of EDOT over pre-synthesized $Ti_3C_2T_x$ in the presence of PSS. The prepared nanocomposite was dip-coated on a PI surface possessing interdigitated electrodes. Furthermore, the prepared sensing film was dried in a vacuum desiccator to analyze ammonia monitoring. Li et al. (2020) fabricated PANI/$Ti_3C_2T_x$, a chemiresistive-type flexible ammonia sensor, using *in situ* polymerization of PANI in the presence of pre-synthesized $Ti_3C_2T_x$ nanosheets on PI substrate. Initially, a mixture of aniline and pre-synthesized $Ti_3C_2T_x$ was prepared. Furthermore, an appropriate amount of oxidant APS (ammonium persulfate) and hydrochloric acid (dopant) was added under constant stir. On achieving pale yellow color of solution, indicating polymerization of aniline, PI substrate with Au-interdigitated electrodes was dip into it, to form sensing film. The films were further dried in vacuum oven and further analyzed for ammonia sensing performance (Li et al., 2020).

11.9.1.2 Structure of $M_3X_2T_x$-P Nanocomposites

Li et al. (2020) reported forming a hierarchical structure of Ti3C2Tx nanosheets and PANI nanoparticles accompanied by PANI dendritic nanofiberson nanoparticles. The presence of dendritic nanofibers increases the gap between the lamellar structures increasing the higher effective surface area for the adsorption of ammonia molecules (Li et al., 2020). They have also observed the formation of core–shell type nanocomposite with $Ti_3C_2T_x$ nanosheets wrapped by PANI sheets (Li et al., 2020). Jin et al. (2020) observed that the inclusion of PEDOT: PSS between $Ti_3C_2T_x$ nanosheets increased their interlayer spacing, which supports a higher adsorption of analyte molecules. Thus, the hierarchical lamellar structure of M3X2Tx-P nanocomposites makes them a promising candidate for analyte sensing due to higher interlayer spacing.

11.9.1.3 Ammonia-Monitoring Measurements

Jin et al. (2020) reported measurements done in static mode in which the analyte was introduced in an air-tight sensing chamber through a micro-syringe. The change in resistance was recorded using a digital multimeter until the sensor's resistance became stable in the presence of the analyte. The sensing measurements were done at room temperature (25 ± 5°C) and in relative humidity of 55 ± 5%. Sensing response was defined as the degree of change in resistance of the sensor in the presence of an analyte using Equation11.2. However, Li et al. (2020) used typical sensing apparatus and performed dynamic mode measurements under controlled varying temperature and humidity with exposure to different ppm levels of ammonia.

11.9.1.4 Sensing Characteristics

Ammonia sensing characteristics of $M_3X_2T_x$-P nanocomposites depend on the structure, formation of interfaces, conductivity, and concentration of parent constituents (Riazi et al., 2021). The enhancement in inter-planar spacing and the presence of various functional groups improves the ammonia sensing performance of M3X2Tx-P nanocomposites compared to their parent constituents (Riazi et al., 2021). Similar observations have been made in the case of PEDOT:PSS/$Ti_3C_2T_x$ and PANI/$Ti_3C_2T_x$ nanocomposites in terms of improved sensing response compared to that of its parents' constituents (Li et al., 2020; Jin et al., 2020). Jin et al. (2020) also reported the variation in sensing response, response time and recovery time of composite with the increase of $Ti_3C_2T_x$ concentration as listed in Table 11.1.

It was observed that with an increase of $Ti_3C_2T_x$ content in the nanocomposite, its ammonia-sensing characteristics increase up to 15 wt% and decrease further (Jin et al., 2020). Hence, PEDOT:PSS/$Ti_3C_2T_x$ composite sensor shows the best sensing characteristics toward ammonia for 15 wt% nanocomposites. The reason was ascribed to higher conductivity of 15wt% nanocomposite compared to that of others (Jin et al., 2020). The higher conductivity provides higher chances of oxidation of nanocomposite in reducing ammonia, which enhances the sensing response (Pandey, 2016). Consequently, response time and recovery time were found to decrease. The reason has also been attributed to optimal interlayer spacing at15 wt% of $Ti_3C_2T_x$, which provides a larger effective surface area and more adsorption sites (Jin et al., 2020). They further analyzed the 15 wt% nanocomposite for other sensing characteristics (Jin et al., 2020). However, Li et al. (2020) reported the ammonia sensing performance of prepared nanocomposite without varying the concentration of $Ti_3C_2T_x$. A comparative analysis of ammonia-sensing performance of both studies is listed in Table 11.2.

The sensitivity of PEDOT:PSS/$Ti_3C_2T_x$ was manifold less than that of PANI/$Ti_3C_2T_x$-based ammonia sensor (Table 11.2). It can be ascribed to dendritic growth of nanofibillar PANI, which was not found in case of PEDOT:PSS/$Ti_3C_2T_x$. Among all the polymers, PANI has been most studied for its ammonia sensing characteristics due to its redox nature and tunable conductivity (Kumar et al., 2020; Tanguy et al., 2018). Thus, the presence of PANI in PANI/$Ti_3C_2T_x$ nanocomposite has been attributed to its amplified sensitivity and lower detection limit compared to that of PEDOT:PSS/$Ti_3C_2T_x$. However, a PANI/$Ti_3C_2T_x$-based sensor lacks response and recovery process, which can be ascribed to slow desorption of ammonia through nanocomposite and possess chances of future advancements. Both the sensors have shown selectivity traits toward ammonia compared to other analytes, which can be ascribed to the adsorption energy of ammonia, and the presence of structural defects and surface functional groups, which act as specific adsorption sites for ammonia (Lee et al., 2017; Yu et al., 2015; Ding et al., 2018).

TABLE 11.1
Variation of 100 ppm Ammonia-Sensing Characteristics with Change in $Ti_3C_2T_x$ Concentration in Nanocomposite (Jin et al., 2020)

Weight Percentage of $Ti_3C_2T_x$ (%)	Conductivity (S/cm)	Sensing Response (%)	Response Time (s)	Recovery Time (s)
0 (Pure PEDOT-PSS)	<0.01	~5–7	~160	130
8	~0.02	~12–15	~158	~100
11	~0.02–0.25	~20–25	~140	~80
15	~0.07	36.6	116	~40
20	~0.03	~20	~120	~70
25	~0.02–0.28	~15–18	~120	~100
100 (Pure $Ti_3C_2T_x$)	Not Mentioned	~10–12	~180	~180

TABLE 11.2
Ammonia-Sensing Characteristics of $M_3X_2T_x$-P Nanocomposites

Sensing Performance	PEDOT:PSS/$Ti_3C_2T_x$ (Jin et al., 2020)				PANI/$Ti_3C_2T_x$ (Li et al., 2020)			
Lowest Detection Limit	10 ppm				25 ppb			
Detection Range	10–1000 ppm				0.025–50 ppm			
Sensitivity	4.94% (10 ppm)	33.1% (100 ppm)	~60% (500 ppm)	~95% (1000 Ppm)	0.05 % (25 ppb)			400% (50 ppm)
Response Time	116 s for 100 ppm				~600s			
Recovery Time	40s for 100 ppm				~1400s			
Linear Regression value	0.957 for 10–100 ppm range		0.983 for 100–1000 ppm range		0.997 for 2–10 ppm			
Stability	~33% at 100 ppm for 4 weeks				88% for 35 days			
Temperature	27°C for all measurements				10–40°C Variation			
Humidity	20–90% RH				20–90% RH			
Repeatability	3 cycles				4 cycles			
Selectivity	1.2% 100ppm toluene	4.6% 100ppm Ethanol	14% 100ppm Methanol	3.4% 100ppm Acetone	~10% 10ppm HCHO	5% 10ppm H_2S	~8% 10ppm CO	~8% 10ppm SO_2
Flexibility	~33% at 100 ppm for different bending angle (60°–240°)				~22% at 10 ppm for bending angle (20, 30, 40)° and bending for 100, 200 and 500 times.			

11.9.1.5 Effect of Variation in Humidity and Temperature on the Working of Sensors

The sensitivity of $M_3X_2T_x$-P nanocomposite sensors toward ammonia increases with surge in relative humidity, which is ascribed to formation of NH_4OH or $NH_3{\cdot}H_2O$ (due to reaction of moisture with ammonia; Abdulla et al., 2015; Mogera wt al., 2014). Since the adsorbed water molecules over PANI surface captures electron from it and form H_3O^+ ions, which increases ionic conductivity of PANI, enhancing its sensing response (Abdulla et al., 2015). On exposure to ammonia, the formation of NH4OH or NH3·H2O takes place, which further captures protons from PANI, resulting in a higher sensing response (Mogera wt al., 2014). However, Li et al. (2020) reported that on increasing the relative humidity (RH) from 40% to 90%, the sensitivity of PANI/$Ti_3C_2T_x$ sensor decreases and is found to be lowest at 90% RH. The reason can be attributed to the accumulation of a thin layer of water over the film at adsorption sites, promoting the formation of H3O$^+$ ions (Abdulla et al., 2015). The excessive adsorption of thin water layer hinders the adsorption of ammonia molecules and limits the sensitivity (Li et al., 2020).

With the rise in environmental temperature, the sensing response of $M_3X_2T_x$-P toward ammonia generally decreases linearly. It is due to desorption of H3O$^+$ ions from the polymer surface and competition between desorption and exothermic adsorption of ammonia on polymer surface at elevated temperature (Matsuguchi et al., 2002). Li et al. (2020) studied the ammonia-sensing performance of a PANI/$Ti_3C_2T_x$ sensor in temperature range of 10–40°C. The temperature range was chosen as it best suits agricultural applications (Li et al., 2020). The response was found to decreases linearly with surge in temperature.

11.9.1.6 Monitoring Volatilization of Agricultural Ammonia by $M_3X_2T_x$-P Sensors

Since the measurement of volatilization of ammonia is essential in agriculture (Wei et al., 2018), it is required to evaluate the feasibility of $M_3X_2T_x$-P sensor for detecting the same. It is generally done by comparing the sensing response of sensor with the conventional sulfuric acid adsorption method (SAAM) and Drager nitrogen tube method (DTM) (Li et al., 2020). For general procedure (Li et al., 2020), in SAAM, 1 kg of fresh soil is placed in a large vessel and cultivated under a dark environment at a temperature around 20°C and humidity about 60%. Furthermore, a mixture of urea in water (in 1:2 ratio) is uniformly sprayed over soil. After fertilization, 15 mL of sulfuric acid in a small beaker is placed in a vessel containing soil, and the concentration of NH_4^+ produced due to reaction of soil and sulfuric acid is recorded by flow analyzer in every 24h. In DTM (Li et al., 2020), a dragger nitrogen tube is also placed in similar vessel. The produced ammonia on entering this tube reacts with the reagent present in a tube and turns its color to blue from yellow. The coloring length of the tube is an indicator of the concentration of ammonia produced. For ammonia volatilization measurement, ammonia sensor is placed in vessel near the soil, and measurements were recorded. Li et al. (2020) compared the response of fabricated PANI/$Ti_3C_2T_x$ sensor to average of results obtained from DTM and SAAM. They noticed that with increasing time (in days), the fabricated sensor gave ammonia trends according to an average of DTM and SAAM, which suggests the feasibility of a PANI/$Ti_3C_2T_x$ sensor toward unattended monitoring of agricultural ammonia volatilization (Li et al., 2020).

11.9.1.7 Sensing Mechanism of $M_3X_2T_x$-P Sensors

$M_3X_2T_x$-P sensors are generally semiconducting in nature due to surface termination and nature of M, X, or polymer (Hajian et al., 2018). The ammonia-sensing mechanism generally includes redox reactions between sensing material and analyte, charge transfer, and synergistic effects (Li et al., 2020; Jin et al., 2020). The sensing phenomena of $M_3X_2T_x$-P sensors can be ascribed to two mechanisms.

11.9.1.7.1 Chemisorption of Oxygen

At room temperature, $M_3X_2T_x$-P composite absorbs the oxygen layer and converts it into oxygen anions through trapping electrons from its conduction band (Thirumalairajan et al., 2014; Gund et al., 2019). It causes the creation of an electron depletion layer with more excellent resistance. On exposure to ammonia, electrons of oxygen ions are released back to the composite conduction band (Gao et al., 2020b). It results in a decrease in the electrical resistance of the composite sensor. On reinstating air to the sensing chamber, oxygen ions replace ammonia over the sensor surface, and the original resistance value is restored (Jin et al., 2020). Jin et al. (2020) have also reported the similar phenomena for PEDOT:PSS/$Ti_3C_2T_x$ composite and summarized it in the following chemical reactions:

$$O_2 + e^- \rightarrow O_2^-$$
$$4NH_3 + 5O_2^- \rightarrow 4NO + 5e^- + 6H_2O$$

However, at room temperature, the conversion of oxygen into oxygen anions is very less, this mechanism contributes very little to the sensing phenomena (Jin et al., 2020).

11.9.1.7.2 Physisorption of Ammonia Molecules

It is predominant phenomenon in ammonia sensing through $M_3X_2T_x$-P composite (Li et al., 2020; Jin et al., 2020). The electrons at surface of $M_3X_2T_x$-P composite interact with ammonia and result in an increase of number of charge carriers (Jin et al., 2020). This decreases the resistance of the sensor and surges its conductivity. This phenomenon is also found to be predominant in other conducting polymers and their composites (Wu et al., 2013). Jin et al. (2020) proposed adding $Ti_3C_2T_x$ into PEDOT:PSS also increases the π–π interaction, leading to a large number of charge carriers and surface adsorption sites, which, in turn, enhances its sensing response. However, Li et al. (2020) extended this view to forming a small Schottky junction between p-type $Ti_3C_2T_x$ and p-type PANI in PANI/$Ti_3C_2T_x$ composite. Since the work function of $Ti_3C_2T_x$ was lower than that of PANI, a hole-depletion layer forms in the PANI region at interfaces of PANI/$Ti_3C_2T_x$. On adsorption of ammonia, a decrease in hole concentration widens the depletion region, resulting in increased resistance of PANI, which explains the sensing phenomenon in terms of decrease in resistance of $M_3X_2T_x$-P composite on exposure ammonia (Li et al., 2020).

Thus, the ammonia sensor based on $M_3X_2T_x$-P nanocomposites are flexible (due to solubility and film formation on PI), cost-effective and energy efficient (due to room temperature operation), highly sensitive (optimal structure of nanocomposite due to the inclusion of a polymer between MXene stacking), improved degree of protonation in polymers (due to the introduction of MXene sheets), environmental- and user-friendly (exclusion of heavy metals or toxic substances), rapid (fast response and recovery), and apt for commercial development (linear detection range). However, a lower level of ammonia, such as ppb measurements with improved sensing characteristics and real-time human breath analysis, is still required with profound sensing characteristics and is the current area of research. It can also be noticed that these studies (Jin et al., 2020; Li et al., 2020) as a well open a window toward the detection of the alcohol group since the response toward ethanol and methanol was significant and cannot be ignored. However, the sensing phenomenon for the alcohol group is different as compared to ammonia and can be easily picked out using electronic circuitry (Chaudhary et al., 2016).

11.9.2 M_2X-Based MXene–Polymer Composites

Naguib et al. (Naguib et al., 2013) proposed that M_2X-based MXenes hold a larger specific surface area than M_3X_2 or M_4X_3-based MXenes due to their fewer atomic layers. Thus, such a high effective surface area makes them promising, excellent candidates for gas sensing applications. Furthermore, conducting polymers such as PANI can enhance their sensing characteristics due to an increase in interlayer distance in MXenes. The following section discusses various reports on M2X-based MXene-conducting polymer nanocomposite for ammonia-sensing applications.

11.9.2.1 Nb_2CT_x/PANI Nanocomposite Ammonia Sensors

Niobium carbide–polyaniline (Nb_2CT_x/PANI) nanocomposites were mostly used for ammonia detection due to their large surface-to-volume ratio and high adsorption sites (Wang et al., 2021, 2021). There are two distinct reports on Nb_2CT_x/PANI nanocomposite-based ammonia sensors by Wang et al. (Wang et al., 2021, 2021) In both the reports active parent materials of nanocomposite are Nb_2CT_x nanosheets and polyaniline nanofibers. However, they differ in synthesis technique as one of the nanocomposites is prepared through ex situ technique and one by *in situ* technique.

11.9.2.2 Sensor Fabrication

Ex-situ Nb_2CT_x/PANI nanocomposite (E-NPC): In this method, 2-D ultrathin Nb_2CT_x nanosheets were prepared using typical method of hydrofluoric acid (HF) etching and tetrapropylammonium hydroxide (TPAOH) intercalation method (Wang et al., 2021). Polyaniline nanofibers were separately synthesized over PI substrate with gold interdigitated electrodes through a typical *in situ* polymerization technique (Wang et al., 2021). Furthermore, a layer of Nb2CTx was spray-coated over a PANI-coated PI substrate and dried in an oven at 60°C (Wang et al., 2021). For sensing measurements, the E-NPC sensor was coupled with a facile triboelectric nanogenerator (TENG) operated by a linear motor of fixed frequency of 1 Hz, which acts as a power source to drive the fabricated sensor (Wang et al., 2021).

***In situ* Nb_2CT_x/PANI nanocomposite (I-NPC):** In this communication (Wang et al., 2021), the method for fabricating 22D ultra-thin Nb2CTx nanosheets is the same as in the previous report (Wang et al., 2021). However, the polyaniline was synthesized over Nb_2CT_x nanosheets using an *in situ* polymerization technique (Wang et al., 2021). Furthermore, the nanocomposite was spray coated on PI substrate with gold interdigitated electrodes for sensing measurements.

Thus, both the synthesis routes for sensor fabrication were different and used TENG nanogenerator for the E-NPC sensor. For the comparison, pure PANI and Nb_2CT_x-based sensors were also fabricated using typical processes (Wang et al., 2021).

11.9.2.3 Sensing Measurements

In TENG-coupled E-NPC sensors, on varying the ammonia concentration, the resistance of E-NPC sensor changes, which, in turn, alters the output voltage of external-loading E-NPC sensor driven by TENG (Wang et al., 2021). The gas-sensing response for the E-NPC sensor was defined in terms of the real-time output voltage of the sensor as the percentage degree of change in voltage in the presence of ammonia given by Equation11.2.

All the measurements were done at room temperature (~25°C) under 87.1% relative humidity conditions (Wang et al., 2021). Thus, the percentage change in output voltage through the E-NPC sensor is the sensing parameter during sensing measurements. For I-NPC sensor (Wang et al., 2021), the percentage change in resistance in the presence of ammonia is taken as a sensing parameter given by Equation11.2 at room temperature, pressure, and humidity conditions (62% RH). The real-time monitoring of breath analysis is further done to illustrate the commercial significance of the prepared sensors.

11.9.2.4 Sensing Characteristics

For E-NPC sensor (Wang et al., 2021):

> Wang et al. (Wang et al., 2021) designed four E-NPC sensors with different concentration of Nb_2CT_x in terms of spray volume i.e. E-NPC-1 (0.05 mL), E-NPC-2 (0.1 mL), E-NPC-3 (0.15 mL), E-NPC-4 (0.2mL) and compared it with pure Nb_2CT_x and PANI sensor. The sensing response towards 100 ppm of ammonia for E-NPC-1, E-NPC-2, E-NPC-3, E-NPC-4, Nb_2CT_x and PANI sensor was found to be 197.20 %, 301.31 %, 108.36 %, 48.90 %, 8.15%, and 128.81 %, respectively. Thus, the E-PNC-2 sensor possesses the highest sensing response compared to other sensors due to its full adsorption sites available to ammonia molecules. On further increasing the concentration of Nb2CTx, the adsorption sites were hindered, limiting the sensing response. In terms of output voltage of sensor driven through TENG, the response of E-NPC-2 sensor was found to be highest amongst all (2.57% per ppm) with least response time (105 s), which was attributed to the synergistic effects due to p-n junction effect amongst Nb_2CT_x and PANI. E-NPC-2 sensor was found to exhibit significant linear response with R^2 = 0.9655, with lowest limit of detection as 1 ppm of ammonia with sensitivity 2.87%. E-NPC-2 sensor was also found to be stable in variation with temperature and humidity with consistent sensitivity towards ammonia. Further, the response is found to be recoverable, repeatable and reproducible for many cycles. The response is found to be significant and manifold enhanced compared to that of sensor reports based on PANI-MWCNT-TENG (255% at 100ppm), $Ti_3C_2T_x$ (0.8% at 100 ppm), $Ti_3C_2T_x$-rGO (4.8% at 10 ppm) and PANI-WO_3 (121% at 100ppm). However, the report lacks in demonstrating the selectivity of said sensor.

For I-NPC sensor (Wang et al., 2021):

> In *in-situ* synthesis, there is always possibility of formation of heterojunctions amongst the parent constituents, which further enhances its sensing response compared to that prepared by ex-situ techniques. This observation is again validated by Wang et al. for I-NPC sensor. The fabricated sensor possessed 29.95% sensitivity towards 1 ppm of ammonia, which is almost 14 fold enhance compared to E-NPC-2 sensor. The enhancement in sensitivity is dedicated to better 3-D morphology and interface interaction (heterojunctions) amongst PANI and Nb_2CT_x. The fabricated sensor is detects linearly with R^2 = 0.9619 in the range of 1–50 ppm of ammonia. Further, humid breath ammonia monitoring was performed and the said sensor demonstrated excellent sensitivity (74.68% at 10 ppm), low detection limit (20 ppb), good linear detection range (R^2 = 0.9951), repeatability (3 cycles), good response time (126s), long term stability (tested for 35 days), stability in variation of humidity with poor recovery time (640s). Further, selectivity was tested and the I-NPC sensor

was found to exhibit maximum sensitivity towards 10 ppm of ammonia (74.44%) compared to that of sulfur dioxide (~9%), ethanol (2%), carbon monoxide (almost negligible) and hydrogen di-sulfide (~1%). However, as per reports in literature 9% sensitivity towards sulfur dioxide is significant, but it can be differentiated from response of ammonia using simple circuitry (Chaudhary et al., 2016, 2017, 2022a, 2022b). This is because sulfur dioxide is an oxidizing gas, whereas ammonia is a reducing gas, and the sensing mechanism for both the analytes is entirely opposite to each other (Chaudhary et al., 2017). Thus, the study also opens [the] window for sulfur dioxide sensing using I-NPC sensor.

11.9.2.5 Sensing Mechanism

It is essential to understand the sensing mechanism of various constituents prior to exploring that nanocomposite.

For PANI:

> Pure PANI undergoes protonation or deprotonation on adsorption or desorption of ammonia molecules (Chaudhary et al., 2015a). Thus, there is conversion of PANI between its reducing and oxidizing states, which means between emeraldine salt and emeraldine base. Adsorbed ammonia molecule lends an electron to p-type PANI and forms ammonium ion (NH_4^+), thereby decreasing its resistance due to reduction in hole density (Liu et al., 2017).

For Nb_2CT_x:

> Nb_2CT_x shows an n-type response towards ammonia (Wang et al., 2021). It means its resistance decreases in the presence of ammonia due increase in its electron density.

For Nb_2CT_x/PANI nanocomposites:

> There are three major sensing phenomena of Nb_2CT_x/PANI nanocomposites for ammonia, which includes the following.

11.9.2.5.1 Formation of p-n Junctions

Since Nb_2CT_x is n-type material and PANI is a p-type material, it is speculated that there is formation p-n junction at the interface of the two parent materials in the case of E-NPC (Wang et al., 2021; Liu et al., 2017; Chaudhary et al., 2015a), which results into enhanced sensing response. Thus, a thin depletion layer forms at interface due to diffusion of holes from PANI side and electrons from Nb_2CT_x side. On adsorbing the ammonia molecule, the resistance of composite decreases due to a decrease in hole concentration of PANI side, which broadens the depletion layer (Wang et al., 2021). It is due to the predominance of PANI in the composite due to its high conductivity compared to that of Nb_2CT_x. A similar mechanism is seen in I-NPC; however, the formation of small p-n heterojunctions throughout the nanocomposite, including the interface, enhances its ammonia sensitivity manifold (Wang et al., 2021; Chaudhary et al., 2021c).

11.9.2.5.2 Formation of Hydrogen Bonds

There is the formation of abundant intermolecular hydrogen bonds among PANI and Nb_2CT_x. These hydrogen bonds occupy many hydrophilic functional groups of the sensor, which reduces the active sites for water adsorption (Gao et al., 2020a; Wang et al., 2021).

11.9.2.5.3 Formation of Unique Hetero-Interfacial Functions

Using various DFT calculations, it has been found that there is a formation of unique hetero-interfacial functional groups at the interfaces among the precursors (Naguib et al., 2013; Gao et al., 2020b). Surface functionalization of MXene by various functional groups (–OH, –F or –O) turns them into semiconductors with a bandgap in the range of 0.05–1.8 eV (Gao et al., 2020a). This functionalization surges interfacial charge carrier transfer, further enhancing its sensing response (Wang et al., 2021).

These three phenomena together contribute to the enhanced sensing response of M_2X-based MXene-conducting polymer-based ammonia sensors.

Thus, the studies on MXene–polymer nanocomposites are very less and dedicated to ammonia gas, which can be ascribed to the selectivity of these sensors towards ammonia. The reason can be attributed to the specific adsorption energy of ammonia and the presence of functional groups which are specifically reactive toward ammonia. However, the studies also show the potential to detect other gases, including sulfur dioxide and nitrogen oxides (Wang et al., 2021). Thus, dedicated studies are required and are the current research topic, along with enhancing ammonia-sensing characteristics of MXene–polymer nanocomposite-based sensors.

11.10 MXene–POLYMER NANOCOMPOSITE-BASED CARBON DIOXIDE SENSORS

The emission of carbon dioxide (CO_2) harms the environment, crop production, and human health. Its monitoring at a low ppm level is the foremost concern of the research community (Mulmi et al., 2020). Although the discovery of MXene has also opened a new window for room temperature detection of CO_2, reports on its sensing through MXene sensors are in scarcity (Zhan et al., 2020; Ho et al., 2021; Kausar et al., 2021). Zhou et al. (2020) reported CO_2-sensing performance of ternary nanocomposite of nitrogen-doped MXene, $Ti_3C_2T_x$ (N-MXene), polyethyleneimine (PEI), and reduced graphene oxide (rGO) and compared it with that of pristine $Ti_3C_2T_x$, PEI, $Ti_3C_2T_x$-PEI, rGO-PEI, $Ti_3C_2T_x$-rGO, and N-$Ti_3C_2T_x$.

11.10.1 Fabrication of Sensors

$Ti_3C_2T_x$ was prepared as discussed previously in $M_3C_2T_x$ section through LiF/HF etching (Li et al., 2020). N-$Ti_3C_2T_x$, rGO, PEI, rGO-PEI, $Ti_3C_2T_x$-rGO were prepared using typical processes reported by Zhou et al. (2020). $Ti_3C_2T_x$-PEI was prepared by ex situ technique by simply mixing pre-synthesized constituents in deionized water under nitrogen flow. For the synthesis of ternary N-$Ti_3C_2T_x$-PEI-rGO (NTPG) nanocomposite, pre-synthesized $Ti_3C_2T_x$, rGO, and PEI were mixed under constant flow of nitrogen for 2hours. Furthermore, sensors were fabricated by spray-coating the target solution on silicon oxide (SiO_2)/Si substrate with planar interdigitated electrodes (Zhou et al., 2020). The fabricated sensors are chemiresistive-type sensors with response defined by Equation 11.1.

11.10.2 Sensing Performance

The concentration of CO_2 beyond 1000 ppm indoors is harmful to human health (Hafiz et al., 2014). Thus, the detection of CO_2 at this limit is indeed required at indoor emission centers. Zhou et al. (2020) analyzed CO_2 monitoring performance of NTPG, $Ti_3C_2T_x$, PEI, $Ti_3C_2T_x$-PEI, rGO-PEI, $Ti_3C_2T_x$-rGO, and N-$Ti_3C_2T_x$ sensors toward 1000 ppm under 48% RH at 20°C. The sensing response of $Ti_3C_2T_x$, N-$Ti_3C_2T_x$, and $Ti_3C_2T_x$-rGO was not unstable with high noise, and the rest were shallow, except the NTPG-based sensor (Zhou et al., 2020). Pure PEI is insulating, and its resistance was not measurable below 36% RH (Zhou et al., 2020). However, with a rise in RH, the resistance of PEI decreases due to humidity-activated proton conduction phenomena. PEI-achieved saturation in resistance at 62% RH (Zhou et al., 2020). Pristine rGO and $Ti_3C_2T_x$ do not show any change in resistance at 62% RH. Thus, it has been chosen to do further sensing measurements (Zhou et al., 2020). Furthermore, four ternary NTPG sensors were fabricated by varying PEI concentration (0.0025, 0.005, 0.0075, and 0.01 mg/mL). NTPG sensor with 0.01 mg/mL shows excellent sensitivity of about 7.4% toward 40 ppm of CO_2under 62% RH at 20°C. Further sensing characteristics are listed in Table 11.3.

Zhou et al. (2020) observed no recovery or significantly less recovery on removing CO_2 in dry air. However, if air supplied to the sensing chamber for evacuation of CO_2 was wet with 2.1% RH, the sensing response surges to 9% for 600 ppm and recovery is achieved. It is attributed to superior proton-hopping conduction of PEI due to moisture in air (Zhou et al., 2020). The response is found to be repeatable and linear in two different ranges as listed in Table 11.3. However, the response decreased with the surge in temperature due to a reduction in adsorbed water molecules. The sensing response observed was found to be significant as compared to the reports present in the literature on CO_2 detection through zinc oxide (20% for 200 ppm at 250°C; Kanaparthi et al., 2019), cesium oxide (110% for 150 ppm at 100°C; Zito ey al., 2020), Ru@WS_2 (1.8% for 20 ppm at 25°C; Rathi et al., 2020) and Ag@Cuo/$BaTiO_3$ (120% for 100 ppm at 120°C; Joshi et al., 2017).

11.10.3 Sensing Mechanism

PEI dominates the sensing mechanism in the ternary nanocomposite. PEI possesses primary R-NH_2, secondary R_1R_2-NH, and tertiary $R_1R_2R_3$-N amino groups (Doan et al, 2014). Primary and secondary amino groups of PEI on exposure to CO_2 result in acid–base reactions to give carbamates, whereas tertiary amino groups result in carbonic acid by the dissolution of CO_2 (Doan et al, 2014). In the presence of humidity, bicarbonates are formed instead of carbamates. N-MXene in the nanocomposite due to its hydrophilic nature adsorbs water molecules, which further protonates PEI the charge transfer (Srinives et al., 2015; Doan et al, 2014). The presence of rGO in nanocomposite serves as rapid conduction channels for charge transfer and collection processes (Zhou et al., 2020). Thus, when nanocomposite is exposed to CO_2, the number of free amines decreases, and the mobility of protons reduces, which is amplified by the action of MXene and rGO, resulting in a net increase in the resistance of the sensor (Zhou et al., 2020; Srinives et al., 2015). Zhou et al. (2020) ascribed fast recovery in wet air to hindrance to acid–base reaction due to steric effects offered by PEI. Thus, all three constituents play a vital role in the sensing phenomenon.

Although the reported sensor is cost-effective with significant sensing characteristics to detect CO_2, improvement to enhance response and recovery still requires ordinate attention for commercial development of these sensors.

TABLE 11.3
CO_2-Sensing Characteristics of NTPG Sensor

Sensing Performance	NTPG Sensor (Zhou et al., 2020)			
Lowest Detection Limit	8 ppm			
Detection Range	8–600 ppm			
Sensitivity	~1% for 8 ppm			
Recovery Time	~9 min for 600 ppm			
Linear Regression value	0.974 for 8–40 ppm range		0.984 for 40–600 ppm range	
Stability	~8% at 600 ppm for 50 days			
Temperature	20–50°C for all measurements			
Humidity	32–64% RH			
Repeatability	4 cycles			
Response Time	~8.8 min for 600 ppm			
Selectivity	No signal (8 ppm of H_2S)	0.5% (8 ppm of SO_2)	0.3% (40 ppm of HCHO)	0.18% (8 ppm of NH_3)

11.11 MXene–POLYMER NANOCOMPOSITE-BASED VOCS/VAPOR SENSORS

VOCs have been extensively used as raw materials in various industries and possess adverse effects on human health and the environment (Tung et al., 2017; Tang et al., 2017). Thus, the monitoring of VOCs, such as methanol, ethanol, and acetone, which are highly flammable and toxic, is indeed required in the field of manufacturing industries, personal healthcare industries, and biomedical applications (Tung et al., 2017; Wang et al., 2020; Jalal et al., 2018). Although there are reports on the sensing of VOCs through MXenes, they lack sensitivity due to fewer adsorption sites (Kim et al., 2018; Chen et al., 2020; Li et al., 2021). However, polymer inclusion in MXene enhances adsorption sites and charge carrier transport for better sensing performance (Carey et al., 2021; Riazi et al., 2021). In literature, there are very few reports on MXene–polymer nanocomposites for VOC detection (Wang et al., 2020; Yuan et al., 2018; Zhao et al., 2019; Zhan et al., 2020). However, these reports are majorly dedicated to $M_3X_2T_x$-P nanocomposites to detect various VOCs at room temperature. Yuan et al. (2018) reported the fabrication of 3D MXene–polymer framework of $Ti_3C_2T_x$ poly-vinyl alcohol (PVA) and PEI(3TPP) and evaluated it for detecting various VOCs, including methanol, ethanol and acetone. Wang et al. (2020) developed a dedicated sensor for methanol detection based on $Ti_3C_2T_x$/PEDOT:PSS nanocomposite. On the other hand, Zhao et al. (2019) reported a dedicated ethanol sensor based on $Ti_3C_2T_x$- PANI nanocomposite.

11.11.1 Sensor Fabrication

The fabrication process of the sensor for VOCs sensing is quite similar to that of the ammonia sensor, as discussed previously. Zhao et al. (2019) reported the synthesis of $Ti_3C_2T_x$- PANI nanocomposite at low temperature through *in situ* polymerization as discussed earlier (Li et al., 2020). However, they have kept the synthesis temperature very low (0–5°C) to get a high yield and limiting the damage to the $Ti_3C_2T_x$ structure during synthesis process. The structure of obtained nanocomposite was similar to the lamellar structure shown in Figure 11.2. A flexible sensor was fabricated on interdigitated PET substrate. Wang et al. (2020) fabricated$Ti_3C_2T_x$/PEDOT:PSS nanocomposite using ex situ mixing technique as mentioned in an earlier section (Jin et al., 2020). They have also varied the ratio of PEDOT:PSS to $Ti_3C_2T_x$ in nanocomposite, particularly 10:1,8:1,4:1, 2:1,1:1, and1:2 and made sensing films over interdigitated electrodes. However, Yuan et al. (2018) have used a coupled electrospinning and self-assembly method to develop a 3TPP framework. The 3D polymer framework was fabricated in a typical synthesis by electrospinning the PVA/PEI mixture using sodium dodecyl sulfate (SDS). Fibers of PVA/PEI framework were collected on interdigitated electrode-based PET substrates through electrospinning and dipped in glutaraldehyde overnight for crosslinking nanofibers. Then, it was immersed in MXene solution and dried to obtain a sensor.

11.11.2 Sensing Performance

VOCs' sensing performance for all the fabricated sensors was measured using a chemiresistive mode. The sensing performance of $M_3X_2T_x$-P nanocomposites were found to be enhanced compared to their pristine parent constituents, which can be ascribed to the presence of lamellar structure with enlarged interlayer spacing, improved charge carrier transport and presence of specific functional groups (Zhao et al., 2019; Wang et al., 2020; Yuan et al., 2018). Wang et al. (2020) also observed that the variation in $Ti_3C_2T_x$ concentration in nanocomposites changes their sensing response toward methanol. They found the sensor based on mass ratio 4:1 showed the highest sensing response toward 300 ppm of methanol, ethanol, and acetone among all sensors. It is ascribed to eliminating MXene-MXene connectivity due to the inclusion of PEDOT:PSS between MXene layers. However, in other combination ratios, MXene–MXene connections are predominant,

TABLE 11.4
Sensing Characteristics of $M_3X_2T_x$-P Nanocomposites toward Methanol

Sensing Performance	$Ti_3C_2T_x$/PEDOT:PSS Sensor (4:1) (Wang et al., 2020)	3TTP Sensor (Yuan et al., 2018)
Lowest Detection Limit	180 ppm	50ppb
Detection Range	180–500 ppm	50 ppb—1.5×10^5 ppm
Sensitivity	0.4% at 180 ppm	2.7% for 5 ppm
Recovery Time	>500s	1.7 min
Response Time	~280s	1.5 min
Temperature	RT	RT
Humidity	Ambient	Ambient
Repeatability	Not mentioned	5 cycles
Linear Regression value	Not mentioned	0.14 per ppm
Selectivity	~0.1% to ethanol and acetone	3.6%, 4.4%, and 2.7% for 5ppm of acetone, ethanol, and methanol
Flexibility	Not mentioned	Flexible for 1000 bending cycles

increasing the nanocomposite's conductivity and reducing its sensing response. They have further evaluated sensing characteristics for 4:1 ratio $Ti_3C_2T_x$/PEDOT:PSS nanocomposite toward methanol as listed in Table 11.4.

It was observed that 3TTP-based sensors possess more significant methanol-sensing characteristics than that of other $M_3X_2T_x$-P nanocomposites in terms of response, detection range, flexibility, and selectivity. However, it suffers from extended response and recovery time, which possess further improvement (Yuan et al., 2018). Zhao et al. (2019) recorded relative change in the sensing layer's current on exposure to VOCs instead of recording variation in resistance. They also observed that the fabricated flexible sensor is dedicated to detection of ethanol. The selectivity of fabricated $Ti_3C_2T_x$/PANI sensors toward ethanol was evaluated through DFT studies (Zhao et al., 2019) and was attributed to smallest bond length for the adsorption of ethanol compared to that of methanol, ammonia, and acetone. The sensor was found to be rapid due to very small response and recovery time, flexible, and highly repeatable. However, studies of related stability can also be included to propose its commercial development. Table 11.5 compares the ethanol-sensing response of various $M_3X_2T_x$-P nanocomposites.

Furthermore, acetone-sensing characteristics of $M_3X_2T_x$-P nanocomposites were compared and listed in Table 11.6. It was observed that 3TTP-based sensors possess better acetone-sensing characteristics compared to others, which can be attributed to its 3D framework–type structure with high effective surface area, more adsorption sites, and large reactive functional groups (Yuan et al., 2018).

3TTP-based sensors possess significant sensing responses toward methanol, acetone, and ethanol with improved sensing characteristics owing to its structure and surface chemistry (Yuan et al., 2018). However, it suffers from slow response and recovery process and selectivity. Thus, it requires more functionalization for the formation of dedicated sensors toward specific VOCs.

11.11.3 Sensing Mechanism

The sensing mechanism of $M_3X_2T_x$-P nanocomposites has been explained in terms of its hydrogen bonding capability due to functional groups (Kim et al., 2018). $M_3X_2T_x$-P nanocomposites possess functional group over its surface, which attaches to VOCs through electrostatic interaction and hydrogen bonding (Kim et al., 2018; Riazi et al., 2021). Yuan et al. (2018) observed a fascinating fact

TABLE 11.5
Sensing Characteristics of $M_3X_2T_x$-P Nanocomposites toward Ethanol

Sensing Performance	$Ti_3C_2T_x$/PEDOT:PSS Sensor (4:1) (Wang et al., 2020)	$Ti_3C_2T_x$- PANI Sensor (Zhao et al., 2019)	3TTP Sensor (Yuan et al., 2018)
Lowest Detection Limit	60 ppm	1 ppm	50 ppb
Detection Range	60–500 ppm	1–200 ppm	50 ppb—8 × 10^4 ppm
Sensitivity	0.1% at 300 ppm	1.56% at 1 ppm	4.4 % at 5 ppm
Recovery Time	~500s	0.5s	1.7 min
Response Time	~280s	0.4s	1.5 min
Temperature	RT	RT	RT
Humidity	Ambient	Ambient	Ambient
Repeatability	Not mentioned	140 cycles	5 cycles
Linear Regression Value	Not mentioned	1.56% per ppm	0.17% per ppm
Selectivity	Not selective to ethanol	~20% for 200 ppm of methanol, acetone, and ammonia	3.6%, 4.4%, and 2.7% for 5ppm of acetone, ethanol, and methanol
Flexibility	Not mentioned	~25–27% for 150 ppm for bending angle (0–120)°	Flexible for 1000 bending cycles

TABLE 11.6
Sensing Characteristics of $M_3X_2T_x$-P Nanocomposites toward Acetone

Sensing Performance	$Ti_3C_2T_x$/PEDOT:PSS Sensor (4:1) (Wang et al., 2020)	3TTP Sensor (Yuan et al., 2018)
Lowest Detection Limit	60 ppm	50 ppb
Detection Range	60–500 ppm	50 ppb—3 × 10^5 ppm
Sensitivity	~0.09% at 300 ppm	0.08% at 50 ppb
Recovery Time	~500s	1.7 min
Response Time	~280s	1.5 min
Temperature	RT	RT
Humidity	Ambient	Ambient
Repeatability	Not mentioned	5 cycles
Linear Regression value	Not mentioned	0.1 per ppm (50 ppb to 20 ppm) 3.7 × 10^{-3} per ppm (25–1.5 × 10^5 ppm)
Selectivity	Not selective to Acetone	3.6%, 4.4% and 2.7% for 5ppm of acetone, ethanol, and methanol
Flexibility	Not mentioned	2% for 20ppm for 1000 bending and unbending cycles

that the resistance of 3TTP sensors was found to increase in every VOC regardless of their nature, whether p-type or n-type. It can be attributed to the dominance of MXene in the nanocomposite, which possesses a metallic type of conductivity (Yuan et al., 2018). They have also proposed a mechanism for selectivity of VOCs due to their ability to form strong hydrogen bonding with more electronegative atoms like oxygen atoms and electrostatic attraction with MXene (Yuan et al., 2018). Thus, VOCs, such as acetone, methanol, and ethanol, are easily detected by $M_3X_2T_x$-P nanocomposites. However, polar organic molecules form weak hydrogen bonds and nonpolar organic molecules form no hydrogen bond, restricting their detection through M3X2Tx-P nanocomposites-based sensors (Yuan et al., 2018).

The sensing response of $M_3X_2T_x$-P nanocomposites sensor was compared to other reports in the literature and found to be consistent and enhanced in terms of sensing characteristics (Wang et al., 2020; Zhan et al., 2020; Chen et al., 2020). Zhao et al. (2019) explained the enhanced sensing response of $M_3X_2T_x$-P nanocomposites in terms of three factors, including increased interlayer distance resulting in more adsorption sites, the dominance of MXene and its metallic conductivity, and hydrogen bonding and electrostatic interaction between VOCs and the nanocomposite surface due presence of abundant functional groups. However, these reports lack the study of variation in sensing response with changes in humidity and working temperature. These studies also miss one of the 3Ss stability studies. Thus, more dedicated research is required for the commercial development of VOC sensors based on $M_3X_2T_x$-P nanocomposites.

11.12 CONCLUSION AND FUTURE PROSPECT

MXene–polymer nanocomposites-based sensors open a new window for detecting various gaseous and vapor analytes, including ammonia, sulfur dioxide, oxides of nitrogen, methanol, ethanol, and acetone at room temperature with significant sensing characteristics. The room temperature operation of MXene–polymer nanocomposite sensors makes them energy- and cost-efficient due to the exclusion of micro-heating assemblages. Due to the absence of any toxic material such as heavy metal, these sensors are user-friendly, generating negligible toxic nano-waste. They are rapid in detecting gaseous analytes. However, the response and recovery in the case of VOCs can be further improved. Due to the chemiresistive module, they can be efficiently designed, handled, occupied, and used at every emission site, unlike sophisticated sensors such as electrochemical or spectroscopic sensors. However, the reports on monitoring VOCs and gases through MXene–polymer nanocomposites are in scared except for ammonia. More MXenes and polymer combinations can be explored for fabricating chemiresistor to detect various pollutants and harmful vapors. MXene–polymer nanocomposites possess prospects in the electronic nose and sensor arrays due to their selectivity towards particular analytes at room temperature. They are a potential candidate for the fabrication of intelligent, rapid, energy-efficient, user and environmentally friendly, cost-effective, flexible, miniature, easy to handle, and portable gas and vapor sensors with enhanced three Ss-and five Rs-sensing characteristics.

REFERENCES

Aarya S., et al., 2020. Recent Advances in Materials, Parameters, Performance and Technology in Ammonia Sensors: A Review. *Journal of Inorganic and Organometallic Polymers and Materials*. Volume 30, p.269.

Abdulla S., et al., 2015. Highly Sensitive, Room Temperature Gas Sensor Based on Polyaniline-Multiwalled Carbon Nanotubes (PANI/MWCNTs) Nanocomposite for Trace-Level Ammonia Detection. *Sensors and Actuators B: Chemical*. Volume 221, p.1523.

Aghaei S. M., et al., n.d. Experimental and Theoretical Advances in MXene-Based Gas Sensors. *ACS Omega*. Volume 6(4), p.2450.

Alrammouz et al, R., 2018. A Review on Flexible Gas Sensors: From Materials to Devices. *Sensors and Actuators A: Physical*. Volume 284, p.209.

Anichini C., et al., 2018. Chemical Sensing with 2D Materials. *Chemical Society Reviews*. Volume 47, p.4860.

Bai H., et al., 2007. Gas Sensors Based on Conducting Polymers. *Sensors*. Volume 7, p. 276.

Baker C., et al., 2017. Polyaniline Nanofibers: Broadening Applications for Conducting Polymers. *Chemical Society Reviews*. Volume 46, p.1510.

Carey M., et al., 2021. MXene Polymer Nanocomposites: A Review. *Materials Today Advances*. Volume 9, p.100120.

Chaudhary V., et al., 2015a. Enhanced and Selective Ammonia Sensing Behaviour of Poly(Aniline co-pyrrole) Nanospheres Chemically Oxidative polymerized at Low Temperature. *Journal of Industrial and Engineering Chemistry*. Volume 26, p.143.

Chaudhary V., et al., 2015b. Enhanced Room Temperature Sulfur Dioxide Sensing Behaviour of *in Situ* Polymerized Polyaniline–Tungsten Oxide Nanocomposite Possessing Honeycomb Morphology. *RSC Advances*. Volume 5, p.73535.

Chaudhary V., et al., 2015c. Solitary Surfactant Assisted Morphology Dependent Chemiresistive Polyaniline Sensors for Room Temperature Monitoring of Low Parts Per Million Sulfur Dioxide. *Polymer International.* Volume 64(10), p.1475.

Chaudhary V., et al., 2016. Surfactant Directed Polyaniline Nanostructures for High-Performance Sulphur Dioxide Chemiresistors: Effect of Morphologies, Chemical Structure and Porosity. *RSC Advances.* Volume 6, p.95349.

Chaudhary V., et al., 2017. Effect of Charge Carrier Transport on Sulfur Dioxide Monitoring Performance of Highly Porous Polyaniline Nanofibres. *Polymer Internationl.* Volume 66(5), p.699.

Chaudhary V., et al., 2021a. Advancements in Research and Development to Combat COVID-19 Using Nanotechnology. *Nanotechnology for Environmental Engineering.* Volume 6, p.8.

Chaudhary V., et al., 2021b. Emerging MXene–Polymer Hybrid Nanocomposites for High-Performance Ammonia Sensing and Monitoring. *Nanomaterials.* Volume 11, p.2496.

Chaudhary V., et al., 2021c. Novel Methyl-Orange Assisted Core-Shell Polyaniline-Silver Nanosheets for highly Sensitive Ammonia Chemiresistors. *Journal of Applied Polymer Science.* Volume 138(43), p.51288.

Chaudhary V. et al., 2022a. Review—Towards 5th Generation AI and IoT Driven Sustainable Intelligent Sensors Based on 2D MXenes and Borophene. *ECS Sensor Plus.* Volume 1(1), p.013601.

Chaudhary V. et al., 2022b. Emergence of MXene–Polymer Hybrid Nanocomposites as High-Performance Next-Generation Chemiresistors for Efficient Air Quality Monitoring. *Advanced Functional Materials.* Volume 32(33), p.2112913.

Chaudhary V. et al., 2022c. Assessing Temporal Correlation in Environmental Risk Factors to Design Efficient Area-Specific COVID-19 Regulations: Delhi Based Case Study. *Scientific Reports.* Volume 12, p.12949.

Chaudhary V. et al., 2022d. Low-Trace Monitoring of Airborne Sulphur Dioxide Employing SnO2-CNT Hybrids-Based Energy-Efficient Chemiresistor. *Journal of Materials Research and Technology.* Volume 20, p.2468.

Chen W.Y., et al., 2020. Nanohybrids of a MXene and Transition Metal Dichalcogenide for Selective Detection of Volatile Organic Compounds. *Nature Communications.* Volume 11, p.1302.

Chen X., et al., 2021. MXene/Polymer Nanocomposites: Preparation, Properties, and Applications. *Polymer Reviews.* Volume 61, p.80.

Ding L., et al., 2018. MXene Molecular Sieving Membranes for Highly Efficient Gas Separation. *Nature Communications.* Volume 9, p.155.

Doan et al, T. C. D., 2014. Carbon Dioxide Detection with Polyethylenimine Blended with Polyelectrolytes. *Sensors and Actuators B: Chemical.* Volume 201, p.452.

Fu S., et al., 2019. Some Basic Aspects of Polymer Nanocomposites: A Critical Review. *Nano Materials Science.* Volume 1(1), p.2.

Gao J., et al., 2020a. NH3 Sensor Based on 2D Wormlike Polypyrrole/Graphene Heterostructures for a Self-Powered Integrated System. *ACS Applied Materials & Interfaces.* Volume 12(20), p.38674.

Gao L., et al., 2020b. MXene/Polymer Membranes: Synthesis, Properties, and Emerging Applications. *Chemistry of Materials.* Volume 32(5), p.1703.

Gogotsi Y., et al., 2019. The Rise of MXenes. *ACS Nano.* Volume 13, p.8491.

Gund G. S., et al., 2019. MXene/Polymer Hybrid Materials for Flexible AC-Filtering Electrochemical Capacitors. *Joule.* Volume 3(1), p.164.

Hafiz S. M., et al., 2014. A Practical Carbon Dioxide Gas Sensor Using Room-Temperature Hydrogen Plasma Reduced Graphene Oxide. *Sensors and Actuators B: Chemical.* Volume 193, p.692.

Hajian S., et al., 2018. Impact of Different Ratios of Fluorine, Oxygen, and Hydroxyl Surface Terminations on Ti3C2Tx MXene as Ammonia Sensor: A First-Principles Study. *IEEE SENSORS.* p.1. https://doi.org/10.1109/ICSENS.2018.8589699.

Hasan M. M., et al., 2021. Two-Dimensional MXene-Based Flexible Nanostructures for Functional Nanodevices: A Review. *Journal of Materials Chemistry A.* Volume 9, p.3231.

Ho D. H., et al., 2021. Sensing with MXenes: Progress and Prospects. *Advanced Materials.* Volume 33(47), p.2005846.

Insausti M., et al., 2020. Advances in Sensing Ammonia from Agricultural Sources. *Science of The Total Environment.* Volume 706, p.135124.

Jalal A. H., et al., 2018. Prospects and Challenges of Volatile Organic Compound Sensors in human healthcare. *ACS Sensors.* Volume 3, p.1246.

Jin L., et al., 2020. Polymeric Ti3C2Tx MXene Composites for Room Temperature Ammonia Sensing. *ACS Applied Nano Materials.* Volume 3(12), p.12071.

Joshi S., et al., 2017. Efficient Heterostructures of Ag@CuO/$BaTiO_3$ for Low-Temperature CO_2 Gas Detection: Assessing the Role of Nanointerfaces during Sensing by Operando DRIFTS Technique. *ACS Applied Materials & Interfaces*. Volume 9(32), p.27014.

Kanaparthi S., et al., 2019. Chemiresistive Sensor Based on Zinc Oxide Nanoflakes for CO_2 Detection. *ACS Applied Nano Materials*. Volume 2(2), p.700.

Kausar A., et al., 2021. Polymer/MXene Nanocomposite–A New Age for Advanced Materials. *Polymer-Plastics Technology and Materials*. Volume 60(13), p.1377.

Khan S. B., et al., 2020. *Gas Sensors*. s.l.:IntechOpen.

Kim S. J., et al., 2018. Metallic $Ti_3C_2T_x$ MXene Gas Sensors with Ultrahigh Signal-to-Noise Ratio. *ACS Nano*. Volume 12(2), p.986.

Kumar V., et al., 2020. Advances in Electrospun Nanofiber Fabrication for Polyaniline (PANI)-Based Chemoresistive Sensors for Gaseous Ammonia. *TrAC Trends in Analytical Chemistry*. Volume 129, p.115938.

Lee E., et al., 2017. Room Temperature Gas Sensing of Two-Dimensional Titanium Carbide (MXene). *ACS Applied Materials & Interfaces*. Volume 9(42), p.37184.

Lee E., et al., 2019. Two-Dimensional Vanadium Carbide MXene for Gas Sensors with Ultrahigh Sensitivity Toward Nonpolar Gases.*ACS Sensors*. Volume 4(6), p.1603.

Li D., et al., 2021. Virtual Sensor Array Based on MXene for Selective Detections of VOCs. *Sensors and Actuators B: Chemical*. Volume 331, p.129414.

Liu C., et al., 2017. Enhanced Ammonia-Sensing Properties of PANI-TiO_2-Au Ternary Self-Assembly Nanocomposite Thin Film at Room Temperature. *Sensors and Actuators B: Chemical*. Volume 246, p.85.

Li X., et al., 2020. Toward Agricultural Ammonia Volatilization Monitoring: A Flexible Polyaniline/$Ti_3C_2T_x$ hybrid Sensitive Films Based Gas Sensor. *Sensors and Actuators B: Chemical*. Volume 316, p.128144.

Matsuguchi M., et al., 2002. Effect of NH_3 Gas on the Electrical Conductivity of Polyaniline Blend Films. *Synthetic Metals*. Volume 127(1), p.15.

Mogera U., et al., 2014. Ultrafast Response Humidity Sensor Using Supramolecular Nanofibre and Its Application in Monitoring Breath Humidity and Flow. *Scientific Reports*. Volume 4, p.4103.

Mulmi S., et al., 2020. Solid-State Electrochemical Carbon Dioxide Sensors: Fundamentals, Materials and Applications. *Journal of The Electrochemical Society*. Volume 167, p.037567.

Naguib M., et al., 2013. New Two-Dimensional Niobium and Vanadium Carbides as Promising Materials for Li-Ion Batteries. *Journal of the American Chemical Society*. Volume 135(43), p.15966.

Nazemi et al, H., 2019. Advanced Micro- and Nano-Gas Sensor Technology: A Review. *Sensors*. Volume 19(6), p.1285.

Pandey S., 2016. Highly Sensitive and Selective Chemiresistor Gas/Vapor Sensors Based on Polyaniline Nanocomposite: A Comprehensive Review. *Journal of Science: Advanced Materials and Devices*. Volume 1(4), p.431.

Rathi K., et al., 2020. Ruthenium-Decorated Tungsten Disulfide Quantum Dots for a CO_2 Gas Sensor. *Nanotechnology*. Volume 31(13), p.135502.

Riazi H., et al., 2021. MXene-Based Nanocomposite Sensors. *ACS Omega*, Volume 6(17), p. 11103.

Sheth Y., et al., 2022. Prospects of Titanium Carbide-Based MXene in Heavy Metal Ion and Radionuclide Adsorption for Wastewater Remediation: A Review. *Chemosphere*. Volume 293, p. 133563.

Singh A., et al., 2022. MnO_2-SnO_2 Based Liquefied Petroleum Gas Sensing Device for Lowest Explosion Limit Gas Concentration. *ECS Sensor Plus*. Volume 1(2), p. 025201.

Srinives S., et al., 2015. A Miniature Chemiresistor Sensor for Carbon Dioxide. *Analytica Chimica Acta*. Volume 8, p.54.

Tang Y., et al., 2017. Superwettability Strategy: 1D Assembly of Binary Nanoparticles as Gas Sensors. *Small*. Volume 13, p.1601087.

Tanguy N. R., et al., 2018. A Review on Advances in Application of Polyaniline for Ammonia Detection. *Sensors and Actuators B: Chemical*. Volume 257, p.1044.

Thirumalairajan S., et al., 2014. Surface Morphology-Dependent Room-Temperature $LaFeO_3$ Nanostructure Thin Films as Selective NO_2 Gas Sensor Prepared by Radio Frequency Magnetron Sputtering. *ACS Applied Materials & Interfaces*. Volume 6(16), p.13917.

Timmer B., et al., 2005. Ammonia Sensors and Their Applications—A Review. *Sensors and Actuators B: Chemical*. Volume 107(2), p.666.

Tung T. T., et al., 2017. Recent Advances in Sensing Applications of Graphene Assemblies and Their Composites. *Advanced Functional Materials*. Volume 27, p.1702891.

Wang J., et al., 2020. Volatile Organic Compounds Gas Sensors Based on Molybdenum Oxides: A Mini Review. *Frontiers in Chemistry*. Volume 8, p.339.

Wang L., et al., 2021a. Recent Advances in Multidimensional (1D, 2D, and 3D) Composite Sensors Derived from MXene: Synthesis, Structure, Application, and Perspective. *Small*. Volume 5(7), p.2100409.

Wang S., et al., 2021b. PANI Nanofibers-Supported Nb2CTx Nanosheets-Enabled Selective NH3 Detection Driven by TENG at Room Temperature. *Sensors and Actuators: B. Chemical*. Volume 327, p. 128923.

Wang S., et al., 2021c. Ultrathin Nb2CTx Nanosheets-Supported Polyaniline Nanocomposite: Enabling Ultrasensitive NH3 Detection. *Sensors and Actuators: B. Chemical*. Volume 343, p.130069.

Wang X., et al., 2020. Ti3C2Tx/PEDOT:PSS Hybrid Materials for Room-Temperature Methanol Sensor. *Chinese Chemical Letters*. Volume 31(4), p.1018.

Wei S., et al., 2018. Greenhouse Gas and Ammonia Emissions and Mitigation Options from Livestock Production in Peri-Urban Agriculture: Beijing—A Case Study. *Journal of Cleaner Production*. Volume 178, p.515.

Wong Y. C., et al., 2020. Conducting Polymers as Chemiresistive Gas Sensing Materials: A Review. *Journal of The Electrochemical Society*. Volume 167, p.037503.

Wu L., et al., 2013. Enhanced Sensitivity of Ammonia Sensor Using Graphene/Polyaniline Nanocomposite. *Sensors and Actuators B: Chemical*. Volume 178(1), p.485.

Wu Z., et al., 2020. Self-Powered Sensors and Systems Based on Nanogenerators. *Sensors*. Volume 20(10), p.2925.

Xiao B., et al., 2016. MXenes: Reusable Materials for NH3 Sensor or Capturer by Controlling the Charge Injection. *Sensors and Actuators B: Chemical*. Volume 235, p.103.

Yuan W., et al., 2018. A Flexible VOCs Sensor Based on a 3D Mxene Framework with a High Sensing Performance. *Journal of Materials Chemistry A*. Volume 6, p.18116.

Yu X.-F., et al., 2015. Monolayer Ti2CO2: A Promising Candidate for NH3 Sensor or Capturer with High Sensitivity and Selectivity. *ACS Applied Materials & Interfaces*. Volume 7(24), p.13707.

Zamhuri A., et al., 2021. MXene in the lens of biomedical engineering: synthesis, applications and future outlook. *BioMedical Engineering OnLine*. Volume 20, p.33.

Zhan X., et al., 2020. MXene and MXene-Based Composites: Synthesis, Properties and Environment-Related Applications. *Nanoscale Horizons*. Volume 5, p.235.

Zhao L., et al., 2019. High-Performance Flexible Sensing Devices Based on Polyaniline/MXene Nanocomposites. *InfoMat*. Volume 1, p.407.

Zhou Y., et al., 2020. Humidity-Enabled Ionic Conductive Trace Carbon Dioxide Sensing of Nitrogen-Doped Ti3C2Tx MXene/Polyethyleneimine Composite Films Decorated with Reduced Graphene Oxide Nanosheets. *Analytical Chemistry*. Volume 92(24), p.16033.

Zito etal., 2020. Low-Temperature Carbon Dioxide Gas Sensor Based on Yolk–Shell Ceria Nanospheres. *ACS Applied Materials & Interfaces*. Volume 12(15), p.17745.

12 MXene-Based Nanocomposites in Energy Conversion and Storage Systems

Dana Susan Abraham, Margandan Bhagiyalakshmi, and Mari Vinoba

CONTENTS

12.1 INTRODUCTION

Since the first report on transition metal carbides—MXenes (Naguib et al., 2011) in 2011, the major application explored on MXene was as electrode materials in energy storage systems (ESSs). Well-established ESSs are batteries and supercapacitors whose performance in terms of energy and power density are critically controlled by the electrode materials employed. From the reports, the MXene are uniquely known to possess metallic conductivity, layered structure, and large surface area that allows diffusion/intercalation of any molecules and ionic transport and adsorption on the surface, respectively. Over the decades, MXenes are synthesized only through etching IIIA or IVA group element layers from bulk MAX powders; the desired new skeleton of MXenes structures can be obtained based on the choice of the parent MAX phase.

MAX phases are layered compositions of early transition metals (M), Al or Si (A), and carbon, nitrogen or a combination of the two (X) in the notation $M_{n+1}AX_n$, where n 1–3 (Barsoum, 2000) Until now, a variety of combinations and structures of MAX phase about more than 150 forms are known (Sokol et al., 2019; Nechiche et al., 2017; Lai et al., 2017; Liu et al., 2014). Generally, etchants, such as hydrofluoric acid (HF) or a mixture of LiF and HCl salts in water are employed for removing aluminum from the MAX phase to obtain MXene. Typically, Al is etched from the MAX phase through three main reactions upon HF treatment as shown:

$$Ti_3AlC_2 + 3HF \rightarrow AlF_3 + 3/2\,H_2 + Ti_3C_2 \quad (12.1)$$
$$Ti_3C_2 + 2H_2O \rightarrow Ti_3C_2\,(OH)_2 + H_2 \quad (12.2)$$
$$Ti_3C_2 + 2HF \rightarrow Ti_3C_2F_2 + H_2 \quad (12.3)$$

DOI: 10.1201/9781003164975-12

Equation 12.1 shows the removal of Al, and the reactions, Equations 12.2 and 12.3, describe the reactivity of the bare surface of MXene with the environment to form O, OH, and F on the surfaces (Naguib et al., 2011; Hope et al., 2016; Caffrey, 2018).

MXene-based energy storage systems, especially supercapacitors, are well appreciated for their high energy density. The layered MXene matrix is highly stable even after 1,000 cycles of intercalation and de-intercalation of ions and, hence, provides high energy (Lukatskaya et al., 2013; Ghidiu et al., 2014; Tao et al., 2013; Fan et al., 2018; Wang et al., 2017). Here it is concentrated on this particular property, which demonstrated that MXene as potential electrode material for ESSs. Noteworthy, the majority of the reports are on the application of MXenes as electrode material due to its isolated layers stacked in an orderly manner, forming heterostructures that contribute to electric and electronic properties.

Yet another interesting property of MXene is the mechanical robustness due to the highest Young's modulus (Lipatov et al., 2018), which endows MXenes as potential filler materials in preparation of polymer nanocomposites. Also, the unique properties of MXenes have made them a suitable nanofiller for making high-performance polymer composites that can satisfy electrical, thermal, mechanical, flame-retardant, and other requirements (Li et al., 2019). Researchers in the field of polymers are attracted to the mechanical property of MXene, and as a result, several exciting developments on polymer nanocomposites with MXene and MXene-based nanocomposites as fillers are available for direct application in various sectors (Gao et al., 2020; Feng et al., 2018; Jimmy and Kandasubramanian, 2020). The properties of polymers like sheer versatility, interfacial tension, corrosion resistance, fracture resistance, and stiffness when combined with highly conductive, mechanically stable results in advanced polymer–MXene nanocomposites, which are promising candidates for energy conversion and storage.

The MXene/polymer nanocomposites are generally categorized as filled composites and complexes. In a filled composite, MXene is utilized as a filler or a component in a bulk material, while MXene and polymer are sparsely oriented with each other in a complex. The most frequently researched materials are filled MXene/polymer nanocomposites, thermoplastic, and thermosetting polymer matrix (Carey and Barsoum, 2021). This chapter discusses the application of MXene/polymer nanocomposites in energy conversion and storage systems, emphasizing supercapacitors, batteries, and fuel cell applications.

12.2 MXene/POLYMER NANOCOMPOSITES FOR ENERGY STORAGE AND CONVERSION APPLICATIONS

12.2.1 MXene/Polymer Nanocomposites as Electrode Materials in Supercapacitors

Supercapacitors are hooked on as long-life energy storage devices in the electrical power storage industry. MXene-based electrode materials have great potential for supercapacitors in the light of their outstanding metallic conductivity, high hydrophilicity, high pseudocapacitance, and ease of preparing flexible, freestanding electrode films. The advantage of MXene electrode materials for supercapacitors over conventional electrode materials is that they are easily assembled to produce freestanding, flexible electrodes and devices without binding agents (Vahid Mohammadi et al., 2018).

In situ pyrrole polymerization between MXene layers was illustrated by Boota et al. (2016), leading to a pseudocapacitive electrode with good electrochemical stability. PPy/$Ti_3C_2T_z$ composite exhibited a maximum volumetric capacitance of ≈1000 F cm^{-3}, and its 92% capacitance was retained even after 25,000 cycles. The incorporation of polypyrrole layer between $Ti_3C_2T_z$ monolayers resulted in enhanced electronic conductivity, fast ion transport, and rapid redox reactions due to the shortened diffusion pathway in PPy/$Ti_3C_2T_z$ composite. The synergistic effect of $Ti_3C_2T_z$ layers intercalated with conductive PPy, and the redox reactions between PPy and MXene-enhanced PPy/$Ti_3C_2T_z$ composite capacitance.

Boota and colleagues (Boota and Gogotsi, 2019) demonstrated asymmetric pseudo-capacitors using conducting polymers (CP) and MXene. MXene served as the anode material for the pseudo-capacitor, and reduced graphene oxide sheet confined with PPy, PANI, and PEDOT acted as cathode. All the CP@rGO//MXene offered impressive power and energy densities, as well as excellent cycling properties. PANI-containing asymmetric device showed an extremely high energy density of 17 Wh kg^{-1} and the highest retention in capacitance of 88%, which is one of the best results for both MXene-based and PANI-containing asymmetric devices.

Li et al. (2020b) engineered an asymmetric and flexible pseudocapacitor from wavy-$Ti_3C_2T_x$//rGO/CNT/PANI, as in Figure 12.1. For the fabrication of negative electrodes, the sheets of MXene were initially mixed in water containing sub-microspheres of polystyrene (PS). PS spheres were then removed by heating in Ar atmosphere at 450°C, developed initially as a freestanding, stretchable, and porous film, which was later compressed into a compact wavelike film. Compared to highly aligned $Ti_3C_2T_x$ film, the wavy $Ti_3C_2T_x$ film with poor alignment enabled improved transport of ions, resulting in a significant volumetric capacitance of 1277 F cm^{-3}. rGO/CNT/PANI electrode was used to complement the negative MXene electrode. Asymmetric wavy-$Ti_3C_2T_x$//rGO/CNT/PANI supercapacitors demonstrated 70 Wh L^{-1} of energy density and 111 kW L^{-1} of power density.

Boota et al. (2017) explored the capacitance characteristics of MXene/PFD (polyfluorene-derivative) nanocomposites as a supercapacitor electrode. A higher capacitance was observed for MXene/PFD with quaternary nitrogen groups because of the intercalation effect, resulting in stronger charge transfer interactions at the MXene/PFD interface. This improves proton accessibility, as well as pseudo-capacitance due to polymer. Moreover, they pointed out that the improvement in capacitance was related to the uniformity in dispersion, the polarity of the PFDs and MXene. The dispersion of molecules with more uniformity will have a stronger interaction with MXene, and the capacitance effect will be more noticeable. A maximum volumetric capacitance of 1026 F cm^{-3} was delivered by MXene/PFDs electrodes. The gravimetric capacitance is more than 1.5 times that of the pure $Ti_3C_2T_z$ film (245 Fg^{-1}, 2 mV s^{-1}) and more than twice that of the $Ti_3C_2T_z$ hybrid (Ling et al., 2014).

Qin et al. (2018) utilized a solution processing technique to develop an ultrathin and flexible supercapacitor, $Mo_{1.33}C$ MXene/PEDOT: PSS. It featured an enhanced volumetric capacitance, highpower, and energy density of 568 F cm^{-3}, 19470 mW cm^{-3}, and 33.2 mWh cm^{-3}, respectively. Upon treatment with H_2SO_4, a higher capacitance was observed 1310 F cm^{-3}. A pair of factors is responsible for the increased capacitance and stability: the inclusion of conducting PEDOT between the $Mo_{1.33}C$ MXene layers resulted in greater interlayer spacing and the interfacial redox processes.

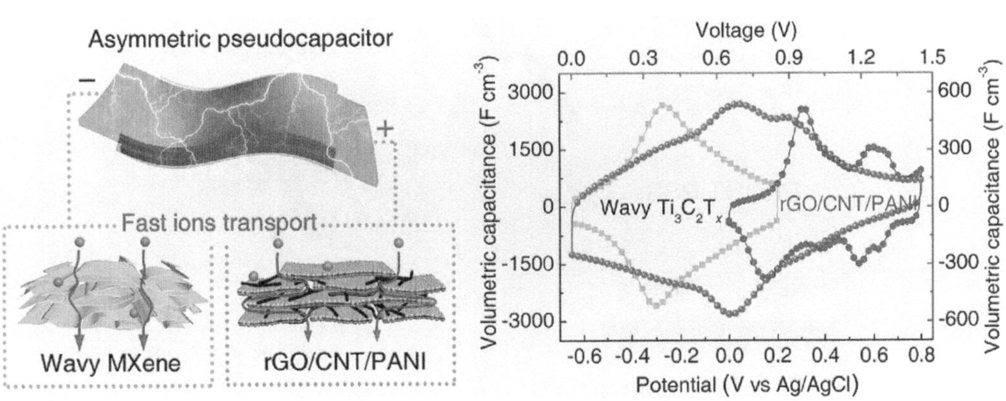

FIGURE 12.1 Schematic representation of asymmetric pseudocapacitor from wavy-$Ti_3C_2T_x$//rGO/CNT/PANI and its electrochemical performance (Li et al., 2020b).

Zhang et al. (2021) described the beneficial effects of tartaric acid as an additive to MXene ($Ti_3C_2T_x$) and in the MXene/polymer composite to form highly conductive and stable functional composites for ultrafast supercapacitors. Tartaric acid was added to MXene during its annealing, resulting in "ta-Mxene". Figure 12.2 shows the effect of tartaric acid on MXene/PEDOT: PSS composite. It was observed that tartaric acid was able to cap and protect the Ti cations present on the defective edges from chemical oxidation, even when the temperatures were elevated. Acidic dispersion of MXene was also enhanced by tartaric acid. In this study, ta-MXene/PEDOT: PSS composites were synthesized using tartaric acid treatment. Along with capping the MXene surface, tartaric acid also induced crosslinking in the MXene/PEDOT: PSS composite, maintaining its structural integrity. The composite also exhibited storage stability for 2 weeks. Along with its chemical and structural stabilizing effects, tartaric acid increased the electrical conductivity by four times, rising to 2,240 S cm^{-1} for ta-MXene/PEDOT: PSS from 552 S cm^{-1} of MXene/PEDOT: PSS.

Qin et al. (2019) investigated the polymerization of organic monomers EDOT and pyrrole by *in situ* electrochemical polymerization (EP) using 2D MXene without employing conventional electrolytes to form flexible solid-state micro-supercapacitors. MXene colloidal solution was observed to serve as a high conductivity solvent during the EP process. Simultaneously, self-assemble into polymer films with a higher content of dopants than conventional electrolytes, resulting in conjugated polymeric MXene films at a molecular level. The pseudo-capacitance and ultra-high-energy capacity of these composite films were found to be 47.4 mF cm^{-2} and 20.05 mWh cm^{-3}, respectively. A significant improvement in both the stability and the rate of the micro-supercapacitors was also achieved. An efficient asymmetric micro-supercapacitors (AMSCs) were developed to improve the energy density and cell voltage with *in-situ* EP composite films with MnO_2; the high energy and power density were found to be 250.1 mWh cm^{-3} and 32.9 W cm^{-3}, respectively. Additionally, the AMSCs showed excellent cyclic stability over 10,000 cycles.

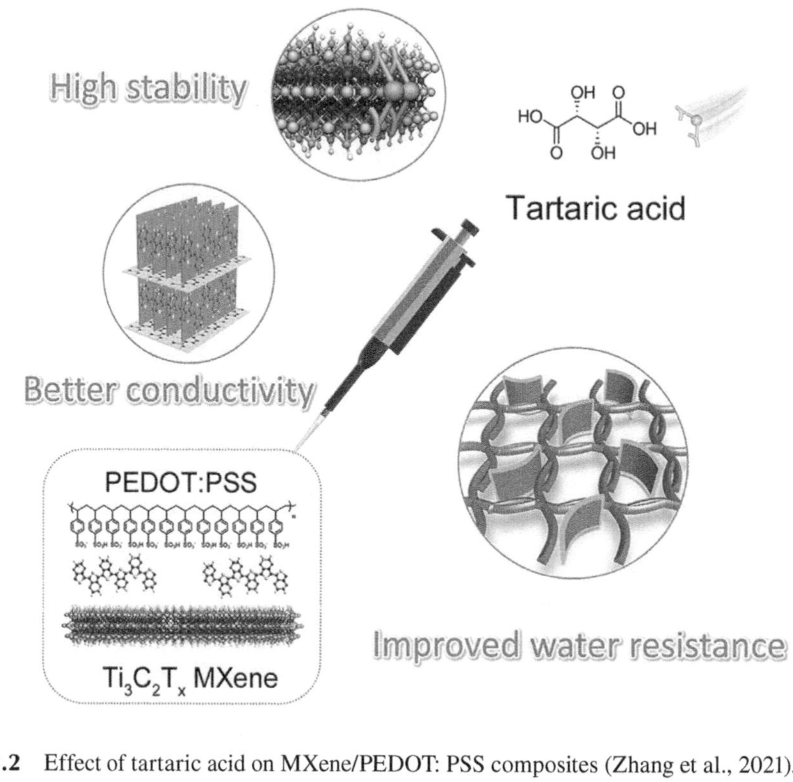

FIGURE 12.2 Effect of tartaric acid on MXene/PEDOT: PSS composites (Zhang et al., 2021).

Wu et al. (2019) fabricated a new decentralized conjugated polymer (PDT) using 2, 6- diaminoanthraquinone (DAQ) and tetrakis(4-bromophenyl) methane (TM), which was coupled with layered MXene by Buchwald–Hartwig coupling. The so-formed electrode displayed an improved areal capacitance of 284 mF cm^{-2}. Furthermore, galvanostatic charge–discharge analysis of PDT/$Ti_3C_2T_x$ film revealed almost 100% capacitance retention after 10,000 cycles. The PDT/$Ti_3C_2T_x$ film displayed excellent electrochemical performance stability, which can be utilized to develop a solid-state supercapacitor with capacitance of 52.4 mF cm^{-2} at 0.1 mA cm^{-2} over 100 cycles. It also proved to be flexible and could withstand 10,000 cycles of static bending at 0–90°. The analysis revealed that when the decentralized chain of PDTs and the $Ti_3C_2T_x$ with better conductivity are combined, electron conduction efficiency, electrochemical activity, stability, and flexibility are increased.

Ajnsztajn et al. (2020) integrated 2D $Ti_3C_2T_x$ nanosheets with poly(9,9-di-n-octylflourenyl-2,7-diyl) (PFO) using a deposition technique, resonant infrared matrix-assisted pulsed laser evaporation (RIR-MAPLE), to generate a transparent supercapacitor electrode material. This technique eliminated the requirement of PFO being soluble in the medium in which nanosheets are suspended. This film composition displayed approximately 20 mF/cm^2 of area capacitance at >75% transmission. The high capacitance can be attributed to $Ti_3C_2T_x$ nanosheet and high transmittance by PFO.

Shao et al. (2018) developed wearable supercapacitors using nanofiber-coated yarn (NCY) electrodes with MXene/polymer composite (PET@MXene NCY). A modified electrospinning procedure was employed to self-wound MXene nanofibers onto polyester (PET). PET@MXene NCY demonstrated desired textile properties, including flexibility, strength, and fabricability. A comparison between graphene and carbon nanotube yarn supercapacitors revealed that they had higher electrochemical performance. In addition, yarn supercapacitors provided a high areal capacitance, enhanced energy density, and high power density of 18.39 mF cm^{-2}, 0.38 μW h cm^{-2}, and 0.39 mW cm^{-2}, respectively. Furthermore, 98.2% of the specific capacitance was retained after 6000 cycles. The high electroactivity of MXene and good mechanical properties of PET offered desirable properties to yarn supercapacitors.

Ren et al. (2018) reported PANI-modified 2D Ti_3C_2 composites by *in situ* polymerization and investigated the electrochemical performance. Incorporating PANI to Ti_3C_2 exhibited enhanced electrochemical performance due to the presence of –NH groups in PANI that facilitated not only faster ion transport and increased electric conductivity. Active interaction between PANI and Ti_3C_2also resulted in improved surface wettability. Hence, more active sites on PANI-Ti_3C_2 composites enhanced the rate of faradaic reactions in the composite. PANI-Ti_3C_2 composites reveal a maximum specific capacitance of 164 F g^{-1} at the scan rate of 2 mV s^{-1} and superior cycling stability because of 96% retention of initial capacitance even after 3000 galvanostatic charge–discharge cycles.

Zhu et al. (2016) synthesized a uniform film of l-Ti_3C_2, to which PPy was intercalated using electrochemical polymerization. They found that when compared to the pristine PPy film, the fabricated supercapacitor showed a 30% enhanced capacitance with a value of 406 F cm^{-3}. Furthermore, the PPy/l-Ti_3C_2 film demonstrated improved cyclic stability; even after 20,000 charge–discharge cycles, the capacitance was almost unaffected. The increased capability and cycling stability were attributed to MXene, which was responsible for assisting ion mobility and charge carrier migration as well as strengthening the PPy backbones. An ultra-thin all-solid-state supercapacitor was also fabricated using this PPy/l-Ti_3C_2 film, which measured an excellent capacitance of 35 mF cm^{-2} and displayed both outstanding deformation tolerance and excellent cycling stability.

Gund et al. (2019) reported porous MXene/conducting polymer hybrids Ti_3C_2/PEDOT: PSS electrode having an interconnected network structure for flexible symmetric electrochemical capacitors. The MXene/polymer nanocomposites exhibited a high capacitance of 24.2 F cm^{-3} at a high frequency of 120 Hz. Additionally, it had long-term durability of 30,000 cycles and could function at 60–10000 Hz. The results of Bode and Nyquist plots evidenced excellent responses to high frequencies of Ti_3C_2/PEDOT: PSS hybrid electrode. The phase angle at 120 Hz is one of the crucial parameters to assess its efficiency as AC filter since an ideal capacitor has a phase angle of –90°. The phase angles and cutoff frequencies of Ti_3C_2/PEDOT: PSS (1:1 wt %) and Ti_3C_2/PEDOT: PSS (1:2 wt %) were 76.1°, and 79.1° at 120 Hz and 1.60, and 1.41 kHz at −45°, respectively. Ti_3C_2/PEDOT: PSS electrodes exhibited a smaller phase angle at low frequencies due to the faradaic effect in the overall capacitive property. The Nyquist plot shows an almost negligible semicircle in the Ti_3C_2/PEDOT: PSS electrode indicating the rapid ionic transport. The Ti_3C_2/PEDOT: PSS electrode film with excellent responses to high frequencies and high volumetric capacitance can be efficiently used in currently available devices as well. Table 12.1 shows a comparative study on the electrochemical performance of various MXene/polymer composite materials used in supercapacitor applications.

12.2.2 MXene/Polymer Nanocomposites in Battery Application

However, despite its excellent electrolyte affinity and electrical conductivity, MXene/polymer nanocomposites are underutilized in battery applications compared to their application in other applications in the field of supercapacitors. Nevertheless, researchers have explored their possible use in whole battery systems comprising electrodes, membrane separators, and electrolytes (Chen et al., 2021a).

Polymer–MXene nanocomposites also find application in lithium–ion storage performance. Chen et al. (2017) developed a combination of d-$Ti_3C_2T_z$ and PEDOT to improve lithium-ion storage performance through enhanced ion transport with increased interlayer spacing for improved lithium–ion uptake. It is noticed that the coulombic efficiency of MXene/PEDOT improved considerably by 53%, while pure MXene had an efficiency of only 38%. Even at 1000 mA g^{-1}, the MXene/PEDOT had a capacity of 71 mAh g^{-1}, which was higher than the pure MXene capacity of 9 mAh g^{-1}.

TABLE 12.1
Comparison of Electrochemical Performance of Different MXene/Polymeric Composite Materials

Material	Specific/Areal/ Volumetric Capacitance	Retention	Energy Density	Power Density	Ref.
PPy/$Ti_3C_2T_z$	1000 Fcm^{-3}(VC)	92%	–	–	(Boota et al., 2016)
$Ti_3C_2T_x$//PANI@ rGO	57 Fg^{-1}(SC)	88%	17 Wh kg^{-1}	–	(Boota and Gogotsi, 2019)
wavy-$Ti_3C_2T_x$//rGO/ CNT/PANI	1277 F cm^{-3}(VC)	–	70 Wh L^{-1}	111 kW L^{-1}	(Li et al., 2020b)
MXene/PFDs	1026 F cm^{-3}(VC)	–	–	–	(Boota et al., 2017)
$Mo_{1.33}C$ MXene/ PEDOT: PSS	568 F cm^{-3}(VC)	–	33.2 mWh cm^{-3}	19470 mW cm^{-3}	(Qin et al., 2018)
ta-MXene/PEDOT: PSS	1.5 F cm^{-3}(VC)	98%	–	–	(Zhang et al., 2021)
PEDOT-MXene	47.4 mF cm^{-2}(AC)	–	250.1 mWh cm^{-3}	32.9 W cm^{-3}	(Qin et al., 2019)
PDT/$Ti_3C_2T_x$	284 mF cm^{-2}(AC)	–	250.1 mWh cm^{-3}	32.9 W cm^{-3}	(Wu et al., 2019)
$Ti_3C_2T_x$/PFO	20 mF cm^{-2}(AC)	–	–	–	(Ajnsztajn et al., 2020)
PET@MXene NCY	18.39 mF cm^{-2}(AC)	98.2%	0.38 μWh cm^{-2}	0.39 mW cm^{-2}	(Shao et al., 2018)
PANI-Ti_3C_2	164 F g^{-1}(SC)	96%	–	–	(Ren et al., 2018)
PPy/l-Ti_3C_2	406 F cm^{-3}(VC)	100%	–	–	(Zhu et al., 2016)
Ti_3C_2/PEDOT: PSS	24.2 F cm^{-3}(VC)	–	–	–	(Gund et al., 2019)

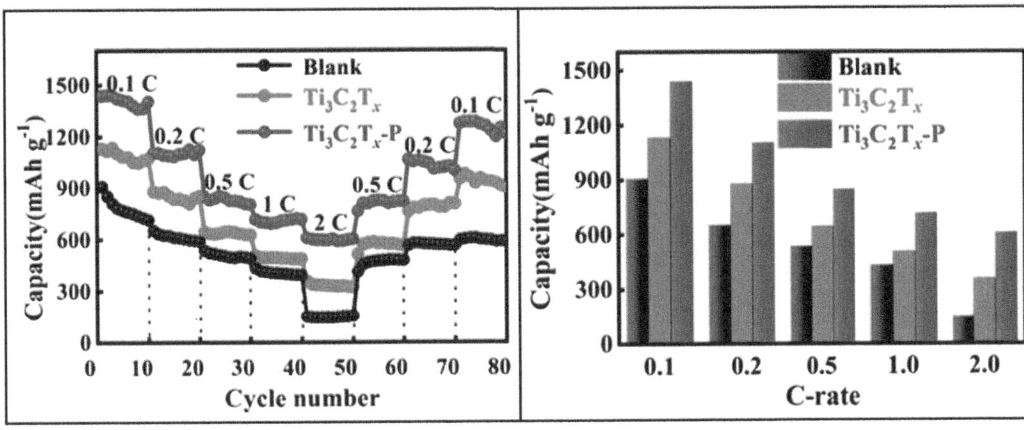

FIGURE 12.3 Cyclic stability of $Ti_3C_2T_x$–PEDOT: PSS hybrid (Li et al., 2020a).

Probably, PEDOT is largely a spacer for MXene layers, enabling better mobility of ions. This report is clear evidence that d-$Ti_3C_2T_z$ as a filler improved Li–ion uptake and is suitable in Li–ion storage devices.

Li et al. (2020a) fabricated $Ti_3C_2T_x$–PEDOT: PSS hybrid to achieve high-performance Li–S batteries. Li–S batteries with PEDOT: PSS experienced increased electrical conductivity and significantly reduced $Ti_3C_2T_x$ stacking. In turn, it promoted more efficient transport of Li^+ ions as well as more sulfur utilization. By incorporating PEDOT: PSS into $Ti_3C_2T_x$ nanosheets, it was also possible to inhibit the shuttle effect and improve the energy storage performance of Li–S batteries. Figure 12.3 shows the cyclic stability of$Ti_3C_2T_x$–PEDOT: PSS hybrid. Consequently, Li–S batteries with $Ti_3C_2T_x$–PEDOT: PSS separator offer superior electrochemical performance with a higher discharge rate of 1241.4 mAh g^{-1}, longer cycle life, and a low decay rate of 0.03% per cycle.

Chen et al. (2021b) synthesized a solid polymer electrolyte (SPE) composed of poly (vinylidene fluoride-co-hexafluoropropylene) filled with poly (methyl acrylate) grafted MXenes (denoted as PVHF/MXene-g-PMA) for high-performance zinc batteries. To facilitate hydrogen bond formation with PVHF as a matrix, the MXenes were grafted with PMA. At room temperature, the resultant SPE exhibited a conductivity (2.69×10^{-4} S cm^{-1}) three times greater than that of the PVHF matrix. A major advantage of the development of SPEs is that they provide significant shelf life for Zinc Ion Batteries. The storage stability studies by galvanostatic charge–discharge cycle illustrated the stable capacitance delivered by PVHF/MXene-g-PMA even after 90 days at a high or low temperature.

Dong et al. (2018) fabricated flexible and integrated sulfur cathode (a-Ti_3C_2-S/d-Ti_3C_2/PP) using alkalized Ti_3C_2 MXene nanoribbon with S/polysulfides host (a-Ti_3C_2-S) and delaminated Ti_3C_2 MXene (d-Ti_3C_2) nanosheets as interlayer on polypropylene (PP) separator, for Li–S batteries. The fabricated electrode achieved a higher capacitance of 1062 mAh g^{-1} at 0.2 Coulomb. Here MXene-based cathode eliminated the need for Al current collector, thus guaranteeing the possibility for next-generation flexible energy storage devices.

Songs et al. (2016) improvised the performance of Li-S batteries by coating $Ti_3C_2T_x$ MXene nanosheets on polypropylene (PP; "Celgard") membrane. Figure 12.4 represents a schematic diagram of Li–S cell with porous PP separator and MXene-modified separator. Powered by improved electrical conductivity and effective trapping of polysulfides, MXene-functionalized lithium–sulfur batteries performed exceptionally well, with a high discharge capacity of 550 mAh g^{-1} after 500 cycles with a capacity decay of only 0.062% per cycle at 0.5 Coulomb. MXenes are thus a prime candidate for improving the electrochemical efficiency of lithium–sulfur batteries. Table 12.2 describes a comparative study on the capacitance of various MXene/polymer composites in a battery application.

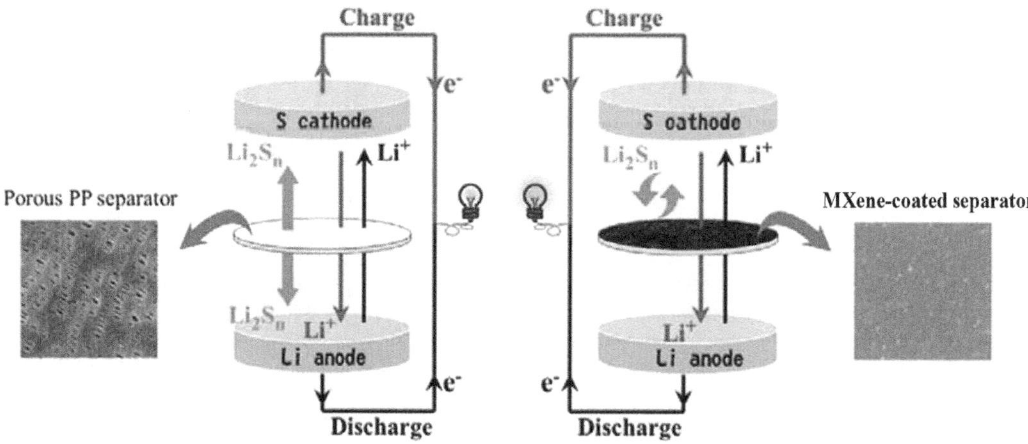

FIGURE 12.4 Schematic diagram of Li–S cell with a porous PP separator and MXene-modified separator (Song et al., 2016).

TABLE.12.2
Comparison of the capacity of Various MXene/Polymer Composites in Battery Application

Material	Capacitance (mAh g^{-1})	Ref.
d- $Ti_3C_2T_z$/PEDOT	71	(Chen et al., 2017)
$Ti_3C_2T_x$–PEDOT: PSS	1241.4	(Li et al., 2020a)
PVHF/MXene-g-PMA	72.3	(Chen et al., 2021b)
a-Ti_3C_2-S/d-Ti_3C_2/PP	1062	(Dong et al., 2018)
$Ti_3C_2T_x$- PP	550	(Song et al., 2016)

12.2.3 MXene/Polymer-Based Electrode Materials in Fuel-Cell Application

Fei et al. (2017) reported the proton conductivity of the $Ti_3C_2T_z$ -PBI composite membrane, finding an application in polymer electrolyte membrane fuel cells. Figure 12.5 shows that $Ti_3C_2T_z$ incorporated into the PBI membrane has enhanced proton conductivities. The conductivity of 3 wt% $Ti_3C_2T_z$ -PBI membranes is in the range of 4.5×10^{-3}–1.4×10^{-2} S cm^{-1}, which is higher than the pristine PBI (8.1×10^{-4}–5.1×10^{-3}). A high activation polarization loss of 0.25 V at 0.1 A cm^{-2} and a fuel-cell voltage loss of 0.6 and 0.4 V from 0.22 to 0.52 A cm^{-2}, was reported for $Ti_3C_2T_z$–PBI membranes. The conductivity of $Ti_3C_2T_z$–PBI membranes with respect to the fuel-cell performance was calculated to be 0.0091 S cm^{-1}. The power density of the 3 wt% $Ti_3C_2T_z$–PBI-based fuel cell is 135 mW cm^{-2} and 209 mW cm^{-2} at 120°C and 150°C, respectively. This study demonstrated the efficiency of $Ti_3C_2T_z$-MXene as a potential filler to obtain $Ti_3C_2T_z$–PBI-based polymer composite membranes, and its application in PEMFC operated at intermediate temperature facilitates enhanced proton conductivity.

For the PEM water electrolysis, Waribam et al. (2021) incorporated crosslinked sulfonated polyether ether ketone (C-SPEEK) with varying amounts of MXene/potassium titanate nanowire (MKT-NW), which acted as a filler for the PEM water electrolysis. Proton conductivity and ion exchange capacity for 12% MKT-NW/C-SPEEK were found to be 0.9971 S cm^{-1} and 1.88 meq g^{-1}, respectively. About 0.180566 mmol of hydrogen gas was produced within 5hours with the 12% MKT-NW/C-SPEEK membrane.

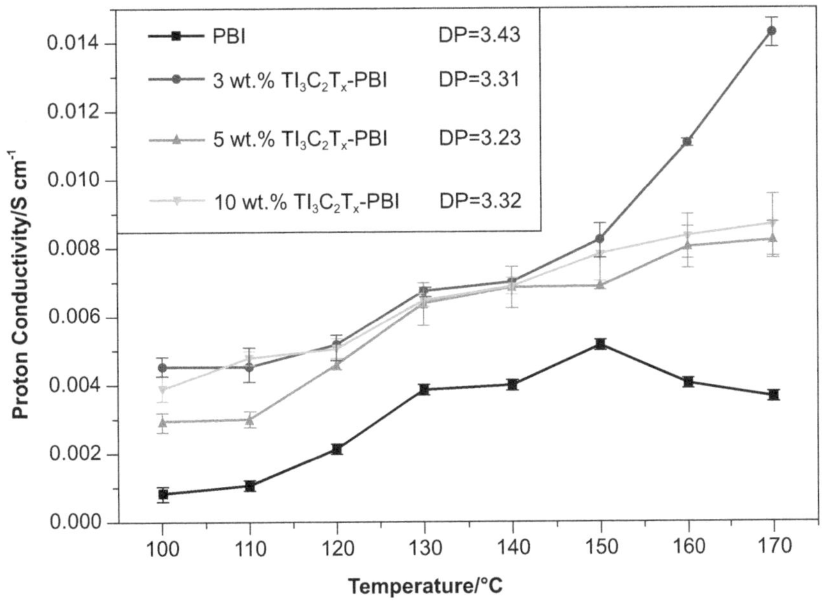

FIGURE 12.5 Proton conductivities of pure PBI and Ti_3C_2Tz-PBI membranes (Fei et al., 2017).

12.3 CONCLUSION

Polymer nanocomposites are widely employed in our daily lives. Compared to the polymer matrix, the characteristics of polymer nanocomposites may be customized and substantially enhanced by incorporating 2D nanofillers. Among various 2D nanofillers, MXene can bestow polymer matrices with favorable traits to meet the growing need for a wider variety and diversity of functional polymers. The particular focus on Ti_3C_2 MXene-based polymer nanocomposites in energy conversion and storage of most of the reports revealed enhancement in electrochemical performance by adding MXene. Thus, it was evident that polymer MXene nanocomposites have a great deal of potential in supercapacitor applications. Although many genuine reports are available on the polymer–MXene nanocomposites, the remarkable properties of these composites, keeping open many opportunities to extract its maximum use for humankind. Nanocomposites with MXene and polymers have been explored sparsely for battery and fuel-cell applications. There has been a vast discovery of conducting polymers and proton exchange polymers, which can be utilized to incorporate in MXene to expand its field of battery and fuel-cell applications. Another challenge that still exists is how to impart the electrochemical advantages of MXenes to polymer–MXene composites for industrial applications. The selection of polymers, synthesis process, and modification of MXenes could be concentrated to unlock the full scientific potential of these nanocomposites in studies related to fuel-cell and battery application for indigenous deliverables.

REFERENCES

AJNSZTAJN, A., FERGUSON, S., THOSTENSON, J. O., NGABOYAMAHINA, E., PARKER, C. B., GLASS, J. T. & STIFF-ROBERTS, A. D. 2020. Transparent MXene-polymer supercapacitive film deposited using RIR-MAPLE. *Crystals*, 10, 152.

BARSOUM, M. W. 2000. The $M_{N+1}A_XN$ phases: A new class of solids: Thermodynamically stable nanolaminates. *Progress in Solid State Chemistry*, 28, 201–281.

BOOTA, M., ANASORI, B., VOIGT, C., ZHAO, M. Q., BARSOUM, M. W. & GOGOTSI, Y. 2016. Pseudocapacitive electrodes produced by oxidant-free polymerization of pyrrole between the layers of 2D titanium carbide (MXene). *Advanced Materials*, 28, 1517–1522.

BOOTA, M. & GOGOTSI, Y. 2019. MXene—conducting polymer asymmetric pseudocapacitors. *Advanced Energy Materials*, 9, 1802917.

BOOTA, M., PASINI, M., GALEOTTI, F., PORZIO, W., ZHAO, M.-Q., HALIM, J. & GOGOTSI, Y. 2017. Interaction of polar and nonpolar polyfluorenes with layers of two-dimensional titanium carbide (MXene): intercalation and pseudocapacitance. *Chemistry of Materials*, 29, 2731–2738.

CAFFREY, N. M. 2018. Effect of mixed surface terminations on the structural and electrochemical properties of two-dimensional $Ti_3C_2T_2$ and V_2CT_2 MXenes multilayers. *Nanoscale*, 10, 13520–13530.

CAREY, M. & BARSOUM, M. 2021. MXene polymer nanocomposites: A review. *Materials Today Advances*, 9, 100120.

CHEN, C., BOOTA, M., XIE, X., ZHAO, M., ANASORI, B., REN, C. E., MIAO, L., JIANG, J. & GOGOTSI, Y. 2017. Charge transfer induced polymerization of EDOT confined between 2D titanium carbide layers. *Journal of Materials Chemistry A*, 5, 5260–5265.

CHEN, X., ZHAO, Y., LI, L., WANG, Y., WANG, J., XIONG, J., DU, S., ZHANG, P., SHI, X. & YU, J. 2021a. MXene/polymer nanocomposites: Preparation, properties, and applications. *Polymer Reviews*, 61, 80–115.

CHEN, Z., LI, X., WANG, D., YANG, Q., MA, L., HUANG, Z., LIANG, G., CHEN, A., GUO, Y. & DONG, B. 2021b. Grafted MXene/polymer electrolyte for high performance solid zinc batteries with enhanced shelf life at low/high temperatures. *Energy & Environmental Science*, 14(6), 3492–3501.

DONG, Y., ZHENG, S., QIN, J., ZHAO, X., SHI, H., WANG, X., CHEN, J. & WU, Z.-S. 2018. All-MXene-based integrated electrode constructed by Ti_3C_2 nanoribbon framework host and nanosheet interlayer for high-energy-density Li–S batteries. *Acs Nano*, 12, 2381–2388.

FAN, Z., WANG, Y., XIE, Z., WANG, D., YUAN, Y., KANG, H., SU, B., CHENG, Z. & LIU, Y. 2018. Modified MXene/holey graphene films for advanced supercapacitor electrodes with superior energy storage. *Advanced Science*, 5, 1800750.

FEI, M., LIN, R., DENG, Y., XIAN, H., BIAN, R., ZHANG, X., CHENG, J., XU, C. & CAI, D. 2017. Polybenzimidazole/Mxene composite membranes for intermediate temperature polymer electrolyte membrane fuel cells. *Nanotechnology*, 29, 035403.

FENG, X.-Y., DING, B.-Y., LIANG, W.-Y., ZHANG, F., NING, T.-Y., LIU, J. & ZHANG, H. 2018. MXene $Ti_3C_2T_x$ absorber for a 1.06 μm passively Q-switched ceramic laser. *Laser Physics Letters*, 15, 085805.

GAO, L., LI, C., HUANG, W., MEI, S., LIN, H., OU, Q., ZHANG, Y., GUO, J., ZHANG, F. & XU, S. 2020. MXene/polymer membranes: Synthesis, properties, and emerging applications. *Chemistry of Materials*, 32, 1703–1747.

GHIDIU, M., LUKATSKAYA, M. R., ZHAO, M.-Q., GOGOTSI, Y. & BARSOUM, M. W. 2014. Conductive two-dimensional titanium carbide 'clay' with high volumetric capacitance. *Nature*, 516, 78–81.

GUND, G. S., PARK, J. H., HARPALSINH, R., KOTA, M., SHIN, J. H., KIM, T.-I., GOGOTSI, Y. & PARK, H. S. 2019. MXene/polymer hybrid materials for flexible AC-filtering electrochemical capacitors. *Joule*, 3, 164–176.

HOPE, M. A., FORSE, A. C., GRIFFITH, K. J., LUKATSKAYA, M. R., GHIDIU, M., GOGOTSI, Y. & GREY, C. P. 2016. NMR reveals the surface functionalisation of Ti_3C_2 MXene. *Physical Chemistry Chemical Physics*, 18, 5099–5102.

JIMMY, J. & KANDASUBRAMANIAN, B. 2020. MXene functionalized polymer composites: Synthesis and applications. *European Polymer Journal*, 122, 109367.

LAI, C.-C., FASHANDI, H., LU, J., PALISAITIS, J., PERSSON, P. O., HULTMAN, L., EKLUND, P. & ROSEN, J. 2017. Phase formation of nanolaminated Mo_2AuC and $Mo_2(Au_{1-x}Ga_x)_2C$ by a substitutional reaction within Au-capped Mo_2GaC and Mo_2 Ga_2C thin films. *Nanoscale*, 9, 17681–17687.

LI, J., JIN, Q., YIN, F., ZHU, C., ZHANG, X. & ZHANG, Z. 2020a. Effect of $Ti_3C_2T_x$–PEDOT: PSS modified-separators on the electrochemical performance of Li–S batteries. *RSC Advances*, 10, 40276–40283.

LI, K., WANG, X., WANG, X., LIANG, M., NICOLOSI, V., XU, Y. & GOGOTSI, Y. 2020b. All-pseudocapacitive asymmetric MXene-carbon-conducting polymer supercapacitors. *Nano Energy*, 75, 104971.

LI, X., YIN, X., LIANG, S., LI, M., CHENG, L. & ZHANG, L. 2019. 2D carbide MXene Ti_2CT_X as a novel high-performance electromagnetic interference shielding material. *Carbon*, 146, 210–217.

LING, Z., REN, C. E., ZHAO, M.-Q., YANG, J., GIAMMARCO, J. M., QIU, J., BARSOUM, M. W. & GOGOTSI, Y. 2014. Flexible and conductive MXene films and nanocomposites with high capacitance. *Proceedings of the National Academy of Sciences*, 111, 16676–16681.

LIPATOV, A., LU, H., ALHABEB, M., ANASORI, B., GRUVERMAN, A., GOGOTSI, Y. & SINITSKII, A. 2018. Elastic properties of 2D $Ti_3C_2T_x$ MXene monolayers and bilayers. *Science Advances*, 4, eaat0491.

LIU, Z., WU, E., WANG, J., QIAN, Y., XIANG, H., LI, X., JIN, Q., SUN, G., CHEN, X. & WANG, J. 2014. Crystal structure and formation mechanism of $(Cr_{2/3}Ti_{1/3})_3AlC_2$ MAX phase. *Acta Materialia*, 73, 186–193.

LUKATSKAYA, M. R., MASHTALIR, O., REN, C. E., DALL'AGNESE, Y., ROZIER, P., TABERNA, P. L., NAGUIB, M., SIMON, P., BARSOUM, M. W. & GOGOTSI, Y. 2013. Cation intercalation and high volumetric capacitance of two-dimensional titanium carbide. *Science*, 341, 1502–1505.

NAGUIB, M., KURTOGLU, M., PRESSER, V., LU, J., NIU, J., HEON, M., HULTMAN, L., GOGOTSI, Y. & BARSOUM, M. W. 2011. Two-dimensional nanocrystals produced by exfoliation of Ti_3AlC_2. *Advanced Materials*, 23, 4248–4253.

NECHICHE, M., CABIOC'H, T., CASPI, E. N., RIVIN, O., HOSER, A., GAUTHIER-BRUNET, V. R., CHARTIER, P. & DUBOIS, S. 2017. Evidence for symmetry reduction in $Ti_3(Al_{1-\delta}Cu_\delta)C_2$ MAX phase solid solutions. *Inorganic Chemistry*, 56, 14388–14395.

QIN, L., TAO, Q., EL GHAZALY, A., FERNANDEZ-RODRIGUEZ, J., PERSSON, P. O., ROSEN, J. & ZHANG, F. 2018. High-performance ultrathin flexible solid-state supercapacitors based on solution processable $Mo_{1.33}C$ MXene and PEDOT: PSS. *Advanced Functional Materials*, 28, 1703808.

QIN, L., TAO, Q., LIU, X., FAHLMAN, M., HALIM, J., PERSSON, P. O., ROSEN, J. & ZHANG, F. 2019. Polymer-MXene composite films formed by MXene-facilitated electrochemical polymerization for flexible solid-state microsupercapacitors. *Nano Energy*, 60, 734–742.

REN, Y., ZHU, J., WANG, L., LIU, H., LIU, Y., WU, W. & WANG, F. 2018. Synthesis of polyaniline nanoparticles deposited on two-dimensional titanium carbide for high-performance supercapacitors. *Materials Letters*, 214, 84–87.

SHAO, W., TEBYETEKERWA, M., MARRIAM, I., LI, W., WU, Y., PENG, S., RAMAKRISHNA, S., YANG, S. & ZHU, M. 2018. Polyester@ MXene nanofibers-based yarn electrodes. *Journal of Power Sources*, 396, 683–690.

SOKOL, M., NATU, V., KOTA, S. & BARSOUM, M. W. 2019. On the chemical diversity of the MAX phases. *Trends in Chemistry*, 1, 210–223.

SONG, J., SU, D., XIE, X., GUO, X., BAO, W., SHAO, G. & WANG, G. 2016. Immobilizing polysulfides with MXene-functionalized separators for stable lithium–sulfur batteries. *ACS Applied Materials & Interfaces*, 8, 29427–29433.

TAO, Y., XIE, X., LV, W., TANG, D.-M., KONG, D., HUANG, Z., NISHIHARA, H., ISHII, T., LI, B. & GOLBERG, D. 2013. Towards ultrahigh volumetric capacitance: graphene derived highly dense but porous carbons for supercapacitors. *Scientific Reports*, 3, 1–8.

VAHIDMOHAMMADI, A., MONCADA, J., CHEN, H., KAYALI, E., ORANGI, J., CARRERO, C. A. & BEIDAGHI, M. 2018. Thick and freestanding MXene/PANI pseudocapacitive electrodes with ultrahigh specific capacitance. *Journal of Materials Chemistry A*, 6, 22123–22133.

WANG, J., TANG, J., DING, B., MALGRAS, V., CHANG, Z., HAO, X., WANG, Y., DOU, H., ZHANG, X. & YAMAUCHI, Y. 2017. Hierarchical porous carbons with layer-by-layer motif architectures from confined soft-template self-assembly in layered materials. *Nature Communications*, 8, 15717 (2017).

WARIBAM, P., JAIYEN, K., SAMART, C., OGAWA, M., GUAN, G. & KONGPARAKUL, S. 2021. MXene potassium titanate nanowire/sulfonated polyether ether ketone (SPEEK) hybrid composite proton exchange membrane for photocatalytic water splitting. *RSC Advances*, 11, 9327–9335.

WU, X., HUANG, B., LV, R., WANG, Q. & WANG, Y. 2019. Highly flexible and low capacitance loss supercapacitor electrode based on hybridizing decentralized conjugated polymer chains with MXene. *Chemical Engineering Journal*, 378, 122246.

ZHANG, M., HÉRALY, F., YI, M. & YUAN, J. 2021. Multitasking tartaric-acid-enabled, highly conductive, and stable MXene/conducting polymer composite for ultrafast supercapacitor. *Cell Reports Physical Science*, 100449.

ZHU, M., HUANG, Y., DENG, Q., ZHOU, J., PEI, Z., XUE, Q., HUANG, Y., WANG, Z., LI, H. & HUANG, Q. 2016. Highly flexible, freestanding supercapacitor electrode with enhanced performance obtained by hybridizing polypyrrole chains with MXene. *Advanced Energy Materials*, 6, 1600969.

13 Biomedical Applications of MXene/Polymer Nanocomposites

Nupur Garg and Farhan J. Ahmad

CONTENTS

13.1 INTRODUCTION

In the class of two-dimensional (2D)-layered materials, MXenes have shown immense potential in the field of healthcare. Attributed to their excellent electrical, thermal and mechanical properties, high surface-to-volume ratio, nanoscale thickness and flexible surface, they are applied to the biomedical field for innovating new methods in therapeutics like biosensing, photothermal or photodynamic therapies, drug delivery, diagnostic and regenerative medicine. Their structure consists of a mixture of different surface functional groups like hydroxyl (–OH), fluoride (–F) or oxygen (–O) groups, that arise during the etching step of the synthesis, with exchangeable cations like water. These give MXenes high hydrophilicity, surface reactivity, good biocompatibility and unique electrical and optical properties. This paves the way for stable MXene suspensions and doping with different hydrophilic polymers. The use of polymer-doped MXenes further increases their versatility to be applied to therapeutics.

With the increasing demand of healthcare and biomedical materials and methods, it is imperative for researchers to come up with innovative techniques and products. MXenes have paved the way for wider research scope in therapeutic materials. MXene/polymer nanocomposites (MX/P-NCs) have been utilized in the development of multifunctional flexible and smart fabrics and devices, wound healing dressings and wearable electronics. MXenes have various surface-active sites, which makes it easier to functionalize with different polymers to form polymer nanocomposites (MX/P-NCs) due to which they display high selectivity and sensitivity. Depending on the choice of polymers, they show high flexibility and can further be developed into a variety of sensors and detectors. Many

DOI: 10.1201/9781003164975-13

research studies have investigated the bactericidal properties of MX/P-NCs that are discussed further in this chapter. Several mechanisms have been proposed for their antibacterial action, including the generation of reactive oxygen species (ROS) causing oxidative stress, interaction with bacterial protein and nucleic acids to cause cell death, inhibition of electron transport, and physical stress with their sharp edges resulting in mechanical damage of the cell membrane and loss of integrity, penetration and leakage of the cell membrane (Rasool et al., 2016). They have been crosslinked with different polymers to cause synergistic action on their antibacterial property due to chemical and physical effects. Polymer doping makes the MXene moiety more biocompatible and enhanced interaction with bacterial cell membranes for better contact.

Apart from these applications, they have been used in diagnostic imaging techniques as contrast agents for better medical image quality that shows diseased tissue in contrast with the normal tissue background, in targeted cancer chemotherapy through photoconversion mechanism and associated hyperthermal treatment and in several on-site novel drug delivery systems by surface associations of various types of receptor ligands, antibodies, functional groups and radical atoms. Their theranostic applicability can be understood by their phototherapeutic ability majorly in cancer chemotherapeutics. The photothermal ablation treatment of cancer has been proved to reduce the tumor tissue apart from diagnosing the cancerous part. They can be used simultaneously for diagnosing the diseased part of the body as well as the release of the therapeutic agent at the target site. Their unique 2D structure is an added advantage to be explored by interacting with different types of polymers and nanoparticles, along with the active agent to form nanocomplexes. They are proving to be scalable for industrial applications despite their specific structure complexity being a challenge (Soleymaniha et al., 2019). This chapter summarizes the important applications in the field of health management and biomedical sector as depicted in Figure 13.1.

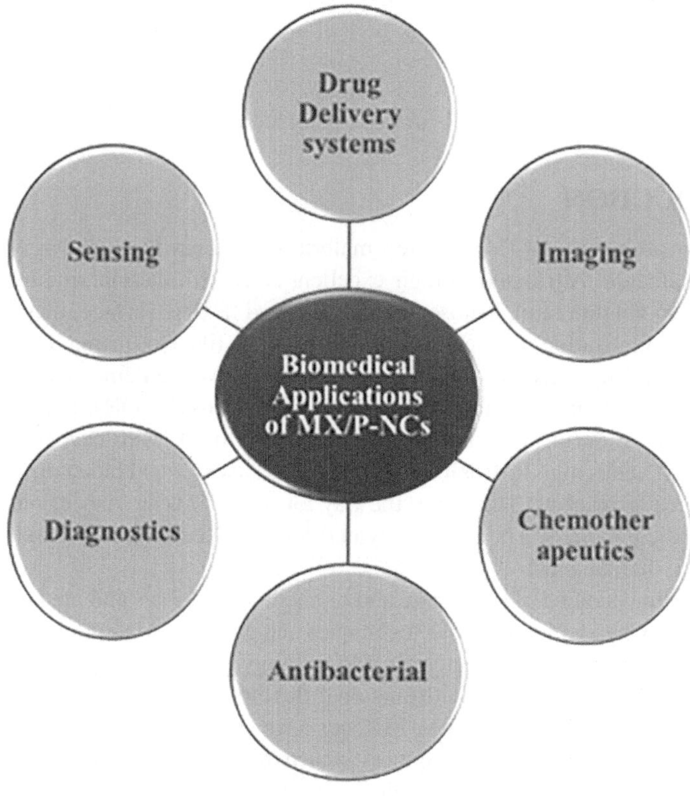

FIGURE 13.1 Biomedical applications of MX/P-NCs.

13.2 BIOMEDICAL PROPERTIES

MXene/P-NCs have several unique properties that make them important constituents to be used in the biomedical sector and disease management. Attributed to their size, structure and ability to crosslink in different ways with different polymers, they present various possibilities to be utilized in therapeutics. Figure 13.2 summarizes the major properties of MX/P-NCs that could be harnessed for biomedical applications.

13.2.1 Surface-to-Volume Ratio

Bestowed with a high surface-to-volume ratio and large specific surface area, surface functionalization with desired groups is much easier and provides a number of polymer crosslinking options. MXenes behave as potent moieties if specific reactive groups are attached to its surface, which helps in its targeted action and in formulating drug delivery systems.

13.2.2 Shape and Morphology

MXene nanosheets comprise 2D thin and flat structures with sharp edges that result from an exfoliation of the layers during synthesizing process. These help in causing stress and physical damage to the bacterial cell membranes. Nanocomposites with desired surface modifications can be obtained by the homogeneous dispersion of exfoliated sheets into desired polymer matrix to reach to the target better. Various electron donors are present and can be doped with desired polymers to achieve targeted results. Their surface morphology acts as a critical factor in therapeutic ability (Soleymaniha et al., 2019).

13.2.3 Hydrophilicity

The presence of their negative surface charge, along with abundant surface functional groups, they can easily interact with water molecules. This surface chemistry is responsible for rendering MXenes hydrophilic due to which they can be formulated to stable suspensions in water or other

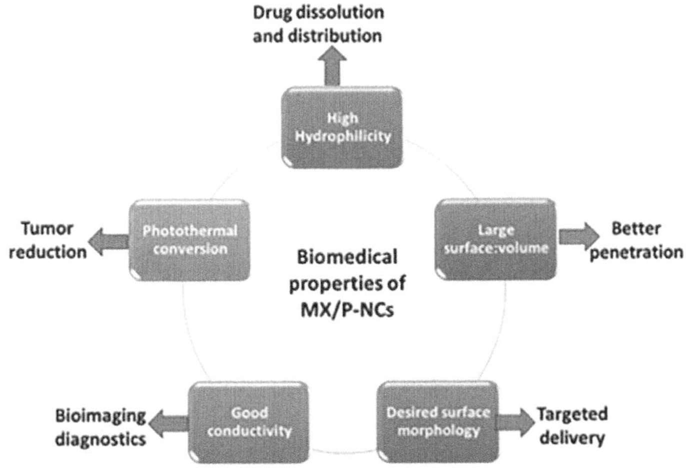

FIGURE 13.2 Biomedical properties of MX/P-NCs.

polar solvents that too at neutral pH. This makes it easier to link and interact with several hydrophilic polymers and biopolymers, making them biocompatible and biodegradable (Chen et al., 2021).

13.2.4 Conductivity

Attributed to their high conductivity, MX-P/NCs are interestingly employed in several bioengineering devices such as sensors, films or fabrics. This property can be exploited in bioimaging and theranostic applications of polymeric nanocomposites. By tracking the metallic conductance, surface functional groups can be identified to know the binding sites for analytes. Gas sensors and molecule sensors can be fabricated for highly sensitive analyte determination. They also have many free charge carriers at their surface due to which they can reflect electromagnetic waves before absorption and, hence, used in detection and imaging. This behavior gives rise to the unique optoelectronic properties of MX-P/NCs, which is exploited in many bioimaging techniques (Carey and Barsoum, 2021).

13.2.5 Photothermal Conversion

Photothermal therapy (PTT) is a novel approach that is done by targeted heating of photothermally active nanocomposites at the cancerous site without damaging the healthy tissues. They act by absorbing electromagnetic radiation in the near-infrared (NIR) window and killing cancer cells by light-to-heat conversion. The efficiency of PTT is based on the size and location of the tumor and the heat generated to reduce it. Several studies have been reported that used MXene hyperthermia as synergistic therapy along with other encapsulated anticancer agents for cancer treatment.

Similarly, photodynamic therapy (PDT) has also been explored for cancer treatment. This therapy acts on cancer cells by generating reactive oxygen species (ROS) from a prodrug using a photosensitizing agent and an activation agent, resulting in cancer cell death through oxidative photodamage. Generally, PDT is used in conjunction with other cancer therapies like chemotherapy or radiotherapy (Lin et al., 2018).

13.3 DRUG DELIVERY

One of the major biomedical applications of MX-P/NCs is their role in novel drug delivery systems. Delivery systems guided, controlled, sustained and targeted release of drugs is inevitable for maximum therapeutic effect with a minimum dose and metabolism in disease therapy. This purpose is attained by exploring the flexibility of MX-P/NCs for polymer–drug interactions. Many types of nanocomplexes can attack the target disease cells efficiently, and MX-P/NCs further enhance the reach, efficiency and therapeutic efficacy of the system by offering different pathways for drug–disease interactions. They have been good carriers for delivering anticancer drugs in cancer chemotherapy. Han et al. investigated the use of SP-modified Ti_3C_2 nanosheet complexes as drug carriers and therapeutic synergistic agents for cancer therapy. These complexes, apart from possessing high drug-loading capacity, also provide pH-responsive and laser-triggered drug release (Han et al., 2018). Similar work was done by Liu and coworkers (2020) by preparing nanocarriers of cobalt nanowires with Ti_3C_2 for synergistic anticancer effect with doxorubicin along with magnetic controlled drug release. A group of researchers (Zhang et al., 2019) fabricated MXene/polyacrylamide nanocomposite hydrogel for higher mechanical gel strength, high drug loading and better sustained released properties. The tumor-targeted therapy using MX/P-NCs have gained a huge interest among researchers. A study group (Xing et al., 2018) developed a Ti_3C_2/cellulose nanocomposite hydrogel that combines the photothermal ability of MXene with increased biocompatibility and sustained drug release properties of cellulose. They reported a 48-hour long-sustained distribution and release of anticancer drug doxorubicin hydrochloride (DOX). This system shows a synergistic therapeutic effect with phototherapy plus chemotherapy against tumor cells. Table 13.1 presentsresearch done on employing MX/P-NCs in drug delivery systems.

TABLE 13.1
Applications of MX/P-NCs in Drug Delivery

Delivery system	Therapeutic Intervention	Results	Reference
2D Ti_3C_2 MXenes	Tumor eradication	High drug loading, pH- and laser-triggered release, synergistic PT	(Han et al., 2018)
Ti_3C_2 cobalt nanowires	Chemo-photothermal therapy	Magnetic/pH responsive/NIR controlled release, PT efficacy	(Liu et al., 2020)
MXene-polyacrylamide hydrogel	Vehicle for enhanced drug release	High mechanical strength, drug loading, flexibility and gel properties	(Zhang et al., 2019)
Ti_3C_2/cellulose hydrogel	Targeting and controlled release for cancer chemotherapy	Sustained drug release, biocompatible, PT tumor scavenging ability with doxorubicin	(Xing et al., 2018)

TABLE 13.2
Applications of MX/P-NCs in Bioimaging

MX/P-NCs	Imaging Technique	Results	Reference
Ti_3C_2 polyoxometalates	MRI/CT	Hyperthermal treatment, diagnostic contrast agent	(Zong et al., 2018)
Ta_4C_3-MnOx-SP	PTT/MRI/CT for tumor	Photothermal hyperthermia, contrast agent	(Dai et al., 2017a)
MnOx/Ti_3C_2-SP	PTT/MRI/CT for tumor	Tumor hyperthermia	(Dai et al., 2017b)
Ta_4C_3-Fe_3O_4	PTT/MRI/CT for tumor	Tumor eradication, photothermal ablation	(Liu et al., 2018)
Ti_3C_2 QDs	PTT for tumor	Multicolor cellular imaging probes	(Xue et al., 2017)
Ti_3C_2 QD/Chitosan	Stem cell–based therapy	Inflammation and tissue repair	(Rafieerad et al., 2019)

13.4 BIOIMAGING

The inherent physicochemical properties of MX/P-NCs, including the conductivity, optic, electronic, magnetic, and the like, present the scope of exploiting them in imaging techniques. Magnetic resonance imaging (MRI) is one such technique that uses a contrast agent that can be guided in the electromagnetic field to give enhanced resolution images. A contrast agent is used to increase the ability of medical imaging to differentiate normal and diseased tissues by providing a background contrast by altering electromagnetism or ultrasonic absorption. They improve the image quality of the diagnostic method. MXenes decorated with different types of magnetic nanoparticles could provide MRI contrasting along with their therapeutic performance at the same time. Few such studies are compiled in Table 13.2. Zong and coworkers (2018) integrated the surface of titanium carbide (Ti_3C_2) MXene nanosheets with polyoxometalates to present MRI or computed tomography (CT) imaging of tumor cells along with their inherent property of hyperthermal action on tumor defining their theranostic applications. Dai *et al.* achieved surface functionalization of tantalum carbide (Ta_4C_3) nanosheets with manganese oxide (MnO_x) nanoparticles to perform as contrast agents for CT/MRI and showed photothermal hyperthermia in tumor suppression so as to use them for PTT. Other work by the same research group (Dai et al., 2017b) showed the efficiency of MnO_x/Ti_3C_2nanosheets as multifunctional theranostic agents in MRI-guided tumor hyperthermia.

Supra-paramagnetic iron oxide nanoparticles have been regarded as excellent contrast agents in diagnostic imaging. Therefore, the surface functionalization of MXenes with these nanoparticles could be utilized to complement these nanosheets with MR-enabled ability. Researchers (Liu et al., 2018) have explored the tantalum carbide MXenes with superparamagnetic iron oxide functionalization for breast cancer theranostic. Xue and coworkers (2017) fabricated biocompatible photoluminescent quantum dots of Ti_3C_2MXenes (MQD) and demonstrated their application in multicolor imaging cellular probes. Another application of MQD was shown by researchers (Rafieerad et al., 2019) by formulating a MQD-chitosan hydrogel with improved physicochemical properties. They demonstrated its scope in translational medicine for tissue repair in inflammatory and degenerative diseases.

13.5 CANCER THERANOSTICS

Several researchers have explored and applied the photothermal properties of various types of MX/P-NCs to cancer therapies majorly by using the inherent hyperthermal ability of the nanosheets along with the therapeutic efficacy of the anticancer drug nanoparticles. By incorporating the anticancer drug in the polymeric 3D porous structure, high drug loading and uptake can be done. In the past couple of years, a number of 2-D nanocomposites were developed for better diagnostic screening (PTT/CT/MRI) and treatment of cancer. This has been achieved due to several merits that MX/P-NCs exhibit, namely, nanosized for enhanced permeation through barriers, tumor penetration and retention, and increased surface area for a wide scope of surface functionalization with different therapeutic moieties. MX/P-NCs have been widely researched for oncological applications. They could be used in different therapies like cancer chemotherapy, radiation therapy, photothermal therapy and photodynamic therapy (PDT). For diagnostic purposes, MX/P-NCS have been utilized as contrast agents in several techniques like MRI, photoacoustic imaging (PAI), CT imaging and sensing. Also, they could be used in cancer theranostics as the delivery systems for anticancer drugs and contrast agents for imaging and treatment simultaneously. Numerous research groups have developed specific MX/P-NCs for cancer diagnostic imaging, some of which are summarized in Table 13.3.

Lin *et al.* first reported the photothermal ablation capability of ultra-thin MXene nanosheets against cancer by developing soybean phospholipid–modified Ti_3C_2 nanosheets for intravenous administration (Lin et al., 2017b). They also demonstrated a similar effect in localized tumor

TABLE 13.3
Application of MX/P-NCs in Cancer Therapies

MX/P-NCs	Therapeutic Intervention	Study Results	Reference
SP-Ti_3C_2	Photothermal ablation	High photoconversion efficiency	(Lin et al., 2017b)
$Al(OH)_4^-$ functionalized Ti_3C_2-DOX	Chemotherapeutic/PTT/PDT	ROS generation, high drug loading, active targeting	(Liu et al., 2017)
Mo_2C-PVA nanoflakes	Photonic tumor hyperthermia	Biocompatible, intense NIR absorption, deep tissue penetration	(Feng et al., 2019)
Nb_2C PVP	Photothermal ablation with two NIR bio-windows	Low phototoxicity, high biocompatibility, enzyme-responsive biodegradability	(Lin et al., 2017a)
Silica coated Ti_3C_2MXene	Synergistic chemotherapy and hyperthermia in hepatocellular carcinoma	High hydrophilicity, controlled drug release, active targeting	(Li et al., 2018)
Ti_3C_2@Au-PEG	Radiotherapy, CT, Photoacoustic therapy	High optical absorption, cancer theranostics	(Tang et al., 2019)

implantable Ti_3C_2/PLGA hybrid. The same group (Lin et al., 2017a) developed niobium carbide MXene nanosheets for near infrared radiation (NIR)–based PTT through two bio-windows NIR-I and NIR-II as shown in *in vivo* xenografts. Liu et al. synthesized multifunctional system of surface-modified Ti_3C_2-DOX for PTT/PDT along with synergistic chemotherapy (Liu et al., 2017). The research on synthesizing novel MX/P-NCs for cancer therapy had gained vast interest in the past couple of years. Feng et al.(2019) developed molybednum carbide (Mo_2C) MX/P-NCs with polyvinyl alcohol (PVA) through surface engineering to increase the biodegradability and biocompatibility of the system. They possessed PT-conversion efficiency in both NIR bio-windows for photonic tumor hyperthermia. Li et al. reported the treatment of hepatocellular carcinoma through PTT and synergistic chemotherapy (Li et al., 2018). They synthesized receptor targeted Ti_3C_2 MXene surface functionalized with a layer of thin mesoporous silica. They showed targeted and controlled drug release, enhanced hydrophilicity and scope for more surface engineering for desired drug-target reach. The chemotherapeutic effect from the surface functionalization and photothermal hyperthermia by the MXene core contributed to the eradication of the tumor as shown by their *in vitro* and *in vivo* results. Tang et al. used gold for surface modification of Ti_3C_2 MX/P-NCs for increasing stability and biocompatibility and maintaining low toxicity. This also enhanced the imaging property using photoacoustic and CT and could be used for radiotherapy in cancer theranostics (Tang et al., 2019).

13.6 ANTIBACTERIAL APPLICATIONS

The antibacterial activity of MX/P-NCs has been well researched and documented by several researchers. The mechanisms, including both physical and chemical factors, are proposed to explain their antimicrobial properties against both gram-positive and gram-negative bacteria. The bacteria-killing ability of MXenes is attributed to the generation of ROS, mechanical damage to the integrity of bacterial membrane by the sharp edges of the 2D MX/P-NCs nanosheets structure, direct penetration and interaction with the sulfur and phosphorus groups of the bacterial protein and DNA, respectively. The high hydrophilicity may enhance the surface contact of bacteria with MXenes leading to the inactivation and killing of bacteria by direct contact. These properties of MXenes may contribute to their use in disinfectant industry since they are biocompatible and biodegradable also. Rasool *et al.* investigated the antibacterial activity of colloidal MXene against gram-negative *Escherichia coli* and gram-positive *Bacillussubtilis* that showed mechanical damage to bacterial cells as well as ROS-dependent oxidative stress to be the major reasons of bacterial cell death (Rasool et al., 2016). Same research group in another study (Rasool et al., 2017) indicated the bactericidal properties of titanium carbide-polyvinylidene fluoride against *E. coli* and *B. subtilis*. Mayerberger et al. prepared nontoxic biocompatible $Ti_3C_2T_z$/chitosan nanofibers antibacterial wound dressing that was found to be effective against *E. coli* and *S. aureus* and showed wound healing properties of the same (Mayerberger et al., 2018). Polymers such as chitosan had already been explored (Chung and Chen, 2008) as a natural bactericide that acts by interacting with the bacterial membrane of *E. coli* and *S. aureus*. They could be explored to have a synergistic action by combining with MXenes. As suggested by Li *et al.*, bactericidal action has been shown by the production of ROS by engineered metal oxide nanoparticles that results in oxidative stress to bacteria (Li et al., 2012). Some of the studies done on MX/P-NCs to show their antibacterial potential are given in Table 13.4.

13.7 BIOSENSING

Several MX/P-NC nanocomplexes have been used to design sensors for detecting various health and biological parameters and disease status by through detection of different analytes, gaseous molecules, contaminants, physical or motion stimuli, among others. The mechanisms behind fabricating MXene biosensors utilize their electrochemical activity, which depends on the intensity

TABLE 13.4
Applications of MX/P-NCs as Antibacterial Agents

MX/P-NCs	Antibacterial Activity	Results	Reference
Colloidal MXene	*E. coli, B. subtilis*	Mechanical damage, oxidative stress	(Rasool et al., 2016)
Ti_3C_2/polyvinylidene fluoride	*E.coli, B. subtilis*	Damage to bacterial cell wall/generation of ROS	(Rasool et al., 2017)
Ti_3C_2Tz/chitosan nanofibers	*E. coli, S. aureus*	Wound healing	(Mayerberger et al., 2018)
Ti_3C_2Tx MXene	*E.coli, B. subtilis*	Physical interaction, cell damage	(Arabi Shamsabadi et al., 2018)

of target signals. When MXene binds the biological target or vice versa, its electronic properties alter leading to generation of current or changes in a detector. With the introduction of high-performance analytical techniques, they have emerged as an advanced platform for biosensing applications. These have been incorporated into the compatible devices for rapid, easy and sensitive detection of specific biological change (Xu et al., 2020). Wearable sensors have been designed to regularly monitor basic health parameters (Xin et al., 2020). With the use of different polymer nanoparticles, their morphology can be controlled and developed to enhance the sensing ability of MXene-based sensors. Rakhi et al. fabricated amperometric glucose biosensor using gold nanoparticle–MXene (Au-NP-MXene) nanocomposite that acts as the matrix for immobilized glucose oxidase enzyme through which amperometric detection could be obtained (Rakhi et al., 2016). Similarly other researchers (Wang et al., 2014; Liu et al., 2015) immobilized hemoglobin on Ti_3C_2-MXene to develop peroxide and nitrite mediator free biosensors. These types of biosensors are used for environmental analysis of different types of analytes. MXenes can also be used to fabricate gas sensors to detect gases like ammonia or acetone to diagnose diseases such as peptic ulcers or diabetes, respectively. Kim et al. (2018) developed a highly sensitive detector of volatile organic compounds that helps in estimating their alarming levels in a diseased state. Lorencova et al. showed $Ti_3C_2T_x$/PtNP nanocomposites deposited on the electrodes can detect small molecules, such as ascorbic acid, dopamine, uric acid and the like, in low limits implying that these can be applied for the development of enzyme-based biosensors (Lorencova et al., 2018). Several body biomarkers can be identified using such sensors. MXene immunosensors based on antigen–antibody interaction have also been developed for high-throughput performance of the analytical test that is cost-effective, portable, highly specific and sensitive (Fang et al., 2020). Kumar and coworkers (Kumar et al., 2018) functionalized MXene nanosheets with aminosilane to provide bioreceptor site for CEA (carcinoembryonic antigen) binding for sensitive detection of cancer biomarker. Table 13.5 summarizes different types of biosensors developed using unique electrochemical and morphological properties of MX/P-NCs.

Chen et al. developed a Ti_3C_2 quantum dots wearable nanosensor for estimating intracellular pH using a ratiometric luminescent probe (Chen et al., 2018). Detection of small quantities of analytes in blood can be done easily through affordable, point-of-care and rapid devices based on the ability of MXenes to interact and bind to different molecules. Zheng et al. (2018) fabricated an MXene-based nanocomposite (MXene/DNA/Pd/Pt) to detect the levels of dopamine whereby DNA is adsorbed on the Pd/Pt nanocomposite–MXene matrix. MX/P-NCs doped with small molecules and metal ions produce similar effect. Peng et al. fabricated a fluorescent nanoprobe based on Ti_3C_2 nanosheets for analysis of human papillomavirus, which is a causative pathogen of cervical cancer (Peng et al., 2019). Attributed to their unique conductive, optical and electronic properties, detection and analysis in readable data are easy to generate.

TABLE 13.5

MXene	Type of Biosensor	Applications	Reference
Au-NP-MXene	Enzyme sensors	Glucose detection	(Rakhi et al., 2016)
$Ti_3C_2T_x$/PtNP	Enzyme sensors	Small biomolecules like dopamine, uric acid, ascorbic acid	(Lorencova et al., 2018)
Ti_3C_2-immobilized hemoglobin	Enzyme sensors	Mediator-free sensor for peroxide or nitrite	(Li et al., 2012· Arabiet al., 2018)
Ti_3C_2-β-hydroxybutyrate dehydrogenase	Enzyme sensors	Diabetic ketoacidosis	(Koyappayil et al., 2020)
$Ti_3C_2T_x$	Gas sensors	Volatile organic compounds	(Kim et al., 2018)
Aminosilane-MXenes	Immunosensors	Anti-Carcinoembryonic antigen cancer biomarker	(Kumar et al., 2018)
MXene-Ti_3C_2Tx	Immunosensors	Monitoring of blood components	(Liu et al., 2019)
$Ti_3C_2T_x$ and polyvinyl alcohol (PVA) hydrogel	Strain sensors	Physical strain sensing	(Zhang et al., 2019)
$Ti_3C_2T_x$-polyacrylamide and polyvinyl alcohol	Strain sensors	Personalized health monitoring	(Liao et al., 2019)
Ti_3C_2 quantum dots	Wearable sensors	Monitoring intracellular pH	(Chen et al., 2018)

13.8 FUTURE PERSPECTIVES

This is to understand that combining MX/P-NCs with various molecular techniques for biomedical applications is still at its infancy. Numerous challenges have been seen in current research progress on applying them to healthcare systems. Their translation applications still need to be exploited and face challenges, like industrial applications or large-scale production, desired surface engineering, biosafety, controlling their behavior and complexity, to name a few. Due to their physicochemical properties, MXene–polymer blend optimization is inevitable. Multidisciplinary approach in their fabrication, to employ more specific, effective and stable surface modifications needs to be addressed in future research. To improve their performances, tailoring and controlling their unique properties are necessary. They have a wide scope in chemotherapeutics and cancer theranostics, which could be explored. Researchers have shown synergistic effects. Despite all these challenges, MX/P-NCs have shown great potential to be applied in biomedical techniques and have a wide scope of research on their bench-to-bedside use.

13.9 CONCLUSION

MXene/polymer nanocomposites have gained interest to be applied to a number of biomedical techniques for better performance, advanced outputs, high-throughput efficiency and target achievability. Their flexible and tunable nature, unique electrochemical and magnetic properties and surface morphology have made them attractive components for applications in chemotherapy, biosensing and bioimaging. Their ultra-thin 2D nanosheet structure has opened several doors to be explored by modifying the surface groups and attachments. Efficient linkage with several types of biopolymers has made them biodegradable and biocompatible, which adds another merit in their use in environmental and healthcare systems. This has enabled them to act as antimicrobial agent. Their hydrophilic and swelling properties have made them excellent carriers in drug delivery systems. The nanocomposites may encapsulate the desired active agent to be unloaded at the target site along with the synergistic action by the MXene moiety itself. This added advantage has been explored in chemotherapeutics. Photothermal conversion is another unique property that converts light to heat and has been efficiently shown by various researchers to reduce tumors. Apart from the versatile applications of MX/P-NCs in theranostics, their applicability in biosensing systems has paved the path for research in a wholly novel domain. Numerous types of biosensors have been fabricated for the detection of biomolecules, analytes and bio-changes. In this way, the development of these multicomponent and multifunctional systems has been shown to achieve novel methods in biomedical and healthcare systems.

BIBLIOGRAPHY

Arabi Shamsabadi, A., Sharifian Gh., M., Anasori, B., Soroush, M., 2018. Antimicrobial Mode-of-Action of Colloidal Ti3C2Tx MXene Nanosheets. ACS Sustainable Chemistry & Engineering 6, 16586–16596. https://doi.org/10.1021/acssuschemeng.8b03823

Carey, M., Barsoum, M.W., 2021. MXene Polymer Nanocomposites: A Review. Materials Today Advances 9, 100120. https://doi.org/10.1016/j.mtadv.2020.100120

Chen, X., Sun, X., Xu, W., Pan, G., Zhou, D., Zhu, J., Wang, H., Bai, X., Dong, B., Song, H., 2018. Ratiometric Photoluminescence Sensing Based on Ti3C2 MXene Quantum Dots as an Intracellular pH Sensor. Nanoscale 10, 1111–1118. https://doi.org/10.1039/C7NR06958H

Chen, X., Zhao, Y., Li, L., Wang, Y., Wang, J., Xiong, J., Du, S., Zhang, P., Shi, X., Yu, J., 2021. MXene/Polymer Nanocomposites: Preparation, Properties, and Applications. Polymer Reviews 61, 80–115. https://doi.org/10.1080/15583724.2020.1729179

Chung, Y.-C., Chen, C.-Y., 2008. Antibacterial Characteristics and Activity of Acid-Soluble Chitosan. Bioresource Technology 99, 2806–2814. https://doi.org/10.1016/j.biortech.2007.06.044

Dai, C., Chen, Y., Jing, X., Xiang, L., Yang, D., Lin, H., Liu, Z., Han, X., Wu, R., 2017a. Two-Dimensional Tantalum Carbide (MXenes) Composite Nanosheets for Multiple Imaging-Guided Photothermal Tumor Ablation. ACS Nano 11, 12696–12712. https://doi.org/10.1021/acsnano.7b07241

Dai, C., Lin, H., Xu, G., Liu, Z., Wu, R., Chen, Y., 2017b. Biocompatible 2D Titanium Carbide (MXenes) Composite Nanosheets for pH-Responsive MRI-Guided Tumor Hyperthermia. Chemistry of Materials 29, 8637–8652. https://doi.org/10.1021/acs.chemmater.7b02441

Fang, L., Liao, X., Jia, B., Shi, L., Kang, L., Zhou, L., Kong, W., 2020. Recent Progress in Immunosensors for Pesticides. Biosensors and Bioelectronics 164, 112255. https://doi.org/10.1016/j.bios.2020.112255

Feng, W., Wang, R., Zhou, Y., Ding, L., Gao, X., Zhou, B., Hu, P., Chen, Y., 2019. Ultrathin Molybdenum Carbide MXene with Fast Biodegradability for Highly Efficient Theory-Oriented Photonic Tumor Hyperthermia. Advanced Functional Materials 29, 1901942. https://doi.org/10.1002/adfm.201901942

Han, X., Huang, J., Lin, H., Wang, Z., Li, P., Chen, Y., 2018. 2D Ultrathin MXene-Based Drug-Delivery Nanoplatform for Synergistic Photothermal Ablation and Chemotherapy of Cancer. Advanced Healthcare Materials 7, e1701394. https://doi.org/10.1002/adhm.201701394

Kim, S.J., Koh, H.-J., Ren, C.E., Kwon, O., Maleski, K., Cho, S.-Y., Anasori, B., Kim, C.-K., Choi, Y.-K., Kim, J., Gogotsi, Y., Jung, H.-T., 2018. Metallic Ti3C2Tx MXene Gas Sensors with Ultrahigh Signal-to-Noise Ratio. ACS Nano 12, 986–993. https://doi.org/10.1021/acsnano.7b07460

Koyappayil, A., Chavan, S.G., Mohammadniaei, M., Go, A., Hwang, S.Y., Lee, M.-H., 2020. Β-Hydroxybutyrate Dehydrogenase Decorated MXene Nanosheets for the Amperometric Determination of β-Hydroxybutyrate. Microchim Acta 187, 277. https://doi.org/10.1007/s00604-020-04258-y

Kumar, S., Lei, Y., Alshareef, N.H., Quevedo-Lopez, M.A., Salama, K.N., 2018. Biofunctionalized Two-Dimensional Ti3C2 MXenes for ultrasensitive Detection of Cancer Biomarker. Biosensors and Bioelectronics 121, 243–249. https://doi.org/10.1016/j.bios.2018.08.076

Li, Y., Zhang, W., Niu, J., Chen, Y., 2012. Mechanism of Photogenerated Reactive Oxygen Species and Correlation with the Antibacterial Properties of Engineered Metal-Oxide Nanoparticles. ACS Nano 6, 5164–5173. https://doi.org/10.1021/nn300934k

Li, Z., Zhang, H., Han, J., Chen, Y., Lin, H., Yang, T., 2018. Surface Nanopore Engineering of 2D MXenes for Targeted and Synergistic Multitherapies of Hepatocellular Carcinoma. Advanced Materials 30, 1706981. https://doi.org/10.1002/adma.201706981

Liao, H., Guo, X., Wan, P., Yu, G., 2019. Conductive MXene Nanocomposite Organohydrogel for Flexible, Healable, Low-Temperature Tolerant Strain Sensors. Advanced Functional Materials 29, 1904507. https://doi.org/10.1002/adfm.201904507

Lin, H., Chen, Y., Shi, J., 2018. Insights into 2D MXenes for Versatile Biomedical Applications: Current Advances and Challenges Ahead. Advanced Science 5, 1800518. https://doi.org/10.1002/advs.201800518

Lin, H., Gao, S., Dai, C., Chen, Y., Shi, J., 2017a. A Two-Dimensional Biodegradable Niobium Carbide (MXene) for Photothermal Tumor Eradication in NIR-I and NIR-II Biowindows. J. Am. Chem. Soc. 139, 16235–16247. https://doi.org/10.1021/jacs.7b07818

Lin, H., Wang, X., Yu, L., Chen, Y., Shi, J., 2017b. Two-Dimensional Ultrathin MXene Ceramic Nanosheets for Photothermal Conversion. Nano Letters 17, 384–391. https://doi.org/10.1021/acs.nanolett.6b04339

Liu, G., Zou, J., Tang, Q., Yang, X., Zhang, Y., Zhang, Q., Huang, W., Chen, P., Shao, J., Dong, X., 2017. Surface Modified Ti3C2 MXene Nanosheets for Tumor Targeting Photothermal/Photodynamic/Chemo Synergistic Therapy. ACS Appl. Mater. Interfaces 9, 40077–40086. https://doi.org/10.1021/acsami.7b13421

Liu, H., Duan, C., Yang, C., Shen, W., Wang, F., Zhu, Z., 2015. A Novel Nitrite Biosensor Based on the Direct Electrochemistry of Hemoglobin Immobilized on MXene-Ti3C2. Sensors and Actuators B: Chemical 218, 60–66. https://doi.org/10.1016/j.snb.2015.04.090

Liu, J., Jiang, X., Zhang, R., Zhang, Y., Wu, L., Lu, W., Li, J., Li, Y., Zhang, H., 2019. MXene-Enabled Electrochemical Microfluidic Biosensor: Applications toward Multicomponent Continuous Monitoring in Whole Blood. Advanced Functional Materials 29, 1807326. https://doi.org/10.1002/adfm.201807326

Liu, Y., Han, Q., Yang, W., Gan, X., Yang, Y., Xie, K., Xie, L., Deng, Y., 2020. Two-Dimensional MXene/Cobalt Nanowire Heterojunction for Controlled Drug Delivery and Chemo-Photothermal Therapy. Materials Science and Engineering: C 116, 111212. https://doi.org/10.1016/j.msec.2020.111212

Liu, Z., Lin, H., Zhao, M., Dai, C., Zhang, S., Peng, W., Chen, Y., 2018. 2D Superparamagnetic Tantalum Carbide Composite MXenes for Efficient Breast-Cancer Theranostics. Theranostics 8, 1648–1664. https://doi.org/10.7150/thno.23369

Lorencova, L., Bertok, T., Filip, J., Jerigova, M., Velic, D., Kasak, P., Mahmoud, K.A., Tkac, J., 2018. Highly stable Ti3C2Tx (MXene)/Pt nanoparticles-modified Glassy Carbon Electrode for H2O2 and Small Molecules Sensing Applications. Sensors and Actuators B: Chemical 263, 360–368. https://doi.org/10.1016/j.snb.2018.02.124

Mayerberger, E.A., Street, R.M., McDaniel, R.M., Barsoum, M.W., Schauer, C.L., 2018. Antibacterial Properties of Electrospun Ti $_3$ C $_2$ T $_z$ (MXene)/Chitosan Nanofibers. RSC Adv. 8, 35386–35394. https://doi.org/10.1039/C8RA06274A

Peng, X., Zhang, Y., Lu, D., Guo, Y., Guo, S., 2019. Ultrathin Ti3C2 Nanosheets Based "off-on" Fluorescent Nanoprobe for Rapid and Sensitive Detection of HPV Infection. Sensors and Actuators B: Chemical 286, 222–229. https://doi.org/10.1016/j.snb.2019.01.158

Rafieerad, A., Yan, W., Sequiera, G.L., Sareen, N., Abu-El-Rub, E., Moudgil, M., Dhingra, S., 2019. Application of Ti3 C2 MXene Quantum Dots for Immunomodulation and Regenerative Medicine. Advanced Healthcare Materials 8, e1900569. https://doi.org/10.1002/adhm.201900569

Rakhi, R.B., Nayak, P., Xia, C., Alshareef, H.N., 2016. Novel Amperometric Glucose Biosensor Based on MXene Nanocomposite. Scientific Reports 6, 36422. https://doi.org/10.1038/srep36422

Rasool, K., Helal, M., Ali, A., Ren, C.E., Gogotsi, Y., Mahmoud, K.A., 2016. Antibacterial Activity of $Ti_3C_2T_x$ MXene. ACS Nano 10, 3674–3684. https://doi.org/10.1021/acsnano.6b00181

Rasool, K., Mahmoud, K.A., Johnson, D.J., Helal, M., Berdiyorov, G.R., Gogotsi, Y., 2017. Efficient Antibacterial Membrane Based on Two-Dimensional Ti 3 C 2 T x (MXene) Nanosheets. Scientific Reports 7, 1598. https://doi.org/10.1038/s41598-017-01714-3

Soleymaniha, M., Shahbazi, M.-A., Rafieerad, A.R., Maleki, A., Amiri, A., 2019. Promoting Role of MXene Nanosheets in Biomedical Sciences: Therapeutic and Biosensing Innovations. Advanced Healthcare Materials 8, 1801137. https://doi.org/10.1002/adhm.201801137

Tang, W., Dong, Z., Zhang, R., Yi, X., Yang, K., Jin, M., Yuan, C., Xiao, Z., Liu, Z., Cheng, L., 2019. Multifunctional Two-Dimensional Core-Shell MXene@Gold

Nanocomposites for Enhanced Photo-Radio Combined Therapy in the Second Biological Window. ACS Nano 13, 284–294. https://doi.org/10.1021/acsnano.8b05982

Wang, F., Yang, C., Duan, C., Xiao, D., Tang, Y., Zhu, J., 2014. An Organ-Like Titanium Carbide Material (MXene) with Multilayer Structure Encapsulating Hemoglobin for a Mediator-Free Biosensor. Journal of the Electrochemical Society. 162, B16. https://doi.org/10.1149/2.0371501jes

Xin, M., Li, J., Ma, Z., Pan, L., Shi, Y., 2020. MXenes and Their Applications in Wearable Sensors. Frontiers in Chemistry 8. https://doi.org/10.3389/fchem.2020.00297

Xing, C., Chen, S., Liang, X., Liu, Q., Qu, M., Zou, Q., Li, J., Tan, H., Liu, L., Fan, D., Zhang, H., 2018. Two-Dimensional MXene (Ti3C2)-Integrated Cellulose Hydrogels: Toward Smart Three-Dimensional

Network Nanoplatforms Exhibiting Light-Induced Swelling and Bimodal Photothermal/Chemotherapy Anticancer Activity. ACS Applied Materials & Interfaces 10, 27631–27643. https://doi.org/10.1021/acsami.8b08314

Xu, B., Zhi, C., Shi, P., 2020. Latest Advances in MXene Biosensors. Journal of Physics: Materials 3, 031001. https://doi.org/10.1088/2515-7639/ab8f78

Xue, Q., Zhang, H., Zhu, M., Pei, Z., Li, H., Wang, Z., Huang, Yang, Huang, Yan, Deng, Q., Zhou, J., Du, S., Huang, Q., Zhi, C., 2017. Photoluminescent Ti3 C2 MXene Quantum Dots for Multicolor Cellular Imaging. Advanced Materials 29. https://doi.org/10.1002/adma.201604847

Zhang, P., Yang, X.-J., Li, P., Zhao, Y., Niu, Q.J., 2019. Fabrication of Novel MXene (Ti3C2)/Polyacrylamide Nanocomposite Hydrogels with Enhanced Mechanical and Drug Release Properties. Soft Matter 16, 162–169. https://doi.org/10.1039/C9SM01985E

Zheng, J., Wang, B., Ding, A., Weng, B., Chen, J., 2018. Synthesis of MXene/DNA/Pd/Pt Nanocomposite for Sensitive Detection of Dopamine. Journal of Electroanalytical Chemistry 816, 189–194. https://doi.org/10.1016/j.jelechem.2018.03.056

Zong, L., Wu, H., Lin, H., Chen, Y., 2018. A Polyoxometalate-Functionalized Two-Dimensional Titanium Carbide Composite MXene for Effective Cancer Theranostics. Nano Research 11, 4149–4168. https://doi.org/10.1007/s12274-018-2002-3

14 Flexible and Wearable Applications of MXene/Polymer-Based Nanocomposites

Heera S and Amal George Cheriyan

CONTENTS

14.1 INTRODUCTION TO MXene/POLYMER-BASED NANOCOMPOSITES

A category of nanocomposites with numerous distinctive properties is the MXene/polymer-based nanocomposite materials. MXene/polymer-based nanocomposites are generally a mixture of bulk MXenes together with randomly dispersed MXenes and polymer components with at least one component in the nano dimension. MXenes are hydrophilic two-dimensional (2D) materials with high electronic density, which makes them metallic. They are derived from the MAX phase or layered ternary $M_{n+1}AX_n$, where 'M' is a transition metal, 'A' is a group IV–V element, and 'X' is carbon or nitrogen. Etching the layer 'A' from the MAX phase gives MXenes. The formula for MXenes is denoted as $M_{n+a}X_nT_z$, where T_z denotes the surface functional group formed from etching. This

DOI: 10.1201/9781003164975-14

functional group enables MXenes to form stable suspensions in polar solvents and results in their hydrophilic nature. The main functional groups are a mixture of oxygen (O), hydroxyl (OH), and fluorine (F). They aid in the betterment of interaction between the fillers and the polymer-based matrix in the nanocomposites. MXenes can even be fabricated with magnetic behavior. MXenes offer a wide range of compositions and have ease of dispersibility. The large surface area and tunable abundant surface functionalities of the MXenes promote their usage as nanofillers. The high electrical and thermal conductivity of MXenes makes the material suitable for numerous sensing applications. The drawbacks of MXenes, which include low flexibility, poor stability in an oxygen atmosphere, and easy restacking, are minimized or even made unnoticed in combination with polymers, possessing high flexibility and stability, in nanocomposites. Along with all the other properties, their antibacterial efficiency, low cost, easy fabrication, and tunable surface terminations have promising benefits in various MXene/polymer nanocomposite applications, especially in wearable applications.

In situ polymerization, ex situ blending, and solution blending are the most common preparation methods for MXenes/polymer nanocomposites. Having the merits of achieving precise control over the composition and forming well-defined polymer structures, ex situ blending is more preferable for fabrication. [1] They combine the aspects of highly elastic-functional polymers and the eccentric properties of MXenes, forming an ingenious group of nanocomposites. The superior mechanical and thermal properties, lightweight nature, flexibility, and increased biodegradability had led to their utilization in a broad range of commercial wearable applications.

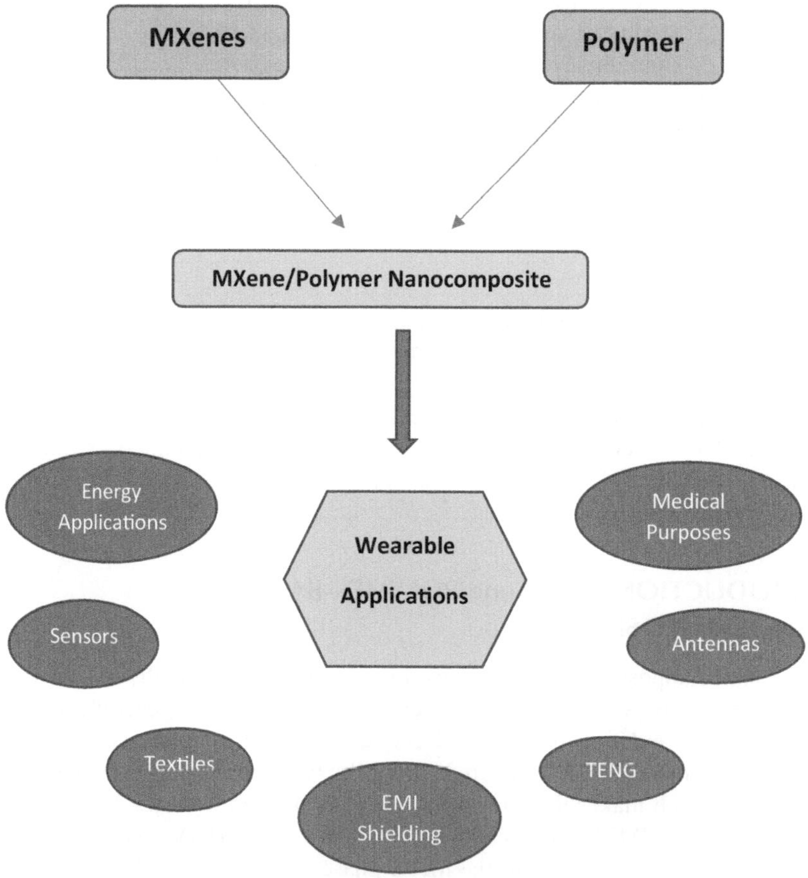

FIGURE 14.1 Schematic representation of wearable applications of MXene/polymer-based nanocomposites.

14.2 CLASSIFICATION OF MXene/POLYMER NANOCOMPOSITES

MXene/polymer nanocomposites are classified into filled composites and complexes. In filled composite materials, MXene is used as a filler in a bulk material, owning to the random dispersion of both MXene and polymer with each other. They are more studied and include both thermoplastic and thermosetting polymers. In a complex material, the dispersion of MXene and polymer is appreciably systematic. They are put together with a specific material structure or as a specific material product. MXene/polymer composite fabric, MXene-coated polymeric fiber, and MXene/polymer composite aerogel, sponge, or foam are all termed as MXene/polymer complexes.

14.3 APPLICATIONS

With the enhanced characteristics of MXene/polymer-based nanocomposites, in comparison with the normal composite materials, they have an extensive variety of usefulness in diverse fields. Biomedical applications, sensing applications, electromagnetic absorption and shielding applications, energy conversion, and storage applications are only a few among them. Each of the applications depends mainly on the constituent components of the composite, their structure, their aspect ratio, the functional groups and the properties.

With the presence of a variety of surface end groups of MXenes, the nanocomposites give unique surface activity, good biocompatibility, hydrophilicity, and exceptional electrical and optical properties. MXenes improve the biocompatibility of polymers, thereby permitting various MXenes/polymer nanocomposites to be used in drug delivery and as antibacterial substances. The antibacterial property of the materials enables them to be used as bandages. The composites are also used in biosensing applications. Conventional composite-based biosensors can be effectively replaced with the new polymer technology in the biosensing applications.

The remarkable features of MXenes like low bandgap, easy functionalization, good conductivity, and rich active sites enable the nanocomposites to be used in high-selectivity and wide-range sensors. There exists a number of sensors based on the MXene/polymer based nanocomposites. Gas, electrochemical, humidity, stress–strain, and photoluminescence sensors are a few among them. [2]

Excellent conductivity, enormous redox cites, and large specific surface area are the preferred properties needed for energy conversion, storage, and other applications. MXene/polymer nanocomposites possessing all the essential qualities can be effectively used for the application. Thus, they have been extensively used as supercapacitor, and in battery applications. MXenes, with their dipole, carrier and defect polarization mechanisms and multi-scattering property between layers, make the nanocomposites suitable material for electromagnetic shielding applications. Among the various traits of MXenes/polymer nanocomposites, the flexible and wearable characteristics are of primary interest and are discussed in the following sections.

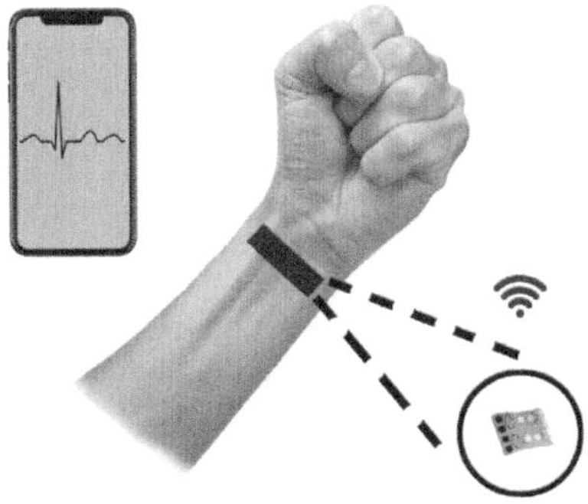

FIGURE 14.2 Pressure sensor attached to the wrist.

14.4 FLEXIBLE AND WEARABLE APPLICATIONS

The MXenes/polymer nanocomposites intertwine the flexibility of polymers along with the various properties of MXenes to engage the material in flexible and wearable applications. [3] This flexible feature of the material drastically eliminated the use of various traditional materials in the respective fields. Wearable devices mainly focus on health treatment and monitoring, sensors, communication and in the field of sports.

14.4.1 General Sensing Applications

14.4.1.1 Pressure Sensors (Piezoresistive Sensors)

The excellent properties of MXenes and their homogeneous dispersion in a suitable polymeric matrix aid the fabrication of highly sensitive wearable MXene-based pressure sensors with a wide tunable sensing range. These kinds of wearable sensors are very effective in the detection of minute bio-signals and hence provide regular health monitoring. Among the various wearable sensors, the MXene-based pressure sensors are the most important, as the sensing capability of the flexible and wearable MXene/polymer-based nanocomposite pressure sensors are unaffected with any deformations and are very much effective in human motion detection in the medical diagnosis.

In a pressure sensor, the pressure signal is converted to an electrical signal. Sensing mechanisms mainly depend on the changes in piezoelectricity, capacitance, piezoresistivity and triboelectricity. Common types of pressure sensors are capacitive, piezoresistive and piezoelectric pressure sensors. The resistance changes are mainly associated with a combination of mechanisms. In a piezoresistive pressure sensor, the pressure signal is converted into resistance signal output concerning the material deformation. Sensing mechanisms rely mainly on the change in seepage path, the tunnel resistance between fillers, and the change in the band structure of fillers. Pressure sensors have to respond quickly and must have fast frequency response, ultra-high sensitivity, wide working range, and high cycle stability. [4] They have to sustain torsion deformation, bending and compression. The ability of a pressure sensor to convert external stimuli into electrical signals is termed its sensitivity It is given by the equation

$$Sensitivity, S = \frac{\frac{\Delta R}{R_o}}{\Delta P}.$$

Gauge factor (GF) is defined as the ratio of the normalized resistance change $\Delta R/R_0$ to the applied strain and is used to evaluate sensitivity. It is given by

$$Gauge\,Coefficient, GF = \frac{\frac{\Delta R}{Ro}}{\varepsilon},$$

where $\Delta R/R_0$ is the normalized resistance change,
ΔP is the pressure change, and
ε is the strain.

For wearable applications, the sensor material must be highly sustainable, deformable, biocompatible, and extremely flexible. The idea of stretchable conductors led to the development of MXene/polymer-based pressure sensors. The sensors, under pressure, undergo a large distance change, causing a reduction in the internal resistance and hence increased conductivity.

Depending on the requirement of the pressure sensors in various wearable devices, the fabrication process varies. One-dimensional single-walled carbon nanotube (SWCNT), MXene sheets and

PVP are incorporated together via solvent evaporation process to form MXene/single-walled carbon nanotube (SWCNT)/polyvinylpyrrolidone (PVP) composite film. In the film material, SWCNT and MXene sheets formed anisotropic bundles and lamellas, respectively. The PVP acted as an adhesive connecting the components. The composite matrix is then coated on a rubber substrate to form a tactile sensor with superior sensitivity, low detection limit, and cyclic stability.

Another kind of wearable pressure sensor is fabricated by directly coating MXene on a suitable flexible skeleton. This idea is further explored and MXene is coated on a cotton textile by dip-coating and a drying method. Together with better electrical properties of MXene, the wavy network nature of the textile substrate provided flexibility and in effect, a sensor with high sensitivity, low response time, and stable response is fabricated. MXene-textile-based pressure sensors can be further explored for their numerous applications as wearable devices. Another choice for a flexible substrate is paper. Tissue paper, with its rough surface, exhibits the properties required for functioning as a good pressure sensor. Since the substrate is completely safe, they are environmentally friendly and biocompatible. This widens its application in wearable devices.

MXene-wrapped sponge with insulated PVA is another sensor material, having the advantage of low cost. Since the substrate, sponge, is common and inexpensive, a high-performance cost-effective wearable sensor device can be fabricated with the material. The working principle of the sensing material can be explained such that, when pressure is applied to the material, the air space between the microfibers got reduced and the contact points are increased providing better conductive pathways. Thus, the changes in external pressure cause changes in resistance of the MXene-sponge and thereby changes in the conductivity. Sensing materials based on mimicking the interlocked human skin structure also proved to be efficient. The sensored signals are enlarged and transferred, leading to better sensing. The microspines on the human skin are carefully mimicked using an abrasive paper to form randomly distributed spines. MXene- and natural microcapsule (NMC)–based flexible biomimetic sensing material is designed under this mechanism for fabricating an effective MXene/NMC pressure-sensitive sensor.

One such sensor with higher conductivity and stretchability is obtained by tuning $Ti_3C_2T_x$ structure to form a distinctive $Ti_3C_2T_x$ MXene nanoparticle–nanosheet network. When subjected to pressure, the nanoparticles migrate resulting in a large change in the resistance and the nanosheets maintain the connectivity of the nanoparticles to the conductive network, thus made capable of withstanding a large strain region. The $Ti_3C_2T_x$/PLA (polylactic acid)–based degradable composites form a wearable pressure sensor with excellent cycle stability, stable response even after 10,000 compression cycles, a low detection limit, low power consumption, and a wide range. They have wide applications in electronic skin and human–machine interfaces.

Apart from resistance-based sensors, capacitive sensors are also developed using MXene/polymer composite as the flexible dielectric material. The composite material enhanced the sensing performance by not only improving the dielectric constant but also by reducing the compression modulus. MXene/poly (vinylidene fluoride-trifluoroethylene) (PVD-TrFE) composite functioned as the dielectric layer and poly (3,4-ethlenedioxythiophene) polystyrene sulfonate (PEDOT:PSS)/PDMS films played the role of an electrode in a wearable capacitive pressure sensor.

14.4.1.2 Strain Sensors

The working principle behind a flexible strain sensor is that it transforms the tensile strain of the sensor material into resistance signal output. Components of a wearable flexible strain sensor need to be highly flexible, with large strain, superior adhesion, and high sensitivity. These characteristics of flexible strain sensors aid in their application in manufacturing wearable electronic devices. A strain sensor can detect any sort of deformation by analyzing the change in resistance or capacitance of the sensing material. The mechanism behind the sensing activity lies in the decrease in conductivity of the composite material with the appearance of deformations, like bending, because of the formation of microcracks, which vanish when the external pressure is released.

Depending on the prerequisite of different applications in wearable devices, MXene/polymer nanocomposites can be designed in different structures like film structures; aerogel structures, with high porosity, low density, and super elasticity; fiber, textile, and planar structures, with large flexibility and mechanical strength; and sponge structures, with very large sensitivity. It is reported that the geometrical structural arrangement of sensing material can affect the sensing range and the sensitivity of the strain sensors to a great extent. This made the scientists explore different structural designs of the materials suitable for designing efficient wearable strain sensors.

This observation leads to the design of MXene/polyaniline fiber (PANIF) nanocomposite-based sensing material with a tile-like stacked hierarchical pattern. Being very vulnerable to getting broken and because of their difficulty to assemble MXene microstructures in a hierarchical pattern, without compromising its conductivity and mechanical properties, negatively charged MXene layers are combined with positively charged PANIF to form the sensing material [5]. Apart from the traditional single-layered sensing material, here the structure is prepared by a layer-by-layer (LBL) assembly. Single-layered pure MXene or pure PANIF sensors were found to have a very low sensitivity and sensing range. This happens because the single layer forms more gaps and microcracks and possesses reduced overlapped area due to layer sliding. This causes the resistance to increase and hence reduces the sensing range and sensitivity.

The resistance of the multilayered structure of MXene and bridged PANIF changes as a result of the change in the overlapped area caused by the reversible slipping of the adjacent layers. Good dispersion and strong interaction between the constituent particles, required for the LBL coating, are present for MXene and PANIF, which made them ideal for fabricating multilayered sensor material.

The fabrication process involves spreading coating of delaminated MXene and PANIF, respectively, on an elastic rubber substrate. The substrate is then stretched to produce microcracks, upon which further coating is carried out and is then released. The material is then characterized and is found that, with the presence of microcracks, reversibly slippage of overlapped layers and spacing propagation, the tile-like hierarchical pattern enhanced the overall sensing range, sensitivity, cyclic stability and ultralow detection limit of the strain sensor occur. The PANIF coated in between the MXene layers acted as a bridge between the MXene layers to guarantee the existence of continuous conductive paths and, hence, improves the overall conductivity of the material. Because of the strong bonding interaction between MXene and PANIF layers, the proper assembling of the constituent particles is made possible, and hence, the fabricated sensor material was found to have an enhancement in stability. With the easiness of fabrication, the nanocomposite can be fabricated in different composition ratios for specific applications.

The tile-like structure also exhibits good response time and recovery time and have better cyclic stability suitable for human motion detection. With these enhancements in properties, the sensor can have wide applications in wearable devices. The specific structural arrangements made the strain sensor precisely detect not only the minute movements like the pulse but also the large motions like bending. The sensor can be directly attached to the skin by medical tapes to detect the motions. This sort of wearable MXene/polymer nanocomposite-based strain sensors can be effectively designed as communication devices for those who are unable to communicate by speaking. The signals can be transmitted wirelessly for effective communication and further human motion monitoring.

Fiber Structure

From the studies of Pu et al., the team developed a good sensing material with fiber structure using a LBL dip-coating process. The multilayered silver nanowire (AgNW)/waterborne polyurethane (WPU)-MXene fiber is fabricated by a successive coating of AgNW/WPU layers and MXene layers on a stretchable substrate, named HPUF. The slippage of the multilayers and the corresponding crack propagation on subjected to a strain resulted in the change in conductivity, thereby resulting in strain sensing. This kind of fiber can be embedded in textiles or fabrics to fabricate numerous wearable devices like a wearable electrical-responsive heater.

Apart from the LBL method, similar fiber structures can be synthesized from other techniques like dip coating and wet-spinning method. $Ti_3C_2T_x$ MXene/polyurethane (PU) composite fiber synthesized from the wet spinning method was found to possess ultra-high sensitivity. The fibers can be woven together to form textiles with good strain-sensing capability, preferred for use in various human health monitoring applications.

Film Structure

Various conductive MXene/polymer composite-based films can be fabricated by different synthesis methods which are determined by the application of the respective material. One such example is the $Ti_3C_2T_x$ MXene/CNT composite film, fabricated by spray-coating technique. The film uses an elastic latex rubber as the substrate upon which MXene and carbon nanotubes (CNTs) are coated successively. The material is characterized by a good detection limit of 0.1%, high sensitivity, and a sensing range. The material is designed as skin attachable and its elasticity can be well employed for widening its application in wearable devices. Another kind of film, $Ti_3C_2T_x$ MXene/graphene-based composite film uses vacuum filtration for its fabrication. As discussed earlier, the strain-sensing mechanism of the material is based on the relative motion of the adjacent layers of the material, which resulted in continuous destruction and maintenance of the conductive path.

Hydrogels

With superior mechanical properties and other characteristics, MXene nanocomposite–based conductive hydrogels are suitable candidates for being used as strain sensors in various wearable electronic devices. Conductive MXene hydrogel (m-hydrogel) with excellent self-adhesive ability, exceptional tensile strength and outstanding self-repairing ability, is prepared by mixing MXene and PVA in a suitable fashion. The basic interactions between the constituent components resulted in exhibiting the superior properties of the material. When subjected to a strain, the m-hydrogels undergo deformation which results in changing contact resistance between the MXene sheets and hence the overall conductivity changes. Hydrogels composed of a combination of 1D nanowires and 2D nanosheets were found to have extraordinary specifications, like fast responsiveness, better linearity, very wide working range, tunable sensing mechanisms, resilience, and superior reproducibility, suitable for application in e-skin.

In contrast with the conventional conductive hydrogels (MNH), which use water as the dispersive medium, MXene-based organohydrogels (MNOH) use an organic solvent as the dispersion medium. The solvent displacement made in MNH enhanced the material with improved self-healing capacity, durability, superior mechanical characteristics, and long-lasting moisture retention with excellent antifreezing properties. With the presence of water in the conventional hydrogels, they suffer evaporation and freezing at room temperature and subzero temperatures, respectively, thereby leading to poor shelf life. As an example, Wan and coworkers developed a kind of MNH by incorporating MXene nanosheets into polyacrylamide (PAAm) and PVA hydrogel. The nanosheets acted as conductive fillers in the hydrogel. The as-prepared MNH is soaked in ethylene glycol (EG) to synthesize MNOH. On immersing in EG, MNH replaces a portion of water molecules and becomes enhanced in its properties to form MNOH. The incorporation of organic solvent aids the material to exhibit better flexibility, self-healing ability, stable moisture retention for 8 days and an antifreezing property even at −40°C. Thus, certain wearable devices, which have to be performed well even at low temperatures and preferred to have non-drying requirements and good sensitivity, can be fabricated with MNOH.

High conductivity, stretchability, and a wide strain-sensing range of MXene/PDAC nanocomposite membranes make them a potential candidate for being used in strain sensors. Strain sensors made of MXene/polymer nanocomposite materials can be used in the manufacture of wearable electronics to identify the extent of deformation of their components, and in biometric sensors. These nanocomposites also allow easy adhesion on other substrates and hence widen the range of applications. [6]

From the finite element method simulations (FEM) for designing piezoelectric pressure sensors, it is observed that the overall properties of the pressure sensor got improved by the introduction of a biomimetic interlocked structure by assembling the natural micro-capsules in Ti_3C_2 nanosheets. This structure significantly enhanced the sensitivity of the sensor. The improvement in the performance, flexibility and mechanical deformability arises from the resultant force distribution adjustments in the biocomposite films.

Gel

A kind of piezoresistive strain sensor is fabricated by the combination of MXenes with a sponge and rGO aerogel. The sensor was found to have a quick response time, broad sensing range and outstanding durability over 10,000 cycles and thus found applications in observing even the tiny strain. The wearable and flexible MXene-sponge/polyvinyl alcohol NW (MSP)–based strain sensors on connecting series to a circuit attached with a Bluetooth system are capable of converting the changes in current and resistance into wireless electromagnetic signals. The device responds well to arterial pulses and can detect various diseases.

The main challenges in developing wearable strain or pressure sensors lie in the development of materials with higher flexibility, sensitivity, wide tunable strain and low dimensions and in the accurate integration of these materials in the nanodevices. The wearable sensors require miniaturization of the device components, mechanical durability and superior flexibility. For wearable, long-time monitoring, self-powered sensors and devices, the circuit must contain flexible energy storage units for the effective performance of the devices. Such units compensate for the power dissipation that occurs in wireless devices. By overcoming these challenges, wearable strain sensors with enhanced performances can be fabricated for analyzing robotic systems, human activities, and many more.

14.4.1.3 Humidity Sensors

Humidity sensors work on the principle of changing the material properties of the nanocomposites caused by hydration and dehydration of water molecules. With proper surface functionalization, humidity or water molecules can be easily absorbed by the material. With the multilayer structure of MXene/polymer nanocomposites, the absorbed water molecules are inserted between the layers and result in an increase in the thickness and distance between the layers. This causes a change in the tunneling resistance of the material layers and can be measured. This type of sensor is of great importance since humidity greatly affects the medical field, food industry, environment, fabric, and much more. Thus, a wearable humidity sensor will be very effective in maintaining and detecting the required amount of humidity in the surroundings. Only those materials with good responses under high and low humidity must be chosen as the fillers and matrices of the nanocomposites.

This criterion is satisfied by, as for example, MXene/PAM (polyacrylamide)-based humidity sensors. They were found to have a wide range of linear sensitive humidity sensing. [7] One such suitable choice for a multifunctional-flexible humidity sensor is MXene/AgNWs-decorated composite silk textile. The fabrication carried out by vacuum-assisted LBL method initially fabricates a leaf-like nanostructure using MXene and AgNWs. The structure provided the substrate material with high conductivity and also maintained the air permeability of the initial textile substrate. Apart from having high sensitivity for water molecules, the hybrid material shows good electromagnetic interference (EMI) shielding capability, which made them a very suitable candidate for multifunctional wearable applications.

Some composites help in monitoring human breathing, clearly help diagnose the humidity variations and help provide timely treatment. The humidity sensors have hope in the fabrication of certain textiles with high sensitivity to humidity (Figure 14.3).

FIGURE 14.3 Textile-based sensors.

14.4.1.4 Gas Sensors

MXene/polymer-based nanocomposites are also very efficient in measuring the presence of certain harmful gases in the atmosphere, thereby give an implication of any sudden accidents. The large number of surface-active sites and the abundant terminal groups on the nanocomposite allow gas adsorption. The gas adsorption/desorption causes the carrier transfer between the composite substrate and the adsorbate gas. This kind of substitution of molecules causes a resistance change of the sensor material, giving a suitable implication of the external conditions.

Materials with much less response time and recovery time, a wide detection range and sensitivity to a variety of gases must have opted for the sensor manufacture. One such type of harmful toxic gas pollutant is volatile organic compounds, VOCs. For sensing the presence of VOCs, a wearable MXene gas sensor is fabricated by a drop-casting method. The process involves platinum sputtering and drop-casting of MXene on a wearable polyimide film. When exposed to VOCs, the numerous terminal groups of MXene reduce the conductivity of MXene to the semiconductor level. The fabricated gas sensor displays a p-type sensing behavior and monitors a variety of gases at room temperature. Since NH_3 molecules possess large absorption energy, they have the fastest response time for detection.

Nonpolar gases like hydrogen and methane can be detected at room temperature by V_2CT_x/polyimide-based gas sensors. The sensor fabrication involves the drop-casting of V_2CT_x MXene solution on a flexible polyimide surface. The rich surface oxygen groups of MXene solution in the sensor made them sense hydrogen and methane, even if having a low composition.

$Ti_3C_2T_x$ MXene/GO composite fiber can be fabricated via the wet-spinning method. The material exhibits high sensitivity toward NH_3 gas. With its outstanding mechanical strength, the material can be woven onto textiles, suggesting extended applications as wearable devices. Similar NH_3 gas-sensing material can be synthesized with cationic polyacrylamide (CPAM)/$Ti_3C_2T_x$ MXene nanocomposites. Here, CPAM plays a dual role in forming hydrogen bonds with NH_3 and in enhancing the mechanical strength of the MXene composite film. [8]

$Ti_3C_2T_x$-PVA/PEI MXene/polymer-based nanocomposites are capable of engineering wearable gas sensors. Here, the MXene component helps in adsorbing polar molecules, through hydrogen bonding while the porous structure of the polymer facilitates a large specific surface area for further adsorption of the gas molecules. With the mutual benefits of the MXenes and the polymers, MXene/polymer-based nanocomposites are very appreciable materials for wearable gas-sensing applications. They can also be used in environmental monitoring.

Yet another type of wearable gas sensor, which senses biomolecules, includes those used in the medical field. $Ti_3C_2T_x$/polymer-based flexible gas sensors monitor the presence of certain components in the human breathe such as monitoring acetone for diabetic care and ammonia for lung discomfort, thereby providing early medical diagnosis and better treatment.

14.4.1.5 Epidermal Sensors

MXene/polymer nanocomposites can be designed as wearable soft epidermal sensors for monitoring human actions. The sensing response of the sensor is given by

$$sensing\, response = \frac{(I - I_o)}{I_o} \times 100\% = \frac{\Delta I}{I_o}\%,$$

where ΔI denote the change in current at a particular pressure and
I_o denotes the initial current without any external force.

Corresponding to each pressure, the sensing response displays a distinctive nature. When the sensors got pressed or their pressure increased, the contact area of the sensor increases resulting in shortening the conductive pathway for ion transfer and efficient electron transport. By this principle, a minute or a large-scale human motion can be effectively detected and monitored. This kind of epidermal sensor can be attached to the wrist for monitoring hand movements and the pulse, for monitoring chest muscle movements during breathing, analyzing fingers movements during bending, and even monitoring the movements during swallowing. By connecting the wearable sensor to a wireless transmitter, the sensing reports are transmitted to a mobile phone, and we can utilize the benefits of wireless monitoring of human motions. This type of sensor has abundant applications, primarily in the field of personal health care.

One such composite which acts suitable for the purpose is the MXene nanocomposite organohydrogel. The impressive self-adhesive, moisture retention, self-healable, and conductive nature of the composite serve as a suitable applicant for many sustainable flexible and wearable electronic applications. The composite is a combination of MXene nanosheet network and polymers, in a dispersion medium like water/glycerol. The initial gel state and long-lasting moisture retention properties can be explained by the presence of glycerol in the polymer matrix. As the amount of glycerol increases, the rate of water evaporation decreases. Glycerol has hygroscopicity and low vapor pressure and forms hydrogen bonds with water molecules, thereby hindering the evaporation of water from the material. With this property, the material can cope with comparatively more bending or twisting under normal temperatures without any breakage. [9] The MXene nanocomposite organohydrogels show self-healing capability, which can be attributed to the presence of dynamic PBA (poly(butyl acrylate))–catechol bonds in the material. The material properties of the as-prepared nanocomposites and the self-healed (after cutting) nanocomposites do not show any large fluctuation. An epidermal sensor made of similar material can have a long lifetime with better accuracy.

The nanocomposites also possess a self-adhesive capability, resulting from the presence of catechol groups in PDA and the hydroxyl groups of glycerol. This ability supports the adhesion of the nanocomposite on different surfaces, but the adhesion strength varies with the selection of the surface. This property enables the nanocomposite to get effectively coated on suitable substrates and thereby increases the range of application of the nanocomposite. All the properties of the MXene nanocomposite organohydrogel facilitate its feasibility of use as epidermal sensors, electronic devices, soft robots, electronic skins and the like. The capabilities of flexible and wearable sensors can be well explored in a good deal of potential uses like in textile, electronic products, electronic skin, automobiles, spacecrafts and much more.

From the discussion, it can be concluded that most of the sensing materials are biocompatible and are environment friendly. This made the material to be applied directly onto the skin or woven into fabrics for wearable applications. Wearable devices, like smartwatches, with a combination of different sensors, have great potential in communication and in the medical field. E-skin developed using different sensors can be very well exploited in future developments in the field of robotics, artificial intelligence, and human–machine interaction. Enhancement of any sensing material basically relies on improving its GF and working range. Morphological changes in the conductive networks made by rational changes and structural optimization are major factors for enhancing the GF and working range of the sensing material. There are many popular strategies that helps achieve better GF. Different structural geometries, like biomimetic interlocked micro/nanostructure, whisker structure, hierarchical structure, homogeneous strain distribution design, microcrack propagation, layer slippage mechanism, overlapped area, length, density, cracked area and width, are the major factors improving the GF. Another aspect determining the performance of a sensor is its working range. The working window is mostly determined by the stretchability of the material. The constituents of the sensing material, the nature of microstructures, substrate, type of sensor and even the fabrication process influences the stretchability of the material. Stretchability can be adversely affected by the strong interactions between the MXene layers. The incorporation of 1D conductive nanomaterials can enhance the stretchability of strain sensors.

Apart from high GF and working window, a sensor must exhibit good data linearity. Data linearity depends on the variation of microstructure morphology when subjected to external stimulus. Once the microstructure undergoes a nonhomogeneous change in its structure upon stretching, the data will be responded to in a nonlinear pattern. For a sensing material to be impressive, it must have high GF, a wide working range, and appreciable data linearity. Thus, the selection of composite material, microstructure geometry and the interactive forces among the active materials are the key points to be noted while designing a sensor for a specific application (Figure 14.2).

14.4.2 Electromagnetic Absorption and Shielding Applications

With the growth of the electronic industry, instrumentation and telecommunication field, a kind of pollution called EMI has emerged. This can take place even inside electronic gadgets densely packed with multiple electronic components. The unwanted interference from other radiations leads to improper functioning of the devices and degrade device performance. EMI can harm not only the gadgets but also the humans, causing many health issues. This demands the need for proper shielding from EMI. The blocking of electromagnetic radiations, either by reflection or by absorption by suitable barrier materials, is called EMI shielding. The shielding material has to be conductive or magnetic and prevents the penetration of high-energy radiations [10]. The common reason for EMI is electrostatic discharge. EMI can be hazardous almost in every field, including automobiles, spacecraft, military safety and even in common household systems.

The shielding materials have to be properly selected depending on the kind of radiation (depending on its energy, the device, etc.) for which the shielding is imparted on. Since MXenes are materials with an appreciable mechanical strength and heterogeneous layer structure with various electromagnetic wave loss mechanisms, they became one among the best candidates for electromagnetic shielding. For the application of electromagnetic shielding, MXene has to be incorporated with certain reinforcing materials to improve their properties. Polymers are a suitable choice for reinforcing materials since it not only improves the overall mechanical property of the composite but at the same time also enhances its conductivity. On being used with polymers, the material can make use of many favorable properties of polymers to form suitable flexible and wearable high-efficiency electromagnetic absorption and shielding materials. The high electrical conductivity and multiple internal reflection characteristics of the MXene/polymer nanocomposites resulted in the high EMI performance of the system. The surface activity of the composites easily enables the incorporation of certain additives used for increasing the conductivity. The strong absorption capacity of the MXene is made used in the composites to provide them with good electromagnetic absorption characteristics. Since reflection can again increase the radiation pollution, materials with good absorption are preferred for the shielding application, which can be well suited by the MXene/polymer nanocomposites. [11] MXene/polymer nanocomposites, with their significant mechanical properties and noncorrosive nature, can be a suitable material for shielding applications. The flexibility, low fabrication cost, lightweight nature, high conductivity and reduced corrosion make the nanocomposites a suitable alternative for the heavy-weighted-corrosive metals. Not only does the flexibility of the composite permit its use in many applications, but the biodegradability of the polymer matrix also encourages its application as an environment-friendly material.

The radiations are initially reflected from the composite's surface, creating a high impedance mismatch at the air–MXene interface and the remaining got absorbed by the composite material. The interlayer structure of MXene enables the radiations to undergo multiple internal reflections. During each reflection, the radiation is absorbed by the composite because of the strong interaction with the high-electron-density MXene layers and the reflection continues until most of the radiations are absorbed by the composite.

Having high absorption and internal reflections of radiation, Ti_2C/PVA nanocomposite materials are a good choice for the shielding application. The preparation methods largely influence the conductivity and strength of the prepared nanocomposites. The dispersibility of MXenes in the polymer matrix can be improved by *in situ* polymerization and the conductivity by thermal reduction. On annealing MXenes, the MXene surface groups get eliminated, causing an increase in internal dipole and electron mobility. These, in turn, increase the electromagnetic energy dissipation capability of the nanocomposites. Thus, wearable MXene/polymer-based nanocomposites with good mechanical properties and absorption and high shielding efficiency (SE) can be achieved by intertwining the properties of MXenes, like high conductivity, interlayer reflection between polymer and MXene, defect and dipole polarization conductance loss and of polymers, like corrosion resistance, lightweight, biocompatible and flexible nature.

MXene/PEDOT:PSS nanocomposites-based film exhibits EMI SE of 42.10dB at 11.1μm. The mechanism of wave attenuation lies primarily on the various polarization processes happening at the interface between MXene and PEDOT:PSS. Similarly, a highly efficient, ultra-flexible EMI shielding material can be fabricated by nanocomposite elastomer $Ti_3C_2T_x$ MXene (m-MXene) based material. The delaminated, multilayered MXene for the purpose was prepared by the selective etching of aluminum using HCl/LiF solution from the Ti_3AlC_2 powder, followed by specific heating and sonication. Along with the superior conductivity, hydrophilic nature and surface reactivity, the MXene network enhances the EMI shielding by possessing high conduction loss, multiple internal reflections and surface polarization effects on MXene networks, compared with other electronegative polymer-based nanocomposites.

Another such productive EMI shielding material developed by Ma and his coworkers is based on aramid nanofiber and $Ti_3C_2T_x$ MXene/silver nanowire (ANF-MXene/AgNW). The increase in the nanocomposite composition increases the effective shielding of the material. The presence of MXene layers and the corresponding ohmic losses drastically improves the shielding effect, even at low composition, than the traditional counterparts. Hence, these materials will immensely help shield wearable electronic devices from EM waves to a great extent.

The MXene/polymer composite–based EMI shielding materials can be integrated on textiles for better shielding applications. The textile fabrication can be explained with the example of MXene/polypyrrole (PPy): polyethylene terephthalate (PET)–based composite. PPy is synthesized via *in situ* polymerization method and is used to prepare a stable PPy/MXene ink. A PET substrate is coated with a PPy/MXene solution by dipping the substrate into the ink solution. The composite-coated fabric is then treated with hydrophobic silicon. The dipping-drying process continues for 10 cycles, and the resulted product displays very high conductivity, approximately 1000 Sm^{-1}, as needed for an excellent EMI shielding. It is found that, as the soaking time increases, the EMI shielding effect and electrical conductivity also increase further. The coating with hydrophobic silicon enhances the material's durability by providing it with a water-resistive nature.

Wearable EMI shielding devices based on different composite materials were explored and the results compared. Using the LBL method, MXene/AgNW solution is spray-coated on a silk textile and formed a leaflike structure. The material with 120-μm thickness exhibits an EMI SE of 54 dB.

Composite materials with high EMI SE and water-repellent character can be synthesized by the sol-gel method. Keeping in mind the need for lightweight, flexible materials for wearable applications, low density, hydrophobic, porous MXene-based foam and aerogel composite are suitable choices for wearable EMI devices. The foam is so fashioned by treating MXene films with hydrazine. The process involves the reaction of hydrazine with the MXene hydroxyl group. During the process, CO or CO_2 gas is evolved, which supports pore formation. The process expands the MXene layers, which leads to the formation of a lightweight foam possessing a well-defined cellular structure.

14.4.3 Heater

Excellent electrical conductivity and flexible nature of MXene-based composite promote its application as wearable heaters. With its superior electrical-thermal heat conversion efficiency, Park et al. [19] developed an electric–thermal heater via a one-step coating process. The heater material is based on MXene-coated PET threads, having the properties of shape adaptable nature, smooth processability, admirable mechanical reliability and good joule heating performance. The fiber can be sewed by a common commercial sewing machine, which simplifies the wearable device fabrication procedure, thereby extending the wearable application. The characteristic properties of the material were found to vary with coating time. The increase in coating time resulted in a decrease in fiber resistance by approximately 300 ohms and an increase in maximum steady-state temperature to 128.6°C.

Similarly, utilizing the high thermal conductivity of MXenes, wearable photo-to-thermal heaters are developed with MXene composites. MXenes display an internal light-to-heat conversion efficiency of almost 100%. It is observed that, even with ultra-low content of the hybrid materials, the temperature can be increased rapidly to approximately 111 ± 2.6°C. This result leads to the development of AgNP/MXene hybrid materials. The materials can be well exploited in fabricating photon captors, molecular heaters and energy transformers.

14.4.4 Triboelectric Nanogenerators

The nanogenerator technology deals with the conversion of mechanical or thermal energy to electricity. It is mainly of three types, namely piezoelectric, triboelectric and pyroelectric nanogenerators. Piezoelectric and triboelectric nanogenerators convert mechanical energy to electricity whereas pyroelectric nanogenerators convert thermal energy into electricity. Triboelectric nanogenerators (TENGs) are based on coupled triboelectrification and electrostatic induction for the conversion of energy. It suitably harvests the kinetic energy of humans to produce electricity. For the purpose, the two electrodes are made in contact in different configurations depending on the needs.

Consider a self-powered wearable wristwatch, as an example of MXene-based wearable and flexible electronic device. The device is fabricated with a 2D MXene-based TENG, integrated with a micro-supercapacitor (MSC) along with a suitable encapsulation material-like silicone. The first step associated with the device working is the separation of charges. Positive charges are accumulated on the human skin while negative charges are induced on the silicon surface. Thus, the potential difference induced between the two layers appears as the input of the associated bridge rectifier and hence allows the electrons to flow from the silicone layer to charge the supercapacitor integrated with the circuit. This self-charging TENG unit powers the digital wristwatch and other similar wearable electronic devices.

To achieve this large potential difference, materials with high triboelectronegativity are preferred for the manufacture of TENGs. Since PTFE, the most triboelectronegative material can be used in single electrode mode operation only their application is limited. With the emergence of MXenes, possessing high electrical conductivity and surface electronegativity, they became a suitable candidate for being used not only in MSC but also as the TENG material, which can be operated in the two-electrode mode.

The high flexibility, transparency, stretchability, and biocompatible nature of polymers led to them being actively used in nanogenerators. To compensate for the lower triboelectronegativity of the polymers, the polymers are incorporated with MXenes to form suitable composites, as required by the nanogenerators. The high conductivity and electronegative surface of MXene/polymer nanocomposites (from the negatively charged functional groups) encourage the composite to be used in nanogenerators. The 3D-MXene/PDMS nanocomposites work as a negative tribo-material in the TENG device. The triboelectric performance of a TENG assembled with 3D MXene/PDMS, shows a higher output voltage than that of pure PDMS-based TENG. Incorporating MXenes into the PDMS matrix induces micro-capacitors in the dielectric PDMS polymer. With the inclusion of the nanofiller, the dielectric is constant, and hence, the capacitance of the nanocomposite get enriched. This sort of improved dielectric constant increases the capacitance and surface charge density of the composite thereby improving the TENG's performance. The 3D-MXene/PDMS composite–based TENG gives a high output current, which is many times greater than that of the pure PDMS, even seven times at optimum conditions, which resulted from the significantly lowered electrical resistance of the nanocomposite. Capacitors can be connected to the circuit and are employed for electrical energy storage. This implies the application of TENG as mechanical power source for driving portable electronic devices. [12]

The superior electronegativity of MXene-based composites is well exploited in the manufacture of wearable TENG with high EMI shielding. This leads to the fabrication of PPUH-m-MXene nanocomposite elastomer-based TENG [13]. The required TENG is synthesized by initially preparing a PPUH polymer matrix and crosslinking $Ti_3C_2T_x$ MXene, modified by PDAC polymer, on the polymer substrate. The prepared TENG is designed to function like a self-powered, wearable sensor with intrinsic healing property, which can sense even delicate physical motions.

Improving the triboelectric charge density of the TENG became the main challenge encountered while developing MXene-based TENGs. The triboelectric charge density can be enhanced by improving the effective contact area. This can be achieved by introducing nano- or microstructures, like pyramid arrays or nanowires, in the triboelectric materials. Since TENGs produce irregular AC output pulses, they need to be stored for providing a steady DC signal for functioning self-powered wearable electronic devices. Solving these challenges will greatly influence to improve the practical self-powered applications of the TENGs. [14]

14.4.5 Wearable Antennas

MXene/polymer nanocomposite-based flexible printed antennas are their infancy regardings their application in wearable electronics. This technique allows to "coat" antennas even on the building walls. High conductivity and flexibility required for the antenna material forced us to consider the composite material as a good choice. Depending on the application, the nanocomposite material can be spray-applied, inkjet printed or screen-printed on suitable substrates. Even with low material quantity, an appreciable amount of communication can be achieved. The lightweight nature and flexibility of the material do not sacrifice the device performance, it is almost comparable with their traditional counterparts. [15] The low cost, ease of fabrication, environmental-friendliness of the material, flexibility, and noncorrosive nature enable the material to be used in sustainable local area communications. Stretchable antennas based on S-MXene are a good choice as wearable antennas. The antenna's resonance frequencies are strain-dependent and are altered between 1.575 to 1.375 GHz. Based on the studies, MXene and SWCNT (S-MXene)–based composites were found to have the ability for being used in EMI shielding and for wireless communication.

14.4.6 Textiles

Sensors printed on fabrics serve many applications, like medical monitoring or communication. For this, the electrodes of sensors have to be suitably placed in or on the textile. Printing of the electrode material allows to achieve the requirement. MXene/polymer nanocomposites can be a suitable choice for the electrode material with their high sensitivity and flexibility.

An effective way to assemble wearable smart devices is by 3D printing. MXene-based inks can be used for the printing process. On comparing with the traditional fibers, these smart fibers are multifunctional, have programmable patterns, and respond to numerous external stimuli.

14.4.7 Biomedical Application

Being biocompatible, MXene/polymer nanocomposites have immense applications in the medical field. The various surface functional groups, hydrophilicity, biocompatible nature, and many other remarkable features enable them to be used in the respective field.

14.4.7.1 Wound Healing Materials

With a biocompatible nature, low cytotoxicity, antimicrobial ability, and good mechanical properties, MXene/polymer-based nanocomposites can be a very good choice for using as wound healing materials. They mostly include bandages and sutures. The adhesiveness of the nanocomposites with other materials makes the surge or the bandages to get easily and completely attached to the wound and provide better blood clotting. The flexible nature and adhesive properties of the nanocomposite offer structural rigidity to ensure the utmost safety to the wounded area and faster healing. The $Ti_3C_2T_x$/CS nanocomposites are found to be very much effective against *E. coli* and *S. aureus*. These characteristics of the nanocomposite enable the composite's service as wearable bandages and surgical sutures and provide better tissue regeneration. [16]

14.4.7.2 Tissue Regeneration

A material used in tissue repair may have to be placed inside the body for a certain period, so it must not react against the body tissues and must not be toxic. Polymer nanocomposites with biodegradable polymers are more effective for the purpose. The MXene/polymer nanocomposite material, with its mechanical properties withstands any degradation of the nanocomposite, hence supports its long life in the body. In the case of providing bone support, the nanocomposites offer strength similar to that of the bones and possess a biocompatible nature and for this reason, the nanocomposites affirm to be one of the most suitable counterparts for the motive. $Ti_3C_2T_x$/PLA nanocomposite membranes are effectively used in bone regeneration.

14.4.7.3 Drug Delivery

A well-developing area in the medical field is targeted drug delivery. This becomes very crucial in the treatment of many diseases, mainly in cancer treatment. In targeted drug delivery the drug will reach or act only at the specified area of infection and does not affect the surrounding healthy tissues. With the presence of various surface functional groups, the MXene/polymer nanocomposites can be properly surface modified and functionalized with the drug material for the selected area drug delivery practice. The intercalated structure of the nanocomposites allows the easy loading of the drug in the composite and imparts them with good drug-loading capacity. By making use of the magnetic property of the nanocomposites, the drug assimilated MXene/polymer nanocomposite can be more excellently tuned and hence permits the controlled release of drugs and targeted drug delivery.

Apart from having good tensile strain sensitivity, high conductivity, significant stretchability and speedy self-repairing ability, MXene/PVA-based flexible sensors excellently bond with human skin. The sensor can easily perceive the postures, body movements and facial expressions, thus having the potential to drastically improve the health care systems. [17,18]

14.4.7.4 Biosensors

Wearable MXene-based electrochemical biosensors measure a wide variety of biomolecules in the body and make continuous evaluations for a long time for the proper monitoring of the body's response to medicines and of the patient's health. For example, a wearable sweat-based biosensor patch composed of Ti_3C_2/PB composite analyses biomolecules present in the sweat. Similarly, a novel bio-electrochemical MXene-based sensor evaluates the human body condition during hemodialysis treatment. The sensor combines the initial dialysis of the whole blood with subsequent detection and makes a conclusion regarding the patient's health. Such MXene-based biosensors simultaneously determine the presence of alacetaminophen (ACOP) and isoniazide (INZ) with better efficiency than the conventional ones.

14.5 CONCLUSION

MXene/polymer nanocomposites are of immense importance because of their unique properties. A great number of studies verified their applications in a diverse variety of fields and proved to be of immense help in improving human life. Among many other enhanced properties, the high conductivity and large surface area drive extreme attention for the material. This makes the MXene/polymer nanocomposites superior to any other nanocomposite materials. Wearable technologies have promising potential in next-generation technologies. MXenes enhanced the electrochemical performance of wearable devices. Their flexible nature permits its integration into portable devices and wireless communication. The predominant peculiarities of the nanocomposites had verified its applications in flexible and wearable device fabrication. Each MXene composition has distinct characteristics and enables its usage in individual applications. Employing the MXene/polymer nanocomposite material initiated the successful fabrication of low cost-highly advanced and sophisticated sensors, communication systems, environmental safety monitoring devices, medical equipment and so on.

Numerous challenges that hinder the further application of MXenes have to the suitably eliminated for harvesting better utilization of the material. Suitable MXene-based electrode materials have to the designed to improve the power density and volumetric energy of MXene-based devices. Further studies need to be carried out to understand the physiochemical characteristics of MXenes. Only by understanding the physics behind the 2D structure, proper manipulation of the MXene structure can be made accordingly. Research has to be done to analyze the easiest but efficient fabrication method to design flexible-wearable MXene-based devices. Studies based on reducing the oxidation degradation of the MXene-based composite have to be explored further so as to improve its durability and stability. Since the complex structures strongly influence a particular application, the mechanism behind those structures has to be studied further for effective device fabrication; that is, the relationship between microstructure and device performance; microstructure and material property, fabrication method and application, different fabrication methods and more have to be explored further. Future developments must focus on different fabrication techniques, improved comfortability for human wearing and effective human–computer interaction or wireless communication. The MXene/polymer-based nanocomposite materials have substantial valuable properties and have to be studied further to inspect their more wearable applications.

REFERENCES

[1] S. Heera, et al., 2021. *Gamma irradiated silver-anthracene nanocomposites for luminescent material fabrications.* s.l., AIP Conference Proceedings. https://doi.org/10.1063/5.0039944

[2] Beatrice, et al., 2018. Polymer nanocomposites with different types of nanofiller. In: *Nanocomposites—Recent Evolutions.* s.l.:s.n. https://doi.org/10.5772/intechopen.81329

[3] Chang Ma, et al., 2021. Flexible MXene-based composites for wearable devices. *Advancd Functional Materials.* 31(22/2009524).

[4] Dandan Lei, et al., 2020. Research progress of MXenes-based wearable pressure sensors. *APL Materials* 8.

[5] Mingyuan Chao, Yonggang Wang, Di Ma, Xiaoxuan Wu, Weixia Zhang, Liqun Zhang, Pengbo Wan, 2020. *Wearable MXene nanocomposites-based strain sensor with tile-like stacked hierarchical microstructure for broad-range ultrasensitive sensing.* Elsevier.

[6] Hossein Riazi, et al., 2021. *MXene-based nanocomposite sensors*. ACS Omega.

[7] Ming Xin, et al., 2020. MXenes and their applications in wearable sensors. *Frontiers in Chemistry*. https://doi.org/10.3389/fchem.2020.00297

[8] Lingfeng Gao, et al., 2020. *MXene/polymer membranes: Synthesis, properties, and emerging applications*. ACS Publications.

[9] Xiaoxuan Wu, et al., 2020. A wearable, self-adhesive, long-lastingly moist and healable epidermal sensor assembled from conductive MXene nanocomposites. *Journal of Materials Chemistry C*. 8, 1788–1795.

[10] S. B. Kondawar, P. R. Modak, 2020. Theory of EMI shielding. In: *Materials for Potential EMI Shielding Applications*. s.l.: Elsevier, pp. 9–25.

[11] Cao, W., Ma, C., Tan, S., et al., 2019. Ultrathin and flexible CNTs/MXene/cellulose nanofibrils composite paper for electromagnetic interference shielding. *Nano-Micro Lett*. 11, 72 (2019). https://doi.org/10.1007/s40820-019-0304-y

[12] Dezhao Wang, et al., 2020. Multifunctional 3D-MXene/PDMS nanocomposites for electrical, thermal and triboelectric applications. *Composites Part A: Applied Science and Manufacturing* 130.

[13] Anon, 2021. Ultraflexible, highly efficient electromagnetic interference shielding, and self-healable triboelectric nanogenerator based on $Ti_3C_2T_x$ MXene for self-powered wearable electronics. *Journal of Materials Science & Technology*. 100, p. 1–11.

[14] Neng Li, et al., MXenes: An emerging platform for wearable electronics and looking beyond. *Matter*. 4, 377–407.

[15] Z. Hamouda, et al., 2018. Magnetodielectric nanocomposite polymer-based dual-band flexible antenna for wearable applications. *IEEE Transactions on Antennas and Propagation*. 66, pp. 3271–3277.

[16] R. Bayan, N. Karak, 2020. Polymer nanocomposites based on two-dimensional nanomaterials. In: *Two-Dimensional Nanostructures for Biomedical Technology*. s.l.: Elsevier, pp. 249–279.

[17] B. L. Gray, 2018. *Polymer Nanocomposites for Flexible and Wearable Fluidic and Biomedical Microdevices*. s.l.: IEEE.

[18] Sithara P. Sreenilayam, et al., 2021. MXene materials based printed flexible devices for healthcare, biomedical and energy storage applications. *Materials Today*, 99–131.

[19] Park et al., 2020. Functional fibers, composites and textiles utilizing photothermal and joule heating. *Polymers*. 12(1), 189. https://doi.org/10.3390/polym12010189

15 Theory, Modeling, and Simulation of MXene/Polymer Nanocomposites

Benjamin Tawiah, Sarkodie Bismark, and Charles Frimpong

CONTENTS

15.1 INTRODUCTION

New two-dimensional (2D) transition metal carbides called MXene hold unlimited potential for future applications in many science and technology areas. The unique characteristics of MXenes such as conductivity, strength, hydrophilicity, and electromagnetic interference (EMI) shielding give a possibility to create new high-tech multifunctional composite materials. As an example, conductive MXene/polymer nanocomposites with these unique properties could be used for structural health monitoring in an airplane wing or a wind turbine blade. The complexity in the degradation process of composites filled with nano-inclusions brings a challenge for experimentalists.

For mechanical properties, the incorporation of MXene with polymeric materials, such as epoxy resins, exhibits an improved fracture toughness and impact strength. The influence of MXene on the mechanical, electrical, and tribological properties of various polymeric systems have been thoroughly investigated (Malaki and Varma, 2020; Wyatt et al., 2021; Aghamohammadi et al., 2021; Ji et al., 2020). Particularly, the tensile strength and moduli, flexural strength and flexural moduli, and hardness have been investigated in a number of studies (Hatter et al., 2020; Sliozberg et al., 2020; Guo et al., 2019; Chen et al., 2021). The interest in MXene polymer composites is due to factors such as high surface area and the compatibility between such nanostructures. When compared to plain polymers, this characteristic is thought to improve the mechanical and thermal properties of polymer nanocomposites, as demonstrated by greater tensile strengths and tensile moduli. The mechanical properties of nanocomposites reinforced with MXene are known to be influenced by the suitability and reactivity of polymers and other fillers, as well as associated production procedures and processing variables. The crosslinking density of the resultant nanocomposites is improved by the high compatibility between polymer matrices and other elements (Aakyiir et al., 2020). Matrix and MXene compatibility can also affect adhesion and interfacial bonding. By enhancing interfacial contacts between matrices and MXene interlayers, the MXene interlayers are altered leading to enhanced MXene compatibility (Aakyiir et al., 2020; He et al., 2021).

Different modeling approaches have been utilized to anticipate the outcomes of nanocomposites in order to understand the progression of damage and identify mechanical characteristics.

DOI: 10.1201/9781003164975-15

Such techniques include finite element (FE) modeling, which is useful in finding the mechanical properties of a single MXene nanosheet and interface layer between nanofillers and polymer matrix using a strategy of inverse finite element modeling under static and dynamic loading. The characterization of modeling and simulation is not simple because it encompasses knowledge in a variety of fields, including fluid dynamics, composites, and computer science to succeed. It has wide applications in energy systems, chemical engineering, composite science, and fabrication, as well as many areas, including logistics and social management. The entire process may be viewed as a microcosm of systems/control theory, numerical inquiry, computer science, artificial intelligence, and/or operations research, depending on the context, modeling, and simulation method. Modeling and simulation (M & S) gradually encompass all of the aforementioned fields. M & S have recently been tipped as the future computing archetype. M & S challenge covers the analysis and design of complicated dynamical systems to the construction of abstract models derived inductively from real-world observations. In principle, M & S design models deductively derive the basis from prior knowledge to build a completely new system that satisfies certain design goals. Real-world problems are frequently solved through an iterative combination of analysis and design. The atomic scale, microscale to mesoscale components of a nanocomposite, and their interactions are explored using M & S. M & S are primarily concerned with the behavior of dynamic systems, including physical and nonphysical systems. Many other theoretical and computational models applied in polymer MXene nanocomposites have been discussed in this chapter.

15.2 THE CONCEPT OF MODELING AND SIMULATION

M & S are the physical and/or logical representations of various constituents within a system to understand their interactions and trepidations virtually to generate data that can help determine decisions or make predictions about systems. The model is generally comparable to a real system, which aids the product design analyst in predicting the impact of system modifications. To put it another way, modeling is the process of developing a profile that represents a system's whole set of attributes. It is a diagrammatic representation of individual units and their interactions within a component (system) that is based on sound theory. This phenomenon could result in known or inferred material properties that could be further manipulated by changing the equation variables to further enhance the properties of the system (in this case the composite). Simulation is the operation of a model in terms of time and space. A performance analysis of an existing or proposed system is obtained when the model is subjected to variable equations within time and space. To put it another way, simulation is the process of studying a system's performance using a model or testing something using a model.

M & S, according to Bratley et al. (2011), can be described as the act of driving a virtual prototype system with appropriate inputs and scrutinizing the resulting outputs. It has been further defined as the process of creating a model of a conceptual system and utilizing it to run experiments in order to understand the system's performance and/or assess various management methods and decision-making processes. The object of M & S includes, but not limited to, prediction, proof, discovery, performance assessment, education, and training. Manufacturing processes, social systems, corporate organizations, government systems, computer systems, ecological and environmental systems, and other multifaceted/complicated processes, systems, and designs may all benefit from M & S. This method has been used in a variety of multidisciplinary study areas, including decision-making processes, integrated product team management, new product creation, and general organizational management. M & S for complex sociotechnical systems have recently emerged as a viable study topic (Zhi et al., 2018)

In the grand scheme of things, models serve two purposes. The first is diagnostic: In this sense, the model is used to develop and test the procedures/practices that explain the observations in this system. In such a situation, the model decides if the processes are well understood and adequately represented. In general, these models and simulations are research tools used to figure out the nature of unexplained phenomena or poorly defined processes.

The model's second function is prognostic, which means it is used to make predictions. In these roles, any sort of model can be utilized. M & S can be applied in many different areas in science and engineering. The various concepts and classifications of theoretical models and simulations of MXene/polymer nanocomposites have been outlined in this section. The rudimentary concept of modeling and simulation include the object, base model, system, experimental frame, simulation, lumped model, verification, and validation.

15.2.1 Definition of Terms

Object: An object in M & S is a real-world entity that is analyzed to help understand how a model behaves. The goal of M & S operations is revealed through the object.

Base Model: A conjectural explication of object attributes and (their) behavior that is valid throughout the model is referred to as a base model.

System: A system is a clear thing that exists under certain parameters in the real world.

Experimental frame: It's a road map for studying nanocomposite systems in a real-world laboratory. This framework includes both the experimental circumstances and the study aims. Two sets of variables make up the basic experimental frame, that is, frame input variables (FIV) and frame output variables (FOV), that correspond to the model terminals. The FIV is in charge of matching the inputs to the system or model, while the FOV is in charge of matching the output values. It's crucial to note that an experiment may or may not affect the operation of the system (by influencing its input and parameters). As a result, the experimentation environment might be considered a separate system (which may, in turn, be modeled by a lumped model). Experimentation also entails making a measurement-based observation.

Simulation of a lumped model: This procedure follows a set of rules (as Petri net, differential-algebraic equations, or bond graph). It produces simulation results with variable input/output behavior (Park et al., 2004). In such circumstances, both symbolic and numerical simulation strategies can be applied. Virtual experimentation is a simulation that imitates real-world physical experimentation and allows one to offer answers to important questions concerning the characteristics of a nanocomposite. The System–Experiment/Model–Virtual Experiment sequence is more basic because of the homomorphic connection between the model and the system. As a result, creating a model of a real system and re-creating its properties should provide similar results to doing a real-world experiment followed by analysis and codification of the findings (Vangheluwe et al., 2002). The guarantee that results drawn from M & S can be trusted is one of the most essential criteria for such a system. The two separate processes of verification and validation are linked to the creation of this confidence.

Verification: The process of connecting or comparing two or more things (results) to ensure their accuracy is known as verification. Verification is generally done in M & S by comparing the dependability of a simulation program and the lumped model. This is accomplished by comparing experimental data to simulation findings inside an experimental frame. If the findings do not match, the model is declared invalid. The simulation model must be verified and validated before it can be used in a real-world situation (Zhi et al., 2018).

Various variables are utilized in M & S. The system state variables are one of these variables; they consist of a set of data that is used to represent the internal development of the system at any given moment in time. It's vital to keep in mind that certain of the system state variables don't change over time. Although in a continuous-event model, the values change at predetermined points. Differential equations are commonly used to specify system state variables, and the values of these equations vary with time.

15.3 THEORETICAL MODELS OF MXene/POLYMER NANOCOMPOSITES

The form of the MXene/polymer nanocomposite is determined by the interaction between the polymer entropy generated by the limitations between the two neighboring plates and the potential enthalpy gains. Any gains emanating from the auspicious interaction between the polymer and fillers interface in most cases determine the morphology of the nanocomposite. Different polymers have different interactions with MXene; that is, some factors such as the polymer molecular structure and orientation can influence the interactions with MXenes. In other words, MXene surfaces could form intercalated hybrids with polymers based on their compatibility. Many means have been employed to modify polymers and MXene to interact well to attain the expected nanocomposites (Zhou et al., 2021; Yu et al., 2021). Theoretical approaches for explaining how the interaction between the matrix, MXene, organic ligands, and clay surfaces, as depicted in Figure 15.1, influences the nanocomposite's phase behavior (Ginzburg, 2019).

For atoms, molecules, and clusters of units, the molecular-scaled approach focuses on molecular dynamics, the Monte Carlo technique, and molecular mechanics. Micro-scaled approaches are utilized to bridge the gap in molecular- and meso/macro-scaled technologies. Structure formation, bulk material flow, and the bonding interaction between matrices and fillers have all been shown to be influenced not just at the subatomic level but also at the micro level. Micro-scaled techniques typically fill the gap between molecular- and meso/macro-scaled methods. In the development of structures, the bulk flow of materials, and the bonding interface between matrices and fillers, it has been demonstrated that elements within a composite are impacted not only at the molecular level but also at the microscopic level. Brownian dynamics (BD), dissipative particle dynamics (DPD), lattice Boltzmann method (LBM), time-dependent Ginsburg-Landau theory (TG-LT), and dynamic density functional theories (DDFT) are some of the approaches for studying microscopic structures and interactions of composite elements within a system that have been established (Salamand Dong, 2019a; Quaresimin et al., 2012).

Under the conditions of both molecular and nanomaterial structures with homogeneity at various scaled levels, the continuous process is the most common technique among the meso-/macroscale

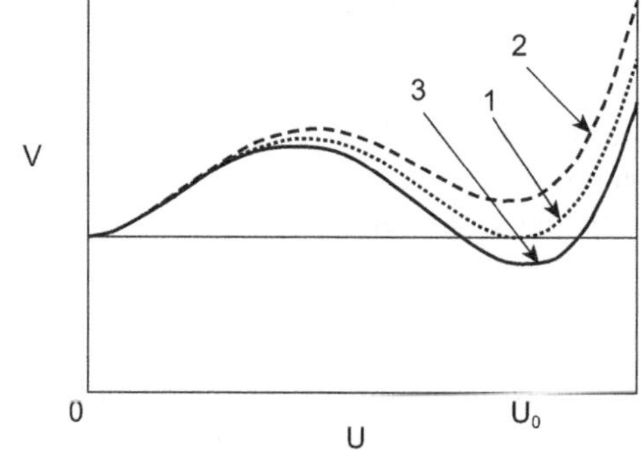

FIGURE 15.1 A representation of kink model of filler intercalation between polymer with a dimensional indication (a) and generated chart in (b).

Source: Reprinted with permission from (Salamand Dong, 2019b).

approaches of modeling. This method examines the deformation of nanocomposite structures caused by external forces, as well as the effects on stress and strain. Micromechanical models, equivalent-continuum models, self-consistent models, and finite element analysis are among the continuum approaches evaluated through modeling and numerical simulations. Salam et al. (Salam and Dong, 2019a) have done a lot of work on using micromechanical models to predict the tensile characteristics of nanocomposites based on the influence of individual elements. The influence of individual components on the usage of micromechanical models in estimating the tensile characteristics of nanocomposites has been widely studied. Six theoretical models have been used to determine the tensile moduli of various polymer nanocomposites, including the Voigt and Reuss models, Hirsch model, Halpin–Tsai (H-T) model, modified H-T models, Hui–Shia (H-S) model, and laminate model. The Danusso–Tieghi (D-T) model, Nicolais–Nicodemo (N-N) model, Lu model, and Turcsányi–Pukànszky–Tüds (T-P-T) model are all used to estimate the tensile strengths of most nanocomposites (Salam and Dong, 2019a).

The rule of mixture (ROM) is the most essential concept in micromechanical modeling of polymer particle composite systems (Mustafa et al., 2015). Filler properties such as shape, morphology, size, aspect ratio, and good interfacial adhesion may be transferred to fillers and matrices using ROM, which is based on the premise that both fillers and polymer matrices are linearly elastic [20]. The simple rule of mixture (SROM) does not give a correct extrapolation for the transverse Young's modulus and in-plane shear modulus, although it is one of the earliest and the simplest technique of micromechanics analysis for estimating the mechanical behavior of unidirectional plies. Therefore, the modified rule of mixture (MROM) proposed by Halpin manages to overcome this shortcoming (Mustafa et al., 2015; Madan et al., 2020)

In general, the Voigt and Reuss models are used to study and simulate composite systems, with the upper and lower boundaries of ROM, respectively (Salamand & Dong, 2019b). In what is known as a parallel model, the Voigt model (VM) includes the amalgamation of total individual components in the composite subject to the average elastic modulus and the volume percentage of each component (Salam and Dong, 2019a; Rossikhin and Shitikova, 2018). In addition, the VM model assumes that an isostrain condition exists in the upper bound for both fillers and matrices, resulting in a change in the modulus of the composites (Bulíček et al., 2012) as presented in Equation 15.1:

$$E_C = E_p\phi_P + E_m(1-\phi_P). \tag{15.1}$$

The Reuss or series model, on the other hand, is known as the inverted ROM or lower bound ROM. The Reuss model, unlike the VM model, is founded on the conjecture that continuous stress arises in both polymer matrices and fillers. Equations 15.2 to 15.4 represent the Reuss model:

$$E_C = \frac{E_p E_m}{E_M\phi_P + E_P(1-\phi_P)} \tag{15.2}$$

$$\varnothing_P = \frac{W_P}{W_P + (\rho_P / \rho_M)(1-W_P)'} \tag{15.3}$$

$$\rho_{blend} = \rho_1\varnothing_1 + \rho_2\varnothing_2, \tag{15.4}$$

where '$\varnothing$' indicates the filler volume proportion in composites with Wp representing the weight fraction of fillers. The density of matrix blends (ρ_{blend}) can be determined as per Equation 15.4, which is usually subject to influence by the relative densities (ρ_1 and ρ_2), and the analogous volume fractions ($\varnothing_1$ and $\varnothing_2$) of the two components in the blended matrices. More so, E_c, E_m, and E_p are the moduli of composites, matrices, and fillers, respectively. In such polymer blends (bio-epoxy blends) the findings must be established experimentally using tensile testing on polymer blends with varying filler amounts.

It's worth noting, however, that the elastic modulus of fillers varies (E_p) can differ based on different MXene dispersion statuses in the polymer matrix in the bulk nanocomposite. Also, in the case of intercalated MXene structures, the appraisal of effective modulus of MXene will take a different approach from that of exfoliated compound mix MXene structures, which can be determined by applying ROM in Equation 15.5.

$$(E_{\text{intercalation}} = E_{\text{MMT}}\phi_{\text{MMT}} + E_{\text{gallery}}(1-\phi_{\text{MMT}}), \tag{15.5}$$

$$if\ E_{\text{gallery}} << E_{\text{MMT}},$$

$$\left(E_{\text{intercalation}} \approx E_{\text{MMT}}\phi_{\text{MMT}} \approx \left(\frac{d_{\text{MMT}}}{d_{002}}\right)E_{\text{MMT}}\right)', \tag{15.6}$$

by which ϕ_{MMT} and E_{MMT} are the volume of MXene nanosheets and the elastic modulus of MXene, respectively, whereas E_{gallery} is the interlayer modulus. Once the interlayer modulus is considerably less than the MMT modulus ($E_{\text{gallery}} \ll E_{\text{MMT}}$), the E_{gallery} will have an insignificant influence on the modulus of the intercalated MXene nanosheets.

As a result, equation may be used to calculate the modulus of interposed MXene composites (Equation 15.6). The limited influence of the reduced-order model (ROM) on the elastic moduli and volume fractions of the composite elements is a disadvantage. More so, critical factors like MXene filler shape, orientation, and filler 3D position are not of any value to the system, which more or less dents the quality of elastic moduli approach to evaluating composite materials. This deficiency can, however, be taken care of with the integration of modulus reduction factor (MRF) in the modified rule of mixture (MROM). This is a typical occurrence when employing MRF with a filler aspect ratio (α), where irregular filler alignment in composite materials may be explained using Equation 15.7:

$$E_c = E_P\phi_P(\text{MRF}) + E_M(1-\phi_P). \tag{15.7}$$

As indicated in Equations 15.8 through 15.10, MRF is calculated using two distinct types of flake-like fillers in nanocomposites: Riley's rule, Padawer's rule, and Beecher's rule.

$$(\text{MRF}) = 1 - \frac{\ln(u+1)}{u} \tag{15.8}$$

Padawer and Beecher's rule:

$$(\text{MRF}) = 1 - \frac{(\tanh u)}{u} \tag{15.9}$$

$$u = \frac{1}{\alpha}\sqrt{\frac{\phi_P G_M}{E_P(1-\phi_P)'}} \tag{15.10}$$

The shear modulus of the matrix is represented by G_M, whereas α denotes the aspect ratio of fillers in composite systems. In this instance, the MRF of nanocomposites is investigated by fitting similar experimental data, and it was discovered that MRF = 0:66 can well match experimental data for modulus prediction when the filler in the composite is less than 6 percent by volume fraction.

Hirsch proposed a model for determining the elastic modulus of nanocomposites that was based on a simple combination of the series and parallel models given in Equation 15.11 (Kalaprasad et al., 1997; Hirsch et al., 2012):

$$E_C = x(E_m(1-\phi_P)+E_P\phi_P)+(1-x)\frac{E_P E_M}{E_M\phi_P + E_P(1-\phi_P)'}, \quad (15.11)$$

where x is an experimental constant that governs the stress transmission between nanofillers and composite matrices. This is based on the precision of curve fitting with realistic investigation data (0 ×1). In calculating the elastic moduli of nanocomposites based on the filler orientation in the nanocomposites, x is seen as an important constraint. With reference to volume fraction, shape, and orientation of fillers, the H-T model gives a realistic approximation for elastic moduli of unidirectional composites. This model is frequently used for constant or fragmentary fillers with a variety of alignments, such as fiber or flake-like nanofillers. The elastic modulus of composites is expressed as follows:

$$E_C = E_M\frac{1+\xi\eta_L\phi_P}{1-\eta_L\phi_P}, \quad (15.12)$$

where E_C, E_M, and E_P are the elastic moduli of composites, matrix, and fillers, respectively, and η_L is determined as follows:

$$\eta_L = \frac{E_P / E_m - 1}{E_P / E_m + \xi}. \quad (15.13)$$

As a result, ξ is a constant in composites, conditional on the geometry and aspect ratios of the fillers. Alternatively, ξ can also be represented by the following formulae:

$$\xi = 2\alpha = 2\left(\frac{l}{t}\right)\text{for loggitudinal modulus}(E_{11}) \quad (15.14)$$

$$\xi = 2\text{ for transverse modulus}(E_{11}) \quad (15.15)$$

The letters l and t stand for the length and thickness/depth of dispersed fillers in composites, respectively. When ξ is very big ($\xi \longrightarrow \infty$), the H-T model equals ROM, which is a parallel model, as shown in Equation 15.1. Once ξ is very small ($\xi \longrightarrow 0$), the H-T model transforms into an inverse of ROM, which is the same as the common series model found in Equation 15.2. As a result, the H-T model is believed to be flexible when it comes to anticipating scenarios in which the ROM is between lower and higher bounds.

The Danusso–Tieghi (D-T) Model proposes a link between rigid matrix-based composite mechanical strength and volume percentage. The D-T model posits that the matrices and fillers have no adhesion; therefore, there is no load transfer from matrices to fillers. As a result, the overall load is considered to be the same as what matrices alone can carry. This approach is used for regular or irregular filler distribution. The strength of composites may be calculated using a simple equation given in Equation 15.16:

$$\sigma_c = \sigma_m(1-\psi), \quad (15.16)$$

where σ_c and σ_m are the composite and matrix tensile strengths, respectively, and ψ is the cross-sectional area fraction. Because of the random orientation of the fillers in this model, the area fraction of the matrix must be constant for every cross section across the matrix, implying that the area fraction of the matrix's cross section is equal to the volume fraction of fillers ($\psi \approx \phi$p; Salam and Dong, 2019a).

Based on the preceding D-T model, another theoretical model Nicolais and Narkis (N-N) substitutes the volume fraction with a power-law function as follows:

$$\sigma_e = \sigma_m (1 - a\phi_P^b), \tag{15.17}$$

where a and b are constants in the composites that are influenced by particle morphological characteristics. Under laboratory conditions, this can be difficult to regulate, making this model burdensome. As a result, the N-N model has been explored for cubic-shaped fillers with evenly dispersed spherical particles, based on the premise that no adhesion exists between the matrix and the fillers. In this model, it is assumed that the continuous phase's minimum cross section is perpendicular to the applied stress. As a result, the strength of composites is determined using Equation 15.18:

$$\sigma_c = \sigma_m (1 - 1.21\phi_P^{2/3}). \tag{15.18}$$

The N-N model is perfect for estimating composites' lower-bound strength. As a result, the strength of composite material might be assumed to be the same as that of the matrix or pristine polymer ($\sigma_c \approx \sigma_m$), which cannot be the reality, and therefore, this model is ineffective.

Other theoretical models, such as the T-P-T model, provides a semi-empirical equation for extremely resilient particle-matrix interfacial bonding using a simple hyperbolic function that describes the changes in the filler cross section with filler concentration (Bharath Kumar et al., 2016). With this model, the exponential component's functionalization is exposed to unforeseen dependency on the matrix strength and filler volume fraction with this approach. As a result, the T-P-T model can only be used to represent composites containing spherical and anisotropic particles.

Theoretical models, in addition to continuum and atomic-scale computing methods, have contributed to a better appreciation of composite systems. The extended Takayanagi model, for example, was used to estimate the tensile strength of polymer/carbon nanotube nanocomposites (PCNTs; Zare and Rhee, 2019a). To produce the predictions, this model assumed that the interphase between the polymer matrix and nanoparticles was strengthened and the percolating efficiencies were high. The model was able to properly predict the average levels of the percolation threshold, interphase thickness, and strength for the two typical PCNTs based on this information (Behera et al., 2019). To further validate the model, mathematical frameworks were paired with an actual research using transmission electron microscopy to analyze and decrease the risk of agglomeration in nanocomposites by carefully changing the volume fraction, aspect ratio, and specific surface area of nanoparticles (Zare and Rhee, 2019b). Another research extended the Takayanagi model into two versions, assuming that the system had networking and dispersion of components above the percolation threshold (Loos and Manas-Zloczower, 2013). The enlarged model ignored the crucial reinforcing and percolating impacts of interphase zones in the modulus because it ignored the interphase areas around distributed and networked nanoparticles. As a result, the nanocomposites reinforcing the effectiveness of the networked nanoparticles were found to be over the percolation threshold (Behera et al., 2019).

The interphase areas around the dispersed MXene nanosheets are given greater weight in the expanded model. To put it another way, the expanded Takayanagi model closely resembles the filler network's reinforcing efficiency (Zare and Rhee, 2020; Zare and Garmabi, 2017) as shown in Equation 15.19.

$$E = \frac{\phi_N (1-\phi_f) E_d E_N + \phi_N (\phi_f - \phi_N) E_m E_N + (1-\phi_N)^2 E_d E_M}{(1-\phi_f) E_d + (\phi_f - \phi_N) E_m} \tag{15.19}$$

The tensile modulus of the nanocomposites is determined by a tensile test and is denoted by the letter E. The volume fractions of nanofiller and networked particles are represented by the letters f

and N, respectively, while the tensile moduli of the dispersed nanofiller, filler network, and polymer matrix are represented by the letters E_d, E_N, and E_m, respectively. It's worth noting that the sum of f and N always equals to one (Zare and Rhee, 2020).

To estimate the tensile moduli and strength of nanocomposites, a mathematical framework of H-S laminate model and H-T models has been proposed. According to this hypothesis, the content of components that can influence the nanocomposites are filler type, dispersion status, orientation, morphological features, and filler-matrix interfacial bonding (SalaM & Dong, 2019b; Affdl and Kardos, 1976). Regardless of the model employed, the orientation of the individual components and their manner of interfacial contact with the polymer matrix have an essential influence in determining the composite's characteristics. In general, the random orientation of dispersed fillers has a big influence in forecasting nanocomposites' elastic moduli. Interestingly, there were some changes between the two models used, although they were not significant. The findings of experimental work at a particular percentage of filler inclusion were reflected in the H-S model, but the H-T model revealed values that were similar to those of the experimental study.

15.4 M & S OF MXene/POLYMER NANOCOMPOSITE

In research and industrial contexts, computational approaches based on electronic to continuum levels are frequently used to investigate various material reactions in defined surroundings and space. While M & S are well-developed fields in metallurgy, ceramics, and polymeric materials, the science of simulating composites is still in its early stages. The challenge in using computational tools to examine composites stems mostly from the fact that

(1) composite materials are mixtures of two or more different materials with similar or dissimilar properties (e.g., metal/ceramic, metal/polymer, ceramic/polymer), necessitating precise designation of not only the discrete constituents in the model but also their complex interfacial interactions, and
(2) the limitless potential for developing unique composite materials by combining several conventional materials, which necessitates precise material knowledge and the appropriate parameters to assure redolent computational insight into any composite product changes.
(Behera et al., 2019; Stukowski, 2014)

With the continuum modeling approach, finite element modeling (FEM) is frequently used to simulate composite materials depending on the length and time scale (Behera et al., 2019). Other modeling approaches, including atomic-scale methodologies, DFT, and molecular dynamic (MD) simulations, are utilized to ensure a complete knowledge of experimental systems and to develop better material models for greater length and time-scale simulations.

FEM is effective for resolving constitutive relations in dynamic structures that are susceptible to a variety of solid-state processes. For analytical systems that are constrained by various geometrical restrictions, FEM is more beneficial. The ability to address complicated geometries associated with different material characteristics in the various components of the system being built is a benefit of FEM-centered simulation. In flat rolling, the constitutive equations relate primarily to the material's stress and strain rates, with the altered material assumed to be homogenous and isotropic. In this scenario, FEM is primarily used to generate detailed stress and strain profiles in the unit cell (matrix, particle, and the contact between them), which are essential to understanding the mechanical behavior of the composites. In a study by Jakšić et al. (2020), the electromagnetic field distribution in polymeric composite structures, as well as the spectrum distribution of their reflection and absorption coefficients were determined using FEM. The theoretical model was compared to an experimental result and the original models for the nanocomposites, and the results are shown in Figure 15.2. The result reveals the effectiveness of the simulated model in predicting the output of nanocomposites.

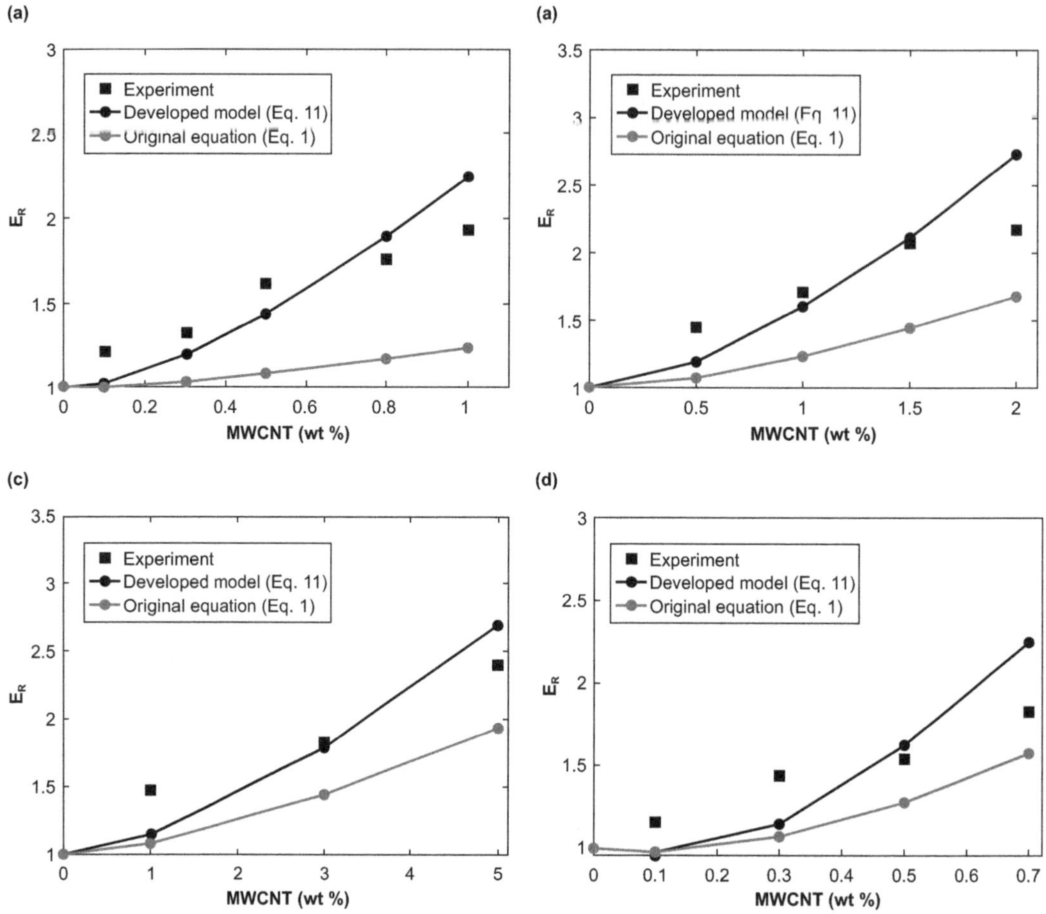

FIGURE 15.2 Relative modulus experimental data, as well as calculations of original and derived models for nanocomposites.

Source: Reprinted with permission from Jakšić et al. (2020).

Some models are built on the assumption that the constituents deform elastically in response to the restricted effective stress level delivered to the system. The spring stiffness layer approach might be used to mimic a typical impairment in a unidirectional nanocomposite, including interface de-bonding. This could be used to simulate the interface of matrix-nanomaterial reinforcement. When the finite layer is replaced in the spring layer model by a short thickness interface, which in this context denotes the transition between different phases, such as MXene and polymer matrix, the MXene nanosheets' interaction with the matrix is thought to be extremely stable and seamless, allowing for the transfer of any stress and strain level. FEM was utilized to simulate 3D hybrid nanocomposites to test this theory, and it was confirmed that stress was transferred from the matrix to the MXene polymer interface (Maghsoudi-Ganjeh et al., 2019). Similarly, Tao Huang et al. (2019) emphasized the usefulness of FEM in examining stress–strain perturbations in nanocomposite studies (Behera et al., 2019). Kilikevičius et al. (2021) investigated composite materials composed of an epoxy matrix and 2D nanosheets of graphene and MXenes. The graphene–matrix interphase properties used in the computational analysis were studied using an inverse modeling approach.

A computer model based on the micromechanical finite element technique was created to explore the mechanical characteristics of such hybrid polymer composites reinforced with graphene and

MXene nanosheets (Behera et al., 2019). In this system, geometrical models of 3D representative volume elements (RVEs) with different volume percentages of graphene and MXene inclusions were used, having different aspect ratios, and diverse alignment configurations created using Digimat-FE (Extreme Engineering, MSC. Software GmbH, Munich, Germany). The RVEs assess the effects of particles' aspect ratio, shape, clustering, orientation (randomly arranged or aligned), volume fraction in total, and, separately for different particles, on the mechanical behavior of nanocomposite.

The RVEs with randomly placed inclusions could be built as cubes, while the RVEs with aligned inclusions could be built as rectangular cuboids, aligning the inclusion in the *x*–*z* plane. As MXene/polymer matrix interface has a momentous effect on the mechanical behavior of composite materials reinforced; therefore, the approach of effective interface models could be adopted where the thin layer, surrounding the MXene inclusions, is formed with specific properties. Typical RVEs used in Kilikevičius et al. (2021) and Zeleniakiene et al. (2018) research are presented in Figure 15.3.

The created RVEs are then imported to the commercial finite element software Abaqus FEA (Dassault Systemes, Vélizy-Villacoublay, France) to develop a computational model and carry out the simulation tasks. The periodic boundary conditions are opted in Digimat-FE (Extreme Engineering, MSC. Software GmbH, Munich, Germany) and imported along with the geometrical models to Abaqus FEA (Dassault Systemes, Vélizy-Villacoublay, France). The RVEs are then subjected to uniaxial tensile loading along the *x*-axis direction. The RVEs are meshed using the 3D 4-node linear tetrahedron element (C3D4) type. The experimental stress–strain curve of epoxy is inserted in the program, and the multilinear hardening plasticity model is considered to define the response to the mechanical loading.

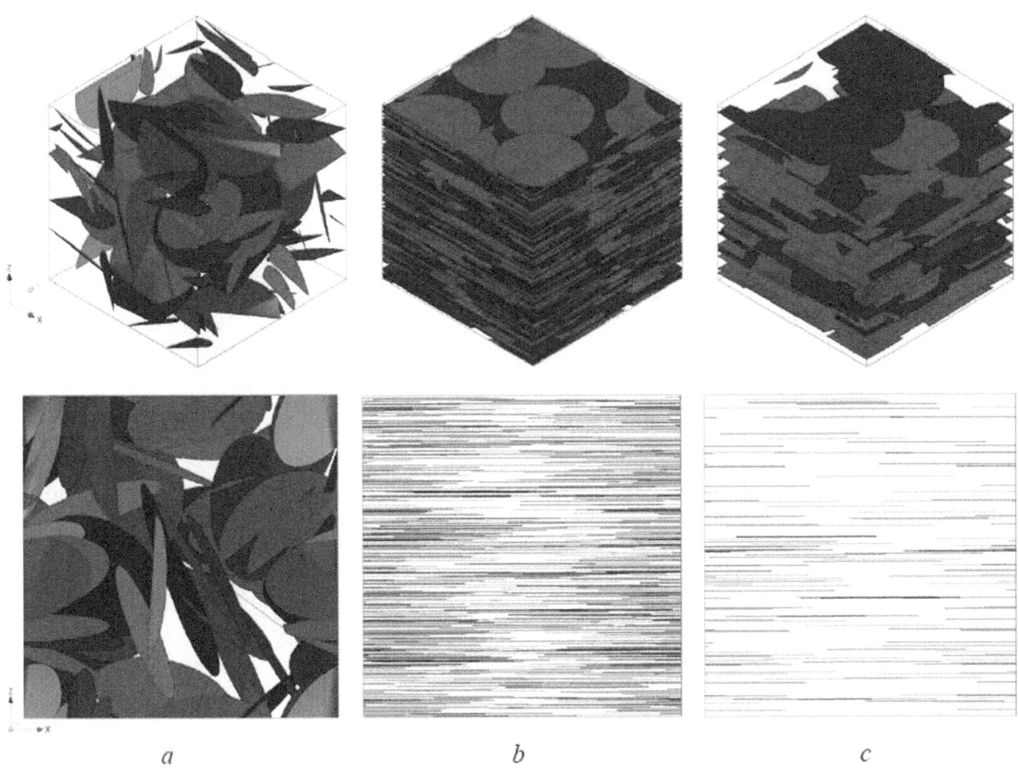

FIGURE 15.3 Some of RVEs of MXene/graphene/epoxy polymer composites: random 0.1% (a); aligned 1% (b); aligned 0.2% (c).

Source: Reprinted with permission from Zeleniakienė et al. (2018).

To investigate the impact of MXene/epoxy interface characteristics on the suggested nanocomposite's strength, several values of the young modulus (E_{MX}) and strength of the interface could be applied. The optimum principal stress benchmark is usually applied for the simulation of matrix and interface cracking with different values. In this regard, the effect of the MXene inclusion in the nanocomposite toward the resistance or damage of the nanocomposite during the simulation would be revealed, delivering reliable and stable results. Figure 15.4 shows a computational model of the effect of randomly placed MXene, and epoxy on the stress distribution studied (Kilikevičius et al., 2021) demonstrates crack upon an increase in strain, while Figure 15.5 shows the stress distribution on aligned MXene/polymer nanocomposites model.

Also, based on the computational model, the damage evolution of the MXene/polymer interface could be obtained. Such evolution in the RVEs was identified in Kilikevičius et al. study (2021). As observed in Figure 15.6, in the beginning, the MXene/epoxy interfaces start to fail (Figure 15.6e). High-stress concentrations at the edges of the nano-reinforcements were detected as shown in

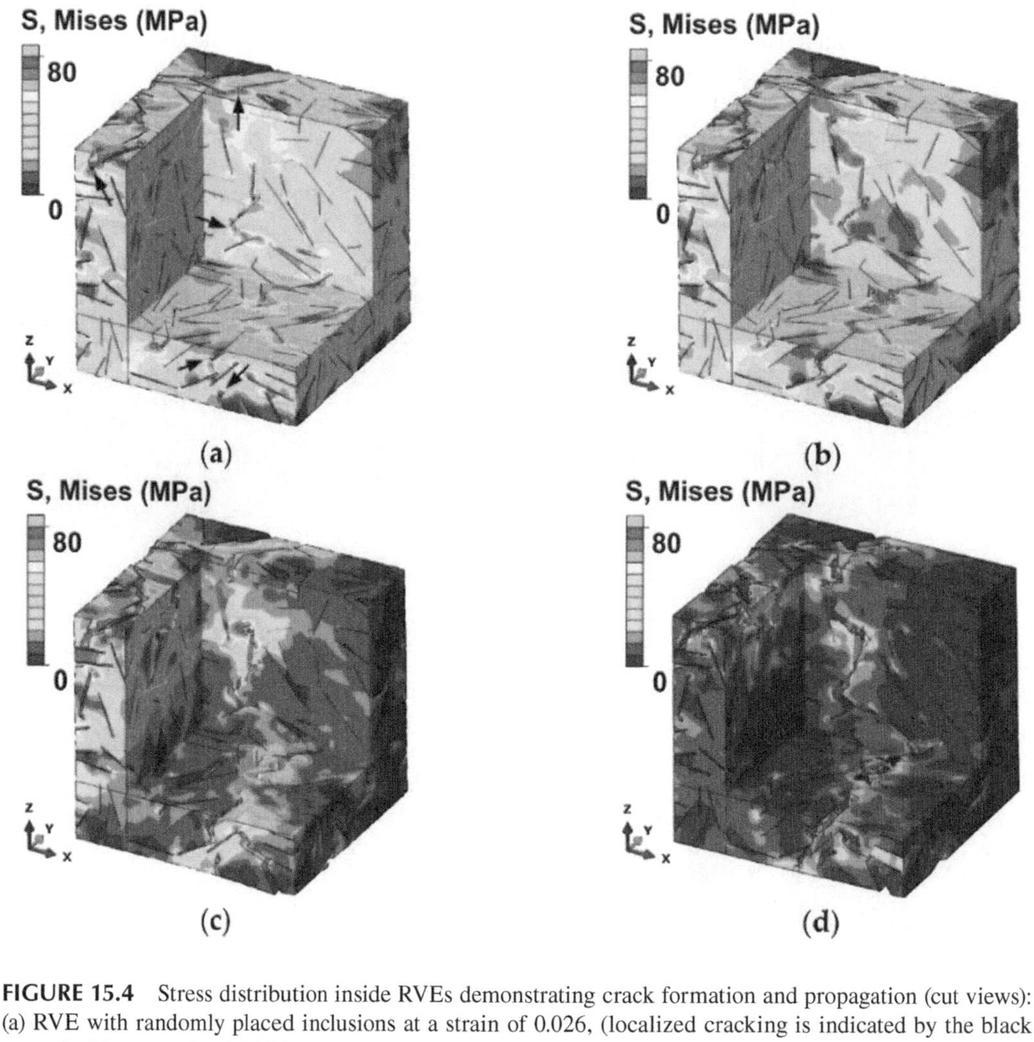

FIGURE 15.4 Stress distribution inside RVEs demonstrating crack formation and propagation (cut views): (a) RVE with randomly placed inclusions at a strain of 0.026, (localized cracking is indicated by the black arrows); (b) at a strain of 0.029; (c) at a strain of 0.031; (d) completely fractured at a strain of 0.038.

Source: Reprinted with permission from Kilikevičius et al. (2021).

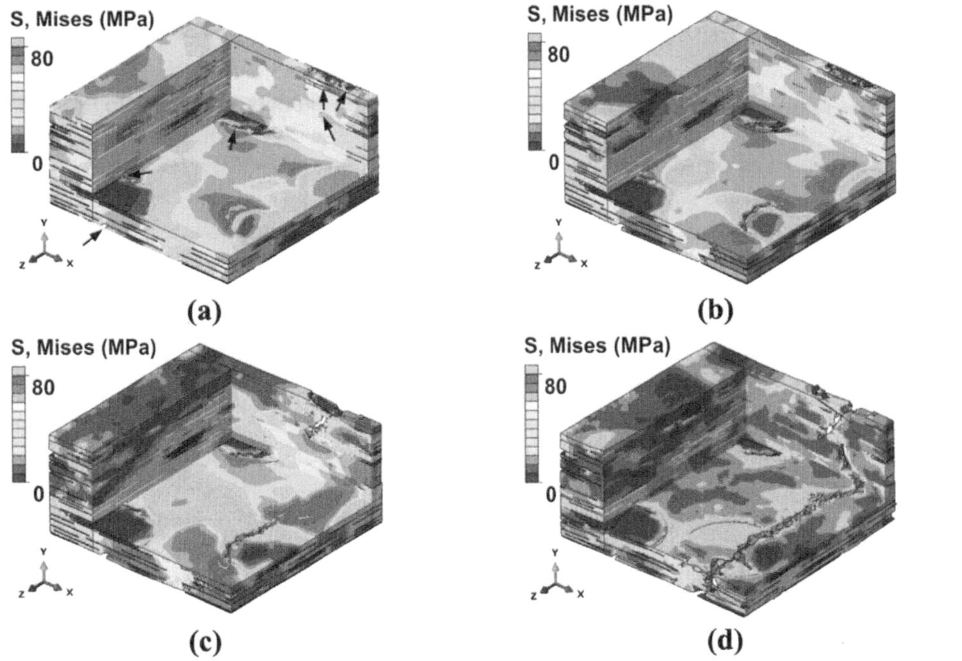

FIGURE 15.5 Stress distribution inside (a) the RVE with aligned MXene inclusions at a strain of 0.032, (localized cracking is indicated by the black arrows) at a strain of 0.036; (b) at a strain of 0.039; (c) at a strain of 0.041; (d) completely fractured at a strain of 0.045.

Source: Reprinted with permission from Kilikevičius et al. (2021).

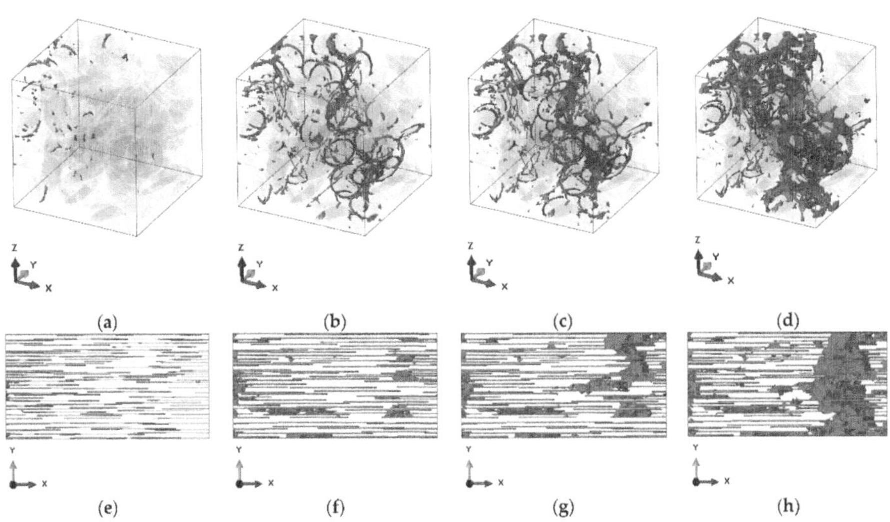

FIGURE 15.6 Damage evolution in the RVEs (damage is highlighted in red). (a–d) The RVE with randomly placed inclusions (the interface layers are hidden for better crack visualization, the MXenes are shown in transparent green, while the graphene inclusions are shown in transparent blue): (a) at a strain of 0.016; (b) at a strain of 0.026; (c) at a strain of 0.029; (d) at a strain of 0.038; (e–h) the RVE with aligned inclusions (the MXenes are shown in green, while the graphene inclusions are shown in blue): (e) at a strain of 0.011; (f) at a strain of 0.036; (g) at a strain of 0.039; (h) at a strain of 0.045.

Source: Reprinted with permission from Kilikevičius et al. (2021).

Figures 15.4a and 15.5e resulting in the formation of localized cracking at the edges of the nano-reinforcements in both RVEs (see Figure 15.6a–e). In the RVE with randomly placed inclusions, matrix damage was observed at a strain value of 0.016 (Figure 15.6a). As the strain increases, the main crack begins to form as shown in Figures 15.4b and 15.6b and propagates to the form shown in Figures 15.4c and 15.6c. A complete fracture of the RVE with randomly placed inclusions was observed at a strain of 0.038 (Figure 15.4d). In contrast, in the case of the RVE with aligned inclusions, it was observed at a strain of 0.045 (Figure 15.5h). Finally, it was observed that upon complete fracturing of the composite model (Figure 15.6d, h), the stress dropped, and crack pinning, and deflection of the epoxy matrix were noted in the fractured RVEs (Figures 15.4d and 15.5h).

Figure 15.7 shows that aligned MXene and graphene inclusions have better effective Young's modulus and tensile strength due to the geometrical orientation of the inclusions. Furthermore, as shown in Figure 15.7b, the influence of the aspect ratio of aligned inclusions reveals distinct effective Young's moduli corresponding to different aspect ratio values of the MXene nanosheets. Figure 15.8 indicates that the MXene/epoxy interface has no effect on the effective Young's modulus of the nanocomposite, but it does enhance the tensile strength in both aligned and randomly inserted MXene composites.

For the nanoscale computational work, molecular dynamics (MD) simulations could be performed using the open-source like LAMMPS software package (Sandia National Laboratories, Albuquerque, NM). The MD modeling procedure was initially built using a fixed-bond force field to perform crosslinking reactions of epoxy monomers (Al Mahmud et al., 2021). Subsequently, a transition simulation into an efficient reactive force field with the parameterization was consequently performed to all the M & S models, followed by equilibration to prepare for the mechanical property prediction at the nanoscale level (Liu et al., 2011). Three-dimensional periodicity was considered for all the M & S models to mimic the bulk behavior of the material. To analyze the mechanical and other essential aspects of complex polymer nanocomposite structures, multifaceted modeling design and analysis, including unit cell modeling and homogenization approach, are commonly used. The molecular and physical relationships between the nanomaterial fillers and the neighboring polymer matrix should be given special consideration because the interface plays a significant role in stress transfer from most polymers to nanomaterials and thus has the potential

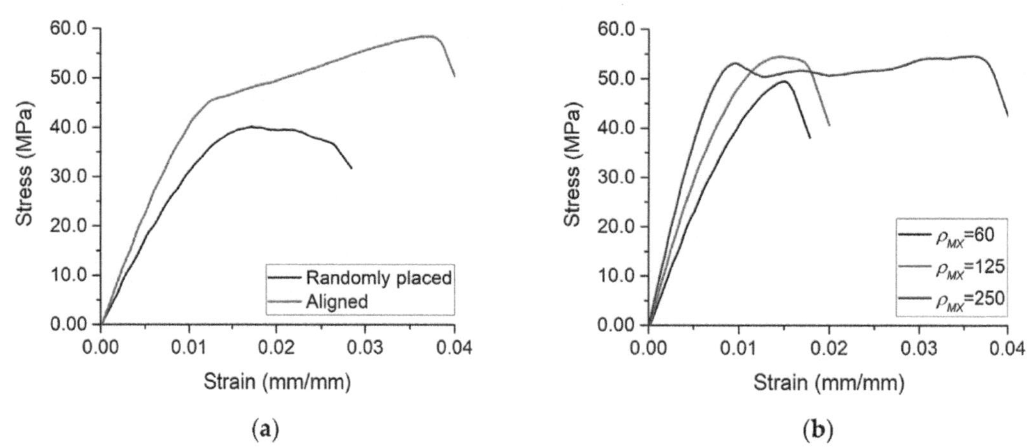

FIGURE 15.7 The influence of (a) geometrical orientation of the inclusions at $\rho_G = 500$, $\rho_{MX} = 125$, $f_G = 0.1\%$, $f_{MX} = 1.6\%$, and $E_{MX} = 0.5E_m$; (b) the aspect ratio of aligned MXene inclusions at $\rho_G = 500$, $f_G = 0.1\%$, $f_{MX} = 3.2\%$, and $E_{MX} = 0.5E_m$.

Source: Reprinted with permission from Kilikevičius et al. (2021).

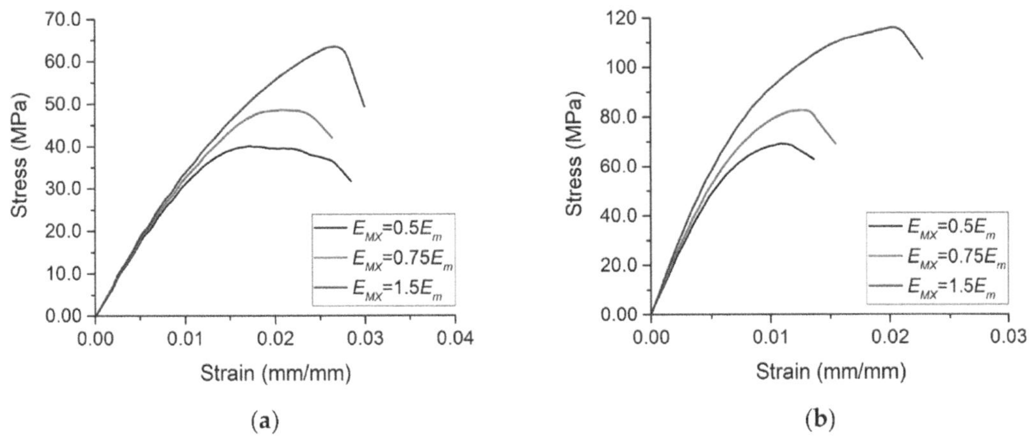

FIGURE 15.8 The influence of the MXene/epoxy interface mechanical properties (a) randomly placed MXenes, $\rho_G = 500$, $\rho_{MX} = 125$, $f_G = 0.1\%$, and $f_{MX} = 1.6\%$; (b) aligned MXenes, $\rho_G = 500$, $\rho_{MX} = 125$, $f_G = 0.1\%$, and $f_{MX} = 6.4\%$.

Source: Reprinted with permission from Kilikevičius et al. (2021).

to have a substantial influence on the mechanical properties. In this context, inverse modeling has been used to assess the interfacial properties of 2D nanomaterials such as graphene and various clay variations in nanocomposites (Monastyreckis et al., 2020a; Kumar et al., 2020). However, there haven't been any modeling studies tailored toward scrutinizing the stress transfer at the MXene/polymer interface currently (Monastyreckis et al., 2020b).

M & S are researcher's method of managing and investigating complex systems before practical experimentation in certain cases. Today's comprehensive models and simulations are so complicated that they require teams of scientists to participate in modeling activities in order for them to be effective. In computation and interpretation, the consequences of model complexity are realized. Almost all modeling findings, particularly theoretical models, are matched with experimental data in composite design to evaluate the viability of relevant theoretical models used as a guide for large-scale manufacture.

15.5 MERITS AND DEMERITS OF M & S OF MXene/POLYMER COMPOSITES

M & S have wider advantages, especially over experimental work, which has greatly broadened its usage in many different fields. Simulation allows for a thorough evaluation of items, systems, and structures in order to make the necessary modifications before they are physically created (Johanns et al., 2013). The mathematical model's elements can be substantially modified in order to examine the many facets of the process. Model restrictions can also be changed such that the simulation produces the most exact copy of reality as possible. The simulation also serves as a mechanism for postulating or analyzing future enhancements. They enable comprehension of how a system should behave without working on real-time systems, allowing problems to be solved before real systems are deployed. M & S allow the alteration to the system and its influence on the output without having to engage with real-time/real-world systems (Robinson, 2008). Different configurations could be applied to determine the best system requirements for the achievement of optimum results. They also enable the performance of tailback analysis that may cause a delay in the experimental work process. M & S allow researchers to understand all the interactions of the various constituents in a nanocomposite and analyze their effects, unlike certain complex systems that are not easy to understand their interaction at a time. Because the real product is not actually processed, mathematical or computationally created M & S can minimize the cost representation of a system entity, phenomenon, or risk of life-cycle activities. Furthermore, new policies, processes, and procedures may be tested independently of the real-world system.

Notwithstanding the vast advantages of M & S, they present some disadvantages. Simulation can be time-consuming and sometimes difficult to translate into the practical experimental world. The simulation process can be costly in terms of computational resources and their availability. Furthermore, creating a usable model is an art that necessitates subject expertise, training, and experience. In addition, due to the limitations in computational resources used for molecular dynamics modeling, computational studies are usually constrained to a few nanoseconds for type molecular dynamics models of organic materials (a few thousand atoms). This could lead to significant uncertainty in the predicted material properties due to the small statistical sampling of the molecular structure and the presence of significant thermal fluctuations at nonzero temperatures.

15.6 CONCLUSION

Some theoretical or computational models employed in simulation experiments at the atomic to the mesoscale level of nanocomposites were discussed in this chapter. Many theoretical approaches with a series of equations have been devised to simulate realistic system nanocomposites. M & S have been established as a veritable strategy for predicting the outcome of a nanocomposite, which could be verified and validated with other modeling methods or the real experimental data to ensure the accuracy of virtual experimental data obtained in the simulation. As there is limited literature on M & S of MXene/polymer nanocomposites, the exploitation of the effect of MXene, polymer, and other constituents in nanocomposites on their properties or functionalities would be a platform for extensive examination of newly developed MXene/polymer nanoparticles. In comparison to experimentally produced nanocomposites, this might broaden the understanding of MXene/polymer nanocomposites' behavior and applications in a short spate of time.

REFERENCES

AAKYIIR, M., YU, H., ARABY, S., RUOYU, W., MICHELMORE, A., MENG, Q., LOSIC, D., CHOUDHURY, N. R. & MA, J. 2020. Electrically and thermally conductive elastomer by using MXene nanosheets with interface modification. *Chemical Engineering Journal*, 397, 125439.

AFFDL, J. C. H. & KARDOS, J. L. 1976. The Halpin-Tsai Equations: A Review. *Polymer Engineering & Science*, 16, 344–352.

AGHAMOHAMMADI, H., AMOUSA, N. & ESLAMI-FARSANI, R. 2021. Recent Advances in Developing the MXene/Polymer Nanocomposites with Multiple Properties: A Review Study. *Synthetic Metals*, 273, 116695.

AL MAHMUD, H., RADUE, M. S., CHINKANJANAROT, S. & ODEGARD, G. M. 2021. Multiscale Modeling of Epoxy-Based Nanocomposites Reinforced with Functionalized and Non-Functionalized Graphene Nanoplatelets. *Polymers*, 13, 1958.

BEHERA, R. K., PINISETTY, D. & LUONG, D. D. 2019. Modeling and Simulation of Composite Materials. *JOM*, 71, 3949–3950.

BHARATH KUMAR, B. R., ZELTMANN, S. E., DODDAMANI, M., GUPTA, N., UZMA, GURUPADU, S. & SAILAJA, R. R. N. 2016. Effect of Cenosphere Surface Treatment and Blending Method on the Tensile Properties of Thermoplastic Matrix Syntactic Foams. *Journal of Applied Polymer Science*, 133.

BRATLEY, P., FOX, B. L. & SCHRAGE, L. E. 2011. *A Guide to Simulation*, Springer Science & Business Media. New York, NY: Springer.

BULÍČEK, M., MÁLEK, J. & RAJAGOPAL, K. R. 2012. On Kelvin-Voigt Model and Its Generalizations. *Evolution Equations & Control Theory*, 1, 17.

CHEN, H., ZHENG, Z., YU, H., QIAO, D., FENG, D., SONG, Z. & ZHANG, J. 2021. Preparation and Tribological Properties of MXene-Based Composite Films. *Industrial & Engineering Chemistry Research*, 60.30, 11128–11140.

GINZBURG, V. V. 2019. Recent Developments in Theory and Modeling of Polymer-Based Nanocomposites. *In:* ANDRIANOV, I. V., MANEVICH, A. I., MIKHLIN, Y. V. & GENDELMAN, O. V. (eds.) *Problems of Nonlinear Mechanics and Physics of Materials*, Cham: Springer International Publishing.

GUO, Y., ZHOU, X., WANG, D., XU, X. & XU, Q. 2019. Nanomechanical Properties of Ti3C2 Mxene. *Langmuir*, 35, 14481–14485.

HATTER, C. B., SHAH, J., ANASORI, B. & GOGOTSI, Y. 2020. Micromechanical Response of Two-Dimensional Transition Metal Carbonitride (MXene) Reinforced Epoxy Composites. *Composites Part B: Engineering*, 182, 107603.

HE, S., SUN, X., ZHANG, H., YUAN, C., WEI, Y. & LI, J. 2021. Preparation Strategies and Applications of MXene-Polymer Composites: A Review. *Macromolecular Rapid Communications*, 2100324.

HIRSCH, M. W., SMALE, S. & DEVANEY, R. L. 2012. *Differential Equations, Dynamical Systems, and an Introduction to Chaos*, Cambridge: Academic Press, an inprint of Elsevier Science, MA.

HUANG, T., ZHAN, M., PEI, Y., XIANG, N., YANG, F., LI, Y., GUO, J., CHEN, X. & CHEN, F. 2019. Establishment of Constitutive Relationships for Laminated Composites Considering the Variation of the Microhardness with the Strain in the Heterostructure Layers and Bonding Regions. *JOM*, 71, 3962–3970.

JAKŠIĆ, Z., OBRADOV, M., TANASKOVIĆ, D., JAKŠIĆ, O. & VASILJEVIĆ RADOVIĆ, D. 2020. Electromagnetic Simulation of MXene-Based Plasmonic Metamaterials with Enhanced Optical Absorption. *Optical and Quantum Electronics*, 52, 83.

JI, Z., ZHANG, L., XIE, G., XU, W., GUO, D., LUO, J. & PRAKASH, B. 2020. Mechanical and Tribological Properties of Nanocomposites Incorporated with Two-Dimensional Materials. *Friction*, 8, 813–846.

JOHANNS, K., LEE, J., GAO, Y. & PHARR, G. 2013. An Evaluation of the Advantages and Limitations in Simulating Indentation Cracking with Cohesive Zone Finite Elements. *Modelling and Simulation in Materials Science and Engineering*, 22, 015011.

KALAPRASAD, G., JOSEPH, K., THOMAS, S. & PAVITHRAN, C. 1997. Theoretical Modelling of Tensile Properties of Short Sisal Fibre-Reinforced Low-Density Polyethylene Composites. *Journal of materials science*, 32, 4261–4267.

KILIKEVIČIUS, S., KVIETKAITĖ, S., MISHNAEVSKY, L., OMASTOVÁ, M., ANISKEVICH, A. & ZELENIAKIENĖ, D. 2021. Novel Hybrid Polymer Composites with Graphene and MXene Nano-Reinforcements: Computational Analysis. *Polymers*, 13, 1013.

KUMAR, A., SHARMA, K. & DIXIT, A. R. 2020. A Review on the Mechanical and Thermal Properties of Graphene and Graphene-Based Polymer Nanocomposites: Understanding of Modelling and MD Simulation. *Molecular Simulation*, 46, 136–154.

LIU, L., LIU, Y., ZYBIN, S., SUN, H. & GODDARD, W. 2011. ReaxFF-/g: Correction of the ReaxFF Reactive Force Field for London Dispersion, with Applications to the Equations of State for Energetic Materials. *The Journal of Physical Chemistry. A*, 115, 11016–22.

LOOS, M. R. & MANAS-ZLOCZOWER, I. 2013. Micromechanical Models for Carbon Nanotube and Cellulose Nanowhisker Reinforced Composites. *Polymer Engineering & Science*, 53, 882–887.

MADAN, R., BHOWMICK, S. & SAHA, K. 2020. A Study Based on Stress-Strain Transfer Ratio Calculation Using Halpin-Tsai and MROM Material Model for Limit Elastic Analysis of Metal Matrix FG Rotating Disk. *FME Transactions*, 48, 204–210.

MAGHSOUDI-GANJEH, M., LIN, L., WANG, X., WANG, X. & ZENG, X. 2019. Computational Modeling of the Mechanical Behavior of 3D Hybrid Organic–Inorganic Nanocomposites. *JOM*, 71, 3951–3961.

MALAKI, M. & VARMA, R. S. 2020. Mechanotribological Aspects of MXene-Reinforced Nanocomposites. *Advanced Materials*, 32, 2003154.

MONASTYRECKIS, G., MISHNAEVSKY JR, L., HATTER, C., ANISKEVICH, A., GOGOTSI, Y. & ZELENIAKIENE, D. 2020a. Micromechanical Modeling of MXene-Polymer Composites. *Carbon*, 162, 402–409.

MONASTYRECKIS, G., MISHNAEVSKY, L., HATTER, C. B., ANISKEVICH, A., GOGOTSI, Y. & ZELENIAKIENE, D. 2020b. Micromechanical Modeling of MXene-Polymer Composites. *Carbon*, 162, 402–409.

MUSTAFA, G., SULEMAN, A. & CRAWFORD, C. 2015. Probabilistic Micromechanical Analysis of Composite Material Stiffness Properties for a Wind Turbine Blade. *Composite Structures*, 131, 905–916.

PARK, C.-S., AUGENBROE, G., MESSADI, T., THITISAWAT, M. & SADEGH, N. 2004. Calibration of a Lumped Simulation Model for Double-Skin Façade Systems. *Energy and Buildings*, 36, 1117–1130.

QUARESIMIN, M., SALVIATO, M. & ZAPPALORTO, M. 2012. Strategies for the Assessment of Nanocomposite Mechanical Properties. *Composites Part B: Engineering*, 43, 2290–2297.

ROBINSON, S. 2008. Conceptual Modelling for Simulation Part I: Definition and Requirements. *Journal of the Operational Research Society*, 59, 278–290.

ROSSIKHIN, Y. A. & SHITIKOVA, M. The Fractional Derivative Kelvin–Voigt Model of Viscoelasticity with and Without Volumetric Relaxation. *Journal of Physics: Conference Series*, 2018. IOP Publishing, 012069.

SALAM, H. & DONG, Y. 2019a. Theoretical Modelling Analysis on Tensile Properties of Bioepoxy/Clay Nanocomposites Using Epoxidised Soybean Oils. *Journal of Nanomaterials*, 2019, 4074869.

SALAM, H. & DONG, Y. 2019b. Theoretical Modelling Analysis on Tensile Properties of Bioepoxy/Clay Nanocomposites Using Epoxidised Soybean Oils. *Journal of Nanomaterials*, 2019, 4074869, 20 pp.

SLIOZBERG, Y., ANDZELM, J., HATTER, C. B., ANASORI, B., GOGOTSI, Y. & HALL, A. 2020. Interface Binding and Mechanical Properties of MXene-Epoxy Nanocomposites. *Composites Science and Technology*, 192, 108124.

STUKOWSKI, A. 2014. Computational Analysis Methods in Atomistic Modeling of Crystals. *JOM*, 66, 399–407.

VANGHELUWE, H., DE LARA, J. & MOSTERMAN, P. J. An Introduction to Multi-Paradigm Modelling and Simulation. Proceedings of the AIS'2002 conference (AI, Simulation and Planning in High Autonomy Systems), Lisboa, Portugal, 2002. 9–20.

WYATT, B. C., ROSENKRANZ, A. & ANASORI, B. 2021. 2D MXenes: Tunable Mechanical and Tribological Properties. *Advanced Materials*, 33, 2007973.

YU, B., YUEN, A. C. Y., XU, X., ZHANG, Z.-C., YANG, W., LU, H., FEI, B., YEOH, G. H., SONG, P. & WANG, H. 2021. Engineering MXene Surface with POSS for Reducing fire Hazards of Polystyrene with Enhanced Thermal Stability. *Journal of Hazardous Materials*, 401, 123342.

ZARE, Y. & GARMABI, H. 2017. Predictions of Takayanagi Model for Tensile Modulus of Polymer/CNT Nanocomposites by Properties of Nanoparticles and Filler Network. *Colloid and Polymer Science*, 295, 1039–1047.

ZARE, Y. & RHEE, K. Y. 2019a. Evaluation of the Tensile Strength in Carbon Nanotube-Reinforced Nanocomposites Using the Expanded Takayanagi Model. *JOM*, 71, 3980–3988.

ZARE, Y. & RHEE, K. Y. 2019b. A Simulation Work for the Influences of Aggregation/Agglomeration of Clay Layers on the Tensile Properties of Nanocomposites. *JOM*, 71, 3989–3995.

ZARE, Y. & RHEE, K. Y. 2020. Development of Expanded Takayanagi Model for Tensile Modulus of Carbon Nanotubes Reinforced Nanocomposites Assuming Interphase Regions Surrounding the Dispersed and Networked Nanoparticles. *Polymers (Basel)*, 12.

ZELENIAKIENĖ, D., GRIŠKEVIČIUS, P., MONASTYRECKIS, G. & ANISKEVICH, A. 2018. *Finite Element Simulation of Mechanical Properties of Novel Polymer Composites Reinforced with Mxene Nanosheets*. Conference: ECCM18 – 18th European Conference on Composite Materials, Athens, Greece, https://az659834.vo.msecnd.net/eventsairwesteuprod/production-pcoconvin-public/0178396fbe7a408dab32578b3943ff36

ZHI, Y.-R., YU, B., YUEN, A. C. Y., LIANG, J., WANG, L.-Q., YANG, W., LU, H.-D. & YEOH, G.-H. 2018. Surface Manipulation of Thermal-Exfoliated Hexagonal Boron Nitride with Polyaniline for Improving Thermal Stability and Fire Safety Performance of Polymeric Materials. *ACS omega*, 3, 14942–14952.

ZHOU, Y., LIN, Y., TAWIAH, B., SUN, J., YUEN, R. K. & FEI, B. 2021. DOPO-Decorated Two-Dimensional MXene Nanosheets for Flame-Retardant, Ultraviolet-Protective, and Reinforced Polylactide Composites. *ACS Applied Materials & Interfaces*, 13, 21876–21887.

16 Future Perspective of MXene-Filled Polymer Nanocomposites

Y C Goswami and Bhavya Pandey

CONTENTS

16.1 INTRODUCTION: AN OVERVIEW OF THE CURRENT STATUS OF POLYMER NANOCOMPOSITES

The modernization of the world brings a lot of challenges associated with sustainable development, energy-efficient systems, optimization of the use of materials, waste management, issues of potable water, and environmental problems, among others [1–2]. Such challenges always motivate us to develop new advanced energy-efficient materials, as a solution. As a result, the beginning of the 21st century witnessed the growth of many advanced materials at the nano level [3–4]. Zero- (0D), one- (1D), and two-dimensional (2D) nanomaterials are being developed with a lot of hope to be used in many day-to-day applications. Of course, such materials have their specific characteristics that were explored with limited dimensions. However, such developments still have enormous potential to incorporate more materials and combinations of materials. Among other materials, very recently discovered new materials MXenes, is a new class of 2D nanostructure materials that exhibit exceptional kind of conductivity, hydrophilicity. MXenes' heterostructures are made up of more than a single element (e.g., C and Ti, in the case of $Ti_3C_2T_x$, and Ti, Mo, and C, in the case of double-transition metal carbides similar to graphene, which consist of only one type of element, i.e., carbon) with randomly distributed functional groups on the surface. MXenes' heterostructures are represented by the $M_{n+1}X_nT_x$, where M is transition metal, X is nitrogen/carbon, and T is a functional group like fluorine, hydroxyl, or oxygen. MXenes are layered ternary metal carbides, carbonitrides, and nitrides. At Drexel University, titanium carbide

DOI: 10.1201/9781003164975-16

(Ti_3C_2) was the first MXene discovered, and since then, studies are going on to understand the different aspects of MXene. MXenes can replace many existing materials [5]. It has found its application in conversion, environment and catalysis, separation membranes, energy storage and conversion, electronics, medicine, and optics [6].

Recently, various polymer nanocomposites have also been developed and emerging as an important class of materials with a lot of future promises for high-performance potential applications and hence attracted intensive research interest. Polymer-based nanocomposite material consists of more than one discontinuous phase, distributed in one continuous phase. They have a great impact on the field of engineering due to their interfacial tension and orientation of dispersed phase inclusions like particles, flakes, laminates, and fibers [7].

Polymer nanocomposites have gained a lot of attention recently, due to their advantage of being used in combination with functional polymers with various types of nanofillers such as clay, inorganic/organic particles, silica, metals, and carbon-based nanoparticles as a filler [8–9], However, despite their excellent properties, polymer nanocomposites have limitations of nonuniform distribution, high viscosity, and fast agglomeration, which need to be addressed in future developments [10]. Recently, polymer-based composites filled with MXene have also been reported with a high potential to enhance the properties and functionalities of nanocomposites [6]. The MXene-filled polymers provide great strength and conductivity due to an increase in surface termination in the process of recombination. MXene-filled polymers are gaining a lot of attention in today's world due to their highly attractive electrochemical and electrical properties compared with other 2D materials like graphene [11]. High electrical conductivity accelerates the heterogeneous electron transfer rate. Because of their tendency of binding with different aimed biomolecules, the presences of many elements together add up again [12] in the heterogeneous catalytic reaction. This also improves distinct physicochemical properties rather than individual components. The properties include easy dispersibility, toughness, excellent barrier performances, improved solvent, good thermal resistance, inflammability, excellent photothermal conversion, excellent electromagnetic shielding, and stores charge capacity in comparison with other nanocomposites [13]. The extraordinary properties of these composites make them titled as wonder materials for future generations. The materials have excellent potential to be used in almost all material applications in life. In this chapter, we discussed, in brief, the future aspects and applications of MXene polymer nanocomposites.

16.2 MXene-BASED POLYMER NANOCOMPOSITES APPLICATIONS: AN OUTLINE

MXene shows a rare combination of high conduction and is put to work on active surfaces [14]. One of the most important advantages of MXene-filled polymer is improvements in fracture toughness. As the surface morphology has drastically modified, with a functional bonding of MXene with polymers, the composites exhibit excellent mechanical properties [15]. In addition to surface-modification reinforcement, a hybrid filler onto the surface of a 2D material to obtain hybrid-reinforced polymers is also reported. The large surface area of MXene helps in developing strong interaction between the Ti and O atoms in Ti_2C and Ti_3C_2 than in the F- or OH-terminated MXene. The O-terminated MXene is highly rigid, and this highly rigid bonding is responsible for the bonding between Ti–O, which is greater than in the Ti–F, and Ti–OH cases in terms of providing strength [16]. In composites, this results in excellent mechanical strength and good thermal stability. Thermodynamics calculations predicted high stiffness for a$Ti_4C_3Ti_2C$, Ti_3C_2 MXenes [5]. Due to its high mechanical strength, MXene-filled polymers are emerging good candidates for supercapacitor application. MXenes give metallic to semiconducting behavior, this can be found out by the electronic structures of MXenes [17]. The OH and F groups in MXene impact the electronic structure of a pristine MXene system. Each F or OH group can only receive one electron from the surfaces.

However, the O group differs from the F and OH groups because it demands two electrons from the surfaces to be stable [18]. $M_{n+1}X_n$-filled polymer gives controlled conductivity with an effective enhancement [19].

The nonmagnetic ground states of the majority of MXene result in a strong covalent bond between the transition metal and the X element. However, some MXenes also exhibit magnetic nature due to unpaired electrons in the spin split *d*-orbitals. Intrinsic point defects can be used to improve the magnetic properties of MXenes. It has also been reported that bare TiC samples exhibit paramagnetic behavior, similar to the samples with F- or S-based terminations, but if the sample contained two kinds of terminations, then its behavior was shifted to ferromagnetic/paramagnetic. Therefore, the magnetic properties of $Ti_3C_2T_x$ can be altered via modulation of the surface terminations [20]. Polymer nanocomposite with controlled magnetic properties has a lot of applications.

MXene-filled polymer nanocomposites exhibit extraordinary optical properties measured through calculation of the imaginary part of the dielectric function tensor [21]. These properties include optical transparency, plasmatic behavior, and efficient photothermal conversion. The ability of MXenes to interact with light in different ways has drawn the attention of the researcher [22].

The property of saturable absorption of MXenes makes it appropriate in the application of ultrafast laser. It was also found that $Ti_3C_2T_x$ can reach similar performance and broadband wavelengths ranging from 1550nm to 1620nm can be produced. This property indicates its potential applications in the signal and communication area [23]. The studies of nonlinear optical properties of MXene-filled polymers are in their initial stage and are emerging as multiple and outstanding applications of the material. More interaction can be done on the interaction of MXene with light.

MXene-polymer nanocomposites have their applications in the industry due to outstanding and different properties. It has potential in storage and transfer of energy, biomedical applications, and sensor applications. Many applications have been developed; still, a lot of work is going on with various modifications concerning improvement in functionalization, the inclusion of a new variety of MXene and polymers. The future of various devices based on MXene-filled polymer is full of applications and innovations.

16.3 EXISTING APPLICATIONS AND FUTURE PERSPECTIVES OF MXene–POLYMER NANOCOMPOSITES

Because of surface terminations, numerous compositions, and high hydrophobicity, MXene-filled polymers exhibit high thermal, electrical, and mechanical properties. Due to excellent mechanical, electrical, magnetic, and optical properties and effective control over that in polymer composites, it has been termed wonder materials for future applications. MXenes and their various films show excellent conduction with metals and great flexibility, due to which they are important for flexible electrochemical devices [24–28]. There are numerous applications of MXene filled Composites that have already been started and have large scope for future improvements.

16.4 FUTURE SCOPE OF CONDUCTIVE APPLICATIONS

The drastic change in the electrical conductivity of MXene-based polymers makes them suitable for many conductive applications [29–30]. An increase in the porosity of hybrid foam with high electromagnetic wave attenuation is also useful for modern applications like conducting fabrics. In the future, MXene, cellulose nanocrystal can be woven into fabrics materials. Three-dimensional printing of MXene frames with tunable electromagnetic interference also has excellent potential as 3D printing that might be very useful in industrial, as well as medical organ, printing applications [14, 30]. MXene-filled polymer patterns with a thermochromic polymer like polydimethylsiloxane can be used for tunability in colors from blue to red under the treatment with high-intensity electromagnetic radiations [31].

MXene-filled polymers have also shown excellent results in EMI shielding that use to protect against incoming or outgoing electromagnetic frequencies (EMF). EMI shield helps in protecting electronic devices from EMI. It may replace traditionally fabricated shields from metal sheets. Although metal shields are effective in terms of shielding but have the disadvantages of being heavy and bulky to design and high susceptibility to corrosion. Its high density restricts its use in broader applications in small and smart electronic products [32–33]. Recently, MXene-based polymers received vast attention to using EMI shielding field because of their exceptional electrical conductivity, hydrophilicity, and chemical activity.

In the recent future with fifth-generation (5G) mobile networks, the use of electronic devices will be raised exponentially, so it will be necessary to take measures for utmost care about people's physical health. For this, EMI and radiation need to be controlled [34]. MXene-filled polymers give hope to develop EMI shields with lightweight and corrosion-resistant materials. With the continuous increase in the use of portable and wearable devices, flexibility and excellent mechanical stability are new essential development directions for EMI shielding materials [35–37]. MXene-filled polymers are the future materials to cope with new development trends of electronic components [38–40]. The cost of the MAX phase, MXenes are relatively high at this stage, and it is vitally important to realize a low-cost industrial production of MXenes.

16.5 FUTURE SCOPE OF TRIBOLOGICAL APPLICATIONS

The efficiency of any machinery and its lifetime are always a matter of concern. As in modern society, friction, wear of materials, and lubrication have become increasingly crucial [41]. Industry concern is always about the minimization of friction, less use of nonrenewable energy in transportation and industrial sectors. Research on the development of excellent lubricant materials to combat friction and wear of materials is of great importance.

MXene-based polymers have shown outstanding tribological properties as a promising lubricant and as an excellent additive to compositions for friction reduction, wear of the materials, and thermal stability. MXene largely improved the antifriction properties of base oil [42]. In a similar study published by Zhang et al. 1.0 wt% of 10–20 nm, Ti_3C_2 nanosheets effectively improved the friction and antiwear properties of base oil [43]. Liu et al. reported low friction with 0.8wt% of highly exfoliated Ti_3C_2 added to base oil [44].

Titanium carbide (Ti_3C_2) MXene-filled polymer is also a promising material for solid lubrication. Zhang et al. reported on the synthesis of Ti_3C_2/ultra-high molecular weight polyethylene composites that reduced the friction wear due to adhesive behavior [45]. Ti_3C_2 nanosheets/copper composites with a very low wear rate and about half the reduction in the coefficient of friction were reported by Mai et al.[46]. Hu et al. reported $Ti_3C_2T_x$/Al composites with a coefficient of friction 0.2 in dry sliding test, which is twice lower than that of bare Al [47]. Recently, Yin et al. also developed Ti_3C_2/nanodiamond polymer composites that exhibited ultra-wear resistance at room temperature on rubbing against a polytetrafluoroethylene (PTFE) ball in the air. The combination of the rolling action of nanodiamond with the slipping and intercalation of MXene provides an effective reduction in wear [48]. Zhang et al. predicted an increase in the friction with Ti and O vacancies, through the increase in surface roughness. The –O-terminated Ti_3C_2 and the –OH terminations of Ti_3C_2 further reduced its interlayer friction [49]. Both surface terminations and intercalated water in $Ti_3C_2T_x$ MXene reduced the bond strength, resulting in a lower friction force under dry on stainless steel surfaces [49]. Ti_3C_2 MXene-filled polymers can also act as a solid lubricant on different substrates at the nanoscale [50]. All these studies have shown the great potential of MXene-filled polymers as solid lubricants; however, no or very few studies reported superlubricity an using MXene-filled combination. MXenes are quite reactive toward the environment particularly in water [51–52], therefore conducting measurements should be performed in an inert atmosphere.

In the future, superlubricity with MXene-filled polymers will be developed with an additional benefit of reduced wear. MXene/graphene composite coating reduces the abrasion of MXene while its superlubricity behavior remains unchanged.

16.6 FUTURE SCOPE FOR SENSING APPLICATIONS

MXene-filled polymers exhibit high metallic conductivity, tunable bandgap, excellent detection, and sensing applicability. MXene-filled polymers can be developed as a variety of sensors specifically electrochemical, gas, and photoluminescence sensors. MXene-filled polymers also show a good response under low humidity and high humidity conditions. Polyacrylamide (PAM) and MXene were laminated to form a double-layer sensor, and with this, the wide-range linear sensitive humidity sensing could be well analyzed [53]. The development of flexible and wearable sensors can also be developed using MXene/polymer composite sensors. It has been reported that a degradable PLA and $Ti_3C_2T_x$ were blended to form a wearable pressure sensor [54]. It was found that the sensor had excellent cycle stability and after 10,000 compression cycles. The response and recovery were reported 11ms and 25ms, respectively. Hu et al. reported Ti_3C_2 Nanosheet composite aerogels as wearable pressure sensors [55]. MXene-filled polymers can be developed as excellent stress sensors in the future because they have very good bending strength and because of their electronic properties. It can also be used to detect long-term stress and strain in the civil construction of the roadways, bridges, and buildings. In the future, MXene-filled polymers come up with major applications as physical sensors.

16.7 FUTURE SCOPE FOR ENERGY STORAGE APPLICATIONS

It has been found that the electrical properties of MXenes are like those of metals and semiconductors, which is due to their composition of elements and groups on the surface terminal [56]. Ideally, pure MXenes are metallic. But they can be turned into semiconductors. Similar to other 2D materials, both organic/inorganic molecules and various ions can spontaneously perform intercalation between MXene layers, which provides a larger surface area for charge storage and gives a way to increase the conductivity of layered materials by increasing the carrier concentration. MXenes has a wide application in electronics because of the like flexibility and metal-like electric conductivity [57–58]. MXene layers with a large interlayer spacing and intercalation of different-sized cations in the MXene-based polymers make them the best candidate for the energy storage device. In the future, in spite of using lithium, there are a variety of cations available, for example, sodium, potassium, magnesium, and others, that can be intercalated between the MXene layers for the development of hybrid devices. In the future, conducting polymers shall also be explored to develop flexible devices. The provision of an additional route for electron conduction by keeping suitable interlayer separations may also be useful to transparent and flexible energy sources. Hence, contributing to the improvement of electrochemical performance. The effect of interlayer distance in energy storage devices and MXene hybrids would lead to a useful path for further investigations towards energy storage applications.

16.8 FUTURE SCOPE OF MXene-POLYMER NANOCOMPOSITES IN BIOMEDICAL STORAGE APPLICATIONS

MXene has different end groups on its surface, which helps it react differently on the surface, excellent hydrophilicity, amazing biocompatibility, and remarkable electrical and optical properties. Thus, MXene has a great source for biomedical purposes [59].

MXenes might be used in many biomedical applications, like antimicrobial activities, drug delivery, diagnostic, sensing, and 3D printing of organs [60–62]. MXene has found its application in many biomedical fields but is still in its growing stage in this field. However, the performance of MXenes mainly focuses on the electrical and optical properties so far. In the future, more biomedical applications shall be developed on magnetic properties besides magnetic resonance imaging (MRI). A better understanding of the mechanism, improving their properties, and using them more appropriately. More applications of MXenes shall be explored in biomedical fields, such as acoustic dynamics therapy. MXene/polymer nanocomposites need to be further explored and developed because they are rarely used in this field. MXene-based polymers have great potential for surface modification and functionalization, so in the future they can find their application in photothermal and chemo-photothermal therapy [63–64]. Due to low toxicity, MXene does not affect the environment much. MXene can also find its application in targeted anticancer drug delivery [65–66]. MXene can find its major application in cancer diagnosis.

16.9 FUTURE SCOPE FOR WEARABLE SMART DEVICES STORAGE APPLICATIONS

The first generation of wearable and smart devices already begins in the form of smart wearable displays, smart garments, and watches with health monitoring systems., this paves a path for future wearable electronics. However, there are many challenges like durability, performance in comparison to existing conventional devices, and a truly flexible nature [67–69]. The extraordinary capacitance, metallic conductivity, superior hydrophily, and rich surface functionality make MXene–polymer nanocomposites the best candidate for promising flexible electrodes. The intercalation of polymers into MXene layers will promote the molecular-level coupling between MXene and polymer molecules. This condensed combination is supposed to effectively improve the strength and flexibility of MXene polymers. Many polymers have been reported, however, PVA demonstrates enormous potential as the candidate reinforcement to prepare MXene-based film electrodes because of the high solubility in water and the affluent hydroxyl groups along the molecular chain, which can be used to form hydrogen bonds between PVA and MXene, with rich electronegative oxygen and fluorine groups. Research on high-quality flexible energy devices has now become the core area of research and product development. Supercapacitors, as energy storage devices, attract attention due to their high power densities and long cycle life span. devices, especially in flexible supercapacitors. The outstanding conductivity of MXenes polymers results in higher power density than semiconductor-type metal oxides, making them suitable for rapid-charge function from the current consumer electronics [70–72]. Promising active sites in MXene support a lot for excellent photocatalytic and hydrogen evolution reactions.

Nowadays foldable phones are already available in the market. MXene-filled polymer as a material has high flexibility, so it can find its future application in the flexible and transparent electronic devices, such as mobile phone, smartwatches, and the like, can be cost-effective also [75].

16.10 FUTURE SCOPE FOR MEMBRANE SEPARATION

Desalination of water is one of the most potential areas of research, due to the scarcity of potable water. In various methods of desalination, membrane desalination is gaining a lot of interest, in which minerals and salts are removed from water by passing through a semipermeable membrane.

The ideal desalination and water treatment membranes should have high flux, high selectivity, stability, and resistance against fouling and chlorine-like chemical [76]. In the field of the membrane, the separation of MXene has drawn attention. MXene-based membranes have shown various potential in separation applications, such as water treatment, osmotic energy conversion, gas separation, and the like [77]. MXenes are considered skeleton materials for preparing lamellar-structure membranes. different types of nanomaterials such as are added to fabricate mixed-matrix membranes with MXenes and MXenes are used as coating materials on membrane supports consisting of various materials such as anodic aluminum oxide, polyvinylidene fluoride, and polycarbonate [78]. The membrane's rejection of inorganic salts is below 23% with flux above 432 L m^{-2} h^{-1} at 0.1 MPa, which indicates that the membrane can be used with extremely high efficiency in dye desalination and wastewater treatment [79].

Recent reports show that MXenes have emerged as one of the amazing materials with potential applications in desalination; however, more still may need to be addressed to exploit their remarkable properties [80–81]. It is still a need for an effective long time storing technique for MXene solution without oxidation. Low yield, control in aggregation, low-cost methods need to be developed. The bottom-up design of an efficient, and environment-friendly system for the production of MXene polymers will be helpful in opening a new door for possible applications [82–83]. Surface modification in polymer-based composites will help to improve their stability, biocompatibility, and recyclability. MXene–polymer-based composites as a solution emerging and have been considered as future desalination technology. Despite challenges, MXene–polymer composites have assured an era of the next-generation 2D nanomaterials water purification and environmental remediation.

16.11 CHALLENGES IN THE DEVELOPMENT OF APPLICATIONS

MXene–polymer nanocomposites have found their application in industries but are still in their primary stage. The results that have been reported are commendable, but the problem related to the production is not resolved. The current studies do not focus on the application in industries [84–86]. Material property improvements cannot be realized without the use of colloidal MXene-filled polymers at scale. Their adoption by industries has challenged the face of the same barriers found in the field of graphene [87]. Figure 16.1 shows a summary of properties, applications, and future scope.

In addition to the challenges taken into consideration from a commercial or industrial point of view, the other issues might be, cost, safety, and end of life. Production costs and then dispersing are quite expensive. Stability during storage and transport must also be considered; the dispersion must be stable for a sufficiently long period, during transport and for storage. Toxic and highly flammable solvents would be avoided. Finally, the end of life and disposal must also be considered [88–89].

16.12 CONCLUSION

MXene-filled polymers have extraordinary conducting, mechanical, optical, and tribological properties. The rare combination of these properties and superior control over these using various ratios, constituents, and functionalization make them suitable for various applications in energy devices, sensing devices, electronic devices, low-friction materials, and medical and environmental applications. Recent research and the future scope of MXene-filled polymers in conducting applications like conducting fabrics, 3D printing, medical organ printing, 5G mobile networks applications, EMI shielding, and tribological applications like super-lubricant materials. Sensing applications, like flexible and wearable sensors, efficient gas sensors, storage devices, and separation membranes, have been envisioned in this chapter. Although MXene-filled polymers are emerging as an excellent candidate for many applications, still, challenges like large-scale production, nontoxic, bottom-up production; effective control on agglomeration; and oxidation are still to be addressed so that more efficient devices can be developed in large-scale environmentally friendly ways. A lot of hope exists in the future of transparent, flexible MXene-filled polymers that shall serve society in almost all domains of day-to-day life.

ACKNOWLEDGMENT

Authors are thankful to ITM University Gwalior, for providing library and lab facilities.

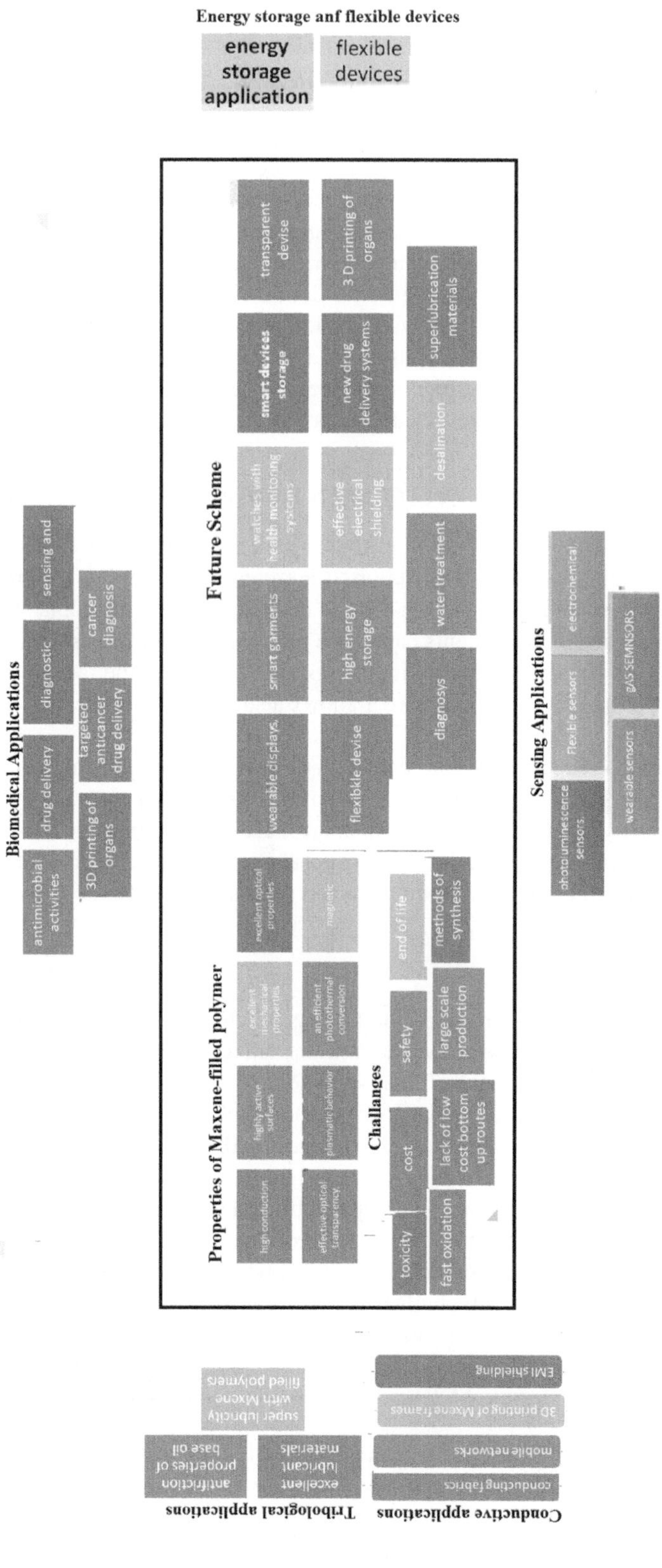

FIGURE 16.1 Properties, applications, and future scope of MXene-filled nanocomposites.

REFERENCES

[1] Bisauriya, R., Verma, D., and Y. C. Goswami Y., (2017). Optically important ZnS semiconductor nanoparticles synthesized using organic waste banana peel extract and their characterization. *Journal of Materials Science: Materials in Electronics*, 29(3), pp. 1868–1876.

[2] Kumar, N., Purohit, L., and Goswami, Y., (2016). Spin coating of ZnS nanostructures on filter paper and their characterization. *Physica E: Low-dimensional Systems and Nanostructures*, 83, pp. 333–338.

[3] Gahlaut, U., Kumar, V., and Goswami, Y., (2020). Enhanced photocatalytic activity of low-cost synthesized Al-doped amorphous ZnO/ZnS heterostructures. *Physica E: Low-dimensional Systems and Nanostructures*, 117, pp. 113792.

[4] Kumar, V., Rajaram, P., and Goswami, Y., (2017). Sol-gel synthesis of SnO2/CdS heterostructures using various Cd:S molar ratio solutions and its application in photocatalytic degradation of organic dyes. *Journal of Materials Science: Materials in Electronics*, 28(12), pp. 9024–9031.

[5] Papadopoulou, K. A., Chroneos, A., Parfitt, D., and Christopoulos, S.-R. G. (2020). A perspective on MXenes: Their synthesis, properties, and recent applications. *Journal of Applied Physics*, 128(17), pp. 170902.

[6] Bogue, R. (2011). Nanocomposites: A review of technology and applications. *Assembly Automation*, 31(2), pp. 106–112.

[7] Sreekala, M.S., Kumaran, M.G., Joseph, R., and Thomas, S. (2001). Stress-relaxation behavior in composites based on short oil-palm fibers and phenol-formaldehyde resin. *Composites Science and Technology*, 61(9), pp. 1175–1188.

[8] Mittal, V. (2009). Polymer layered silicate nano composites: A review. *Materials*, 2(3), pp. 992–1057. doi: 10.3390/ma2030992

[9] Banda-Cruz, E. E., Flores-Gallardo, S. G., and Rivera-Armenta, J. L. (2017). Study of the dispersion of Cloisite 10A in recycled polyethylene terephtalate by extrusion. *DYNA*, 84(200), pp. 107–111.

[10] Julkapli, N. M., Bagheri, S., and Sapuan, S. M. (2015). Multifunctionalized carbon nanotubes polymer composites: Properties and applications. *Eco-Friendly Polymer Nanocomposites*, pp. 155–214.

[11] Garg, R., Agarwal, A., and Agarwal, M. (2020). A review on MXene for energy storage application: Effect of interlayer distance. *Materials Research Express*, 7(2), pp. 022001.

[12] Shahzad, F., Zaidi, S. A., and Naqvi, R. A. (2020). 2D Transition metal carbides (Mxene) for electrochemical sensing: A review. *Critical Reviews in Analytical Chemistry*, 1(18). doi: 10.1080/10408347.2020.1836470.

[13] Garg, R., Agarwal, A., and Agarwal, M., (2020). A review on MXene for energy storage application: Effect of interlayer distance. *Materials Research Express*, 7(2), pp. 022001.

[14] Zhi, W., Xiang, S., Bian, R., Lin, R., Wu, K., Wang, T., and Cai, D. (2018). Study of Mxene-filled polyurethane nanocomposites prepared via an emulsion method. *Composites Science and Technology*. doi: 10.1016/j.compscitech.2018.10.026.

15] Bai, Y., Zhou, K., Srikanth, N., Pang, J. H., He, X., and Wang, R. (2016). *RSC Advances*, 6, pp. 35731.

[16] Magnuson, M., Halim, J., and Näslund, L.-Å. (2018) Chemical bonding in carbide MXene nanosheets. *Journal of Electron Spectroscopy and Related Phenomena*, 224(27). doi: 10.1016/j.elspec.2017.09.006

[17] Khazaei, M., Arai, M., Sasaki, T., Chung, C.-Y., Venkataramanan, N. S., Estili, M., Sakka Y., and Kawazoe, Y. (2013). Novel electronic and magnetic properties of two-dimensional transition metal carbides and nitrides. *Advanced Functional Materials*, 23, pp. 2185.

[18] Khazaei, M., Ranjbar, A., Arai, M., Sasaki, T., and Yunoki, S. (2017). Electronic properties and applications of Mxenes: A theoretical review. *Journal of Materials Chemistry C*, 5(10), pp. 2488–2503.

[19] Hart, J. L., Hantanasirisakul, K., Lang, A. C., Anasori, B., Pinto, D., Pivak, Y., van Omme, J. T., May, S. J., Gogotsi, Y., and Taheri, M. L. (2019). Control of MXenes' electronic properties through termination and intercalation. *Nature Communications*, 10(1), pp. 2041–1723.

[20] Sun, W., Xie, Y., and Kent, P., (2018). Double transition metal Mxenes with wide band gaps and novel magnetic properties. *Nanoscale*, 10(25), pp. 11962–11968.

[21] Yorulmaz, U., Ozden, A., Perkgoz, N. K., Feridun, A., Sevi, C. (2016). Vibrational and mechanical properties of single layer MXene structures: A first-principles investigation. *Nanotechnology*, 27(33), pp. 335702.

[21] Heyd, J., Scuseria, G., and Ernzerhof, M. (2003). Hybrid functionals based on a screened Coulomb potential. *The Journal of Chemical Physics*, 118(18), pp. 8207–8215.

[22] Frey, N., Wang, J., Vega Bellido, G., Anasori, B., Gogotsi, Y., and Shenoy, V., (2019). Prediction of synthesis of 2D metal carbides and nitrides (Mxenes) and their precursors with positive and unlabeled machine learning. *ACS Nano*, 13(3), pp. 3031–3041.

[23] Jiang, X., Liu, S., Liang, W., Luo, S., He, Z., Ge, Y., Wang, H., Cao, R., Zhang, F., Wen, Q., Li, J., Bao, Q., Fan, D., and Zhang, H. (2018). Broadband nonlinear photonics in few-layer Mxene Ti3C2Tx (T = F, O, or OH). *Laser & Photonics Reviews*, 12(2), pp. 1870013.

[24] Naguib, M., Unocic, R.R., Armstrong, B.L. et al. (2015). Large-scale delamination of multi-layers transition metal carbides and carbonitrides "Mxenes." *Dalton Transactions*, 44(20), pp. 9353–9358.

[25] Chaudhari, N. K., Jin, H., Kim, B., San Baek, D., Joo, S. H., and Lee, K. (2017). Mxene: An emerging two-dimensional material for future energy conversion and storage applications. *Journal of Materials Chemistry A*, 5(47), pp. 24564–24579.

[25] Gogotsi, Y., Anasori, B. (2017). The Rise of Mxene. *ACS Nano*, 13(8), pp. 8491–8494. doi: 10.1021/acsnano.9b06394. PMID: 31454866.

[26] Müller, K., Bugnicourt, E., Latorre, M., Jorda, M., Echegoyen Sanz, Y., Lagaron, J., Miesbauer, O., Bianchin, A., Hankin, S., Bölz, U., Pérez, G., Jesdinszki, M., Lindner, M., Scheuerer, Z., Castelló, S., and Schmid, M. (2017). Review on the processing and properties of polymer nanocomposites and nanocoatings and their applications in the packaging, automotive and solar energy fields. *Nanomaterials*, 7(4), pp. 74.

[27] Saini, P. (2015). Intrinsically conducting polymer-based blends and composites for electromagnetic interference shielding: Theoretical and experimental aspects. *Fundamentals of Conjugated Polymer Blends, Copolymers and Composites*, pp. 449–518. doi: 10.1002/9783527659647.ch10

[28] Saini, P. (2013). Electrical properties and electromagnetic interference shielding response of electrically conducting thermosetting nanocomposites. *Thermoset Nanocomposites*, pp. 211–237.

[29] Khan, W., Sharma, R., and Saini, P. (2016). Carbon nanotube-based polymer composites: Synthesis, properties and applications. *Carbon Nanotubes—Current Progress of their Polymer Composites*. doi: 10.5772/62497

[30] Li, S., Meng Lin, M., Toprak, M. S., Kim, D. K., and Muhammed, M. (2010). Nanocomposites of polymer and inorganic nanoparticles for optical and magnetic applications. *Nano Reviews*, 1(1), pp. 5214. doi: 10.3402/nano.v1i0.5214

[31] Wu, X., Tu, T., Dai, Y., Tang, P., Zhang, Y., Deng, Z., Li, L., Zhang, H., and Yu, Z., (2021). Direct ink writing of highly conductive Mxene frames for tunable electromagnetic interference shielding and electromagnetic wave-induced thermochromism. *Nano-Micro Letters*, 13(1). doi: 10.1007/s40820-021-00665-9

[32] Wang, C., Murugadoss, V., Kong, J., He, Z., Mai, X., Shao, Q., Chen, Y., Guo, L., Liu, C., Angaiah, S., and Guo, Z. (2018). Overview of carbon nanostructures and nanocomposites for electromagnetic wave shielding. *Carbon*, 140, pp. 696–733. doi: 10.1016/j.carbon.2018.09.006

[33] Huangfu, Y., Liang, C., Han, Y., Qiu, H., Song, P., Wang, L., Kong, J., and Gu, J. (2019). Fabrication and investigation on the Fe3O4/thermally annealed graphene aerogel/epoxy electromagnetic interference shielding nanocomposites. *Composites Science and Technology*, 169 pp. 70–75. doi: 10.1016/j.carbon.2018.09.006

[34] Kumar, P. (2019). Ultrathin 2D nanomaterials for electromagnetic interference shielding. *Advanced Materials Interfaces*, 6(24), pp. 1901454. doi: 10.1002/admi.201901454

[35] C. Liang, Ruan, K., Zhang, Y., and Gu, J. (2020). Multifunctional flexible electromagnetic interference shielding silver nanowires/cellulose films with excellent thermal management and joule heating performances. *ACS Applied Materials & Interfaces*, 12(15), pp. 18023–18031. doi: 10.1021/acsami.0c04482

[36] Jia, L.-C., Yan, D.-X., Liu, X., Ma, R., Wu, H.-Y., and Li, Z.-M. (2018). Highly efficient and reliable transparent electromagnetic interference shielding film. *ACS Applied Materials & Interfaces*, 10(14), pp. 11941–11949.

[37] Sun, Y., Luo, S., Sun, H., Zeng, W., Ling, C., Chen, D., Chan, V., and Liao, K. (2018). Engineering closed-cell structure in lightweight and flexible carbon foam composite for high-efficient electromagnetic interference shielding. *Carbon*, 136(4), pp. 299–308. doi: 10.1016/j.carbon.2018.04.084

[38] Fu, J., Yun, J., Wu, S., Li, L., Yu, L., and Kim, K. (2018). Architecturally robust graphene-encapsulated Mxene Ti2CTx@polyaniline composite for high-performance pouch-type asymmetric supercapacitor. *ACS Applied Materials & Interfaces*, 10(40), pp. 34212–34221. doi: 10.1021/acsami.8b10195

[39] Liang, C., Qiu, H., Song, P., Shi, X., Kong, J., and Gu, J. (2020). Ultra-light Mxene aerogel/wood-derived porous carbon composites with wall-like "mortar/brick" structures for electromagnetic interference shielding. *Science Bulletin*, 65(8), pp. 616–622.

[40] Liu, J., Zhang, H.-B., Sun, R., Liu, Y., Liu, Z., Zhou, A., and Yu, Z.-Z. (2017). Hydrophobic, flexible, and lightweight Mxene foams for high-performance electromagnetic-interference shielding. *Advanced Materials*, 29(38), p. 1702367.

[41] Yang, J., Chen, B., Song, H., Tang, H., and Li, C. (2014). Synthesis, characterization, and tribological properties of two-dimensional Ti3C2. *Crystal Research and Technology*, 49(11), pp. 926–932.

[42] Huang, S., Mutyala, K.C., Sumant, A.V., and Mochalin, V.N. (2021). Achieving superlubricity with 2D transition metal carbides (Mxenes) and Mxene/graphene coatings. *Materials Today Advances*, 9, pp. 100133.

[43] Zhang, X., Xue, M., Yang, X., Wang, Z., Luo, G., Huang, Z., Sui, X., and Li, C. (2015). Preparation and tribological properties of Ti3C2(OH)2 nanosheets as additives in base oil. *RSC Advances*, 5(4), pp. 2762–2767.

[44] Liu, Y., Zhang, X., Dong, S., Ye, Z., and Wei, Y. (2016). Synthesis and tribological property of Ti3C2T X nanosheets. *Journal of Materials Science*, 52(4), pp. 2200–2209.

[45] Zhang, D., Ashton, M., Ostadhossein, A., van Duin, A.C.T., Hennig, R.G., and Sinnott, S.B. (2017). Computational Study of Low Interlayer Friction in Tin+1Cn (n = 1, 2, and 3) Mxene. *ACS Applied Materials & Interfaces*, 9(39), pp. 34467–34479.

[46] Mai, Y.J., Li, Y.G., Li, S.L., Zhang, L.Y., Liu, C.S., and Jie, X.H. (2019). Self-lubricating Ti3C2 nanosheets/copper composite coatings. *Journal of Alloys and Compounds*, 770, pp. 1–5.

[47] Hu, J., Li, S., Zhang, J., Chang, Q., Yu, W., and Zhou, Y. (2019). Mechanical properties and frictional resistance of Al composites reinforced with Ti3C2Tx Mxene. *Chinese Chemical Letters*, 31, pp. 996–999.

[48] Yin, X., Jin, J., Chen, X., Rosenkranz, A., and Luo, J. (2019). Ultra-wear-resistant Mxene-based composite coating via *in situ* formed nanostructured tribofilm. *ACS Applied Materials & Interfaces*, 11(35), pp. 32569–3257.

[49] Rosenkranz, A., Grutzmacher, P.G., Espinoza, R., Fuenzalida, V.M., Blanco, E., Escalona, N., Gracia, F.J., Villarroel, R., Guo, L., Kang, R., Mucklich, F., Suarez, S., and Zhang, Z. (2019). Multi-layer Ti3C2Tx-nanoparticles (Mxenes) as solid lubricants—role of surface terminations and intercalated water. *Applied Surface Science*, 494, pp. 13–21.

[50] Rodriguez, A., Jaman, M.S., Acikgoz, O., Wang, B., Yu, J., Grützmacher, P.G., Rosenkranz, A., and Baykara, M.Z. (2021). The potential of Ti3C2TX nano-sheets (Mxenes) for nanoscale solid lubrication revealed by friction force microscopy. *Applied Surface Science*, 535, pp. 147664.

[51] Huang, S., and Mochalin, V.N. (2019). Hydrolysis of 2D transition-metal carbides (Mxenes) in colloidal solutions. *Inorganic Chemistry*, 58(3), pp. 1958–1966.

[52] Natu, V., Hart, J.L., Sokol, M., Chiang, H., Taheri, M.L., and Barsoum, M.W. (2019). Edge capping of 2D-Mxene sheets with polyanionic salts to mitigate oxidation in aqueous colloidal suspensions. *Angewandte Chemie International Edition*, 58(36), pp. 12655–12660.

[53] Naguib, M., Saito, T., Lai, S., Rager, M.S., Aytug, T., Parans Paranthaman, M., Zhao, M.-Q., and Gogotsi, Y. (2016). Ti3C2Tx(Mxene)–polyacrylamide nanocomposite films. *RSC Advances*, 6(76), pp. 72069–72073.

[54] Jiao, E., Wu, K., Liu, Y., Lu, M., Hu, Z., Chen, B., Shi, J., and Lu, M. (2021). Ultrarobust Mxene-based laminated paper with excellent thermal conductivity and flame retardancy. *Composites Part A: Applied Science and Manufacturing*, 146, pp. 106417.

[55] Hu, Y., Chen, Z., Zhuo, H., Zhong, L., Peng, X., and Sun, R. (2019). Advanced compressible and elastic 3D monoliths beyond hydrogels. *Advanced Functional Materials*, 29(44), pp. 1904472.

[56] Hart, J.L., Hantanasirisakul, K., Lang, A.C. et al. (2019). Control of Mxenes' electronic properties through termination and intercalation. *Nature Communications*, 10(1), pp. 522

[57] Zhou, J., Yu, J.L., Shi, L.D. et al. (2018). A conductive and highly deformable all-pseudocapacitive composite paper as supercapacitor electrode with improved a real and volumetric capacitance. *Small*, 14(51), pp. 1803786. doi: 10.1002/smll.201803786

[58] Yan, C.S., Fang, Z.W., Lv, C. et al. (2018). Significantly improving lithium-ion transport via conjugated anion intercalation in inorganic layered hosts. *ACS Nano*, 12(8), pp. 8670–8677, 109.

[59] Chen, X., Zhao, Y., Li, L., Wang, Y., Wang, J., Xiong, J., . . . Yu, J. (2020). Mxene/polymer nanocomposites: Preparation, properties, and applications. *Polymer Reviews*, 1–36. doi: 10.1080/15583724.2020.1729179

[60] Pham, V.P., and Yeom, G.Y. (2016). Recent advances in doping of molybdenum disulfide: Industrial applications and future prospects. *Advanced Materials*, 28(41), pp. 9024–9059.

[61] Anasori, B., Lukatskaya, M.R., and Gogotsi, Y. (2017). 2D metal carbides and nitrides (Mxenes) for energy storage. *Nature Reviews Materials*, 2(2). www.nature.com/articles/natrevmats201698

[62] Lin, H., Chen, Y., and Shi, J. (2018). Insights into 2D Mxenes for versatile biomedical applications: Current advances and challenges ahead. *Advanced Science* (Weinh), 10, pp. 1800518.

[63] Zhang, J., Zhao, Y., Guo, X., Chen, C., Dong, C.-L., Liu, R.-S., Han, C.-P., Li, Y., Gogotsi, Y., and Wang, G. (2018). Single platinum atoms immobilized on an Mxene as an efficient catalyst for the hydrogen evolution reaction. *Nature Catalysis*, 1(12), pp. 985–992.

[64] Cao, W., Ma, C., Mao, D., Zhang, J., Ma, M., and Chen, F. (2019). Mxene-reinforced cellulose nanofibril inks for 3D-printed smart fibres and textiles. *Advanced Functional Materials*, 29(51), pp. 1905898.

[65] Cao, W., Ouyang, H., Xin, W., Chao, S., Ma, C., Li, Z., Chen, F., and Ma, M. (2020). A stretchable highoutput triboelectric nanogenerator improved by Mxene liquid electrode with high electronegativity. *Advanced Functional Materials*, 30(50), pp. 2004181.

[66] Wang, K., Lou, Z., Wang, L., Zhao, L., Zhao, S., Wang, D., Han, W., Jiang, K., and Shen, G. (2019). Bioinspired interlocked structure-induced high deformability for two-dimensional titanium carbide (Mxene)/natural microcapsule-based flexible pressure sensors. *ACS Nano*, 13(8), pp. 9139–9147.

[67] Huang, J., Li, Z., Mao, Y., and Li, Z. (2021). Progress and biomedical applications of Mxenes. *Nano Select*, 2(8), pp. 1480–1508.

[68] Ma, J., Zheng, S., Das, P., Lu, P., Yu, Y., and Wu, Z.-S. (2020). Sodium ion microscale electrochemical energy storage device: Present status and future perspective. *Small Structures*, 1(1), pp. 2070003.

[69] Wen, Z., Yeh, M.-H., Guo, H., Wang, J., Zi, Y., Xu, W., Deng, J., Zhu, L., Wang, X., Hu, C., Zhu, L., Sun, X., and Wang, Z.L. (2016). Self-powered textile for wearable electronics by hybridizing fiber-shaped nanogenerators, solar cells, and supercapacitors. *Science Advances*, 2(10), pp. 1600097.

[70] Liu, Z., Xu, J., Chen, D., and Shen, G. (2015). Flexible electronics based on inorganic nanowires. *Chemical Society Reviews*, 44(1), pp. 161–192.

[71] Xue, Q., Sun, J., Huang, Y., Zhu, M., Pei, Z., Li, H., Wang, Y., Li, N., Zhang, H., and Zhi, C. (2017). Recent progress on flexible and wearable supercapacitors. *Small*, 13(45), p. 1701827.

[72] Gogotsi, Y. (2000). Graphite polyhedral crystals. *Science*, 290(5490), pp. 317–320.

[73] Xiong, Z., Liao, C., Han, W., and Wang, X. (2015). Mechanically tough large-area hierarchical porous graphene films for high-performance flexible supercapacitor applications. *Advanced Materials*, 27(30), pp. 4469–4475.

[74] Yang, Q., Cui, S., Ge, Y., Tang, Z., Liu, Z., Li, H., Li, N., Zhang, H., Liang, J., and Zhi, C. (2018). Porous single-crystal NaTi2(PO4)3 via liquid transformation of TiO2 nanosheets for flexible aqueous Na-ion capacitor. *Nano Energy*, 50, pp. 623–631.

[75] Boota, M., Pasini, M., Galeotti, F., Porzio, W., Zhao, M.-Q., Halim, J., and Gogotsi, Y. (2017). Interaction of polar and nonpolar polyfluorenes with layers of two-dimensional titanium carbide (MXene): Intercalation and pseudocapacitance *Chemistry of Materials*, 29(7), pp. 2731.

[76] Huang, L., Ding, L., and Wang, H. (2021). Mxene-based membranes for separation applications. *Small Science*, 1(7), pp. 2100013. doi: 10.1002/smsc.202100013

[77] Goh, P.S., and Ismail, A.F. (2018). A review on inorganic membranes for desalination and wastewater treatment. *Desalination*, 434, pp. 60–80.

[78] Al-Hamadani, Y. A. J., Jun, B.-M., Yoon, M., Taheri-Qazvini, N., Snyder, S. A., Jang, M., and Yoon, Y. (2020). Applications of Mxene-based membranes in water purification: A review. *Chemosphere*, pp. 126821. doi: 10.1016/j.chemosphere.2020.126821

[79] Han, R., Ma, X., Xie, Y., Teng, D., and Zhang, S. (2017). Preparation of a new 2D Mxene/PES composite membrane with excellent hydrophilicity and high flux. *RSC Advances*, 7(89), pp. 56204–56210. doi: 10.1039/c7ra10318b

[80] Salim, O., Mahmoud, K.A., Pant, K.K., and Joshi, R.K. (2019). Introduction to MXenes: Synthesis and characteristics. *Materials Today Chemistry*, 1(14), pp. 100191. doi: 10.1016/j.mtche m.2019.08.010

[81] Khazaei, M., Mishra, A., Venkataramanan, N.S., Singh, A.K., and Yunoki, S. (2019). Recent advances in MXenes: From fundamentals to applications. *Current Opinion in Solid State and Materials Science*, 23(3), pp. 164–178.

[82] Ihsanullah, I. (2020). MXenes (two-dimensional metal carbides) as emerging nanomaterials for water purification: Progress, challenges and prospects. *Chemical Engineering Journal*, 388, pp. 124340.

[83] Sinopoli, A., Othman, Z., Rasool, K., and Mahmoud, K.A. (2019). Electrocatalytic/photocatalytic properties and aqueous media applications of 2D transition metal carbides (MXenes). *Current Opinion in Solid State and Materials Science*, 23(5), p. 100760.

[84] An, H., Habib, T., Shah, S., Gao, H., Patel, A., Echols, I., Zhao, X., Radovic, M., Green, M. J., Lutkenhaus, J. L., H. *et al.* (2019). Water sorption in Mxene/polyelectrolyte multilayers for ultrafast humidity sensing. *ACS Applied Nano Materials*, 2(2), pp. 948–955.

[85] Guo, Y., Zhong, M., Fang, Z., Wan, P., and Yu, G. (2019). A wearable transient pressure sensor made with Mxene nanosheets for sensitive broad-range human–machine interfacing. *Nano Letters*, 19(2), pp. 1143–1150.

[86] Hu, Y. *et al.* (2019). Biomass polymer-assisted fabrication of aerogels from Mxenes with ultrahigh compression elasticity and pressure sensitivity. *Journal of Materials Chemistry. A, Materials for Energy and Sustainability*, 7(17), pp. 10273–10281.

[87] Naguib, M., Unocic, R.R., Armstrong, B.L. et al. (2015). Large-scale delamination of multi-layers transition metal carbides and carbonitrides "Mxenes." *Dalton Transactions*, 44(20), pp. 9353–9358 62.

[88] Johnson, D.W., Dobson, B.P., and Coleman, K.S. (2015). A manufacturing perspective on graphene dispersions. *Current Opinion in Colloid & Interface Science*, 20(5–6), pp. 367–382.

[89] Yang, H. *et al.* (2017). Layered PVB/Ba 3 Co 2 Fe 24 O 41/Ti 3 C 2 Mxene composite: Enhanced electromagnetic wave absorption properties with high impedance match in a wide frequency range. *Materials Chemistry and Physics*, 200, pp. 179–186.

Index